An American Physical Therapy Association Anthology

GAIT

Basic Research

VOL. 1

This anthology is a compilation of articles
originally published in the March/April 1936
through October 1992 issues of *Physical Therapy*.

ISBN # 0-912452-88-9

© 1993 by the American Physical Therapy Association. All rights reserved.

For more information about this and other APTA publications, contact the American Physical Therapy Association, 1111 North Fairfax Street, Alexandria, VA 22314-1488.

[Order No. P-93]

Table of Contents

Foreword vii

Descriptions and Requirements

The Base of Support in Stance: Its Characteristics and the Influence of Shoes Upon the Location of the Center of Weight / Frances A Hellebrandt, Doris Kubin, Winifred M Longfield, L E A Kelso (Nov/Dec 37) 1

The Mechanics of Walking: A Clinical Interpretation / Jacquelin Perry (Sep 67) 5

Metabolic Energy Cost of Unrestrained Walking / Raymond L Blessey, Helen J Hislop, Robert L Waters, Daniel Antonelli (Sep 76) 29

Functional Ambulation Velocity and Distance Requirements in Rural and Urban Communities: A Clinical Report / Christy S Robinett, Mary A Vondran (Sep 88) 35

Dynamic Biomechanics of the Normal Foot and Ankle During Walking and Running / Mary M Rodgers (Dec 88) 39

Foot Trajectory in Human Gait: A Precise and Multifactorial Motor Control Task / David A Winter (Jan 92) 49

The Dual-Task Methodology and Assessing the Attentional Demands of Ambulation with Walking Devices / David L Wright, Tammy L Kemp (Apr 92) 61

Trunk Kinematics During Locomotor Activities / David E Krebs, Dennis Wong, David Jevsevar, Patrick O Riley, W Andrew Hodge (Jul 92) 71

Methods of Studying Gait

Methods of Studying Gait / Gary L Smidt (Jan 74) 81

A Method of Measuring the Duration of Foot-Floor Contact During Walking / Gena M Gardner, M Patricia Murray (Jul 75) 87

Evaluation of a Clinical Method of Gait Analysis / Darlene D Boeing (Jul 77) 93

A Light-Emitting Diode System for the Analysis of Gait: A Method and Selected Clinical Examples / Gary L Söderberg, R H Gabel (Apr 78) 97

Footprint Analysis in Gait Documentation: An Instructional Sheet Format / Marion Shores (Sep 80) 105

Suggestions from the Field: A Portable Feedback Gait Apparatus for Five Segments of Footfall / Debra Warshal, Mike Jacobs, Wynne Lee, Miguel Garcia (Oct 81) 111

Methods of Measurement in Soviet Gait Analysis Research, 1963-1974 / Leopold G Selker (Feb 76) 115

Clinical Determination of Energy Cost and Walking Velocity via Stopwatch or Speedometer Cane and Conversion Graphs / David H Nielsen, David G Gerleman, Louis R Amundsen, David A Hoeper (May 82) 121

Use of the Krusen Limb Load Monitor to Quantify Temporal and Loading Measurements of Gait / Steven L Wolf, Stuart A Binder-Macleod (Jul 82) 127

Absorbent Paper Method for Recording Foot Placement During Gait: Suggestion from the Field / Bertha H Clarkson (Mar 83) 137

A Clinical Method of Quantitative Gait Analysis: Suggestion from the Field / Kay Cerny (Jul 83) 139

Assessing the Reliability of Measurements from the Krusen Limb Load Monitor to Analyze Temporal and Loading Characteristics of Normal Gait / Patricia B Carey, Steven L Wolf, Stuart A Binder-Macleod, Raymond L Bain (Feb 84) 141

Gait Analysis Techniques: Rancho Los Amigos Hospital Gait Laboratory / JoAnne K Gronley, Jacquelin Perry (Dec 84) 147

Objective Clinical Evaluation of Function: Gait Analysis / R Keth Laughman, Linda J Askew, Robert R Bleimeyer, Edmund Y Chao (Dec 84) 155

Effects of Electromyographic Processing Methods on Computer-Averaged Surface Electromyographic Profiles for the Gluteus Medius Muscle / Rob F M Kleissen (Nov 90) 163

The Foot and Ankle

The Feet in Relation to the Mechanics of Human Locomotion / R Plato Schwartz, Arthur L Heath (Mar/Apr 36) 169

The Vertical Pathways of the Foot During Level Walking: I. Range of Variability in Normal Men / M P Murray, B H Clarkson (Jun 66) 173

The Vertical Pathways of the Foot During Level Walking: II. Clinical Examples of Distorted Pathways / M P Murray, B H Clarkson (Jun 66) 179

The Base of Support in Stance: Its Characteristics and the Influence of Shoes Upon the Location of the Center of Weight / Frances A Hellebrandt, Doris Kubin, Winifred M Longfield, L E A Kelso (Nov/Dec 37) 189

Walking Patterns of Healthy Subjects Wearing Rocker Shoes / Margery J Peterson, Jacquelin Perry, Jacqueline Montgomery (Oct 85) 193

Reliability of Kinematic Measurements of Rear-Foot Motion / Michael J Mueller, Barbara J Norton (Oct 92) 199

The Knee and Hip

Hip Motion and Related Factors in Walking / Gary L Smidt (Jan 71) 207

Knee Flexion During Stance as a Determinant of Inefficient Walking / David A Winter (Mar 83) 221

Bilateral Analysis of the Knee and Ankle During Gait: An Examination of the Relationship Between Lateral Dominance and Symmetry / Lori A Gundersen, Dianne R Valle, Ann E Barr, Jerome V Danoff, Steven J Stanhope, Lynn Snyder-Mackler (Aug 89) 225

The Upper Extremity

Patterns of Sagittal Rotation of the Upper Limbs in Walking: Study of Normal Men During Free and Fast Speed Walking / M P Murray, S B Sepic, E J Barnard (Apr 67) 237

Upper-Extremity Muscular Activity at Different Cadences and Inclines During Normal Gait / Raymond E Hogue (Sep 69) 251

Age-Related Changes in Walking Performance

Development of Gait at Slow, Free, and Fast Speeds in 3- and 5-Year-Old Children / Ruth Rose-Jacobs (Aug 83) 261

Gait Cycle Duration in 3-Year-Old Children / Darlene S Slaton (Jan 85) 271

Comparison of Gait of Young Women and Elderly Women / Patricia A Hagemen, Daniel J Blanke (Sep 86) 277

Comparison of Gait of Young Men and Elderly Men / Daniel J Blanke, Patricia A Hageman (Feb 89) 283

Biomechanical Walking Pattern Changes in the Fit and Healthy Elderly / David A Winter, Aftab E Patla, James S Frank, Sharon E Walt (Jun 90) 289

Relationships Between Physical Activity and Temporal-Distance Characteristics of Walking in Elderly Women / Carol I Leiper, Rebecca L Craik (Nov 91) 297

The Influence of Assistive Devices on Walking

Weight-Distribution Variables in the Use of Crutches and Canes / Metta L Baxter, Ruth O Allington, George H Koepke (Apr 69) 311

System of Reporting and Comparing Influence of Ambulatory Aids on Gait / Gary L Smidt, M A Mommens (May 80) 317

Elbow Moment and Forces at the Hands During Swing-Through Axillary Crutch Gait / Marc Reisman, Ray G Burdett, Sheldon R Simon, Cynthia Norkin (May 85) 325

Effects of Selected Assistive Devices on Normal Distance Gait Characteristics / Chukwuduziem U Opara, Pamela K Levangie, David L Nelson (Aug 85) 329

Energy Expenditure of Ambulation Using the Sure-Gait Crutch and the Standard Axillary Crutch / Annette L Annesley, Monica Almada-Norfleet, David A Arnall, Mark W Cornwall (Jan 90) 333

Energy Cost, Exercise Intensity, and Gait Efficiency of Standard Versus Rocker-Bottom Axillary Crutch Walking / David H Nielsen, Joan M Harris, Yvonne M Minton, Nancy S Motley, Jeri L Rowley, Carolyn T Wadsworth (Aug 90) 339

The Influence of Loading on Gait

Effect of Load and Carrying Position on the Electromographic Activity of the Gluteus Medius Muscle During Walking / Donald A Neumann, Thomas M Cook (Mar 85) 347

Influence of Body Weight Support on Normal Human Gait: Development of a Gait Retraining Strategy / Lois Finch, Hugues Barbeau, Bertrand Arsenault (Nov 91) 355

Foreword

The large number of articles in *Physical Therapy* and its predecessor journals reflect our profession's interest in gait. This volume focuses on studies that span topics as diverse as descriptions of gait in healthy subjects, age-related differences, effects of assistive devices, and kinematic and kinetic descriptions.

The articles span not just topics but also decades, and although collectively they demonstrate how vital gait is to our profession's function, they also show how elusive the subject may be. There is much to be studied and much to be described about human walking. There is much to be learned. Which deviations may be adaptive, and which maladaptive? This anthology should provide clinicians and students with a remarkable wealth of knowledge not just about gait, but also about the strategies we have used to study gait. In another anthology entitled *Gait—Volume 2: Clinical Practice*, articles relating to the measurement of gait in patients have been collected.

There are several purposes for an anthology such as this one. Clearly a key function is to present between one set of covers all the material on a topic area to facilitate the reader's ability to see all that has been published on a topic. But I would be remiss as an editor if I did not encourage the reader also to pause and ponder other issues. Although on the one hand I am impressed by the remarkable programmatic publication efforts of some of our authors, and the great effort that many authors made years ago in contributing to our body of knowledge, I also am impressed by our growth. The articles, therefore, reflect not just the growth of knowledge, but also the growth in sophistication in the way we physical therapists examine issues. That is not to suggest that what is new is always better than what is old, but the observation should serve as a warning to all readers to look cautiously at all that lies within the pages of this volume.

With time, our views often change. With time, just as the enthusiasm for a topic may change, so might standards for acceptable research and documentation. Although we have included no articles in this anthology that have glaring errors, we also have made no effort to alter the historical record by re-editing or re-reviewing papers. All articles relating to the subject have been published regardless of whether they would meet the standards of the Journal today. This compendium, therefore, is rich in content and diverse in quality. It is as if we have presented before the interested reader a rather extensive smorgasbord of regional cuisine. Some of the offerings you may like better than I, and some neither of us may like, but only by offering the array can we represent the cuisine as it is rather than as we would like it to be. Like any smorgasbord, however, there can be many schemes for the organization of the presentation. In this anthology, articles have been placed in sections based on content, and then organized within those sections based on both content and chronology.

On behalf of the Journal, I encourage you to read what lies between the covers and to share my appreciation for those authors whose work has increased our understanding of human walking.

Jules M Rothstein, PhD, PT
Editor, *Physical Therapy*
January 1993

Basic Research

The Base of Support in Stance*

Its Characteristics and the Influence of Shoes Upon the Location of the Center of Weight

Frances A. Hellebrandt, Doris Kubin, Winifred M. Longfield and L. E. A. Kelso

According to Morton[1] the ease and economy with which man meets the prime mechanical disadvantages of the biped stance represents the acme of his physical evolution. The unfavorable factors are the height of man's center of gravity and the smallness of his base of support. Morton postulates that as long as the line of gravity falls within a certain sector of the area of underpropping, rotatory stresses will be equilibrated by increase in tone without resorting to phasic contraction of the antigravity muscles. He locates the center of gravity of a subject standing in a comfortable posture as perpendicularly above a point between the feet in the region of the navicular bones. Steindler[2] believes there is an inherent tendency for each individual to maintain his center of weight at a constant point in relation to the structural supports of his body. However, we[3] observed that sway is inseparable from the upright stance and that the center of weight shifts incessantly during the maintenance of a natural comfortable posture.

The object of this study was (1) to observe the base of support characteristics in stance; (2) to project the average location of the shifting center of weight into the base of support; and (3) to determine the effect of shoes upon the location of the center of weight and upon postural stability.

METHODS

1. *Recording the Base of Support.* All observations were made in Morton's foot position of greatest stability in stance. In this posture the toes turn out so that the lateral diameter of structural support approximates the antero-posterior. The position was standardized by the insertion of a 30-degree wedge between the heels which were made to abut against a low vertical ledge. The subject stood upon absorbent paper after first moistening the soles of the feet with a concentrated solution of potassium permanganate. The contact area of the feet with the ground was recorded during five minutes of quiet standing. The resulting footprint was then analyzed in accord with the suggestions embodied in Morton's description of the functionally important diameters of support. The total area of underpropping was measured in square centimeters with a planimeter after introducing tangents to the heel, the lateral border of the feet and the toes. The diameter of antero-posterior stability was taken as the vertical distance between the heel tangent and a line through the heads of the first metatarsals, the latter being placed midway between the two in the presence of inequality in the length of the feet. The distance between the fifth metatarsals measured the lateral support. The accessory margin of anterior stability was taken as the distance between the first metatarsal line and the great toe tangent.

2. *Projecting the Center of Weight into the Base of Support.* Figure 1 illustrates the method and equipment. The subject stands upon the uppermost of two platforms placed at right angles to each other and supported on steel wedges to minimize friction. One knife-edge of each beam rests upon a hundred-pound platform scale modified to graphically record changes in weight upon a constant-moving kymograph. The subject faces the antero-posterior scale which records all forward and backward oscillations of the center of weight. A simultaneous record of sway in the transverse vertical plane is made by the auxiliary lever of the lateral scale. Knowing the weight of the subject, the distance between the knife-edges, the indications of the scales and the tares, the location of the center of gravity in the two vertical orientation planes may be determined at any instant by equating moments. Since the position of the feet is fixed in regard to the knife-edges in terms of which the center of gravity is located, the latter may be projected into the imprint of the base of support.

3. *Determining the Average Position of the Shifting Center of Weight.* The secondary of a slotted core transformer excited from a constant potential source was suspended from each auxiliary scale lever. Since the position of the scale lever is directly proportional to the former, the involuntary vibratory motion of the center of weight may be electrically integrated by placing the secondary of each transformer in series with the current coil of a rotating watt hour meter. From the reading of the meters the average load on the antero-posterior and lateral scales may be obtained and by simple calculation converted into the mean location of the center of gravity in the two cardinal vertical orientation planes.

4. *Modifying the Size of the Base of Support and the Height of the Center of Gravity by Wearing Shoes.* A series of consecutive 3 minute records was made standing in a comfortable posture in bare feet, wearing shoes with low (1.0 to 2.5 cm.), medium (2.8 to 4.5 cm.) and high (4.8 to 7.0 cm.) heels. Sufficient rest was allowed between each standing period to eliminate fatigue. Heel height was estimated by the difference in the tallness of the subject measured with and without shoes.

RESULTS AND THEIR DISCUSSION

1. *The Characteristics of the Base of Support in Stance.* Observations were made on 88 professional students in physical education accustomed to vigorous exercise and possessing, by that functional criterion, essen-

---

*From the Department of Physiology, University of Wisconsin, Madison, Wisconsin, supported in part by a grant from the Wisconsin Alumni Research Foundation.

ACKNOWLEDGMENTS—Our thanks are due Profs. Trilling and Bassett for the use of certain of the facilities of the Physical Education Department, Genevieve Braun for help in the conduct of the experiments and O. H. Brogdon for assistance in the preparation of the manuscript.

tially normal feet. The data are presented in Table I. These young adult women show a 63 per cent larger margin of static security laterally than antero-posteriorly in the foot posture of Morton, disregarding the accessory area of support which is functionless during stance in the average subject. The greatest lateral diameter more nearly approximates total foot length. Considerable variability was evident in the diameters studied, especially in the relative importance of the accessory area of support which is a margin of safety called into activity when the center of gravity shifts too far forward in the antero-posterior plane.

TABLE I

Characteristics of the Base of Support in Stance

	Mean	Standard Deviation	Coefficient of Variability
Number of subjects—88			
Antero-posterior foot support	16.43 ± 0.05 cm.	0.75 ± 0.04	4.6
Lateral foot support	26.78 ± 0.12 cm.	1.63 ± 0.08	6.08
Accessory foot support	6.10 ± 0.04 cm.	0.49 ± 0.02	8.03
Total area of underpropping	502.0 ± 3.37 sq. cm.	46.89 ± 2.38	9.3
Foot length	23.09 ± 0.07 cm.	1.00 ± 0.05	4.3

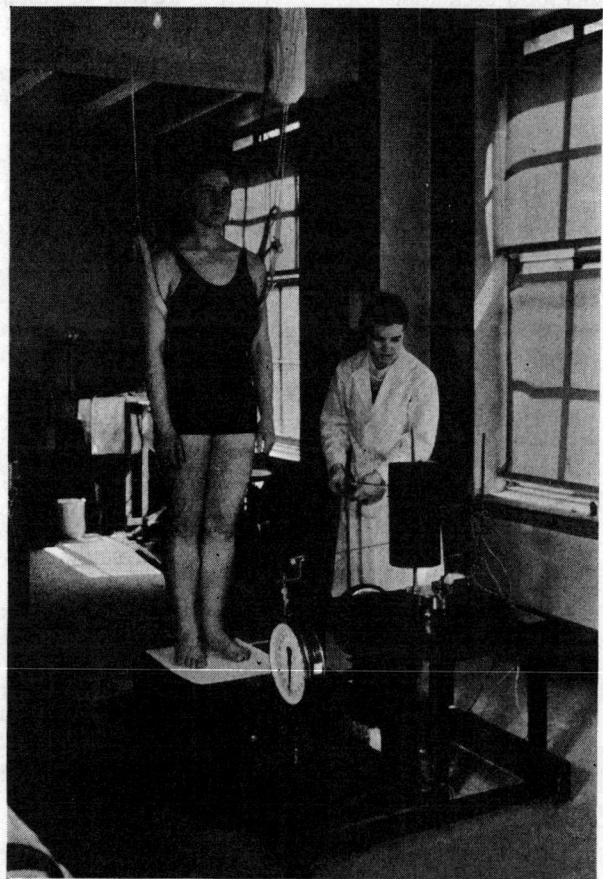

Fig. 1.—Photograph of the apparatus developed for the graphic registration of concurrent shifts of the center of gravity in the antero-posterior and transverse vertical planes during the maintenance of the upright stance. The subject assumes a natural comfortable posture with the gaze fixed upon a designated point. The arm straps are loosely adjusted and offer no restraint. Their function is protective in the event of syncope.

2. *The Location of the Center of Weight in the Base of Support.* The area of structural support in the foot position of Morton forms a trapezoid bounded by the intersection of the heel and lateral tangents and the extension of the horizontal line passing through the heads of the first metatarsals. The geometric center of the base of support may be approximately located by constructing an antero-posterior median from the midpoints of the heel tangent and the metatarsal line, and another, perpendicular to the first, bisecting it. The intersection of these medians marks the center of the active support. The projected average position of the center of weight within the base of support may then be mathematically related to the geometric center of the trapezoidal area approximating the superficial extent of the underprop space. The distance from the geometric center of the base to the center of gravity projection is called the eccentricity, $\frac{eAP}{\frac{1}{2} AP}$ and $\frac{eT}{\frac{1}{2} T}$ being the eccentricity ratios evaluating deviations as regards the antero-posterior and transverse areas respectively. (eAP = the perpendicular distance from the center of gravity to the transverse median. $\frac{1}{2} AP$ = half the length of the antero-posterior median. eT = the perpendicular distance from the center of gravity to the antero-posterior median. $\frac{1}{2} T$ = half the length of the lateral median.) Figure 2 illustrates the above points.

To support the center of gravity and uphold the body in the erect posture, the straight line joining the center of weight with the center of the earth must fall within the area enclosed by the feet. The average position of the center of weight falls relatively close to the geometric center of the base. That it is moderately eccentric was indicated by our[4] observation that the upright stance was asymmetrical in the majority of a group of 445 young adult women. The weight fell preponderantly to the left in a comfortable stance. However, the mean location of the center of weight is such that the margins of static security are roughly equal on all sides when the functionless accessory area of support is disregarded. We have also demonstrated that, as Steindler believes, there is an inherent tendency for each individual to maintain his center of gravity at a constant point in relation to his structural supports. The standard procedure for the graphic study of the trajectory of the center of gravity during three minutes of quiet standing in a natural and comfortable stance was followed on different days. The utmost care was exercised to assume the same posture. The involuntary sway of one of our subjects was unusually great. The maximum displacements of her center of gravity forward, back, right and left during a typical three-minute period of standing are projected into the base of support in Fig. 3. The maximal forward sway brings the center of weight dangerously near the fore extremity of the antero-posterior diameter of structural foot support. The subject was forced to grip with the toes to keep from falling forward, even though the center of gravity never moved beyond the heads of the first metatarsals. Backward motion was less great. The remarkable feat is that, in spite of involuntary postural sway of this extraordinary magnitude, the average position of the center of gravity on eight different days was confined to an area of 1.5 sq. cm. in a base of support measuring approximately 500 sq. cm. The data are illustrated in Fig. 3. Six consecutive trials on the same day yielded essentially identical results, the average centers of weight now being still more closely clustered. However capricious the behavior of the center of gravity may seem during so-called "quiet" standing, its oscillations are so accurately balanced that the average relation of the center of gravity to the base of support remains remarkably constant at least in vigorous, healthy young adults with a good equilibratory and kinesthetic sense.

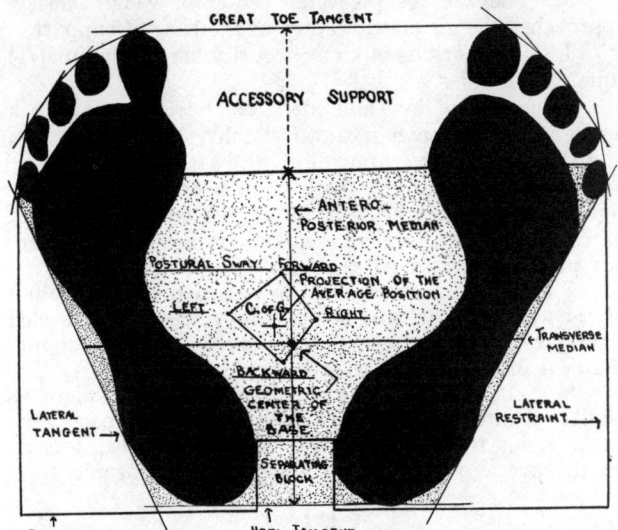

Fig. 2.—*Planographic reproduction of the base of support of a normal young adult woman illustrating the slight eccentricity of the average position of the center of weight during three minutes of comfortable standing. The peripheral points FBRL indicate the maximal shift in the center of weight forward, back, right and left. The foot guide is removed before a record is made. If the projected center of weight falls anterior to the transverse median, the AP eccentric value is positive; if it falls posterior to it, it is negative. Similarly, if the projected center of weight falls to the right of the antero-posterior median, the lateral eccentric value is positive; if it falls to the left, the value is negative.*

3. *The Effect of Shoes upon the Location of the Center of Weight and upon Postural Stability.* Fifty-one experiments were performed on a homogeneous group of 12 vigorous young adult women, habitually wearing low-heeled shoes. The area of underpropping varied with the heel height of the shoe being worn. The low-heeled shoes provided an average increase of 10.5 per cent in the total supporting area. The base of support furnished by shoes of medium heel height was slightly smaller than that of the bare feet. The highest heeled shoes decreased the contacting foot area on an average by 15.1 per cent.

The subjects responded similarly. In quiet standing without shoes the average position of the center of weight was close to the geometric center of the base of support. When a low-heeled shoe was worn, the center of weight remained in approximately the same position with respect to the new base. Heels of moderate height similarly failed to significantly shift the average location of the center of weight. When shoes with the highest heels to which the subject was accustomed were worn, the projection of the center of gravity into the base of support continued to fall near the geometric center of the area of underpropping in 7 of the 12 subjects. In the remaining 5, the center of weight was thrown forward. The data indicate that the center of gravity when projected vertically, typically occupies a position near the geometric center of the base even when the heels are elevated, raising the center of gravity, and further reducing the already small base of support. To achieve this, the multijointed body must redistribute its parts, thus nullifying the derangement caused by uplifting the heels. This may be accomplished in a variety of ways, the most common being a compensatory flexion of the knees or hyperextension of the lumbo-dorsal spine, the latter conveying the upper portion of the body backward and thus balancing the forward pitch caused by elevating the heels.

Since the position of the projected center of gravity remained nearly constant when shoes were worn, involuntary postural sway was observed in order to determine the influence upon stability of augmenting the chief mechanical impediments to the orthograde stance, raising the center of gravity and narrowing the base of support. The margins of the area within the base allocated to shifts in the center of weight were found by projecting the maximum oscillations in the coronal and sagittal planes. Low-heeled shoes had little effect on postural stability, the area of sway being in some instances slightly decreased in magnitude in consonance with the enlargement of the base. An augmentation of the involuntary postural sway was evident with moderate elevation of the heels. In all but one subject sway was excessive when shoes of greatest heel height were worn. The degree of pertubation may be unduly accentuated by the fact that the subjects were unaccustomed to standing in high-heeled shoes, but the tendency to become less steady with each infringement of the security of a body already in unstable

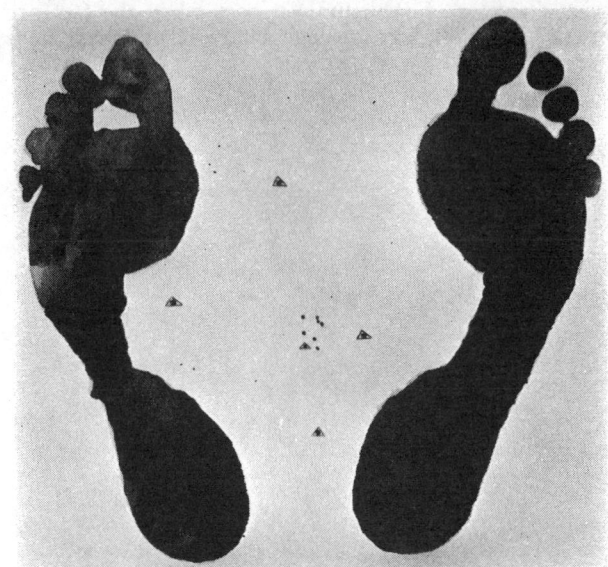

Fig. 3.—*A photographic reproduction of the base of support in stance. The central cluster of eight dots represents the projection of the average location of the center of gravity during three minutes of standing in a natural comfortable stance on eight different days. The maximal shift of the center of weight to and fro in the two vertical orientation planes on the day differentiated by the triangular outline of the average spot is represented by the outlying similarly enclosed data.*

equilibrium might be deduced on *a priori* grounds. Table II contains a typical set of data illustrating the trend of the results. From its examination it is evident that even though very high-heeled shoes heighten the center of gravity, lessen the base of support and magnify the unintentional postural sway, the greatest area within the pedal foundation to which shifts in the center of weight are apportioned is only a fraction of the total underprop space. A large margin of safety still remains, affording protective security in all planes.

Table II—Showing the Influence of Shoes With Heels of Various Heights upon the Base of Support, the Location of the Projected Center of Weight and Postural Stability.

Heel Height	Total Base of Support	Eccentricity in the Vertical Orientation Planes		Area of Maximal Sway
cm.	sq. cm.	Antero-Posterior	Transverse	sq. cm.
0.0	426.9	+ .129	— .063	4.4
2.5	429.9	+ .098	— .052	3.3
4.0	380.9	+ .027	— .014	5.5
5.2	369.3	.000	— .129	19.1

If the projected center of weight falls anterior to the transverse median or to the right of the antero-posterior median, the eccentric value is positive; if it falls posterior to or to the left of the respective medians, it is negative.

Summary and Conclusions

The characteristics of the base of support in stance were studied on young adult women possessing functionally normal feet. The average position of the incessantly shifting center of weight was located within the area of underpropping, and disturbances in static security induced by shoes were measured. From the evidence presented it may be concluded that:

1. The transverse diameter exceeds the antero-posterior diameter of structural support when the feet are separated by a 30-degree wedge, but closely approximates total foot lengths.

2. Irrespective of the magnitude of the involuntary shift in the center of weight which is inseparable from the upright stance, young adults of good equilibratory and kinesthetic sense maintain the *average* location of the center of gravity in a highly constant relation to the structural diameters of the base of support.

3. The average projected center of weight closely approaches the geometric center of the base of support.

4. The margins of static security are approximately equal on all sides.

5. Shoes of low and moderate heel height have a negligible effect upon postural stability and the location of the center of weight within the base of support.

6. Shoes with extremely high heels may either insignificantly effect the location of the center of weight in respect to the base of support or move it forward, but postural sway now becomes excessive.

7. It is suggested that the augmented vascillation of the center of weight is obligatory when high-heeled shoes are worn because they intensify the natural mechanical instability of the orthograde stance.

8. The maximal shifts in the center of weight as projected into the base of support rarely encroach seriously upon the limits of the underprop area, leaving on the average a relatively wide margin of safety.

Bibliography

1. Morton, D. J.: The Human Foot. Columbia University Press, N. Y., 1935.
2. Steindler, A.: Mechanics of Normal and Pathological Locomotion in Man. Charles C. Thomas, Springfield, Illinois, 1935.
3. Hellebrandt, F. A.: Standing as a Geotropic Reflex. In press.
4. Hellebrandt, F. A., R. H. Tepper, G. L. Braun and M. C. Elliott: The Location of the Cardinal Anatomical Orientation Planes Passing Through the Center of Weight in Young Adult Women. In press.

Basic Research

THE MECHANICS OF WALKING

A Clinical Interpretation

JACQUELIN PERRY, M.D.

WALKING IS FREQUENTLY described as a simple act of falling forward and catching oneself. If this is so, why does the person who has recovered some 50 to 60 degrees of knee flexion and has good quadriceps strength, still limp following casting for a fractured femur? Why does the hemiplegic patient walk poorly when he can flex and extend his paretic extremity quite well? Answers to these kinds of questions have stimulated numerous persons throughout the past century to investigate the actual mechanics of walking.

Improved instrumentation and close teamwork between medicine and engineering permitted Inman, Eberhart and their associates [1,2,3] to define and expand the works of earlier investigators.[4,5,6] As a result, they have provided temporal and qualitative relationships, as well as more explicitly delineated, the fundamental components of walking. These components are the arcs of joint motions, se-

---

Dr. Perry is Orthopaedic Surgeon, Chief, Spinal and Hip Deformity Service, and Director of Kinesiological Research at Rancho Los Amigos Hospital, Downey, California 90242.

quence of muscle actions, and rates of body advancement, trunk alignment and ground force reactions.

Subsequent to these studies valuable additions, confirmations and refinements have been contributed independently by Murray [7,8] and Sutherland.[9] The efforts of these many investigators have clearly identified the complex mechanics of walking. Modern prosthetic practice depends heavily on these data, and some of the information has been applied to tendon transplant surgery,[10] but little if any of the data have been incorporated in the management of the many other disabilities which constitute the bulk of orthopedic practice. This omission appears to result from fragmentary publication, presentation of data in obscure reports, and adherence to unfamiliar terminology and strange frames of reference. Scientific investigators find significance in value changes of each variable independently, whereas the clinician attributes significance only to those factors which demonstrably influence the patient's ability to perform. This is not to say that one is more important than the other, but to note the way different persons use data. For the clinician to utilize the scientist's findings, the data must be reinterpreted into functional terminology and concepts. Reinterpretation of the data on walking is the purpose of this paper.

FUNCTIONAL TASKS OF WALKING

Though sometimes used for recreation, walking is basically a means of travel from one place to another—a way of reaching a position to see, to hear, to perform a manual task. Therefore, walking is of secondary importance and should require only minimal amounts of time and energy. In addition, the body needs a smooth ride so as to avoid jarring the sensitive tissues that comprise the brain, the heart and other vital organs, if top performance is to be retained. Smoothness and energy economy are best accomplished by a wheel traveling over an even surface.[3] But wide variations in both natural and man-made terrain are every day experiences. These obstacles are better handled by the versatility of multi-jointed limbs.

To closely approximate the smoothness and energy economy of a wheel, and yet retain the ability to accommodate irregular terrain, man's two lower extremities pass through an intricate series of muscle and joint actions. During the course of travel three functional tasks are accomplished: (1) forward progression, (2) alternately balancing the body over one limb and then the other, and (3) repeated adjustment of relative limb length. Each has its own mechanical demands and responses (Table 1).

TABLE 1
FUNCTIONAL TASKS OF WALKING
Forward Progression
  Shock Absorption
  Momentum Control
  Forward Propulsion
Single Limb Balance
Limb Length Adjustment

Forward Progression

The multitude of actions related to advancing the body in a smooth and economical manner may be grouped into three functions: absorption of the shock related to a rapid transfer of weight on to the forward foot; control of momentum that threatens the stability of the limb as a weight-bearing structure; and, generation of sufficient force to carry the body forward. By clever utilization of momentum to assist in shock absorption and in propelling the body forward, the work requirements are minimized in all three of these tasks. The details of these accomplishments will be discussed as the total walking cycle is analyzed.

Single Limb Balance

To advance by the use of two limbs, the individual must be able to balance the body over one limb while swinging the other. Without such balance (or an adequate substitute) he cannot walk.

When standing in the traditional erect posture, the trunk is well centered between the two supporting limbs. As soon as one foot is lifted to take a step the body becomes grossly off-balance because of the loss of one of these supports (Fig. 1A). The person would fall unless (1) there is a massive holding force from the hip abductor muscles, and (2) he shifts laterally over the weight-bearing foot. Both actions are utilized in normal walking.

The normal person shifts his weight prior to attempting to take a step. In fact, he appears incapable of lifting his foot without this shift. Persons with normal proprioception and muscle control, but who lack adequate hip

MECHANICS OF WALKING

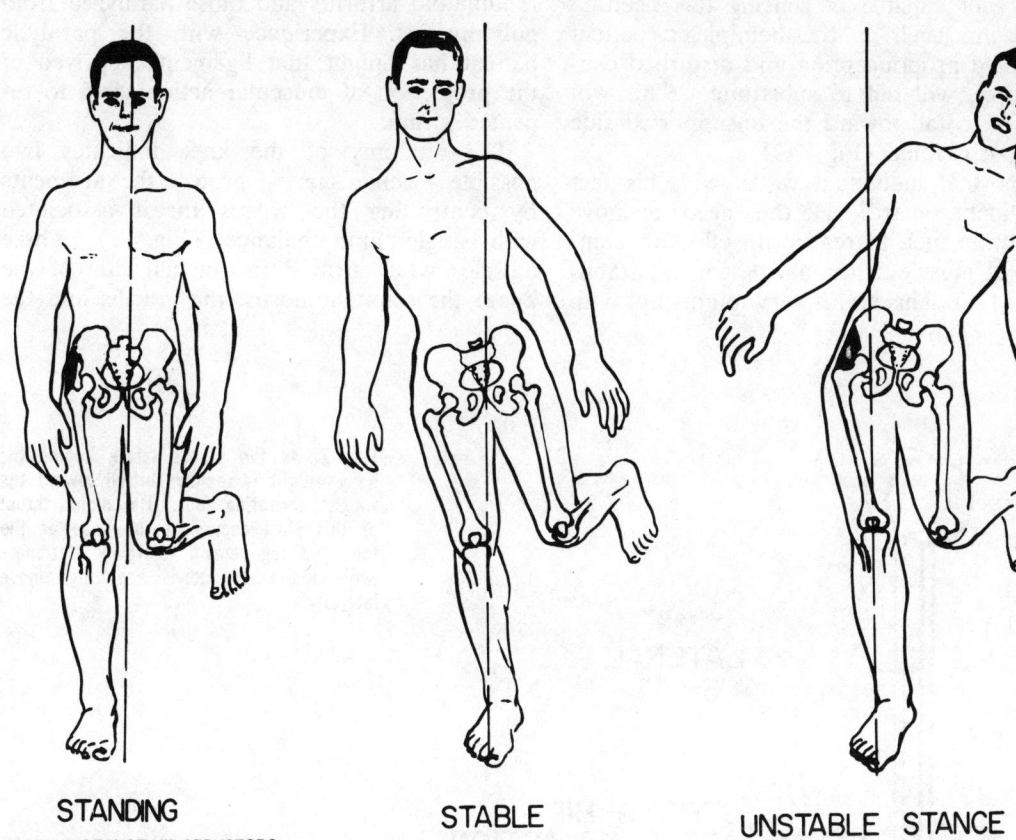

STANDING
NORMAL STANCE HIP ABDUCTORS
STRONGLY ACTIVE

A.

STABLE
POSTURAL SUBSTITUTION FOR
PARALYZED HIP ABDUCTORS

B.

UNSTABLE STANCE
INACTIVE HIP ABDUCTORS
NO POSTURAL SUBSTITUTE

C.

FIG. 1A. The normal individual balances on one limb by shifting his weight slightly to that side (approximately one inch) and by holding forcefully with the hip abductor muscles.

FIG. 1B. A person with intact sensation but an inadequate abductor mechanism (such as a poliomyelitis patient with paralysis of the gluteus medius-minimus complex) balances on one limb by shifting the trunk laterally in order to substitute trunk weight for the absent abductor force.

FIG. 1C. When a person has inadequate proprioception or body image to substitute for non-functioning hip abductor muscles he will be unable to balance on one limb. Instead he will fall toward the unsupported side (a positive Trendelenberg test).

abductor stability, either because of paralysis of the abductor muscles or mechanical inefficiency at the hip joint (from old fractures or dislocations), substitute by exaggerating the lateral shift of the trunk (Fig. 1B). Patients who are not capable of sensing this need to shift weight, such as the hemiplegic patient with limited proprioception and disturbed central control, will fail to substitute. This will cause him to fall toward the unsupported side as the foot is lifted (Fig. 1C).

The normal individual walks with his feet about 3 inches apart.[7] He thus needs to move over only an inch to realize an effective compromise of muscle action and alignment stability (Fig. 1A). This seems very minor, but with the body weight locking the foot on the ground there is a significant lateral thrust (valgus thrust) on the knee and foot (Fig. 2). This seems to account for the valgus knee deformities which occur so readily in patients with rheumatoid arthritis and those paralyzed from poliomyelitis. Experience with the paralytic patient has taught that ligaments deprived of the protection of muscular action yield to repeated strain.

The anatomy of the knee indicates two possible mechanisms to protect the ligaments by controlling the valgus thrust associated with single limb balance (Fig. 3). Three muscles wrap around the medial side of the knee: the semitendinosus, the gracilis and the

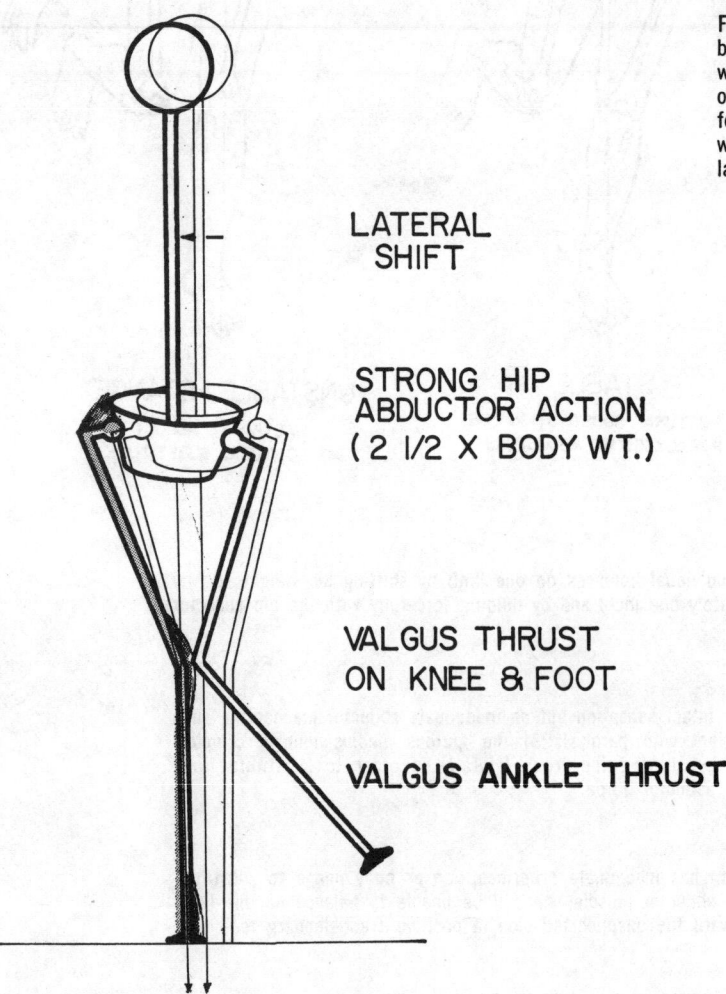

FIG. 2. As the stance phase begins the body weight is rapidly shifted toward the weight accepting foot. The major thrust of this shift occurs at the knee as the foot and leg remain relatively stationary while the trunk, pelvis and thigh move laterally.

Basic Research

MECHANICS OF WALKING

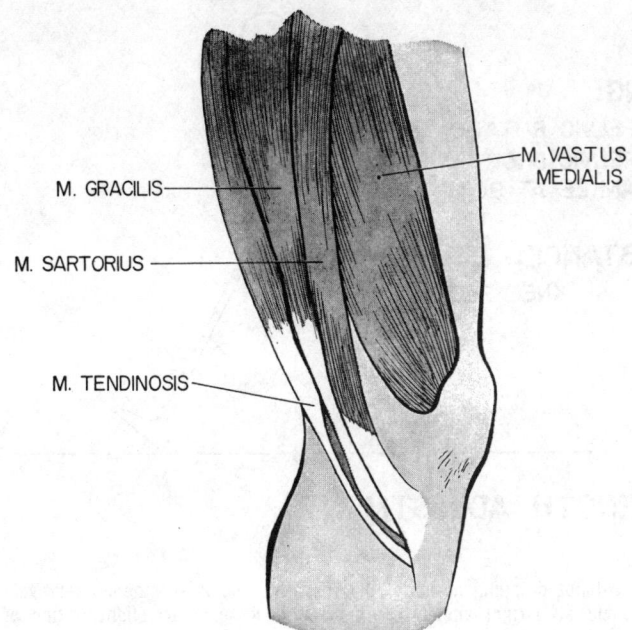

FIG. 3. The four muscles illustrated wrap around the medial side of the knee joint. Since all are active at the beginning of stance it appears that they are actively protecting the medial ligaments from the valgus thrust on the knee that occurs during single limb balance.

sartorius. They all have the common function of knee flexion and medial support. One is a hip extensor, one a hip adductor and the third a hip flexor. These three muscles also have different rotation actions. Thus one could speculate that on weight bearing there is a mechanism to support the knee medially against the valgus thrust while the position of the hip is changing from flexion to extension and from internal rotation to external rotation. The second protective mechanism is probably the vastus medialis. Recent studies indicate that the vastus medialis does not have a special role in knee extension except to prevent lateral dislocation of the patella. But, it may have a very important function in controlling the valgus angulation of the knee as body weight is shifted onto one foot.

A similar valgus stress occurs at the foot. The posterior tibialis muscle which becomes active as soon as weight is borne on the heel, appears to provide protective restraint. Experience with the hemiplegic patient has shown that the medial insertion of the soleus also gives some inversion, and this muscle also comes into play during the first part of weight bearing.

Limb Length Adjustment

Relative lengthening and shortening of the limb is required as the position changes to enable the foot to reach the ground with ease, whether the extremity is directed straight downward, or reaching either forward or backward (Fig. 4). Obviously, the diagonal distance between the trunk and the ground is greater than the vertical distance, and thus, the extremity which is reaching forward to take a step must be longer than the other limb which is providing vertical support. To just drop down as the body passes onto the forward foot is potentially detrimental, as evidenced by the discomfort accompanying such a jarring action. It is also inefficient, as this would cause an abrupt change in direction and hence loss of momentum that otherwise might be utilized for forward travel.

The forward reaching extremity is relatively lengthened by borrowing some of the width of

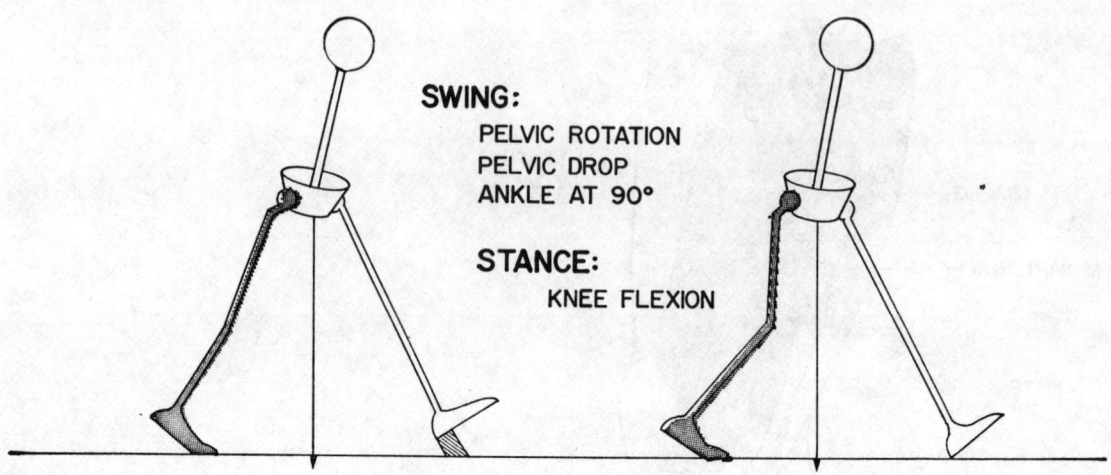

FIG. 4. To reach the desired point of ground contact without dropping abruptly, the reaching limb is lengthened relatively by pelvic rotation, pelvic drop and by holding the ankle at a right angle. The demand is lessened by slight flexion of the stance limb.

the pelvis through rotating the pelvis forward with the reaching limb, and also by allowing the pelvis to drop on that side. Further length is gained from the heel by holding the foot at a right angle. Finally, the total need for length is decreased by slightly flexing the weight bearing knee.

Stress on the Hip

These motions, which accompany the swinging limb, are also creating significant stress on the hip of the stance limb. While bearing the full weight of the body plus the compressive force of the stabilizing abductor and extensor muscles, the hip passes through adduction and internal rotation and swings from flexion to extension—good reason, indeed for even minor discrepancies in the ball and socket contour of the hip joint to cause pain. These multiple stresses also mean that limitations in joint rotation will cause painful tension on the capsule and ligaments, even though the ranges of flexion and extension are still good.

Clinical experience suggests that these stresses can become symptomatic even without x-ray evidence of bony change. Thus the early treatment for such disability is a program to restore the normal ranges of rotation, abduction, and extension. Obviously, painful soft tissues cannot tolerate vigorous challenge, so the exercise program must be gentle and of brief duration. The basic rule to be followed is that if the exercise causes pain, exercise, *per se,* is not contraindicated, but the amount of exercise has been excessive, or the method inappropriate. A person cannot walk unless he can move the limb. Hence, an appropriately graduated exercise program to improve movement is essential.

Stress on the Knee

During rotation at the hip, comparable rotatory forces are active on the knee. The stress will strain the ligaments if there is insufficient muscular strength to protect them. In addition, the person stands with the knee flexed approximately 10 or 15 degrees while he is bearing his full body weight. As a result, support is gained through muscular action rather than ligamentous tension. The quadriceps,

Basic Research

MECHANICS OF WALKING

which grasps three-fourths of the knee joint through its retinaculae, has a very vital role at this time.

THE PHASES OF GAIT

The alternate standing and stepping aspects of walking are technically defined as the *stance* and *swing* phases of gait respectively. Stance begins at "heel-strike" and ends at "toe-off". The limb then swings forward to the next heel-strike. As a means of better identifying related actions, stance has been divided into the periods of heel-strike, mid-stance, and push-off (Fig. 5). Thus the term heel-strike has been given two meanings. It may denote the initial moment of contact between the foot and the ground, or it may refer to the sequel of events resulting from ground contact. The swing phase is often divided into early and late periods.

Each of these intervals contains a complex of activity related to accomplishing a particular task. The nature of these tasks is best identified by the use of functional terminology. Appropriate functional descriptions are: weight acceptance, trunk glide, push, and balance-assist for the stance phase; pick-up and reach during the swing phase (Fig. 5). Each task is a composite of the appropriate components of forward progression, single limb balance and limb length adjustment.

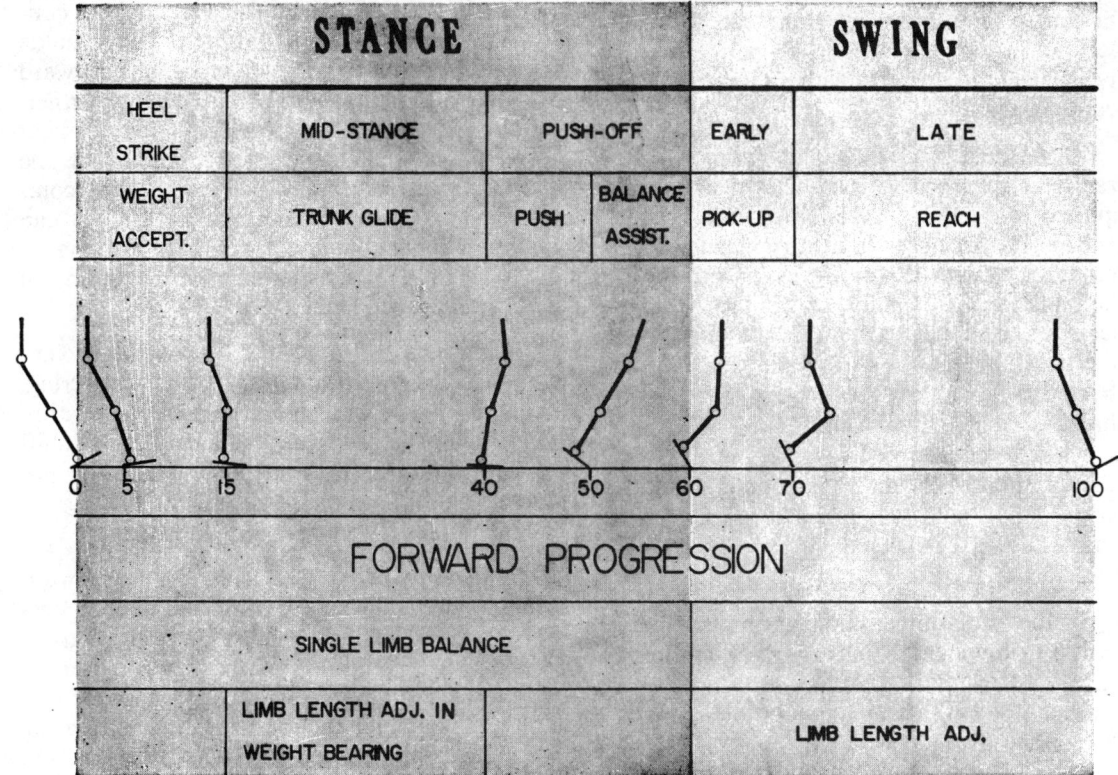

FIG. 5. The relationship between the three basic patterns of action (forward progression, single limb balance and limb length adjustment) and the subdivisions of the walking cycle are presented diagramatically. The subdivisions of the walking cycle have been identified both by time intervals (heel strike, mid-stance, etc.) and by the task to be accomplished (weight acceptance, trunk glide, etc.)

Weight Acceptance

Heel strike represents a moment of great change in demands (Plate 1). Just before the heel touches the ground, the limb was swinging forward quite rapidly as a result of its previous push-off and the active flexion at the hip and knee. To reach its forward position in time, the limb has to travel at approximately five miles an hour. At the same time the body has also received a recent forward push from the other foot so that its traveling speed is about two miles an hour. Consequently, there is considerable forward momentum in effect at the time the heel strikes the ground. Ground contact causes the foot to stop its forward travel abruptly while momentum is still tending to carry the tibia forward. If uncontrolled, this would cause the knee to collapse and the limb would be unable to support weight at the same time weight is being rapidly shifted forward and laterally from the other limb. As a result, the functional demand in the post-heel strike period is effective weight acceptance without impeding forward travel.

When the heel strikes the ground the extremity is stretched forward with the hip flexed approximately 30 degrees, the knee fully extended and the ankle is at a right angle. The weight is transmitted to the ground through the tibia, but the point of contact is at the heel, which is approximately one-third of the foot length behind the axis of the tibia. Through the leverage of this heel length, a downward thrust is created that would cause the foot to slap if it were not controlled. Control is by a rapid response of the ankle dorsiflexors (anterior tibialis and the toe extensors). Their action allows the forefoot to touch the ground gradually without a slap.

At the same time, momentum is carrying the tibia forward so that a rocker motion results which allows total forward progression without any abrupt changes in direction. If this early and rapid forward travel of the tibia were not restrained, however, it would advance to a point where the extremity would become unstable because of excessive knee flexion. This is avoided by prompt action of the soleus and posterior tibial muscles which create a relative plantar flexing force. Under such control the tibia advances gradually and in a manner consistent with the dual demands of forward progression and extremity stability. Direct restraint of knee flexion is also gained by action of the quadriceps.

During this entire period of weight acceptance the body weight is still behind the weight bearing foot and the hip is in flexion. This is an unstable position, for without control the body weight would tend to force the hip into further flexion. Action by the "hamstrings" and the gluteus maximus restrain the tendency toward hip flexion and leads to gradual hip extension.

Immediately following heel strike, two actions are occurring in the line of forward travel. One is rapid plantar flexion of the foot which is controlled by the ankle dorsiflexors. The other is rapid tibial advancement causing knee flexion, which is controlled by the soleus and posterior tibial, acting at the ankle to restrain the tibia while the quadriceps acts directly on the knee to give it support. Hip extension is also essential to control the trunk-thigh relationships. If the ankle is stabilized so the tibia cannot fall forward and the hip is prevented from flexing further, momentum will carry the thigh–body segment forward and extend the knee. Thus while the quadriceps is a very normal and useful component, it is not essential if the patient can maintain a continuous flow in his walking so as to have the assistance of momentum, in addition to tibial stability.

In addition to the forward progression challenges to weight acceptance that are occurring, demands for single limb balance also arise (Fig. 2). The latter necessitates a rapid shift laterally and a strong response from the abductor muscles. It also means protecting the knee and ankle from the valgus thrust.

As a result of the dual demands of forward progression and single limb balance, the customary data charts (Fig. 9) indicate activity in most of the muscles of the lower extremity within the short period following heel strike.[9,11] The data on joint motion [7] indicate decreasing hip extension, increasing knee flexion and quick ankle plantar flexion followed by gradual dorsiflexion. The force charts [12] show very rapid transfer of weight onto the stance foot so that 95 per cent of the weight has been transferred within the first 10 per cent of the walking cycle (approximately 0.1 sec.). All of the work of extremity stabilization and smooth weight transfer is accomplished by the 15 per cent mark.

Basic Research

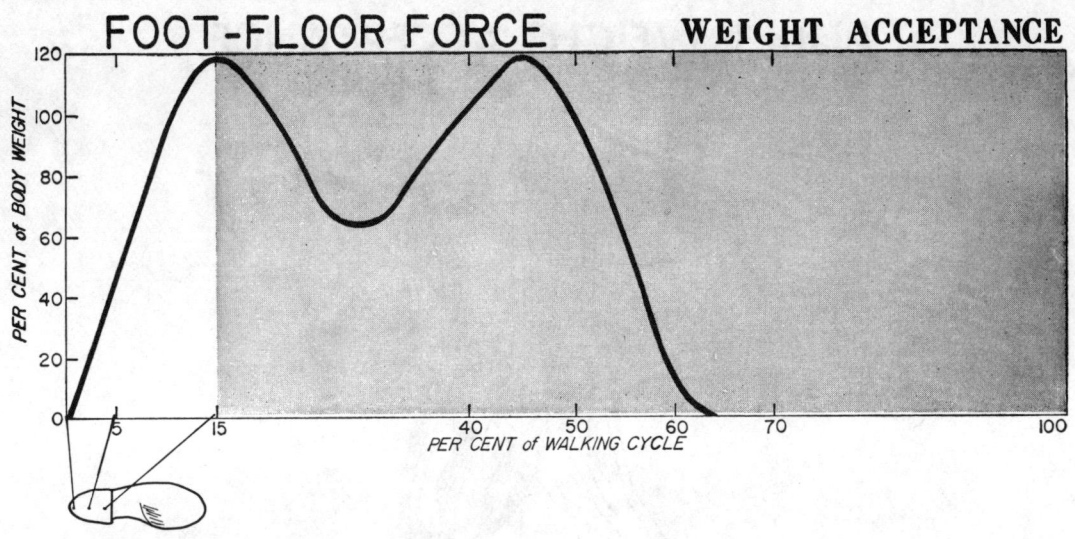

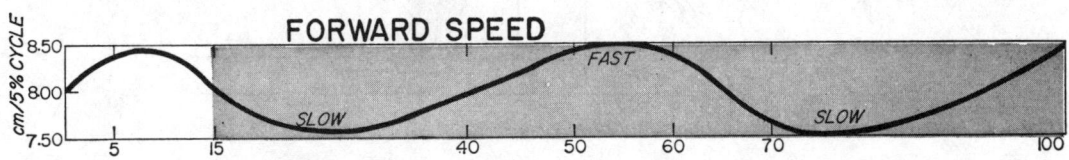

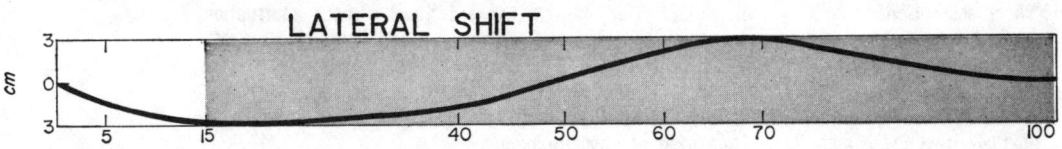

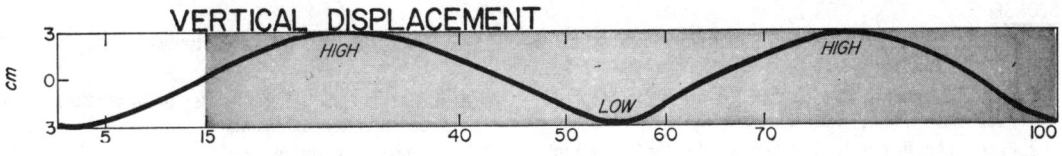

Plate 1. WEIGHT ACCEPTANCE

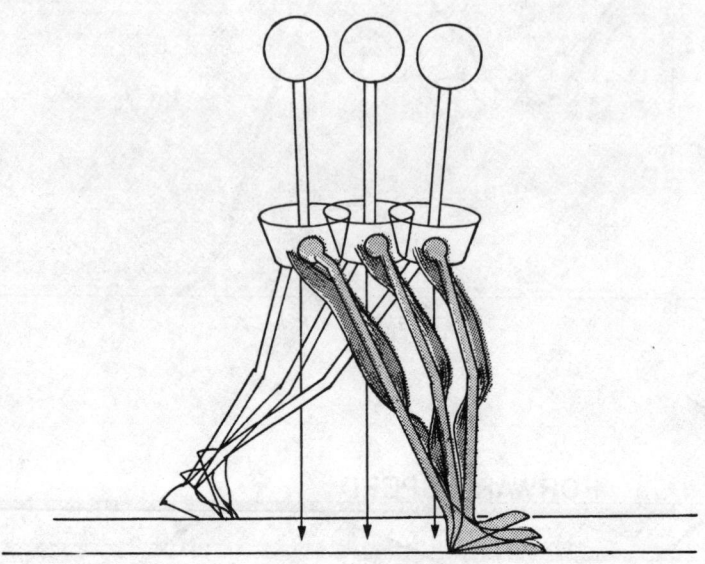

TASK: WEIGHT ACCEPTANCE (INTERVAL: HEEL–STRIKE)
Heel–Strike to Foot–Flat: 0 to 15% of Walking Cycle.

DEMANDS:
1. Shock absorption
2. Limb stabilization
3. Forward travel without interruption
4. Balance on one limb

SITUATION:
1. Strong forward momentum just before heel strike
   a. Body traveling 2 mph (force from push of opposite limb)
   b. Swing limb traveling 5 mph (force from own push plus active hip and knee flexion)
2. Extremity reaching ahead of body
3. Heel strike abruptly stops forward travel of foot; momentum now concentrated on lower leg (tibia)

RESPONSE:

Events
FORWARD PROGRESSION
1. Immediate plantar flexion (due to ground contact of heel, body weight along tibia).
2. Rapid knee flexion to 15° (due to tibial advancement with thigh and trunk aligned behind foot).

3. Hip flexion tendency (due to body weight being behind weight bearing foot).

SINGLE LIMB BALANCE:
1. Tendency to fall away from support limb.

2. Valgus thrust on knee secondary to lateral shift.

3. Valgus thrust on ankle.

Anatomical Activity
1. Restraint by ankle dorsiflexors: anterior tibialis, great and common toe extensors.
2. Knee flexion restrained by:
   a. Tibial advancement restrained by soleus and posterior tibialis
   b. Quadriceps activity
   c. Thigh stabilization through hip extensor activity by semitendinosis, biceps (long head), gluteus maximus.
3. Reversed by hip extensors and forward momentum.

1. Lateral shift of body. Pelvis stabilization by hip abductors: Gluteus medius, gluteus minimus, tensor fascia femoris.
2. Restrained by medial knee muscles: vastus medialis, semitendinosis, gracilis.
3. Restrained by posterior tibialis and medial insertion of soleus.

Basic Research

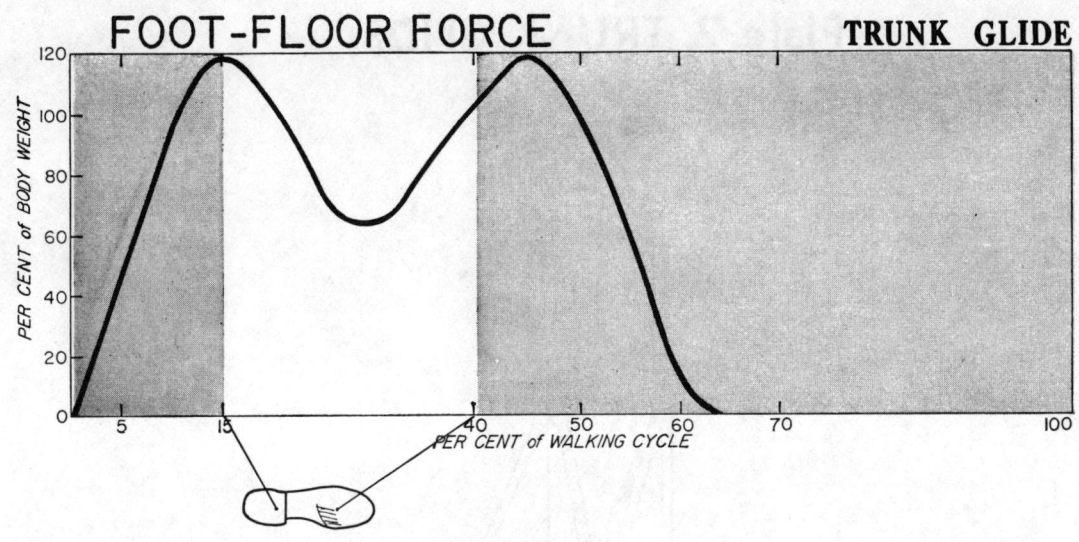

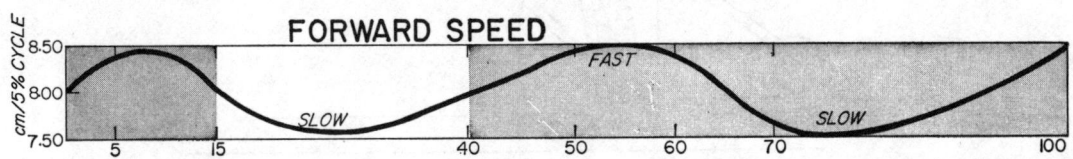

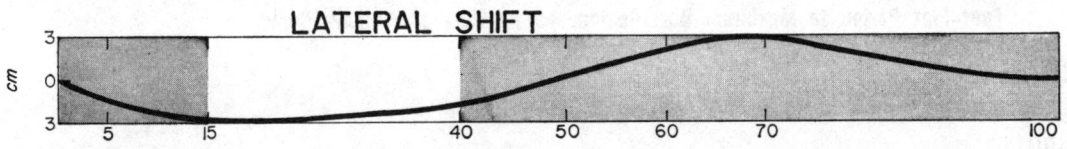

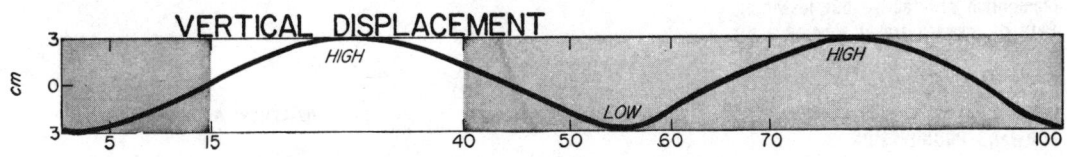

Plate 2. TRUNK GLIDE

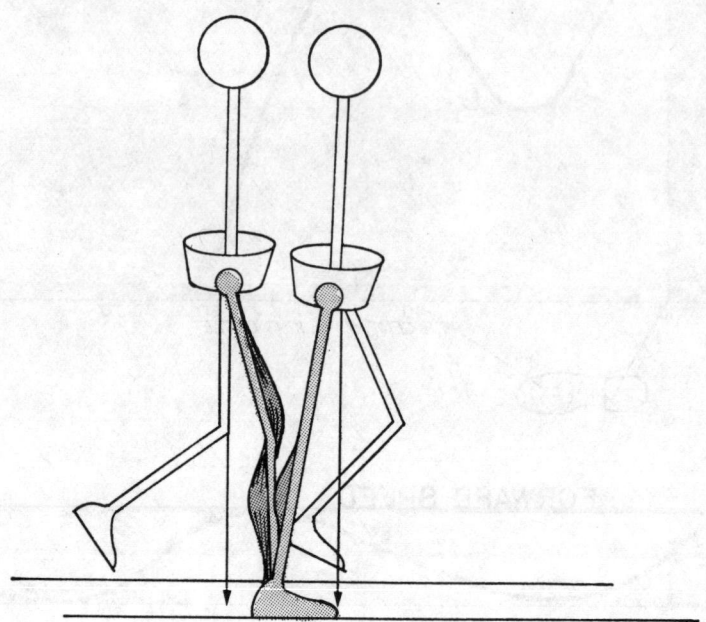

TASK: TRUNK GLIDE (INTERVAL: MID–STANCE)
Foot–Flat Period to Maximum Dorsiflexion: 15 to 40% of Walking Cycle

DEMAND:
Continue forward travel of body over flat foot.

SITUATION:
Complete single limb support has been attained.
Foot flat on the ground.
Extremity stable.
Momentum still active but lessening.
Rate of forward travel slowing a bit.

RESPONSE:

Events	Anatomical Activity
FORWARD PROGRESSION	
1. Momentum carries trunk and limb segments forward over stationary foot.	1. Rate of advancement controlled by tibial restraint: soleus and posterior tibialis activity.
a) Knee extended as thigh advancement over stable tibia.	a) Quadriceps quiet.
b) Hip extended by thigh advancement.	b) Hip extensors quiet.
2. Body weight passes from behind heel to over forefoot.	2. Ankle advances from 5° plantar flexion to 10° dorsiflexion.
SINGLE LIMB BALANCE:	
1. Total single limb support.	1. Continued hip abductor activity.
2. Lateral shift maximum at 20% point, then starts to decrease.	2. Knee stress relieved and protector muscles relaxed.
LIMB LENGTH ADJUSTMENT:	
Other limb swinging forward.	Simultaneous abduction, internal rotation and extension demand on weight bearing hip joint.

Basic Research

MECHANICS OF WALKING

Trunk Glide

Following the great challenge of weight acceptance, there is a quiet period of coasting forward over the flat foot (Plate 2). Extremity stability and balance having been attained in the first period, little active effort is required now. Momentum appears to be the main propelling force as the body glides forward. In the course of this travel body weight changes its alignment from behind the heel to over the forefoot. To attain this position with minimal expenditure of energy, tibial advancement is rigorously controlled by the continued action of the soleus and posterior tibialis muscles. This allows momentum to decrease the demands of the hip and knee so that the hip muscles drop out very quickly. The quadriceps become inactive by the time the thigh has reached the vertical position, and the ankle is in about 10 degrees of dorsiflexion by the time the weight is aligned over the forefoot.

Throughout this period the body is still balanced over one leg. Hence, the hip abductors are still very active. The lateral shift, however, diminishes during the latter part of the interval as the body prepares to approach the other limb. In addition to sustained abduction stability, there is progressive internal rotation (recovery from external rotation) as the pelvis swings forward with the other limb. This accounts for the strong action of the tensor fascia femoris. This gliding period might be considered as a rest between intervals of intense work. It accounts for 25 per cent of the walking cycle, or almost half of the stance phase.

The clinical significance of this period is the need for a range of dorsiflexion at the ankle. If the tibia will not advance about 10 degrees in front of the vertical position the person loses the stabilizing effect of momentum at the hip and knee (Fig. 6A). To stand erect his knee must go into a considerable degree of hyperextension (Fig. 6B). If this is not possible, his only other means of remaining upright is to lean forward at the hips—a posture that requires good strength of the hip extensor muscles or good arm supoprt (Fig. 6C).

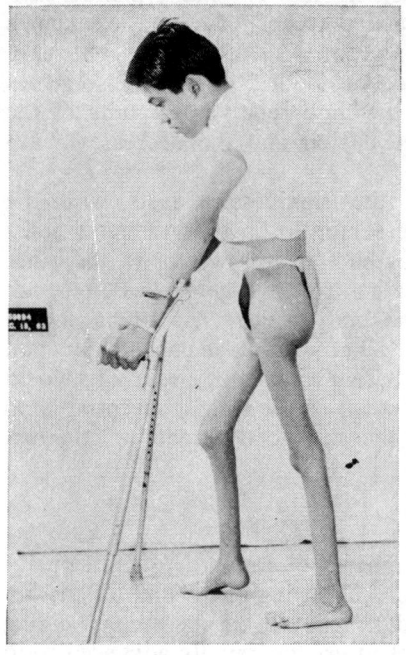
A.

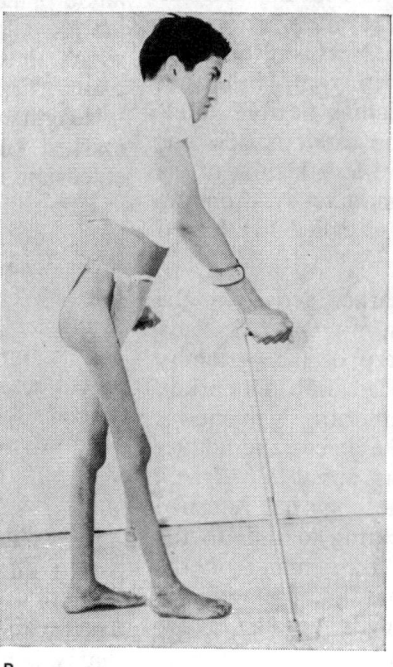
B.

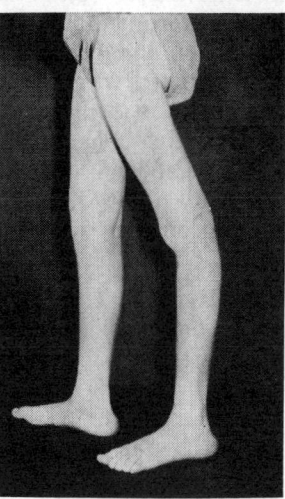
C.

FIG. 6. The patient in A and B has complete paralysis of both lower extremities resulting from poliomyelitis. He has bilateral ankle (pantalar) fusions. The left foot in A (the posterior foot in the photo) was stabilized in 10° of dorsiflexion. This allows him to balance his weight over the forefoot with the hip and knee extended. Minimal crutch support is needed and no deformities have developed. In contrast, the right foot in B was fused in the traditional position of 15° equinus. To bring his trunk forward over the foot on weight bearing requires considerable hip flexion. Lacking hip extensor muscles requires him to put all his weight on his arms. The relative lengthening of the right limb combined with his inability to support weight on it has encouraged hip and knee flexion deformities which add further to his instability. Fig. 6C: Lack of ankle dorsiflexion leads to exaggerated knee hyperextension as the body weight is brought forward of the weight bearing foot.

Push

At the end of the gliding period the body weight is in front of the foot, the knee is extended, and the heel is just rising to support the ankle against the dorsiflexing influence of the body weight that is so far forward (Plate 3). The foot is also preparing to push the body forward again. The rest of the plantar flexors become active. The gastrocnemius, the peroneals and toe flexors join the posterior tibialis and the soleus which continue their activity. The flexion action of the gastrocnemius on the knee is controlled by the forward position of body weight. Being anterior to the foot, it locks the knee in extension (there is no quadriceps activity at this time). As a result, all gastrocnemius action is at the ankle.

The combined push of the seven plantar flexor muscles creates a ground force that exceeds body weight by about 20 per cent. The speed of forward progression is increased and one might say the patient is propelled forward by the plantar flexors pushing against the ground. In the meantime the other extremity has come forward to catch the body weight as it advances. Clinically this means that the plantar flexion force is extremely important for smooth and efficient gait. This also means that the body weight is far in front of the foot and if the person lacks plantar flexion stability he cannot come up on his forefoot. He has lost a component of relative lengthening of his extremity and will accommodate by dropping the hip on that side (the so-called flat footed gait).

During this period of marked activity at the ankle, hip control has been minimal. By having the body weight forward of the extremity the hip is passively pushed into extension. Excessive strain of the anterior ligaments is avoided by the actions of the iliacus and adductor longus. As weight is just about to be transferred to the other foot, the body has returned to the midline and is preparing to shift to the other side. This lessens the demand on the hip abductors. At the same time the adductor magnus and adductor longus become active to control the shift medially.

Balance Assist

Almost immediately after the peak of the push-force, there is a rapid drop in the amount of weight being supported by that foot, yet the toe remains in contact with the ground. This is an interval of double limb support for the weight that is being rapidly transferred from one limb to the other (Plate 4). The continued contact with the floor by the toe of the limb that is discharging its weight, would seem to serve as a means of assisting in balance as the weight is being accepted by the other limb.

The knee flexes rapidly to a position of about 65 degrees. This appears to result largely from release of body weight on the taut gastrocnemius, as the gracilis is the only other flexor muscle active. Subsequent rectus femoris action at the end of this interval restrains the extent of flexion. It only reaches 70 degrees by the end of the next period.

At the same time as the knee is flexing rapidly, the hip is flexing at a more gradual rate as it recovers from the extended position. Again, the active musculature seems to be the adductors. They are probably providing the dual role of flexion as well as restraint of abduction as weight is being transferred across the midline to the other foot. During this interval of hip and knee flexion, toe contact is maintained by a proportional increase in ankle plantar flexion. The many plantar flexors are still active except for the gastrocnemius which has become silent. This extensive plantar flexion of about 20 degrees also serves to lengthen the limb relatively in spite of the marked knee flexion and the decrease in hip extension.

Clinically the significance rests with the ability of the person to have a graduated assist in balance as he transfers weight to the other side. When controlled plantar flexion is not available, the body weight has to be shifted at one time. The abrupt change in direction would be apparent as a limp, and it also would increase the work of the weight-accepting limb because of the increased impact of the exchange.

Pick-up

The final phases of forward progression relate to the forward swing of the limb (Plate 5). The extremity, relieved of its weight-bearing duties, is picked up and rapidly advanced from behind the body to in front of it.

Because the forefoot protrudes several inches beyond the anterior surface of the tibia, the toe is pointing down any time the hip is extended behind the body, unless there is extreme ankle dorsiflexion. This downward pointing is even greater when the knee is flexed. Thus,

Basic Research

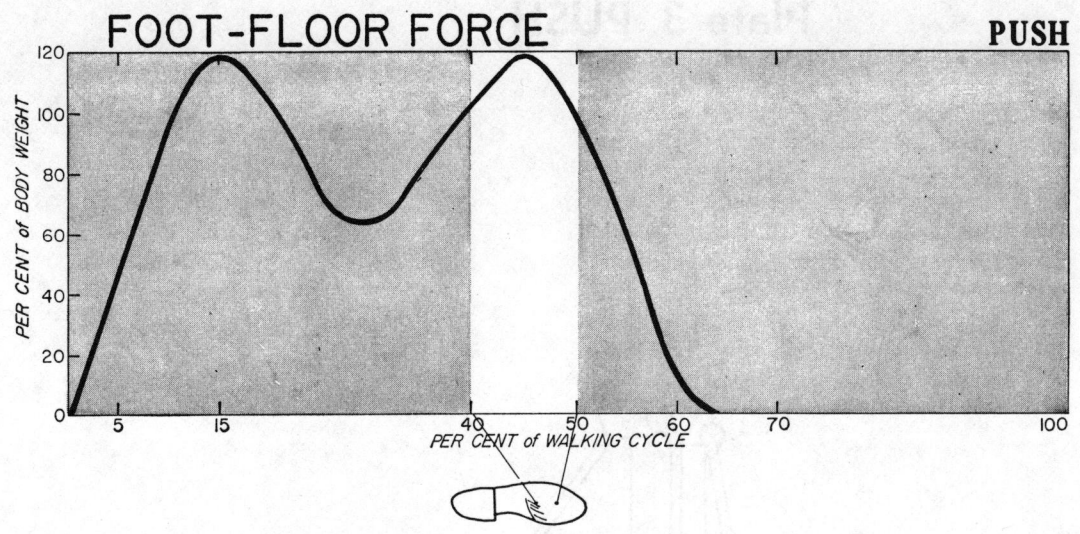

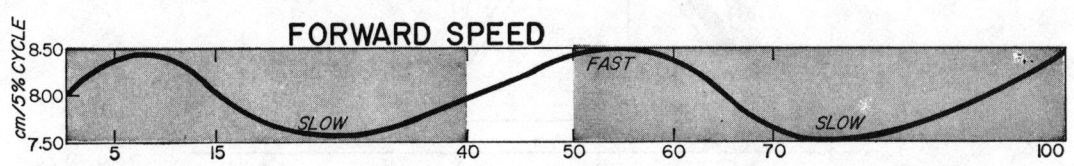

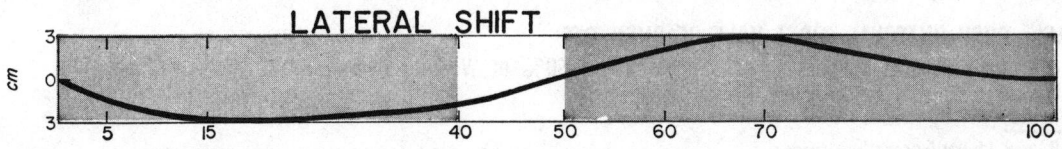

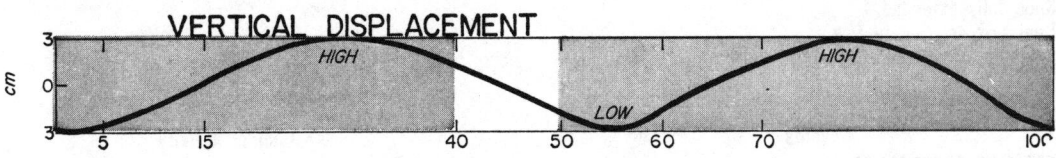

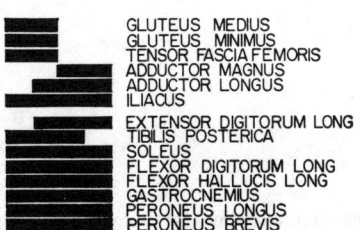

Plate 3. PUSH

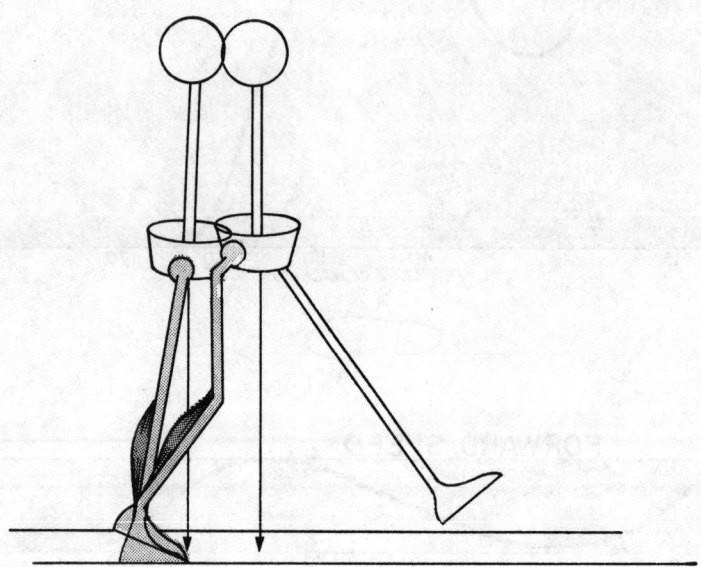

TASK: PUSH (INTERVAL: FIRST HALF OF PUSH–OFF)
Heel–Rise to Maximum Push Force: 40 to 50% of Walking Cycle

DEMAND:
 Renew forward propelling force.

SITUATION:
 Body slightly ahead of foot.
 Knee fully extended.
 Heel just starting to rise.
 Ankle in 10° dorsiflexion.

RESPONSE:

Events	Anatomical Activity
FORWARD PROGRESSION	
1. Body weight tends to pull:	
a. Hip into more extension	1. a. Hip extension restrained by iliacus.
b. Knee into more extension	b. Knee extension restrained by gastrocnemius to 10° flexion.
c. Ankle into more dorsiflexion	c. All seven plantar flexors active: gastrocnemius, peroneus longus and brevis, great and common long toe flexors join soleus and posterior tibialis which continue activity.
2. Create push force.	2. Increased activity of all seven plantar flexor muscles.
SINGLE LIMB BALANCE:	
1. Trunk returns to midline in preparation for weight transfer to other limb.	1. Hip abductors relaxed by middle of period.
2. This creates passive abduction of hip.	2. Shift controlled by hip adductors longus and magnus.

Basic Research

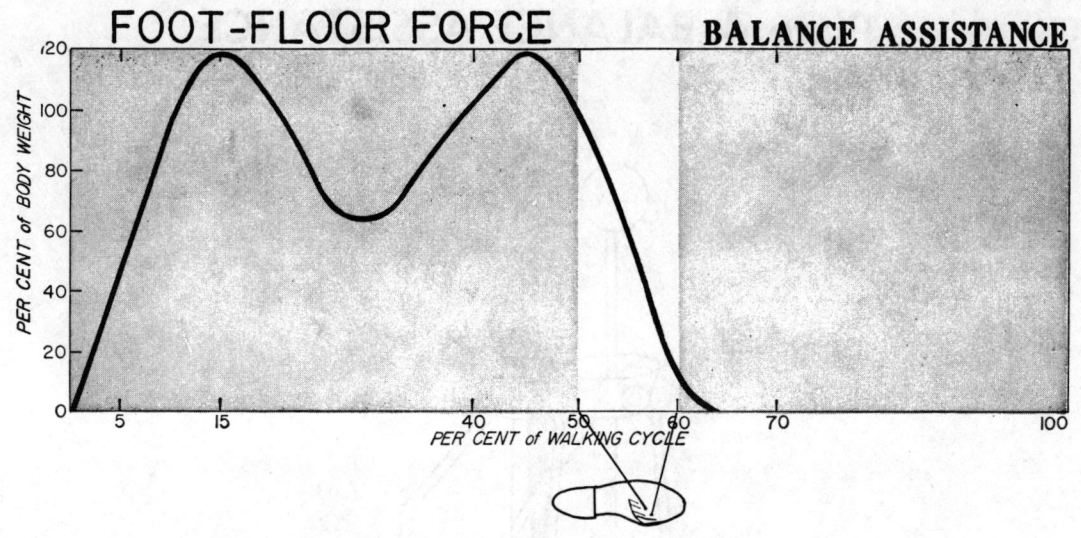

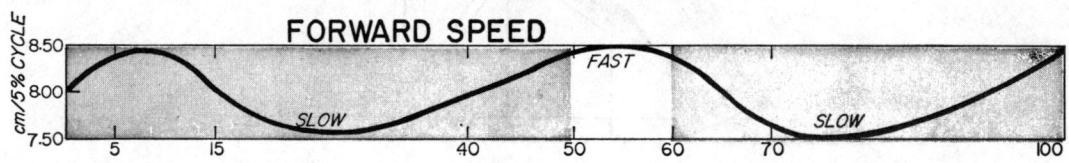

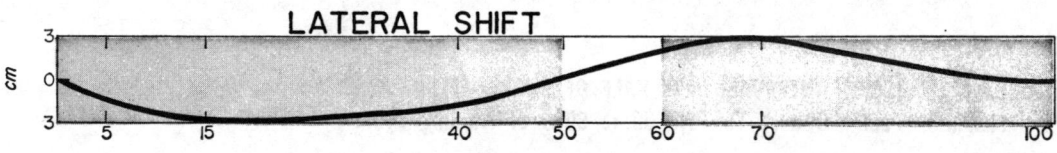

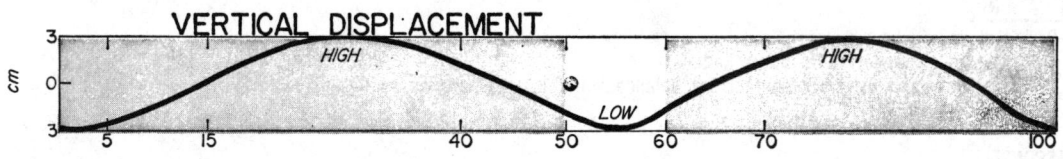

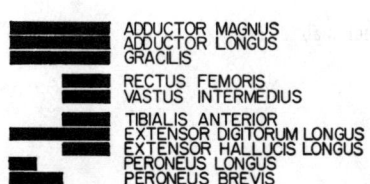

Plate 4. BALANCE ASSISTANCE

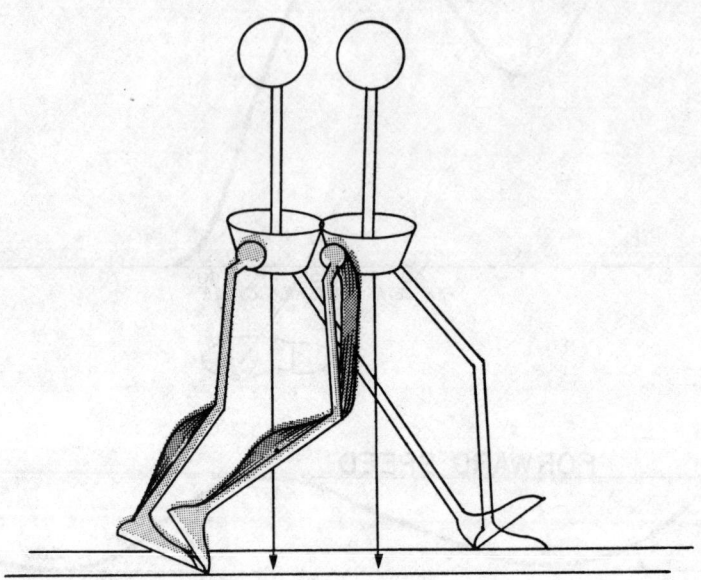

TASK: BALANCE–ASSIST (INTERVAL: LAST HALF OF PUSH–OFF)
Maximum Push Force to Toe–Off: 50 to 60% of Walking Cycle

DEMAND:
Assist body balance as other limb "struggles" to accept weight.

SITUATION:
Period of double limb support.
Weight rapidly transferred to other limb.
Primary limb maintains floor contact for balance while it prepares for swing.
Body well ahead of limb.

RESPONSE:

Events	Anatomical Activity
FORWARD PROGRESSION:	
1. Rapid weight transfer removes resistance at knee and ankle.	1. Rapid and marked passive knee flexion (0 to 50°). No knee flexor muscle activity evident.
2. Floor contact maintained.	2. a. Postural equinus due to forward tipping of tibia by the knee flexion with the hip extended. b. Active plantar flexion: only gastrocnemius and posterior tibialis silent.
	3. Hip extension lessens (−10° to 0°) Adductor longus and magnus active (Let's not quibble whether this is hip joint or pelvis motion).
SINGLE LIMB BALANCE (lateral alignment) Period of double limb support. Weight shifting rapidly across midline to other foot.	Adductors (magnus and longus) restrain lateral shift, hence add stability.

Basic Research

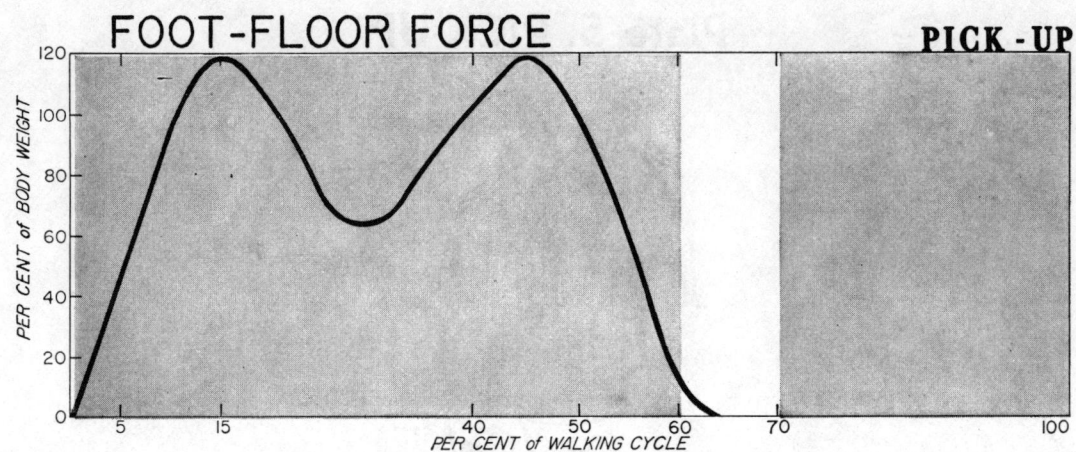

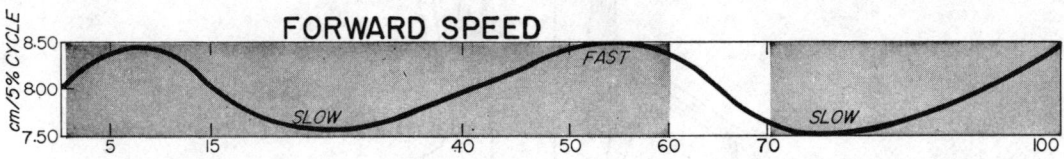

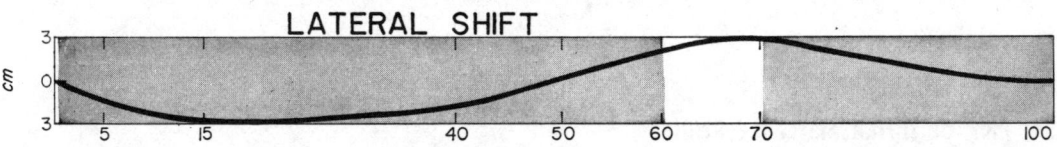

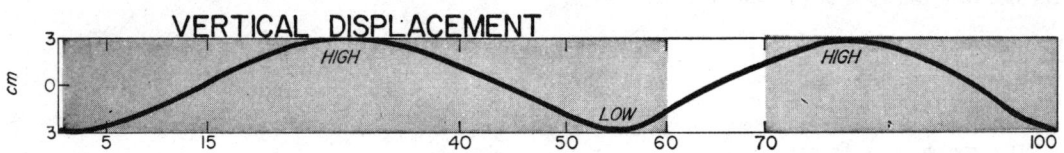

MUSCLE ACTION

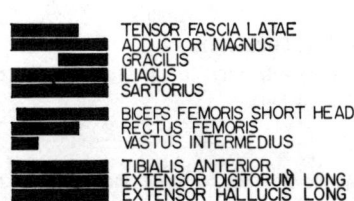

Plate 5. PICK-UP

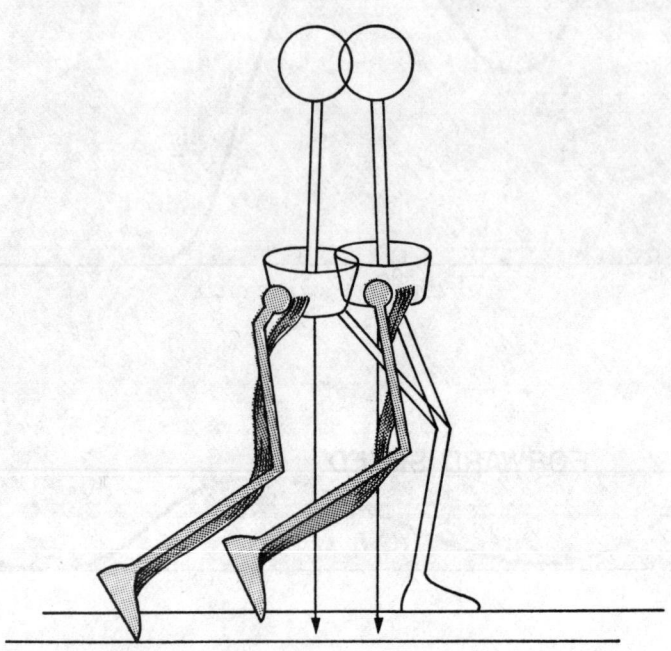

TASK: PICK-UP (INTERVAL: EARLY SWING)
   Toe-Off to End of Knee Flexion: 60 to 75% of Walking Cycle

DEMAND:
   Lift foot from ground in preparation for forward reach.

SITUATION:
   Weight entirely on other limb
   Extremity far behind body axis
   Toe extended down toward ground as a result of:
   1) the marked knee flexion
   2) the length of foot that protrudes beyond the line of the leg
   3) ankle in maximum equinis from assisting balance

RESPONSE:

Events	Anatomical Activity
FORWARD PROGRESSION:	
1. Entire extremity lifted to overcome postural and true equinus.	1. a. Active hip flexion (0° to 5°) by: iliacus, sartorius, tensor fascia femoris.
	b. Active knee flexion (50° to 70°) by: biceps femoris (short head), sartorius.
2. At toe-off, foot posterior and lateral to axis of body.	2. Extremity brought toward midline by Adductor magnus.
LIMB LENGTH ADJUSTMENT:	
1. Limb shortened to aid toe clearance.	1. Pelvis rotates forward from its maximum posterior position.

Basic Research

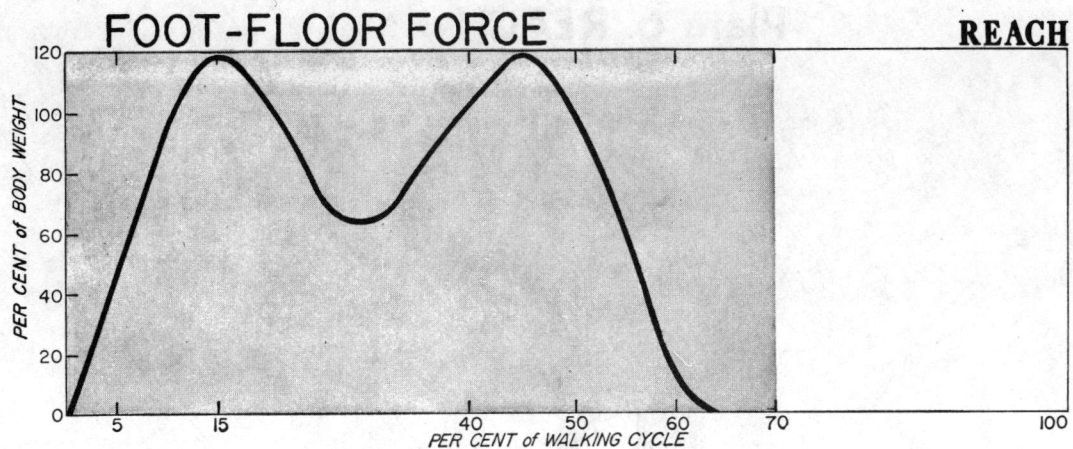

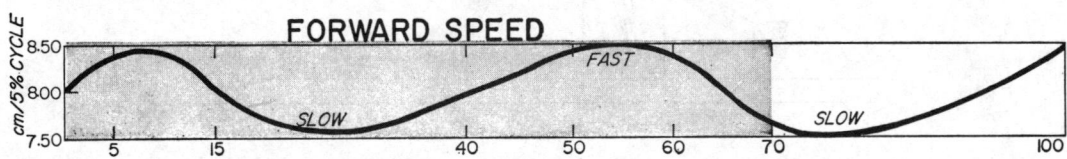

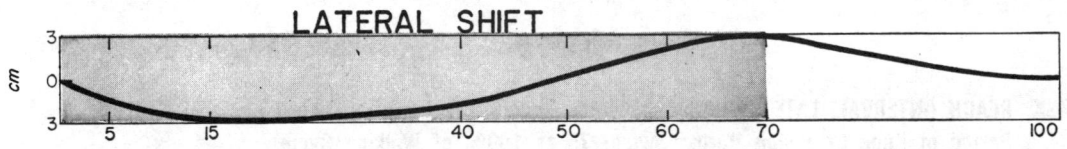

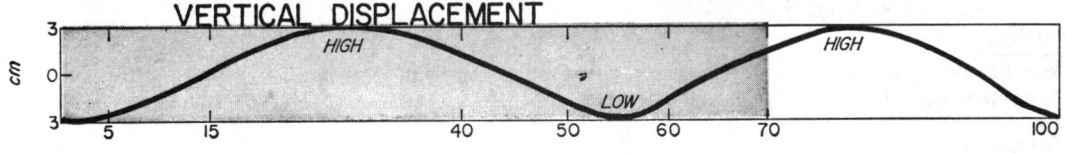

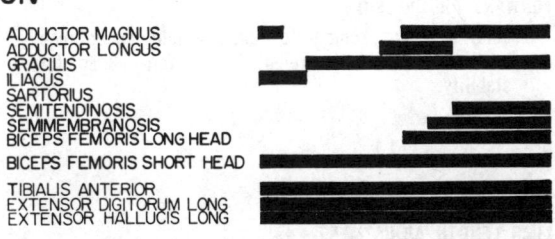

These graphs adapted from data published by Murray [7] and Cunningham.[12]

Plate 6. REACH

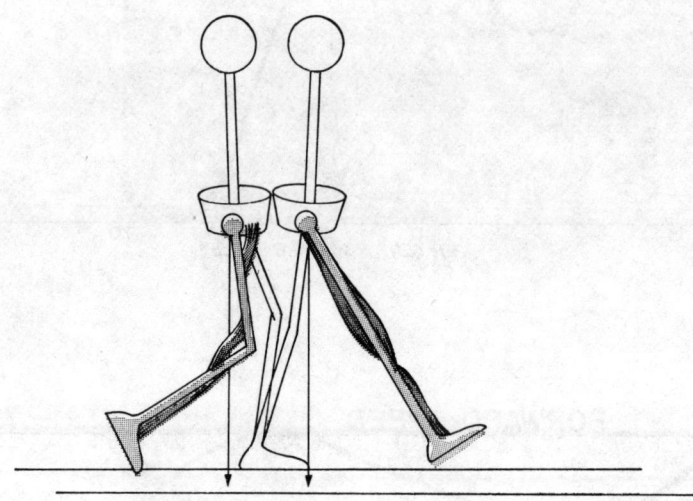

TASK: REACH (INTERVAL: LATE SWING)
  Period of Knee Extension During Swing: 75 to 100% of Walking Cycle

DEMAND:
  Advance foot for next step in forward progression.
  Be ready to receive the advancing body weight.

SITUATION:
  Body traveling forward as a result of previous push and stance activity of other limb.
  Extremity suspended in a flexed posture at every joint.
  Foot still behind axis of body.
  Toe clear.

RESPONSE:

Events	Anatomical Activity
FORWARD PROGRESSION:	
1. Limb advances rapidly to reach weight acceptance position before body weight is too fare ahead for stability.	1. Knee extends rapidly from its 70° flexed posture by relaxation of flexors and pendulum effect. Extensors (Vasti) become active at end of period to maintain full knee extension. Hip flexion increased slightly (to 30°) and maintained by adductors.
2. Toe kept clear of ground.	2. Active dorsiflexion.
LIMB LENGTH ADJUSTMENT:	
1. Limb lengthened.	1. Pelvis continues to rotate with advancing limb. Also drops into further adduction.

MECHANICS OF WALKING

there is relative equinus of the foot even with the ankle in a neutral position.

Consequently, toe clearance is a combination of hip flexion, knee flexion and ankle dorsiflexion. It has always been difficult for investigators to determine the exact moment of toe-off because the decrease in ground force is a gradual lessening of contact during balance assist, and the amount of ground clearance is minimal. For maximum efficiency, no extra energy is expended and the toe just barely clears the ground by about a centimeter. At the same time, the hip and knee motions are smooth continuations of the postures started during the balance assist phase. The only real change is the abrupt shift from active plantar flexion while toe contact is being maintained to active dorsiflexion just after toe clearance has begun. The toe is prevented from dragging primarily by the continuance of the hip and knee flexion motions which lift the whole limb. By the end of the pick-up phase, the knee has flexed to a maximum 70 degrees. The hip and ankle are about mid-way in their flexion course. Because of the extensive knee flexion, the foot is still behind the line of the body.

Knee flexion is accomplished by the short head of the biceps femoris and the sartorius. The iliacus, sartorius, and tensor flex the hip. This is a significant interval clinically because the patient who has inadequate hip or knee flexion will drag his toe despite adequate ankle control. Thus the natural inclination to apply a short leg brace to anyone who drags his toe must be restrained. During the period of pick-up, toe drag is due to the relative equinus of the foot and inadequate elevation of the entire extremity.

Inability to lift the extremity may be the result not only of paralytic difficulties, but also of restrictions of joint motion. The person who has considerable knee stiffness following prolonged immobilization for femoral shaft fractures, hip fusions, and so forth, must hike his pelvis to accommodate for this loss of motion. A fused hip would require even more knee flexion to clear the toe. Seldom do such patients have the excessive dorsiflexion that could substitute because the prolonged immobilization has included the ankle joint as well. Hence, an individual needs a minimum of 70 degrees to clear his toe on smooth ground, more on any rough terrain and over 100 degrees for negotiating stairs.

Reach

Having cleared the ground, the extremity is now stretched forward to prepare for the oncoming task of weight acceptance (Plate 6). This is accomplished by continuing hip flexion to a final position of 30 degrees. The so-called primary flexors of the hip which were active during pick-up are now silent even though hip flexion continues and is maintained. Undoubtedly much of this is momentum from the previous impetus but is also sustained by the gracilis, adductor longus and adductor magnus which are again active. Knee flexion is continued by the short head of the biceps femoris. There is a brief burst of rectus femoris activity at the end of the pick-up period just before the knee starts extending. The quadriceps are otherwise silent until the end of the reach when the vasti become active to maintain this extended posture in preparation for weight acceptance. At the same time the semimembranosus, semitendinosus and long head of the biceps become active, presumably to restrain the forward momentum of the limb at both the hip and knee. Hip flexion restraint assures a logical position for ground contact and knee restraint protects the ligaments from strain.

The foot attains a position of just less than neutral (5° equinus) and maintains it until heel strike. This is under the continued activity of the dorsiflexor muscles (tibialis anterior, extensor digitorum longus, extensor hallucis longus). The clinical problem of toe drop is now directly related to anterior ankle control.

The meagerness of muscular activity through this swing phase indicates the ease of swinging the extremity if there has been an adequate push-off. The greater magnitude of knee flexion range and knee flexor muscle activity, as compared to the hip, also indicates which area has the greatest need. This is borne out clinically for seldom need we be concerned about hip flexor activity unless the person lacks knee control as well. A further aid to clearing the foot is the fact that the weight is on the opposite leg and thus that side of the pelvis is relatively elevated even though it never comes above the horizontal.

As the extremity reaches forward, it is also being relatively lengthened by associated pelvic rotation and tilt (Fig. 4). In the course of the total swing phase from the moment of toe-off until the heel strikes once again, the extremity has recovered from relative lengthening to accommodate the posterior reach, attains its maximally shortened posture as it passes the vertical line and then again lengthens in order to reach forward. The pelvis both lifts and rotates forward as the limb advances to the midline. After this point the pelvis drops while it continues the forward rotation.

SUMMARY

The following points have the greatest clinical significance for normal walking.

(1) As one starts to balance on one limb, there is a definite valgus thrust on the knee and foot as the individual shifts his weight on to the weight-bearing foot (Fig. 2).

(2) By utilizing the pelvis as a means of increasing the relative length of the limb, the process of swinging one extremity from behind to forward carries the hip of the weight-bearing limb through adduction, internal rotation and extension while it is supporting full body weight (Fig. 4). Similar rotatory stresses are made on the knee and foot.

(3) During the forward progression complex of actions, the single greatest factor preventing the knee from buckling on weight acceptance and trunk glide is control of the tibia through strong action of the plantar flexor muscles (Plate 1). The calf muscles must be strong or they must be replaced by contracture, fusion or bracing. Hip extensor control is the second significant factor. Direct quadriceps stabilization of the knee is the least important though, of course, very useful when present.

(4) The body weight is advanced in front of the foot only if approximately 10 degrees of dorsiflexion is available (Plate 2). Otherwise, there will be a strong force to hyperextend the knee, and the hip must be able to support weight in flexion.

(5) There is a postural equinus at the beginning of the swing phase because the entire foot is tipped downward by the marked knee flexion that is present (Plate 5). Consequently, toe drag in this period is due to inadequate pick-up activity of the hip and knee flexors. Later, as the extremity reaches forward for the next step, continued toe drag would be the result of inadequate anterior ankle control.

CONCLUSION

Whenever the patient lacks the required joint range or muscle response (strength or timing), he will limp or have to exaggerate other actions to compensate for the deficiencies.

In treating patients the goal is safe, efficient ambulation. Encouragement to eliminate a limp or walking aid should be given only if the patient can afford the extra effort and tissue strain resulting from compensatory actions.

These goals for the patient can only be achieved if the physician and the therapist have a working knowledge of the mechanics of walking, and can apply these concepts to their evaluation and program planning for the individual patient and his specific problem.

REFERENCES

1. Eberhart, H. D., V. T. Inman, J. B. DeC. M. Saunders, A. S. Levens, B. Bresler and T. D. McCowan, Fundamental Studies of Human Locomotion and other Information Relating to the Design of Artificial Limbs. A Report to the National Research Council, Committee on Artificial Limbs. University of California, Berkeley, 1947.
2. Eberhart, H. D., V. T. Inman, and Boris Bresler, The Principal Elements in Human Locomotion, Chapter 15. In P. E. Klopsteg and P. D. Wilson: *Human Limbs and Their Substitutes*. McGraw-Hill Book Co., New York, 1954.
3. Saunders, J. B. DeC. M., V. T. Inman, and H. W. Eberhart, The Major Determinants in Normal and Pathological Gait. J. Bone Jt. Surg., 35-A:543–558, July 1953.
4. Elftman, H., A Cinematic Study of the Distribution of Pressure in the Human Foot. Anat. Record, 59:481–491, 1934.
5. Hubbard, A. W., and R. H. Stetson, An Experimental Analysis of Human Locomotion. Am. J. Physiol., 124:300–314, 1938.
6. Schwartz, R. P., A. L. Heath, W. Mislek, and J. N. Wright, Kinetics of Human Gait. J. Bone Jt. Surg., 16:343–350, 1934.
7. Murray, M. P., A. B. Drought, and R. C. Kory, Walking Patterns of Normal Men. J. Bone Jt. Surg., 46-A: 335–360, 1964.
8. Murray, M. P., and B. H. Clarkson, The Vertical Pathways of the Foot During Level Walking: I. Range of Variability in Normal Men. J. Amer. Phys. Ther. Assoc., 46:585–589, 1966.
9. Sutherland, David H., An Electromyographic Study of the Plantar Flexors of the Ankle in Normal Walking on the Level. J. Bone Jt. Surg., 48A:66–71, January 1966.
10. Close, J. R. and F. N. Todd, The Phasic Activity of the Muscles of the Lower Extremity and the Effect of Tendon Transfer. J. Bone Jt. Surg., 41-A:189–208, 1959.
11. The Pattern of Muscular Activity in the Lower Extremity During Walking. A Presentation of Summarized Data. Prosthetic Devices Research, Institute of Engineering Research, University of California, Berkeley. Series II, Issue 25, September 1953.
12. Cunningham, D. M., Components of Floor Reactions During Walking. Prosthetic Devices Research Project, Institute of Engineering Research, University of California, Berkeley. Series II, Issue 14, November 1950 (Reissued, October 1958)

Basic Research

Metabolic Energy Cost of Unrestrained Walking

RAYMOND L. BLESSEY, MA,
HELEN J. HISLOP, PhD,
ROBERT L. WATERS, MD,
and DANIEL ANTONELLI, MA

Physiologic factors of metabolic energy cost as well as selected mechanical characteristics of gait are described in a group of 40 presumably normal men and women between the ages of 20 and 60 years. Special emphasis was placed on unrestrained (free cadence) walking to provide a reliable baseline for comparison to persons with physical gait impairments. Velocity of walking was the most important factor in determining oxygen uptake and was independent of age or sex. An empirical equation was established relating oxygen uptake to the speed of walking. The average velocity for men in unrestrained walking trials was 89 meters per minute; for women 74 meters per minute. These differences were related to the greater stride length in men. The average cadence selected by both sexes was 116 steps per minute. None of these factors was age dependent. Of the physiologic measures only systolic blood pressure was age dependent predictably rising with age. The mean heart rate was 103 beats per minute and did not vary significantly between men and women. The mean respiratory rate was 19 per minute; the mean respiratory quotient, 0.85, neither being age or sex dependent. Oxygen uptake values averaged 12.95 ml/kg-min for the population studied and again were neither age nor sex dependent.

Although a number of determinations of the metabolic energy cost of normal walking have been reported, no data have emerged which can be used reliably as standards from which the penalties incurred by persons with physical disabilities may be assessed. Each investigator has employed a slightly different test procedure generally using treadmills or walkways where the speed of walking has been regulated thus imposing environmental velocity artifacts on the results.[1-5] The variety of testing techniques complicates comparison of results because oxygen uptake depends on the speed of walking[1-4, 6, 7] and walking on surfaces other than a treadmill imposes a higher metabolic cost.[8] The availability of baseline data is important for comparison of energy-cost tests on persons with physical gait disabilities who are unable to walk on a treadmill and who can only walk at their own naturally selected velocities. For these reasons, this study was designed to determine the metabolic energy cost of unrestrained walking for normal adult men and women of different ages.

METHOD

Forty adults, twenty men and twenty women, were studied. Five of each sex were in each of the following age groups: 20 to 29 years, 30 to 39 years, 40 to 49 years, and 50 to 59 years. The mean age for the entire group was 39.6 years. None of the participants had a

---

Mr. Blessey is a research physical therapist with the Environmental Health Service, Rancho Los Amigos Hospital, Downey, CA 90242.

Dr. Hislop is professor of Physical Therapy, University of Southern California, Downey, CA 90242.

Dr. Waters is assistant director of the Pathokinesiology Service and chief of Neurological Service at Rancho Los Amigos Hospital.

Mr. Antonelli is chief engineer of the Pathokinesiology Laboratory at Rancho Los Amigos Hospital.

TABLE 1
Mean Values of Anthropometric Data

	N	Age	Height (cm)	Weight (kg)	Leg Length (cm)
Women	20	40.0 ± 13.4[a]	164.9 ± 8.7[a]	62.1 ± 11.2[a]	87.4 ± 5.7[a]
Men	20	39.1 ± 12.5[a]	180.3 ± 9.3[a]	81.0 ± 19.6[a]	96.2 ± 5.9[a]

[a] ± = 1 Standard Deviation

known history of chronic or severe respiratory, cardiovascular, or musculoskeletal disease. Their anthropometric data are presented in Table 1.

PROCEDURE

Oxygen intake was determined by the Douglas bag method using a 2-way valve. The inner diameter of the valve was 26 mm and the inner diameter of the connecting tubing was 40 mm. A small thermistor, sensitive to the differences in temperature between inspired and expired air, was placed in the airway just beyond the mouthpiece to monitor respiration rate. Bipolar leads were placed on the subject's chest to record heart rate. Step frequency was determined by a heel switch taped to the subject's shoe. An aneroid sphygmomanometer was attached to the subject's right arm. The total weight of all the equipment worn including the supporting harness was less than 2.5 kg. Electronic signals from the bipolar chest leads, the thermistor, and the heel switch were transmitted to a receiver* by a FM-FM VHF radiotelemetry system allowing the subjects to walk unimpeded by wires or cables.

Samples of expired air were analyzed by a dry gas method using a paramagnetic oxygen analyzer† and an infrared carbon dioxide analyzer.†† Expired air volumes were measured by a dry gas flow meter. All gas volumes were corrected to standard values for temperature, pressure, and water vapor (BTPS units).

Walking was carried out on a level concrete oval track 60.5 meters in circumference. All subjects normally wore an oxford type shoe during everyday activity. Two controlled cadence tests at approximately 60 and 120 steps per minute were conducted using a metronome. Tests at a free unrestrained cadence and at a maximum walking speed were also done during which cadences were recorded. Each walk lasted approximately five minutes. The first period served as a warm-up during which a steady state condition was achieved (usually within 3 minutes). During the last two minutes the expired air was collected with no interruption of the subject's performance. Heart rate and respiration rate were monitored continuously throughout the 2-minute collection period to ensure that the subject maintained a steady state.

RESULTS

Mechanical Characteristics of Gait

Considerable differences between men and women subjects were observed (Tab. 2). The speed of free cadence walking in the 20 men ranged from 62 to 121 meters per minute (mean, 89 m/min). In comparison, the women walked at lower velocities which ranged from 53 to 95 meters per minute with a mean of 74 meters per minute. These differences were statistically significant ($P < .01$).

The average step frequency for both sexes was 116 steps per minute at the comfortable walking speed (Tab. 2). Mean stride length, however, varied significantly ($P < .01$) between the men (1.54m) and women (1.27m). Stride length values ranged from 1.21 to 2.02 meters for the men and from 1.20 to 1.45 meters for the women.

The stride length of an individual has been noted to vary directly with the step frequency at different walking speeds.[9] This observation was confirmed throughout the range of cadences between 60 and 130 steps per minute (Fig. 1). The men tended to walk with a greater stride length than the women at all step frequencies, with a more pronounced difference at the faster cadences. To determine whether the greater stride length of the

---

* Signatron Model 4200, Biosentry Telemetry, Torrance CA 90510.

† Servomex Type OA 250, Taylor Instruments, San Leandro, CA 94577.

†† Capnograph Godart Type 146, Beckman Instruments, Fullerton, CA 92631.

Basic Research

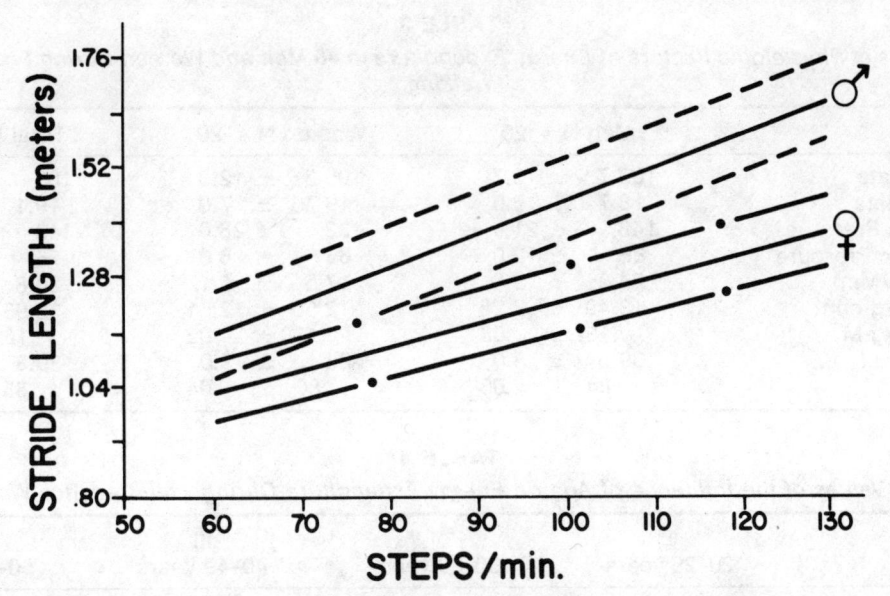

Fig. 1. Regression lines for the relationship between stride length and step frequency in the 20 women and 20 men studied. The difference in stride length between the two groups at a given step frequency tended to increase with the faster cadences.

men was related primarily to their greater body size, stride length was normalized by dividing by both height and leg length. Even after normalization, stride length in men was greater for any given step frequency. The best data fit resulted after normalizing by leg length with the following equations:

$$L = 0.862 + .005C \text{ (women)}$$

$$L = 0.770 + .007C \text{ (men)}$$

In the preceding equations, L represents the ratio of stride length to leg length and C represents cadence recorded in steps per minute.

Among the four age groups between 20 and 60 years, no significant differences in any of the mechanical characteristics of gait studied were found.

Metabolic Energy Expenditure

Expired air samples were collected from all 40 subjects after five minutes of standing in order to obtain a baseline measurement of oxygen consumption. The group had a mean oxygen uptake of 4.05 ml/kg-min ± .76. No statistically significant differences related to sex or age were noted.

The rate of oxygen uptake during free cadence walking was found to be virtually the same for men (13.4 ml/kg-min) and women (12.5 ml/kg-min). Individual values ranged between 7.8 and 17.1 ml/kg-min for the men and between 8.5 and 17.0 ml/kg-min for the women (Tab. 3).

During free cadence walking, the women tended to have a slightly higher mean heart rate, 105 beats per minute, than the men who had 101 beats per minute (Tab. 3). The average respiration rate per minute was 19 for both sexes. The mean values of the systolic and diastolic blood pressures in men (148/85 mm Hg) were somewhat higher than the same measures in women (133/65 mm Hg). The difference in diastolic blood pressure between the two sexes was significant ($P < .02$).

TABLE 2
*Mechanical Characteristics of Free Cadence Walking
Mean Values in 40 Subjects*

	Men (20)	Women (20)	Total
Velocity	89 m/min ± 11.4[a]	74 m/min ± 10.7[a]	82 m/min ± 13.1[a]
Stride Length	1.54 m ± .18[a]	1.27 m ± .11[a]	1.40 m ± .16[a]
Step Frequency	116/min ± 6.4[a]	116/min ± 10.7[a]	116/min ± 8.8[a]

[a] ± = Standard Deviation.

TABLE 3
Mean Values of Physiologic Factors of Energy Expenditure in 40 Men and Women During Free Cadence Walking

	Men N = 20	Women N = 20	Total N = 40
Heart Rate	100.7 ± 13.7	105.3 ± 12.8	103 ± 13.3
Resp. Rate	18.7 ± 6.0	19.45 ± 7.0	19.1 ± 7.0
Systolic Pressure	148 ± 21.0	132 ± 28.0	140 ± 26.0
Diastolic Pressure	85 ± 12.0	65 ± 8.0	75 ± 10.
$\dot{V}$ Air (L/Min)	24.2 ± 3.8	17.5 ± 5.1	20.8 ± 4.5
$\dot{V}_{O_2}$ ml/kg-min	13.40 ± 2.28	12.50 ± 2.51	12.95 ± 2.70
$\dot{V}_{O_2}$ ml/kg-M	.152 ± .03	.170 ± .02	.161 ± .02
% $\dot{V}_{O_2}$ max.	39.5 ± 8.0	38 ± 9.0	39.8 ± 8.0
RQ	.84 ± .03	.86 ± .04	.85 ± .04

TABLE 4
Mean Values of the Influence of Age on Energy Expenditure During Free Cadence Walking

	I 20–29 years	II 30–39 years	III 40–49 years	IV 50–59 years
Heart Rate	103.8 ± 15.6	101.6 ± 13.0	104.4 ± 13.4	102.2 ± 13.1
Resp. Rate	18.0 ± 9.4	18.8 ± 8.0	17.7 ± 9.5	22.2 ± 5.8
Systolic	129.4 ± 24.5	132.4 ± 17.0	146.0 ± 37.4	145.0 ± 18.4
Diastolic	80.9 ± 16.1	77.8 ± 11.2	83.1 ± 7.6	80.1 ± 7.7
$\dot{V}$ Air (L/Min)	20.7 ± 7.2	18.2 ± 4.9	22.6 ± 5.1	21.8 ± 4.5
$\dot{V}_{O_2}$ ml/kg-min	13.5 ± 2.7	12.1 ± 2.7	13.5 ± 1.2	12.7 ± 2.76
$\dot{V}_{O_2}$ ml/kg-M	.165 ± .024	.159 ± .032	.159 ± .019	.163 ± .021
% Pred. O_2 Max.	35.8 ± 5.9	34.8 ± 7.6	41.1 ± 7.1	43.1 ± 10.3
RQ	.84 ± .04	.89 ± .03	.83 ± .04	.87 ± .03

To evaluate the influence of age on energy expenditure, the subjects were divided into the following age groups of ten persons each: 20 to 29 years, 30 to 39 years, 40 to 49 years, and 50 to 59 years.

The mean values of oxygen uptake, heart rate, respiratory rate, and diastolic pressure did not vary significantly with age (Tab. 4). A tendency was noted for an increase in systolic blood pressure during unrestrained walking which could be related to age (Tab. 4). Other factors did not vary with age.

The oxygen uptake was directly related to velocity (Fig. 2). When oxygen consumption (ml/kg-min) was plotted against the square of the walking speed (at cadences between 60 and 130 steps per minute) a linear relationship resulted with a line of best fit described by the equation:

$$E_{O_2} = 7.55 \pm .000811 V^2 \pm 2.33$$

where E_{O_2} is the energy cost expressed in ml O_2/kg-min and V is the velocity in m/min (Fig. 2). Separate regression lines were generated for men and women, however, these did not differ significantly.

The predicted maximum oxygen uptake was calculated for each subject by establishing a regression line for the relationship between oxygen uptake and heart rate at the four different walking speeds, then extrapolating to the predicted maximum heart rate using the formula of Fox: [HR = 220 – age].[10] As expected, the predicted maximum oxygen consumption tended to decrease with age for both sexes (Tab. 5).

For each subject, the absolute value of oxygen uptake was divided by the value of the predicted maximum oxygen uptake to compute the relative energy cost (expressed as a percentage of predicted maximal capacity) of unrestrained walking. The mean values of the relative energy cost of free cadence walking was 39.5 percent for the men and 38.1 percent for the women, which was not a significant difference. The relative metabolic cost of free cadence ambulation, however, tended to increase with age. The differences in mean values between the age groups of 50 to 59 years (43.1%), 20 to 29 years (35.8%), and 30 to 39 years (34.8%) were significant at the .05 level (Tab. 6).

Basic Research

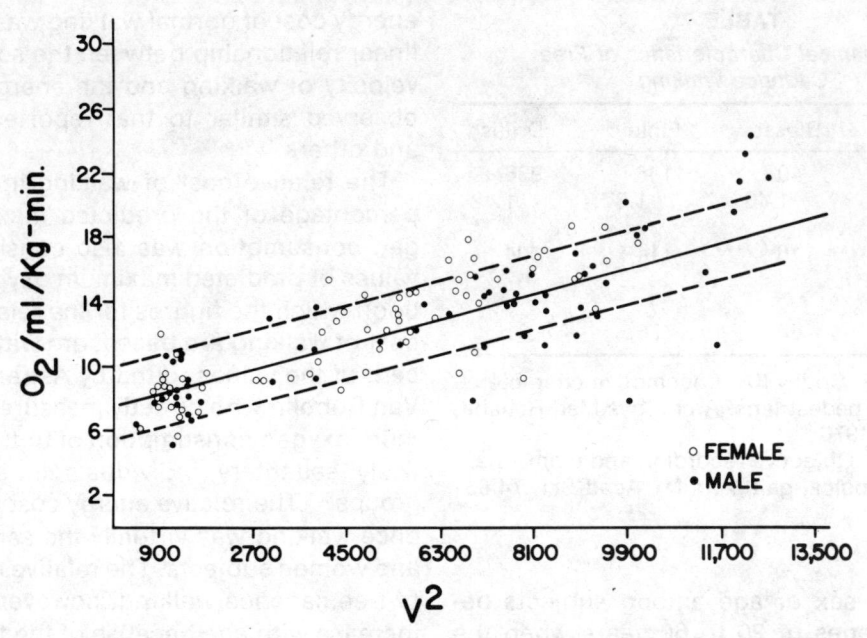

Fig. 2. Individual data for the 40 adults (men and women) studied demonstrating the linear relationship between oxygen uptake (ml/kg-min) and velocity squared. Equation for regression line: $E_{O_2} = 7.55 + .000811V^2$. S.E. = 2.33.

DISCUSSION

The results indicate considerable differences between men and women in velocity, stride length, and step frequency during free cadence walking, but no differences among the subjects that could be attributable to age. The men had an average comfortable walking velocity of 88 meters per minute compared to 74 meters per minute for the women. The major reason for the difference in walking speed was the tendency for the men to take longer strides, since both groups walked at a cadence of approximately 116 steps a minute. The group of forty subjects had a mean velocity of 82 meters per minute, an average cadence of 116 steps per minute, and a mean stride length of 1.40 meters. The group means in this study (as well as the means for men and women) are extremely close to the values reported by both Finley[11] and Drillis[12] who derived their data from more than 2,000 pedestrians walking in selected urban areas unaware they were being observed (Tab. 7). Based on observations of 572 women and 534 men, Finley reported that as a group women selected a slower walking speed (74 m/min) than men (82 m/min). He also found that women had a mean cadence of 116 steps per minute compared to the men of 110 steps per minute. On this basis, we concluded that the subjects in this experiment walked in an unrestrained manner and that their normal gait was not altered by the experimental procedure which was the major objective of the study.

In agreement with previous studies, the metabolic energy cost of walking was inde-

TABLE 5
Predicted Maximum Oxygen Uptake in Four Age Groups (n = 5 in each group)

	Age	$\dot{V}_{O_2}$ Max.
Women	20–29	34.74
	30–39	36.60
	40–49	32.02
	50–59	31.78
Men	20–29	42.70
	30–39	37.30
	40–49	36.98
	50–59	29.60

TABLE 6
Relative Energy Cost of Free Cadence Walking in Men and Women (n = 10 in each group)

Age	Percent $\dot{V}_{O_2}$ Max.
20–29	35.8
30–39	34.8
40–49	41.1
50–59	43.1

TABLE 7
Mechanical Characteristics of Free Cadence Walking

	Blessey	Finley[a]	Drillis[b]
N	40	1106	936
Stride L (m)	1.40	1.37	1.42
Step Freq (steps/min)	116	114	114
Velocity (M/min)	82	78	81

[a] Finley FR, Codey KA: Locomotive characteristics of urban pedestrians. Arch Phys Med Rehabil, 51:423–426, 1970

[b] Drillis RJ: Objective recording and biomechanics of pathological gait. Ann NY Acad Sci, 74:86–109, 1958

pendent of sex or age among subjects between the ages of 20 to 60 years when the data were normalized for different body sizes.[2,3,5] The major factor determining the energy cost of normal walking was velocity. A linear relationship between the square of the velocity of walking and the energy cost was observed similar to that reported by Cotes and others.[1,4,6]

The relative cost of walking in terms of a percentage of the predicted maximum oxygen consumption was also considered. The values of predicted maximum oxygen uptake, upon which the figures for the relative energy cost of walking are based, are within ten percent of the values found by Astrand[13] and by Van Dobelin who directly measured the maximum oxygen consumption of untrained, relatively sedentary individuals in similar age groups.[14] The relative energy cost of free cadence walking was virtually the same for men and women subjects. The relative energy cost of free cadence walking, however, tended to increase with age because of the tendency of the maximum oxygen consumption to decline with age.

REFERENCES

1. Bobbert AC: Energy expenditure in level and grade walking. J Appl Physiol, 15:1015–1021, 1961
2. Booyens J, Keatinge WR: The expenditure of energy by men and women walking. J Physiol [London], 138:165–171, 1957
3. Corcoran PJ, Brengelmann GL: Oxygen uptake in normal and handicapped subjects in relation to the speed of walking beside velocity-controlled cart. Arch Phys Med Rehabil 51:78–87, 1970
4. Cotes JE, Meade F: The energy expenditure and mechanical energy demand in walking. Ergonomics. 3:97–119, 1960
5. Mahadeva K, Passmore R, Woolfe B: Individual variations in metabolic cost of standardized exercises: Effect of food, age, sex and race. J Physiol [London], 121:225–231, 1953
6. Givoni B, Goldman RF: Predicting metabolic energy cost. J Appl Physiol, 30:429–433, 1971.
7. Passmore R, Durnin JV: Human energy expenditure. Physiol Rev, 35:802–840, 1955
8. Daniels F, Vanderbie JH, Wensiman FR: Energy Cost of Treadmill Walking Compared to Road Walking. Natick, MA, US Army Quartermaster Center, EPD Rept. 1953, p 220
9. Lameroux L: Human kinematics of walking. Bull Prosthet Res, 10:3–84, 1971
10. Fox SM, Naughton JP, and Haskell WL: Physical activity and the prevention of coronary heart disease. Ann Clin Res, 3:404–432, 1971
11. Finley FR, Codey KA: Locomotive characteristics of urban pedestrians. Arch Phys Med Rehabil, 51:423–426, 1970
12. Drillis RJ: Objective recording and biomechanics of pathological gait. Ann NY Acad Sci, 74:86–109, 1958
13. Astrand I: Aerobic work capacity in men and women with special reference to age. Acta Physiol Scand [Suppl], 169:8–88, 1960
14. Van Dobelin W: An analysis of age and other factors related to maximum oxygen uptake. J Appl Physiol, 22:934–941, 1967

Basic Research

Functional Ambulation Velocity and Distance Requirements in Rural and Urban Communities
A Clinical Report

CHRISTY S. ROBINETT
and MARY A. VONDRAN

The purposes of this clinical report are 1) to document the distances and velocities that individuals must ambulate to function independently in their community and 2) to demonstrate the differences in travel distances and velocities among communities of various sizes. In seven communities of different sizes, we measured distances from a designated parking space to commonly frequented sites (eg, stores, post offices, banks, and medical buildings). We also measured street widths and the time allowed by crossing signals to cross streets safely. From these data, we calculated the velocities needed to safely cross streets in each community. The study results showed that ambulation distances and velocities vary depending on the size of the community and that for individuals to function independently within their community, they must ambulate at velocities and distances much greater than the ambulation objectives that may be set at most rehabilitation settings. Based on the results of this study, we suggest that rehabilitation centers take distance measurements of communities from which they receive the majority of their patients to more accurately prepare patients for functional independence after rehabilitation discharge.

Key Words: Early ambulation, Functional training and activities, Gait training, Rehabilitation.

Mobility is a primary concern in the rehabilitation process and is a problem addressed by physical therapists in all settings. Mumma found that patients reported mobility as their main functional loss after a stroke.[1] Weerdt and Harrison reported that physical therapists noted lack of active movement and mobility in general as the primary problem in patients with stroke.[2]

In many rehabilitation settings, patients are taught ambulation skills involved in both indoor and outdoor activities and in movement up and down ramps and stairs. Many rehabilitation facilities have developed functional training evaluation forms to evaluate patients' progress. Most evaluation forms include a section pertaining to ambulation skills, but very few of the forms we have encountered include specific skills (eg, shopping, crossing streets with traffic lights, walking at specified velocities, and walking long distances).

---

* C. Robinett, BS, is Staff Physical Therapist, St. Mary of the Plains Hospital, 4000 24th St, Lubbock, TX 79410 (USA).

M. Vondran, BS, is Staff Physical Therapist, Northeast Rehabilitation Hospital, 70 Butler St, Salem, NH 03079.

This article was submitted July 20, 1987; was with the authors for revision 23 weeks; and was accepted March 15, 1988. Potential Conflict of Interest: 4.

The longest distance that patients were required to ambulate in the forms we surveyed was 183 m on a smooth surface, and the fastest velocity required was 18.2 m/min.[3-5] Normal walking velocity is 89 m/min for men and 74 m/min for women.[6]

In both acute care and rehabilitation settings, many of the patients treated in physical therapy have difficulty ambulating and receive gait training as one aspect of their treatment.[1-2] Scranton et al suggested that one of the major obstacles preventing patient independence is difficulty in ambulation.[7] Although attainment of ambulation skill is a major rehabilitative milestone, ambulation skill alone is not enough for an individual to function independently within the community. To be independent in their community, patients must be able to ambulate appropriate distances at adequate velocities and to ascend and descend curbs.[8]

The purposes of this clinical report are 1) to document the distances and velocities that an individual must be able to ambulate to achieve independence within the community and 2) to demonstrate how these distances and velocities differ among rural and metropolitan communities. We believe that rehabilitation centers should develop criteria for functional ambulation that will enable physical therapists to more adequately prepare patients for independence within their community.

MEASUREMENT OF FUNCTIONAL DISTANCES

The six sites we measured in this study were selected by Lerner-Frankiel et al in their article on functional community ambulation.[8] The six sites are places in the community that are frequently visited by individuals in the performance of routine activities: 1) supermarkets, 2) large drugstores, 3) banks, 4) department stores, 5) post offices, and 6) physician's offices. Lerner-Frankiel et al also measured curb heights, commercial and residential street widths, and the amount of time that crossing signals allowed for safely crossing streets.[8]

We measured seven communities and placed them into three categories according to population.[9] *Rural towns* had a population of less than 10,000 persons, *small towns* were classified as those with a population of 10,000 to 40,000 persons, and *cities* had a population of greater than 95,000 persons (Tab. 1).

To obtain our data, we followed the measurement protocol established by Lerner-Frankiel et al (Tab. 2).[8] Because their protocol was geared toward a metropolitan area with large drugstores and shopping malls, we made some modifications to accommodate the facilities in small and rural towns. Many of the small towns we measured lacked shopping malls and had a limited number of large drugstores, so we included smaller pharmacies and department stores not in shopping malls in our measurements. For rural areas without street-crossing signals, the length of time from the green light to the beginning of the yellow light was measured.

To obtain distance measurements, we used a rolling device that measures distance in feet. We obtained the measuring device from police departments or associated facilities within the communities measured. Distances were measured from the closest parking space through a portion of the site and back to the car. A stopwatch was used to time street-crossing signals.

In the cities, we measured three or four different places in each of the site categories. In San Antonio, Tex, for example, four banks and three post offices were measured. In small towns, we measured two to four places in each site category according to availability. In rural towns, one or two places were measured for each site category because a limited number of sites were available for measurement.

REPORT OF FUNCTIONAL DISTANCES AND VELOCITIES

We grouped the data into three groups according to population size of the community. The median and range for velocity and distance were then determined for each type of site and recorded under the respective population category (Tab. 3). For most sites, distances were shorter in the rural and small towns than in the cities. The longest distances

TABLE 1
Communities Categorized by Population Size

Community	Population Size
<10,000 people	
Slaton, Tex	6,804
Lovington, NM	9,727
10,000–40,000 people	
New Braunfels, Tex	22,402
Hobbs, NM	28,794
>95,000 people	
Abilene, Tex	98,315
Fort Worth, Tex	385,164
San Antonio, Tex	785,940

TABLE 2
Measurement Protocol for Individual Locations[a]

Location	Start and Return Point	Goal
Supermarket	Handicapped parking space or space closest to entrance.	Through closest entrance, down half the total number of aisles, exit through check stand.
Large drugstore (if available)	Handicapped parking space or space closest to entrance.	Through closest entrance, down half the total number of aisles, exit through check stand.
Bank	Handicapped parking space or space closest to entrance.	Through closest entrance, into the designated line, and to the farthest teller.
Physician's office or medical building	Handicapped parking space or space closest to entrance.	Through main entrance to elevator. If only one floor, to main lobby.
Post office	Handicapped parking space or space closest to entrance.	Through front entrance, into line, to farthest clerk.
Department store (in a shopping mall, if available)	Handicapped parking space or space closest to desired store.	Through closest entrance of department store, ground floor. Around perimeter plus down middle aisle.
Crosswalk	From one corner to the other, across the street.	At street level, walking from one side of the street to the other. Length of time from "Walk" signal appearance to end of flashing "Don't Walk" signal. If no crossing signal, time from green light to beginning of yellow light.
Curb	From street level to sidewalk level.	Stepping up and down a curb.

[a] Adapted from Lerner-Frankiel et al.[8]

that individuals would need to ambulate were found in supermarkets, where the longest distance recorded was 480 m. Post offices required individuals to walk the shortest distances; the smallest post office distance recorded was 13 m. Little variability existed among the seven communities in curb height, residential crosswalk length, and distances to banks and physicians' offices. The most variability between population categories was found in the velocity required to cross a street safely. The velocities recorded for safe street crossing ranged from 30 to 82.5 m/min. The rural and small towns generally required a slower walking velocity than the cities.

DISCUSSION

The distances obtained from the seven communities are probably all greater than the distances patients are typically trained to ambulate in physical therapy. The study data are conservative, because our measurements involved only half the total number of aisles in drugstores and supermarkets and only the perimeter and middle aisle of department stores. To achieve independence within their community, individuals will at some point need to visit sites such as those measured and will need to ambulate distances sufficient to perform activities at those sites. The minimum safe velocity of 30 m/min for crossing a street is seldom addressed or documented when gait training patients. The data obtained in our study demonstrate the need for inclusion of more appropriate distance and velocity goals in gait training. Velocity goals may need to be revised for elderly individuals, whose walking speed—even for healthy individuals—may not be fast enough to cross streets with signals safely.[10]

TABLE 3
Medians and Ranges of Measurements for Specific Locations in Communities

Variable	Population Size								
	<10,000			10,000–40,000			>95,000		
	Median	Range	n	Median	Range	n	Median	Range	n
Curb height (cm)	17.0	(11.5–20.0)	14	18.0	(14.0–20.0)	9	18.5	(14.0–21.5)	11
Crosswalk									
Commercial distance (m)	15.0	(13.0–17.0)	6	19.5	(13.0–24.0)	5	16.5	(8.5–27.5)	9
Residential distance (m)	11.5	(9.0–12.0)	7	12.5	(9.0–13.0)	8	10.0	(9.0–12.0)	11
Velocity required for safe crossing (m/min)	44.5	(30.0–69.0)	7	58.5	(31.0–69.0)	5	63.5	(42.5–82.5)	11
Location									
Post office (m)	39.0	(13.0–65.5)	2	47.0	(44.5–49.0)	2	60.5	(36.0–84.0)	9
Bank (m)	64.0	(57.0–68.0)	3	51.0	(38.0–85.0)	5	63.0	(30.0–217.0)	10
Physician's office or medical building (m)	45.5	(41.5–61.0)	3	47.0	(19.5–90.0)	3	47.5	(26.5–122.0)	8
Supermarket (m)	230.0	(122.0–289.0)	4	332.0	(145.0–474.0)	5	342.0	(201.0–480.0)	9
Department store (m)	132.0	(62.0–202.0)	2	175.5	(111.0–292.0)	6	327.0	(169.0–505.0)	8
Drugstore (m)	100.0	(32.5–247.0)	4	148.0	(48.0–210.0)	6	139.5	(103.0–213.0)	7

Patients living in communities of different sizes may need to achieve different distance and velocity goals. Required ambulation velocities vary considerably between rural towns and cities. Patients returning to a rural community may not need to achieve the same ambulation velocity as patients returning to a metropolitan area. We generally found that the ambulation distances measured in rural towns were smaller than those in cities, and the distances increased progressively with the increase in population. Patients from rural or small towns need to ambulate shorter distances than patients from cities to be independent in their community. A rehabilitation facility should consider patients' geographical origin and destination when establishing gait-training goals for distances and velocities.

SUMMARY

Distances that individuals need to ambulate to function independently in their community probably exceed the distances that patients are expected to ambulate in physical therapy settings. Velocity is not always an evaluation criterion for patients undergoing gait training but is an important factor for patients functioning outside the rehabilitation center. The data from this clinical report demonstrate that distances and velocities needed for community ambulation vary according to the population size of the community. When establishing ambulation distance and velocity criteria, physical therapists should consider the community to which the patient will return.

Rehabilitation centers should take measurements of distances and velocities of frequently-visited sites for the geographical areas from which they receive the majority of their patients. Physical therapists then could more accurately prepare their patients for independence in the community after discharge. Distance and velocity ambulation goals should be set for all patients, both young and old, who intend to return to an independent active life in their community after rehabilitation.

Acknowledgments. We express our appreciation to Claire Kispert, PhD, and Betsy Littell, PhD, for their advice and assistance on this manuscript. We also thank Kim Osborne, PT, and Kelly Parsons, PT, for their assistance in generating the idea for this article.

REFERENCES

1. Mumma CM: Perceived losses following stroke. Rehabilitation Nursing 11(3):19–24, 1986
2. Weerdt WJG, Harrison MA: Care for stroke patients: Against whose yardstick shall we measure? Physiotherapy 71:298–300, 1985
3. Jette A: State of the art in functional status assessment. In Rothstein JM (ed): Measurement in Physical Therapy: Clinics in Physical Therapy. New York, NY, Churchill Livingstone Inc, 1985, vol 7, pp 148, 153, 157, 158
4. Palmer ML, Toms JE: Manual For Functional Training. Philadelphia, PA, F A Davis Co, 1980, pp 171–182
5. Deniston OL, Jette A: A functional status assessment instrument: Validation in an elderly population. Health Serv Res 15:21–34, 1980
6. Blessey RL, Hislop HJ, Waters RL, et al: Metabolic energy cost of unrestrained walking. Phys Ther 56:1019–1024, 1976
7. Scranton J, Fogel M, Erdman W: Evaluation of functional levels of patients during and following rehabilitation. Arch Phys Med Rehabil 51:1–21, 1970
8. Lerner-Frankiel MB, Vargas S, Brown M, et al: Functional community ambulation: What are your criteria? Clinical Management in Physical Therapy 6(2):12–15, 1986
9. Lane HU (ed): The World Almanac and Book of Facts, 1987. New York, NY, World Almanac, 1986
10. Aniansson A, Rundgren A, Sperling L: Evaluation of functional capacity in activities of daily living in 70-year-old men and women. Scand J Rehabil Med 12:145–154, 1980

Basic Research

Dynamic Biomechanics of the Normal Foot and Ankle During Walking and Running

MARY M. RODGERS

This article presents an overview of dynamic biomechanics of the asymptomatic foot and ankle that occur during walking and running. Functional descriptions for walking are provided along with a review of quantitative findings from biomechanical analyses. Foot and ankle kinematics and kinetics during running are then presented, starting with a general description that is followed by more specific current research information. An understanding of the dynamic characteristics of the symptom-free foot and ankle during the most common forms of upright locomotion provides the necessary basis for objective evaluation of movement dysfunction.

Key Words: Ankle; Foot; Kinesiology/biomechanics, gait analysis; Kinetics; Lower extremity, ankle and foot.

The foot and ankle, by virtue of their location, form a dynamic link between the body and the ground. The foot and ankle are basic to all upright locomotion performed by the human, constantly adjusting to enable a harmonious coupling between the body and the environment for successful movement. The dynamic characteristics of the foot and ankle have been inferred traditionally from cadaveric examination and qualitative clinical assessment. Advancements in biomechanical techniques for dynamic analysis have enabled more quantitative and accurate documentation of foot and ankle function during movement, especially during the process of walking.

The objective of this article is to provide a selected review of quantitative information relevant to the dynamic function of the foot and ankle complex. Although results have often confirmed traditional anatomical assumptions regarding foot and ankle function, they have also contradicted long-accepted theories in certain cases.

The most frequently performed movements of the foot and ankle for healthy people occur during walking. Much research has been conducted in the analysis of walking, and the majority of this article will concentrate on the dynamic biomechanics of the foot and ankle during this activity. A classical description of the biomechanics of gait as found in clinical literature is followed by an overview of quantitative findings that document kinematic and kinetic characteristics during walking. As interest in physical fitness continues to grow, therapists are treating an increasing number of runners, both recreational and competitive. The foot and ankle kinematics and kinetics that occur during running will be presented briefly in the final section. This review includes information relevant to symptom-free individuals.

FOOT AND ANKLE KINEMATICS DURING WALKING

Although the foot has been viewed traditionally as a static tripod or a semirigid support for body weight (BW), it has evolved primarily for walking and is therefore a dynamic mechanism. The body requires a flexible foot to accommodate the variations in the external environment, a semirigid foot that can act as a spring and lever arm for the push off during gait, and a rigid foot to enable BW to be carried with adequate stability. The dynamic biomechanics of the foot and ankle complex that allow successful performance of all these requirements can only be understood when studied in relation to the biomechanics of the lower limb during walking.

The gait cycle (or stride period) provides a standardized frame of reference for the various events that occur during walking (Fig. 1). The *gait cycle* is the period of time for two steps and is measured from initial contact of one foot to the next initial contact of the same foot. The gait cycle consists of two phases: 1) stance (when the foot is in contact with the supporting surface) and 2) swing (when the limb is swinging forward, out of contact with the supporting surface). Along with providing forward momentum of the leg, the swing phase also prepares and aligns the foot for heel-strike and ensures that the swinging foot clears the floor. Stance comprises about 60% of the total gait cycle at freely chosen speeds and functions to allow weight-bearing and provide body stability. Five distinct events occur during the stance phase: heel-strike (HS), foot flat (FF), mid-stance (MS), heel rise (HR), and toe-off (TO).

General Description

An understanding of the various joint axes of the foot and ankle (see articles by Riegger and Oatis in this issue) is essential to the discussion that follows. Figure 1 summarizes these joint motions as they relate to different phases of gait. Numerous authors have contributed to a clinical description of walking kinematics based primarily on observation.[1-7] To understand the movements of the foot and ankle during walking, other portions of the lower extremity must be included.[1] During walking, rotation of the pelvis causes the femur, fibula, and tibia to rotate about the long axis of the limb.[2] The magnitude of this rotational motion increases progressively from pelvis to tibia. For example, during normal walking on level ground, the pelvis undergoes a maximum rota-

---

M. Rodgers, PhD, PT, is Research Health Scientist, Laboratory of Applied Physiology, Wright State University, 3171 Research Blvd, Dayton, OH 45420 (USA), and Veterans Administration Medical Center, 4100 W Third St, Dayton, OH 45428.

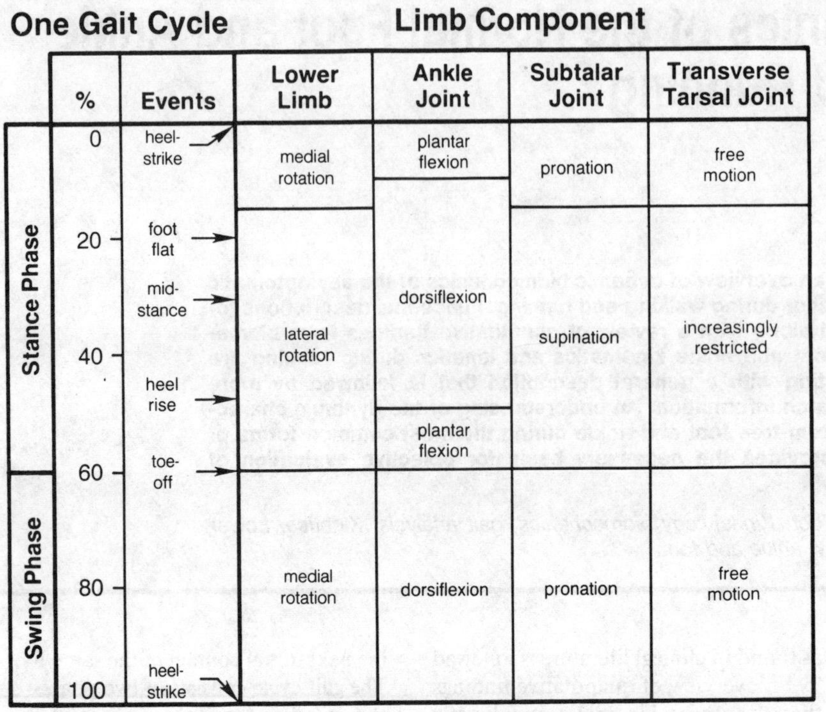

One Gait Cycle			Limb Component			
	%	Events	Lower Limb	Ankle Joint	Subtalar Joint	Transverse Tarsal Joint
Stance Phase	0	heel-strike	medial rotation	plantar flexion	pronation	free motion
	20	foot flat		dorsiflexion	supination	increasingly restricted
	40	mid-stance	lateral rotation			
		heel rise				
	60	toe-off		plantar flexion		
Swing Phase	80		medial rotation	dorsiflexion	pronation	free motion
	100	heel-strike				

Fig. 1. Summary of phases of gait cycle and accompanying motions of lower limb joints. (Adapted from Mann.[6])

tion in each gait cycle of about 6 degrees, and the tibia undergoes a rotation of about 18 degrees in the same period. Generally, the limb rotates medially (internally) during the swing phase and early stance phase and then laterally (externally) until the stance phase is complete and TO has occurred.[3]

At HS, the tibia is rotated medially about 5 degrees from its neutral position, and the ankle joint is either in its neutral position or in slight plantar flexion.[4] According to Perry, compression of the heel pad occurs at HS and is followed by traction on both anterior and posterior calcaneal attachments during terminal stance.[5] Immediately following HS, the foot flexes toward the floor, with the dorsiflexors controlling this plantar motion to prevent the foot from slapping down to the FF position. From HS to just before FF, the increasing medial rotation of the tibia and fibula is transmitted through the ankle mortise to the talus.[6] The medial rotation of the mortise, combined with the plantar-flexed position of the ankle, tends to shift the forefoot medially from its neutral, toe-out position. The heel contact with the ground is lateral to the center of the ankle joint where BW is transmitted to the talus, creating a pronatory moment at the subtalar joint that, in turn, stresses the structures of the medial arch. The talus rotates medially on the calcaneus about the subtalar axis forcing the calcaneus into pronation. According to Wright and associates, the foot quickly pronates, about 10 degrees within the first 8% of stance at an average walking speed.[7] In this pronated position, free motion is available at the transverse tarsal joint so that the foot remains flexible, distal to the navicular and cuboid, and can bend into close contact with the supporting surface.

At the FF position, the lower limb begins to rotate laterally. Because the forefoot is now fixed on the ground, the entire lateral rotation of the ankle mortise is transmitted to the talus. As lateral rotation continues, the foot supinates, increasing stability at the transverse tarsal joint and along the longitudinal arch of the foot. The stability of the transverse tarsal joint is further improved by the increasing body load being carried and by the firm fit of the convex head of the talus into the concave face of the navicular bone.[1,3,6]

When the leg has passed over the foot, the ankle begins dorsiflexion. After HR, the ankle joint moves back into plantar flexion forcing the metatarsophalangeal joints to dorsiflex. Because the plantar aponeurosis wraps around the metatarsal heads, a "windlass" effect takes place that increases tension across the longitudinal arch, further elevating the arch and increasing foot stability. Just before TO, the combination of weight-bearing, windlass effect, and supination ensures that the foot is in a maximally stable position for lift-off. After TO, some authors report that the leg rotates medially, again pronating the foot and unlocking the transverse tarsal joint so that the foot returns to its flexible state for the swing phase of gait.[1,3,6] It should be noted that other authors report that the leg continues to be in lateral rotation throughout mid-swing and that the foot remains supinated throughout swing.[2]

Kinematic Studies

Kinematics refers to the description of motion, independent of the forces that cause the movement to take place. Linear and angular displacements, velocities, accelerations, center of rotation for joints, and joint angles are all examples of kinematics.[8] Kinematic information can be collected using direct measurement techniques (ie, goniometers, accelerometers) and with indirect measurement using imaging techniques (ie, cinematography, high-speed video, stroboscopy). Each technique has advantages and disadvantages that have been described by several authors and will not be detailed in this discussion.[9,10] Instead, the results of selected studies relevant to dynamic biomechanics of the foot and ankle during walking and running will be presented.

Walking cadence and velocity. Many factors affect foot and ankle biomechanics during walking, including the velocity of gait and anthropometric characteristics (ie, limb length). Winter defines *natural cadence*, or *free cadence*, as the number of steps per minute when a subject walks as naturally as possible and reports an average natural cadence range of 101 to 122 steps/min.[11] In general, the natural cadence for women is 6 to 9 steps/min higher than that of men. Foot and ankle kinematic measurements also are directly related to the walking velocity. Studies have documented the changes that occur with increasing speed.[12,13] For this reason, walking velocity must be considered when comparing biomechanical findings.

Displacements—paths of movement. Motion of the heel in walking has been reported by Winter in a study with 14 subjects walking at their natural cadences.[11] Vertical displacement of the heel begins well before TO and reaches maximum upward velocity just before

TO. The heel reaches its highest displacement shortly after TO. Horizontal velocity increases gradually after HR, reaching its maximum late in the swing phase, and then rapidly decreases just before HS. Vertical velocity of the heel slows abruptly at about 1 cm above ground level, after which the heel is lowered very gently to the ground.

The path of the forefoot differs from that of the heel. For the same sample of 14 subjects, Winter reports an initial rise in the forefoot during late push-off and early swing.[11] As the leg and foot are swung forward, the forefoot just clears the ground and then rises to a second peak just before HS. Because the toe is the last part of the foot to leave the ground, and because of the accompanying leg and foot angles, the toe rises to no more than 2.5 cm above the ground and then drops to only 0.87 cm of clearance at mid-swing. As the knee extends and foot dorsiflexes, the toe rises to a maximum of 13 cm just before HS.

Ankle range of motion, foot placement, and arch movement. Ankle-joint angles, foot-placement angles, and arch movement are other kinematic characteristics that have been investigated. Winter reported mean ankle-joint ranges of motion during walking for 19 subjects as a maximum of 9.6 degrees of dorsiflexion and 19.8 degrees of plantar flexion.[11] Murray and associates found that foot-placement angle showed high variability on successive steps of the same foot.[14] A mean value of 6.8 degrees of foot abduction (out-toing) was reported, with the average difference between successive foot angles being 2.4 degrees.

Dynamic arch movement was studied by Kayano using an "electro arch gauge."[15] He found that the medial longitudinal arch lengthens from the vertical force of BW from early stance to FF. It then shortens with the decrease in BW and activation of the arch supporting muscles. As the calf muscles activate for push-off, the arch lengthens again. It finally shortens rapidly because of the windlass action of the plantar aponeurosis as the toes dorsiflex for TO.

FOOT AND ANKLE KINETICS DURING WALKING

General Description

Kinetics is the study of the forces that cause movement, both medially (muscle activity, ligaments, friction in muscles and joints) and laterally (from the ground, active bodies, passive bodies).[8]

A large number of researchers have analyzed muscular activity and ground reaction forces (GRFs) during gait. Joint moments, segmental energy, joint reaction, and pressure distribution beneath the foot during walking have received less attention. The findings from electromyographic studies of the foot and ankle muscles during walking will be presented in the first subsection, followed by findings from force-plate and pressure-distribution studies. Calculated kinetic variables, such as ankle-joint moments and joint reaction forces, will be included in the final subsection.

Electromyographic studies of foot and ankle muscles during walking. Many researchers have investigated the electrical activity of muscles during walking, and Basmajian and Deluca have presented a review of their findings.[16] In general, studies have shown that many of the changes in levels of muscular activity occur at 15% to 20% of the gait cycle (FF), when the foot adapts to the supporting surface.

Winter and Yack have contributed extensively to the literature on EMG during walking.[17] Specific EMG patterns for several of the foot and ankle muscle groups that are active during walking are shown in Figure 2. The tibialis anterior muscle (TA) has its major activity at the end of swing to keep the foot in a dorsiflexed position. Immediately after HS, the TA peaks and generates forces to lower the foot to the ground in opposition to the plantar-flexing GRFs. The TA is the only inverting muscle active during the period

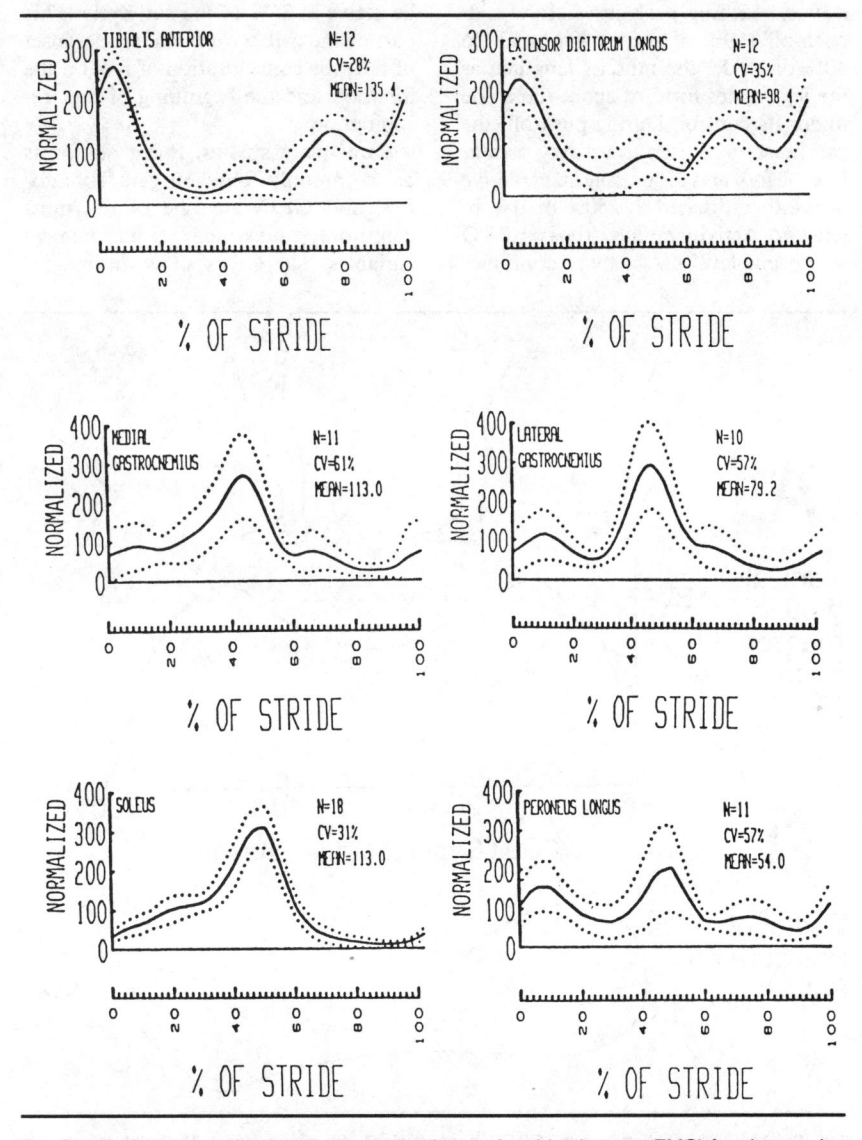

Fig. 2. Electromyographic activity (normalized to each subject's mean EMG) for six muscles during walking. Plots show mean EMG (solid line) and one standard deviation (dotted lines) for samples of varying size. Activity of medial and lateral gastrocnemius muscles is very similar and is combined for discussion in text. (Reprinted with permission.[17])

of maximum everting stress, when BW is completely on the heel. In some individuals, the TA plays a minor role in pulling the leg forward over the foot shortly after FF. A second burst of activity commences at TO and results in dorsiflexion for foot clearance during mid-swing.

The extensor digitorum longus muscle (EDL) has almost identical activity to the TA. It functions to lower the foot after HS and to dorsiflex the foot and toes for clearance during swing. A minor third phase occurs during push-off and appears to be a co-contraction to stabilize the ankle joint.[17]

The gastrocnemius muscle (GA) and soleus muscle (SO) exhibit one major long-duration phase of activity throughout the single-limb support period. It begins just before HS and rises during stance, reaching peak just before mid-push-off (50% of stride). From FF to 40% of stride, the muscles lengthen as the leg rotates forward about the ankle under its control. During push-off, the calf muscles shorten to actively plantar flex the foot and to generate an explosive push-off (estimated at 250% of BW in tension). Activity rapidly drops until TO where low-level GA activity continues into swing, probably showing the GA acting as a knee flexor to cause adequate knee flexion before swing-through.[17]

The peroneus longus muscle (PL) has a small burst of activity during weight acceptance (10% of stride), which appears to stabilize the ankle (possibly as a co-contraction to the TA). A larger burst during push-off (50% of stride) shows the PL acting as a plantar flexor. Low-level PL activity during early swing is likely a co-contraction to the TA to control the amount of foot dorsiflexion and supination.[17]

Other investigators have reported their findings of intrinsic muscle activity in the foot during walking.[18,19] The group of intrinsic muscles covered by the plantar fascia (flexor digitorum brevis, abductor hallucis, and abductor digiti minimi muscles) were shown to be active at 35% of the gait cycle. This part of the gait cycle includes the onset of HR, the concentration of BW on the forefoot, and the beginning of foot resupination.

Force-plate studies. Force platforms are commonly found in gait laboratories, and GRFs are one of the most commonly measured biomechanical variables. The GRFs show the magnitude and direction of loading directly applied to the foot and ankle structures during locomotion. Because the foot and ankle are the first parts of the body involved in contact with the ground during walking, they must be able to withstand and transmit these GRFs. The GRF data also provide information necessary for the calculation of ankle-joint reaction forces, which will be discussed later.

Figure 3 shows a graph of typical vertical GRFs during walking. The magnitude of vertical GRFs has been reported to range from 1.1 to 1.3 times BW, depending on walking speed.[20] Footwear has been shown to attenuate the peak vertical GRF values.[20] A rapid loading rate, often seen in vertical GRFs during the first 25 msec after contact, has been described as a possible contributing factor in joint degeneration.[21]

The force plate provides only one instantaneous measure of force distribution. This measure is called the center of pressure (COP), and it identifies the geometric centroid of the applied force distribution.[8] The path of the COP is created by plotting the instantaneous COP at regular time intervals during the entire stance phase of gait (Fig. 3). Studies of the COP show a normal progression of the path from just slightly lateral to the midline of the heel, along the midline of the foot, up to the metatarsal heads.[20,22] At this point, medial migration occurs so that by TO the COP lies under the first or second toe. This medial migration aspect of the COP path has been described as the most variable among subjects. The COP path is altered by different footwear, as illustrated by the findings of Katoh and associates (Fig. 4).[22]

Pressure-distribution studies. Force-plate systems are limited in the analysis of foot movement because the force information is not specific to foot anatomical locations. For example, the forces recorded may occur underneath both the fore and rear parts of the foot simultaneously so that the COP may fall at some intermediate point, which may not actually be loaded. Pressure-distribution devices provide the specific location of pressures as they occur beneath the moving foot. Recent studies in pressure distribution have revealed new information regarding dynamic foot function during walking.[23-27]

Although a great deal of individual variability exists in foot pressures during walking, the usual location of peak pressure is beneath the heel. A comparison of mean regional peak pressures found

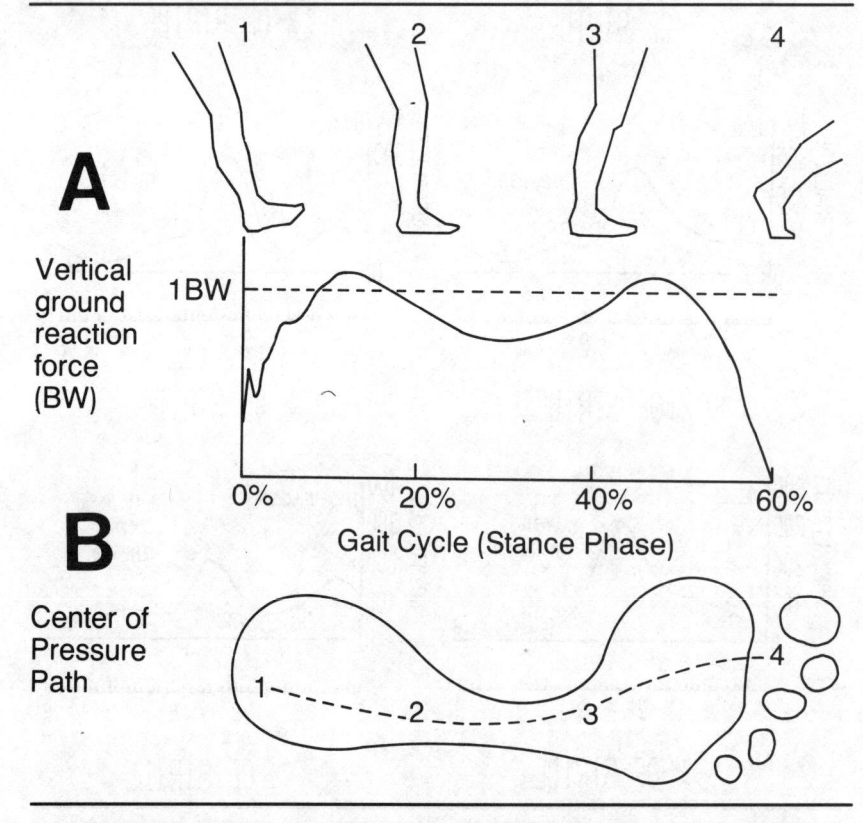

Fig. 3. Ground reaction forces (GRFs) beneath foot during walking: (A) Graph of classic vertical GRF during stance phase of gait cycle (BW = body weight; 1 = heel-strike; 2 = foot flat; 3 = midstance; 4 = toe-off); (B) path of the center of pressure, which represents series of instantaneous centroids of GRF during walking.

Basic Research

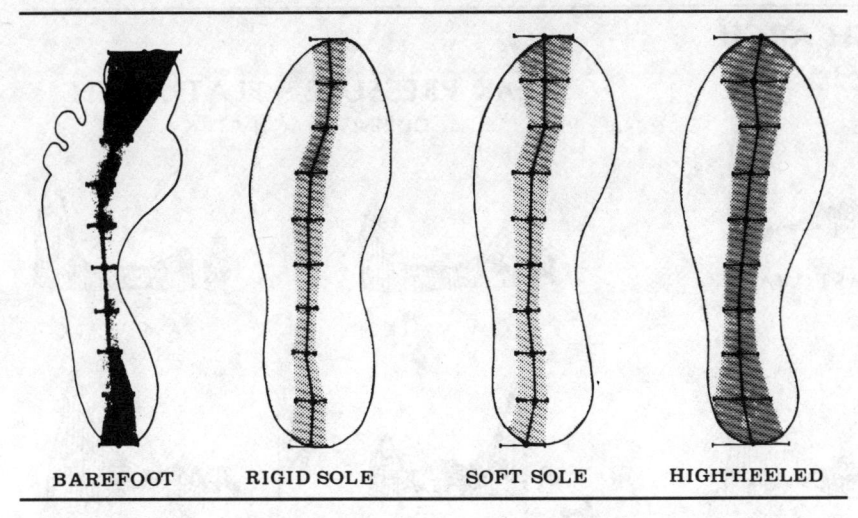

Fig. 4. Mean and one standard deviation for center-of-pressure paths during normal walking in different foot conditions: barefoot and wearing rigid-soled, soft-soled, and high-heeled shoes. (Reprinted with permission.[22])

TABLE
Comparison of Mean Regional Peak Pressures (in Kilopascals) from Pressure-Distribution Studies

Region	Rodgers[23]	Soames[a]	Grieve and Rashdi[b]	Betts et al[c]	Clarke[24]
Hallux	219	400	178	432	...
Medial toes	180	300	...	...	378
Lateral toes	163	200	...	...	160
First metatarsal	245	520	163	353	319
Second metatarsal	336	510	212	392	319
Lateral metatarsals	312	550	151	281	324
Medial midfoot	60	...	68	...	43
Lateral midfoot	103	150	6	...	95
Medial heel	337	780	208	363	443
Lateral heel	333	450	208	363	391

[a] Soames RW: Foot pressure patterns during gait. *J Biomed Eng* 7:120–126, 1985.
[b] Grieve DW, Rashdi T: Pressures under normal feet in standing and walking as measured by foil pedobarography. *Ann Rheum Dis* 43:816–818, 1984.
[c] Betts RP, Franks CI, Duckworth T: Analysis of loads under the foot: Part 2. Quantification of the dynamic distribution. *Clinical Physics and Physiological Measurement* 1(2):113–124, 1980.

by several different investigators is shown in the Table. Differences in values reported result from the variety of techniques and subject samples used by investigators.[23] These pressure-distribution studies have shown that all metatarsal heads are loaded during the stance phase of gait. This finding negates the concept of tripod stance, which would not allow pressure beneath the middle metatarsal heads.

Many variables have been identified that directly influence pressure distribution beneath the foot. Clarke found that with increasing speed, pressures increase and shift medially.[24] The toes contribute more as the walking speed increases. Walking barefoot alters both kinetic and kinematic variables when compared with walking in shoes.[20] Structural characteristics of the foot, such as arch type, also affect pressure distribution.[25] As shown in Figure 5, the more rigid high-arched foot tends to concentrate pressure beneath the heel and forefoot, with minimal pressure beneath the midfoot. This absence of midfoot pressure is present even in the higher loading conditions that occur with increasing speed of locomotion. The flexible flat-arched foot shows more spreading of pressure, including the area beneath the midfoot.

The classic Morton's foot structure, characterized by a second metatarsal head that is placed more distally than the first, has also been shown to influence pressure distribution. Rodgers and Cavanagh reported that second metatarsal head pressures were significantly higher in subjects with Morton's foot when compared with control subjects without Morton's foot.[26] This finding suggests that individuals with a Morton's foot structure may be more prone to second metatarsal pressure problems than individuals with other foot structures. Pressure-distribution studies have also been useful in identifying areas of concentrated pressure that may lead to pressure ulcers for individuals with insensitive feet.[27]

Joint moments and joint reaction forces. Indirect methods have been used to calculate gait kinetics when direct methods are not feasible. These methods are necessary to calculate forces within the joint because force transducers currently cannot be used safely in subjects. Winter[9,11] and Winter and Robertson[28] have made significant contributions in the calculation of joint moments of force and energy patterns during walking. The mean maximum ankle-joint moment (normalized to body mass) generated during walking was found to be a plantar moment of $1.6\ N \cdot m/kg$, occurring between 40% and 60% of the gait cycle. Plantar flexors were found to absorb energy during the early stance and MS phases of the gait cycle as the leg rotates over the foot. Late in stance, these same muscles plantar flex rapidly (producing the plantar moment) and generate an explosive burst of energy (push-off).

As mentioned in the section on force-plate studies, the GRFs during gait are transmitted proximally to the rest of the body through the foot and ankle, compressing each joint along the way. These compressive forces have been shown to contribute to the formation of osteoarthrosis.[21,29] Joint reaction studies of the ankle have been few, probably because this joint demonstrates osteoarthritic changes less often than the hip and knee joints. Stauffer and co-workers have shown ankle-joint compressive forces of about 3 times BW from HS to FF.[30] A further rise to a peak value of 4.5 to 5.5 times BW occurs during heel-off when the plantar flexors are undergoing strong contraction. Seireg and Arvikar have derived maximal ankle-joint reaction forces of 5.2 times BW from mathematical models.[31] Procter and Paul found a peak of 3.9 times BW for ankle-joint reaction force during walking.[32]

Stauffer and associates also reported ankle shear forces of 0.6 times BW in a posterior direction.[30] After HR, talo-

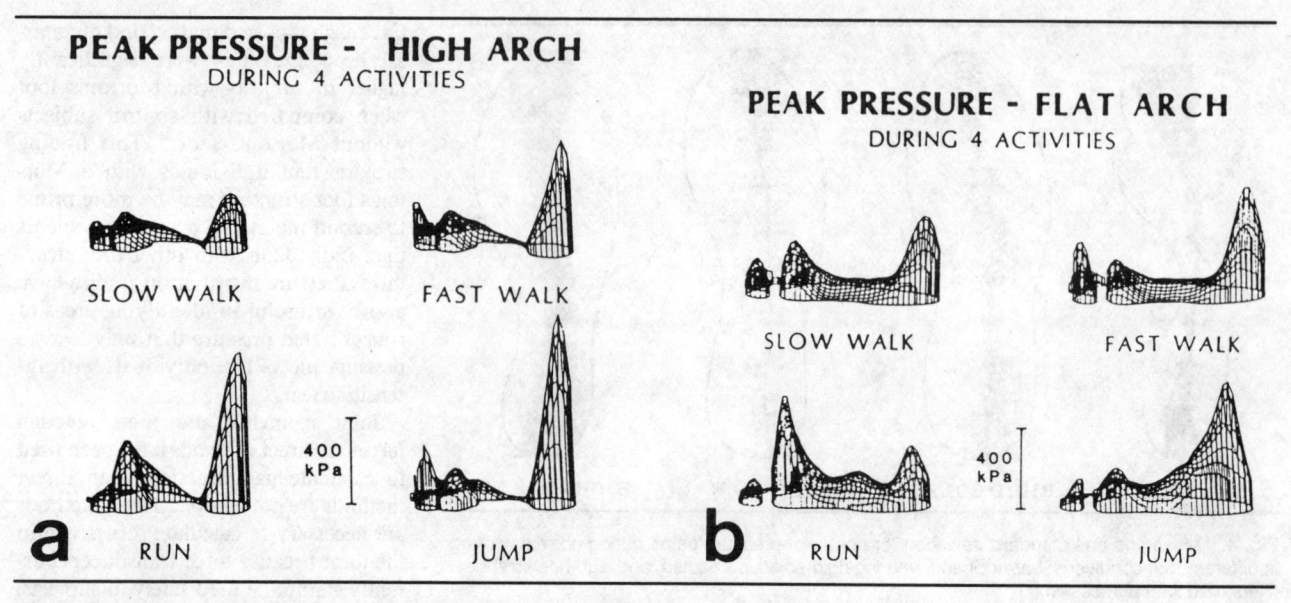

Fig. 5. Pressure-distribution patterns during slow and fast walking, running, and landing from a jump beneath a high-arched (a) and a flat-arched (b) foot. The flat-arched foot shows more spreading of pressure beneath the midfoot region. (Reprinted with permission of Martinus Nijhoff/Dr W Junk Publishers.[25])

crural shear was anterior and reduced to less than half of the previous posterior forces. Subtalar-joint reaction forces have been calculated by Seireg and Arvikar.[31] The peak resultant force in the anterior facet of the talocalcaneonavicular joint was 2.4 times BW and for the posterior facet, 2.8 times BW. Peaks for both locations occurred in the late stance phase of the gait cycle.

FOOT AND ANKLE KINEMATICS DURING RUNNING

A considerable amount of research has been conducted in the area of running biomechanics and is presented in a detailed review by Williams.[33] The position of other body parts and the timing of their movements are basic to an understanding of the motion of the foot and ankle. Although other body parts (primarily the hip and knee) have received most of the attention, several investigators have contributed to a functional description specific to foot and ankle motions during running at moderate speeds.[34,35]

General Description

For the running gait in which HS occurs, initial contact is at the lateral heel with the foot slightly supinated.[34,35] This position results from swinging of the leg toward the line of progression. Slight plantar flexion of the subtalar joint occurs along with supination of the forefoot and calcaneus. The subtalar joint passes from a supinated to a pronated position between HS and 20% into the support phase. The foot remains pronated between 55% and 85% of the support phase. Maximum pronation occurs between 35% and 40% of support phase, approximately the time when total-body center of gravity passes over the base of support. Full pronation marks the end of the absorbing and braking period of support as the foot begins its propulsive period. Maximum ankle dorsiflexion occurs 50% to 55% into the support phase when the center of gravity is forward of the support leg. The foot begins to supinate and returns to the neutral position at 70% to 90% of the support phase. The foot then assumes a supinated position for push-off.[34,35]

Kinematic Studies

Several stride variables that directly affect running kinematics and kinetics have been described by Cavanagh.[36] These variables include stride length at different speeds, optimal stride length, timing of the phases of running gait, and foot placement. Timing of the biomechanical events in running is variable because it depends on running speed, type of shoe, and individual anatomic variations. For example, Kaelin et al reported the interindividual (N = 70) and intraindividual variabilities (20 repetitions each for 6 of the subjects) for several variables during running.[37] The maximum pronation angle during foot-ground contact showed a range of 20 degrees among the subjects, but only 7 to 12 degrees within the same individual. Vertical touchdown velocity of the foot during running varied between 0.64 and 2.3 m/sec among the subjects. Scranton and associates reported an average duration of the support phase for jogging of 0.2 sec and for sprinting of 0.1 sec.[38]

Clinical evaluations have suggested a relationship between pronation of the foot during running and a variety of lower extremity problems such as shin splints and knee pain. Currently, quantitative data do not support the relationship, although this finding may result from inadequate analytical techniques. For example, studies of rear-foot motion have been conducted in two dimensions, although pronation occurs in more than one plane. Clarke and associates have reviewed several different studies of rear-foot movement in running (Fig. 6).[39] They reported an average maximum pronation angle of 9.4 degrees over all studies. The authors suggest that a maximum pronation angle of 13 degrees and total rear-foot motion greater than 19 degrees during running would be considered excessive. Currently, however, no single variable reliably predicts safe rear-foot movement during running.

Basic Research

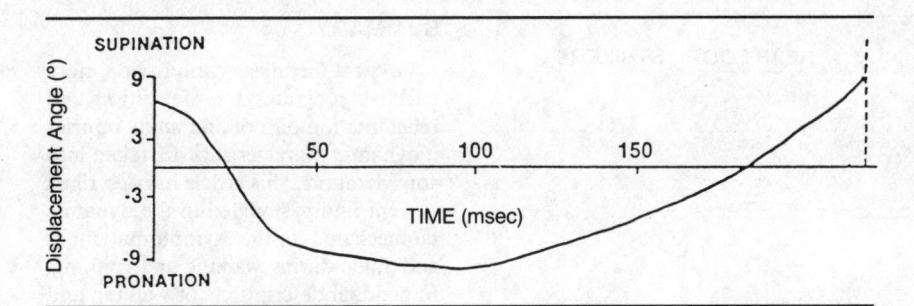

Fig. 6. Curve showing average rear-foot angular displacement during support phase of running based on rear-foot motion studies conducted by various researchers. The foot remains pronated for the majority of the support phase. (Adapted from Clarke TE, Frederick EC, Hamill CL: The study of rearfoot movement in running. In Frederick EC (ed): *Sport Shoes and Playing Surfaces: Biomechanical Properties*. Champaign, IL, Human Kinetics Publishers Inc, 1984, p 180.)

FOOT AND ANKLE KINETICS DURING RUNNING

General Description

Direct measurement of running kinetics poses more difficult technical problems than during the slower speeds of walking gait. Targeting a force plate is more difficult at higher speeds without altering the normal running gait patterns. The faster motion requires more distance for running, and longer cables or telemetry systems therefore must be used for EMG data collection. Treadmill running has been used for EMG data collection, although the pattern of running is different from that seen over natural terrain or on a track. Because of these problems, few researchers have directly measured foot and ankle muscle activity.[33] More research has been conducted in GRFs and pressure distribution during running. Indirect calculations of foot and ankle muscle forces, segmental moments, and joint reaction forces during running have been performed by a few researchers.

Electromyographic studies of foot and ankle muscles during running. Studies have shown that EMG activity increases with running as compared with walking. Miyashita and associates have reported that integrated EMG (IEMG) activity of the TA and GA increases exponentially with increasing speed.[40] Ito et al report that with increasing running speed, the IEMG increased during swing but remained the same during the support phase.[41]

Force-plate studies. Several authors have suggested a link between common running injuries and the impact forces at foot-strike that can occur thousands of times during running.[34,42] Force-plate analysis has shown that peak loading force during running is more than twice that of walking and occurs at least twice as fast. Perry extrapolates that the forces imposed on the supporting tissues would reflect a fourfold increase in strain.[5] Because microtrauma is cumulative, running creates symptoms that do not arise with ordinary walking.

Force-plate data for jogging and running are much more variable from step to step when compared with walking. The pattern and magnitude of the vertical GRFs during running also differ significantly from those that occur during walking. Variables that affect vertical GRF data include touchdown velocity of the heel, position of the foot and lower leg before contact, and movement of these structures during impact.[43] The vertical GRF curve for heel-toe running ("heel strikers") usually shows two distinct peaks: 1) the impact force peak and 2) the active force peak.[44,45] Typical peak vertical GRF values for distance running speeds are 2.5 to 3.0 times BW.

The pattern of force is dependent on the orientation of the foot at initial contact, which is determined by whether the runner is a "forefoot striker," a "midfoot striker," or a "rear-foot striker."[44] Most runners initially contact the ground with the outside border of the shoe, some with the rear lateral border (rear-foot strikers), and some with the middle lateral border (midfoot strikers). Harrison and associates report that mean foot contact time is reduced in forefoot strikers as compared with rear-foot strikers (0.20 vs 0.19 seconds, respectively).[46] Cavanagh and Lafortune also found slightly shorter contact times for the midfoot strikers compared with the rear-foot strikers.[44]

Additional differences in GRF patterns have been described.[44] Rear-foot strikers demonstrate a sharp initial spike in vertical GRF that is generally absent from the midfoot-striker patterns. Midfoot strikers produced two positive peaks in the anteroposterior force during the braking phase. The mean peak-to-peak amplitude for mediolateral (ML) GRF was three times greater in the midfoot strikers than that for the rear-foot strikers (0.35 and 0.12 BW, respectively). These findings indicate that the loading rates within the muscle and joints are affected by the type of initial foot contact during running.

The path of the COP also depends on the type of initial foot contact during running (Fig. 7). Cavanagh and Lafortune found that the COP path for rear-foot strikers followed from the rear lateral border to the midline within 15 msec of contact.[44] The COP path then continued along the midline to the center of the forefoot where it remained for almost two thirds of the entire 200-msec support phase. Midfoot strikers running at the same running speed made initial contact at 50% of shoe length. The COP path then migrated posteriorly as the rear part of the shoe made contact with the ground. This posterior movement coincided with a drop in the AP GRF. When the end of posterior migration was reached, the COP rapidly moved to the forefoot where it remained for most of the support phase.

Pressure-distribution studies. Very little information is available regarding pressure distribution under the foot during running. Pressure patterns during running vary with foot type (Fig. 5). The increased loading that occurs with running remains concentrated under the heel and forefoot in the more rigid high-arched foot. In the more flexible flat-arched foot, the increased load is spread beneath the entire foot, including the midfoot region.[25] Cavanagh and Hennig found that the average peak pressure during the contact phase of running (868.0 kPa) occurred under the heel for a sample of 10 rear-foot strikers.[47] Although pressures were much higher beneath the heel of these rear-foot strikers, more of the contact time was spent on the forefoot.

Muscle forces, segmental impulse, and joint reaction forces. Several investigators have developed mathematical models to predict muscle forces during running. Forces generated by the dorsiflexors and the GA have been calculated by Harrison and associates.[46] They report peak forces in the dorsiflexors of 0.5 times BW, which are active only during the first 10% of the stance phase. The GA generated a substantially greater peak force of 7.5 times BW. Calculations by Burdett revealed that the

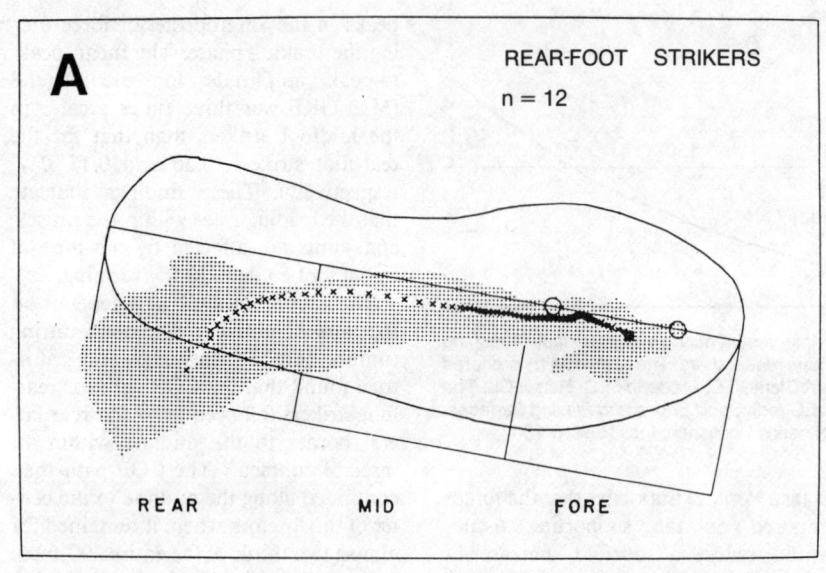

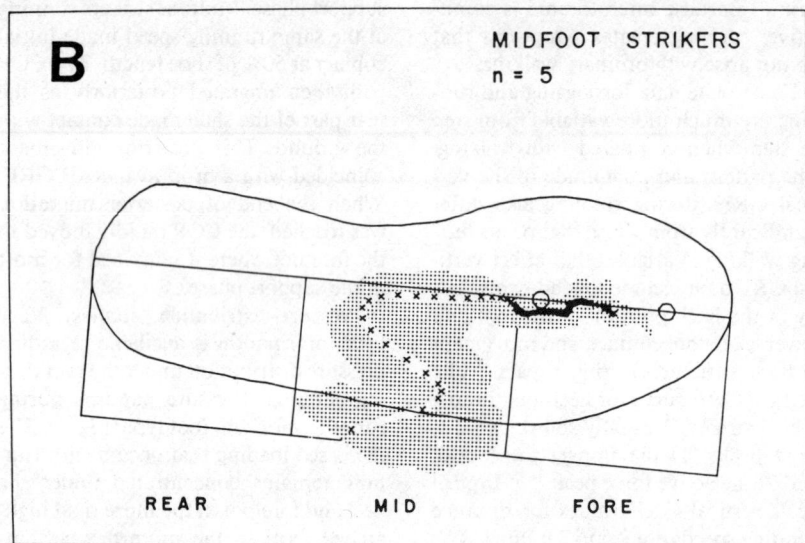

Fig. 7. Comparison of center-of-pressure paths during running for rear-foot (A) and midfoot (B) strikers. (Reprinted with permission.[44])

GA-SO group had the highest predicted force (5.3-10.0 times BW) of the ankle muscle groups.[48] Predicted forces in the tibialis posterior, flexor digitorum longus, and flexor hallucis longus musculature ranged from 4.0 to 5.3 times BW. The peroneus tertius muscle and EDL did not show any predicted force during the stance phase of running.

Impulse is the effect of a force acting over a period of time and is determined mathematically as the integral of the force-time curve.[8] Ae and associates calculated the impulse generated by different body segments during running.[49] The researchers found that the foot generated the largest mean impulse compared with other body segments. This impulse increased with faster running, suggesting that the foot plays an important role in projecting the body and increasing running velocity.

Ankle-joint reaction forces during running have also been calculated by several investigators. Harrison and associates reported maximum ankle-joint reactions of 8.97 and 4.15 times BW for the compressive and shear components, respectively.[46] Burdett predicted that compressive forces on the foot along the longitudinal axis of the leg reached peak values of 3.3 to 5.5 times BW during running.[48] In addition, he reported ML shear forces that ranged from a medial force of 0.8 times BW to a lateral force of 0.5 times BW. Furthermore, the vertical reaction forces and other calculated forces were determined to be about 2.5 times larger in running (at a 4.47-m/sec pace) when compared with walking.

SUMMARY

Physical therapists can provide more effective programs for prevention and rehabilitation of foot and ankle injuries if dynamic characteristics are taken into consideration. This article has described current findings related to the dynamic biomechanics of the asymptomatic foot and ankle during walking and running. Functional descriptions of walking and running biomechanics have been provided along with quantitative findings from current biomechanical studies. Extensive databases are still unavailable for many of the biomechanical variables that affect dynamic foot and ankle motion. As advances in biomechanical methods continue and more clinicians include quantitative techniques in their routine evaluations, however, more insight into dynamic foot and ankle function will be provided.

REFERENCES

1. Inman VT, Mann RA: Biomechanics of the foot and ankle. In Inman VT, Du Vries HL (eds): Surgery of the Foot. St. Louis, MO, C V Mosby Co, 1973, pp 3–22
2. Inman VT, Ralston HJ, Todd F: Human Walking. Baltimore, MD, Williams & Wilkins, 1981
3. Manley MT: Biomechanics of the foot. In Helfet AJ, et al (eds): Disorders of the Foot. Philadelphia, PA, J B Lippincott Co, 1980, pp 21–30
4. Soderberg GL: Kinesiology: Application to Pathological Motion. Baltimore, MD, Williams & Wilkins, 1986
5. Perry J: Anatomy and biomechanics of the hindfoot. Clin Orthop 177:9–15, 1983
6. Mann RA: Biomechanics of the foot. In Bunch WH, et al (eds): Atlas of Orthotics: Biomechanical Principles and Application, ed 2. St. Louis, MO, C V Mosby Co, 1985, pp 112–125
7. Wright DG, Desai ME, Henderson BS: Action of the subtalar and ankle-joint complex during the stance phase of walking. J Bone Joint Surg [Am] 46:361–382, 1984
8. Rodgers MM, Cavanagh PR: Glossary of biomechanical terms, concepts, and units. Phys Ther 64:1886–1902, 1984
9. Winter DA: Biomechanics of Human Movement. New York, NY, John Wiley & Sons Inc, 1979
10. Yack HJ: Techniques for clinical assessment of human movement. Phys Ther 64:1821–1830, 1984
11. Winter DA: The Biomechanics and Motor Control of Human Gait. Waterloo, Ontario, Canada, University of Waterloo Press, 1987
12. Andriacchi TP, Ogle JA, Galant JO: Walking speed as a basis for normal and abnormal gait measurements. J Biomech 10:261–268, 1977
13. Winter DA: Kinematic and kinetic patterns in human gait: Variability and compensating effects. Human Movement Science 3:51–76, 1984
14. Murray MP, Kory RC, Sepic S: Walking patterns of normal women. Arch Phys Med Rehabil 51:637–650, 1970
15. Kayano J: Dynamic function of medial foot arch. Journal of the Japanese Orthopaedic Association 60:1147–1156, 1986
16. Basmajian JV, Deluca CJ: Muscles Alive: Their Functions Revealed by Electromyography, ed 5. Baltimore, MD, Williams & Wilkins, 1985
17. Winter DA, Yack HJ: EMG profiles during normal human walking: Stride-to-stride and inter-subject variability. Electroencephalogr Clin Neurophysiol 67:402–411, 1987

18. Basmajian JV, Stecko G: The role of muscles in arch support of the foot. J Bone Joint Surg [Am] 45:1184–1190, 1963
19. Mann RA, Inman VT: Phasic activity of intrinsic muscles of the foot. J Bone Joint Surg [Am] 46:469–481, 1964
20. Cavanagh PR, Williams KR, Clarke TE: A comparison of ground reaction forces during walking barefoot and in shoes. In Morecki A, et al (eds): Biomechanics VII. Baltimore, MD, University Park Press, 1981, pp 151–156
21. Radin E, Whittle M, Yang KH, et al: The heel strike transient, its relationship with the angular velocity of the shank, and the effects of quadriceps paralysis. In: Proceedings of the American Society of Mechanical Engineers Annual Conference, December 8–12, 1986, pp 121–123
22. Katoh Y, Chao EYS, Laughman RK, et al: Biomechanical analysis of foot function during gait and clinical applications. Clin Orthop 177:23–33, 1983
23. Rodgers MM: Plantar Pressure Distribution Measurement During Barefoot Walking: Normal Values and Predictive Equations. Doctoral Dissertation. University Park, PA, The Pennsylvania State University, 1985
24. Clarke TE: The Pressure Distribution Under the Foot During Barefoot Walking. Doctoral Dissertation. University Park, PA, The Pennsylvania State University, 1980
25. Cavanagh PR, Rodgers MM: Pressure distribution underneath the human foot. In Perren SM, Schneider E (eds): Biomechanics: Current Interdisciplinary Research. Dordrecht, The Netherlands, Martinus Nijhoff/Dr W Junk Publishers, 1985, pp 85–95
26. Rodgers MM, Cavanagh PR: Pressure distribution in Morton's foot structure. Med Sci Sports Exerc, to be published
27. Cavanagh PR, Hennig EM, Rodgers MM, et al: The measurement of pressure distribution on the plantar surface of diabetic feet. In Whittle M, Harris D (eds): Biomechanical Measurement in Orthopaedic Practice. Oxford, England, Clarendon Press, 1985, pp 159–166
28. Winter DA, Robertson DGE: Joint torque and energy patterns in normal gait. Biol Cybern 29:137–142, 1978
29. Radin E, Martin B, Burr DB, et al: Mechanical factors influencing cartilage damage. In Peyron JG (ed): Osteoarthritis: Current Clinical and Fundamental Problems. Paris, France, CIBA-GEIGY Corp, 1985, pp 90–99
30. Stauffer RN, Chao EYS, Brewster RC: Force and motion analysis of the normal, diseased, and prosthetic ankle joint. Clin Orthop 127:189–196, 1977
31. Seireg A, Arvikar RJ: The prediction of muscular load sharing and joint forces in the lower extremities during walking. J Biomech 8:89–102, 1975
32. Procter P, Paul JPL: Ankle joint biomechanics. J Biomech 15:627–634, 1982
33. Williams KR: Biomechanics of running: In Terjung RL (ed): Exercise and Sport Sciences Reviews. New York, NY, Macmillan Publishing Co, 1985, vol 13, pp 389–441
34. Mann RA, Baxter DE, Lutter LD: Running symposium. Foot Ankle 1:190–224, 1981
35. Bates BT, Osternig LR, Mason B: Lower extremity function during the support phase of running. In Asmussen E, Jorgensen K (eds): Biomechanics VI. Baltimore, MD, University Park Press, 1978, pp 31–39
36. Cavanagh PR: The biomechanics of lower extremity action in distance running. Foot Ankle 7:197–217, 1987
37. Kaelin X, Unold E, Stussi E, et al: Interindividual and intraindividual variabilities in running. In Winter DA, et al (eds): Biomechanics IX-B. Champaign, IL, Human Kinetics Publishers Inc, 1985, pp 356–360
38. Scranton PE, Rutkowski R, Brown TD: Support phase kinematics of the foot. In Bateman JE, Trott A (eds): The Foot and Ankle. New York, NY, Thieme Medical Publishers Inc, 1980, pp 195–205
39. Clarke TE, Frederick EC, Hamill CL: The study of rearfoot movement in running. In Frederick EC (ed): Sport Shoes and Playing Surfaces: Biomechanical Properties. Champaign, IL, Human Kinetics Publishers Inc, 1984, pp 166–189
40. Miyashita M, Matsui H, Miura M: The relation between electrical activity in muscle and speed of walking and running. In Vredenbregt J, Wartenweiler JW (eds): Biomechanics II. Baltimore, MD, University Park Press, 1971, pp 192–196
41. Ito A, Fuchimoto T, Kaneko M: Quantitative analysis of EMG during various speeds of running. In Winter DA, et al (eds): Biomechanics IX-B. Champaign, IL, Human Kinetics Publishers Inc, 1985, pp 301–306
42. James SL, Bates BT, Osternig LR: Injuries to runners. Am J Sports Med 6:40–50, 1978
43. Nigg BM: Biomechanical analysis of ankle and foot movement. Med Sci Sports Exerc 23:22–29, 1987
44. Cavanagh PR, Lafortune MA: Ground reaction forces in distance running. J Biomech 13:397–406, 1980
45. Frederick EC, Hagy JL, Mann RA: Prediction of vertical impact force during running. J Biomech 14:498, 1981
46. Harrison RN, Lees A, McCullagh PJJ, et al: Bioengineering analysis of muscle and joint forces acting in the human leg during running. In Jonsson B (ed): Biomechanics X-B. Champaign, IL, Human Kinetics Publishers Inc, 1987, pp 855–861
47. Cavanagh PR, Hennig EM: Pressure distribution measurement: A review and some new observations on the effect of shoe foam materials during running. In Nigg BM, Kerr BA (eds): Biomechanical Aspects of Sport Shoes and Playing Surfaces. Calgary, Alberta, Canada, The University of Calgary Press, 1983, pp 187–190
48. Burdett RG: Forces predicted at the ankle during running. Med Sci Sports Exerc 14:308–316, 1982
49. Ae M, Miyashita K, Yokoi T, et al: Mechanical power and work done by the muscles of the lower limb during running at different speeds. In Jonsson B (ed): Biomechanics X-B. Champaign, IL, Human Kinetics Publishers Inc, 1987, pp 895–899

Basic Research

Research Report

Foot Trajectory in Human Gait: A Precise and Multifactorial Motor Control Task

The trajectory of the heel and toe during the swing phase of human gait were analyzed on young adults. The magnitude and variability of minimum toe clearance and heel-contact velocity were documented on 10 repeat walking trials on 11 subjects. The energetics that controlled step length resulted from a separate study of 55 walking trials conducted on subjects walking at slow, natural, and fast cadences. A sensitivity analysis of the toe clearance and heel-contact velocity measures revealed the individual changes at each joint in the link-segment chain that could be responsible for changes in those measures. Toe clearance was very small (1.29 cm) and had low variability (about 4 mm). Heel-contact velocity was negligible vertically and small (0.87 m/s) horizontally. Six joints in the link-segment chain could, with very small changes ($\pm 0.86°-\pm 3.3°$), independently account for toe clearance variability. Only one muscle group in the chain (swing-phase hamstring muscles) could be responsible for altering the heel-contact velocity prior to heel contact. Four mechanical power phases in gait (ankle push-off, hip pull-off, knee extensor eccentric power at push-off, and knee flexor eccentric power prior to heel contact) could alter step length and cadence. These analyses demonstrate that the safe trajectory of the foot during swing is a precise endpoint control task that is under the multisegment motor control of both the stance and swing limbs. [Winter DA. Foot trajectory in human gait: a precise and multifactorial motor control task. Phys Ther. 1992;72:45–56.]

David A Winter

Key Words: *Kinesiology/biomechanics, gait analysis; Lower-limb trajectory, measurements; Slipping; Tripping.*

Walking is primarily a lower-extremity control activity, and researchers have recognized this by focusing their research on the kinematics and kinetics of the lower limb. The upper body (head, arms, and trunk [HAT]) has received limited attention, and that has dealt mainly with kinematic descriptions.[1] Some recent focus has been placed on the HAT's large inertial load, as it affects balance,[2] and on the HAT's large gravitational load, as it affects collapse.[3] The role of the lower extremity in controlling both balance and collapse was identified as unique stance-phase tasks. The detailed role of the lower extremity in achieving forward progression has been limited, however, to kinematic descriptions and a number of kinetic analyses. Forward progression is essentially a lower-extremity task and begins late in stance during push-off[4] and continues throughout swing. The detailed energetics that decide the magnitude of step length and the precise trajectory of the foot during swing have not been analyzed and were the subject of this research.

Review of Literature

To date, there has been considerable effort focused on the kinematics of the lower limb during normal walking. Joint angle data have most commonly been reported.[5–12] Absolute segment kinematics (linear and angular displacements, velocities, and accelerations) are not commonly reported.[12] Other than the occasional "stick-diagram" plot and a few individual

---

DA Winter, PhD, PEng, is Professor, Department of Kinesiology, University of Waterloo, Waterloo, Ontario, Canada N2L 3G1.

This research was funded, in part, by Grant MT4343 from the Medical Research Council of Canada.

This article was submitted November 26, 1990, and was accepted July 24, 1991.

trajectory plots,[13] there has not been a comprehensive study that has examined the trajectory of the foot (heel and toe), especially critical variables such as toe clearance and heel-contact velocity.

Several energy-related motor patterns have been identified as influencing the magnitude of step length.[14] Because the swing limb constitutes the major energy demand in walking,[15,16] we must look at the mechanical energy-generating and energy-absorbing phases that accelerate and decelerate the lower limb immediately prior to and during swing. Energy generation during push-off by the plantar flexors is the largest single work phase in the gait cycle[4] and is responsible for the upward and forward acceleration of the lower limb. Simultaneous with this plantar-flexor contraction (during 40%–60% of the walking stride), the knee is flexing under the control of the eccentrically acting quadriceps femoris muscle. During late stance (50% of stride), the hip flexors commence a concentric contraction, initiating a "pull-off" power phase that continues past toe-off (TO) into mid-swing (80% of stride). Finally, the major deceleration of the leg and foot is achieved by the hamstring muscles, which contract eccentrically to reduce the foot velocity to near-zero prior to heel contact (HC). What is not known is how these energy-generating and energy-absorbing phases vary as stride length (and cadence) varies in normal level gait.

Methodology

Biomechanical Model

The precision of any task must be considered relative to the number of segments involved, their size and mass, and the number of degrees of freedom. The link chain for the control of the foot during swing begins with the stance foot and proceeds up to the hip, across the pelvis, and down to the distal end of the swing foot/phalangeal segment. This chain can be considered to consist of seven segments (or nine if a phalangeal segment is considered), with 12 major angular degrees of freedom at the ankle, knee, and hip that can influence the displacement of the heel or toe during the swing phase of gait. Figure 1 represents this anatomical model with those important degrees of freedom indicated. For a typical adult male subject (mass=70 kg, height=1.8 m), the length of this chain exceeds 2 m. If we consider the large number of muscles crossing those joints, the end-point control of the heel and toe trajectories is a challenging task.

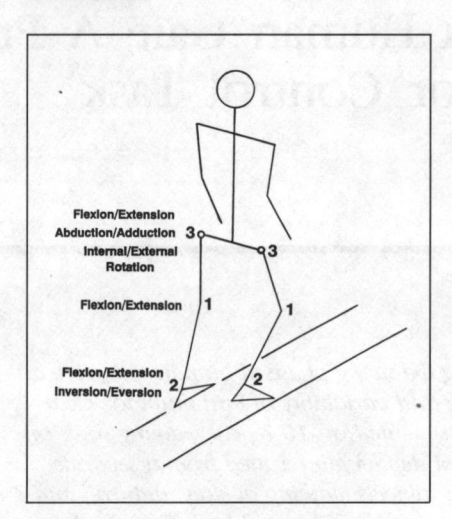

Figure 1. *Stick diagram of link-chain system of seven segments of the support limb, pelvis, and swing limb involved in the control of the toe and heel trajectories. The 12 major degrees of freedom at the six joints that influence those trajectories are indicated.*

Procedure and Subjects

The experimental evidence presented in this report was taken from gait laboratory data collected from young adults. Some analyses were based on individual walking trials, and other analyses were based on repeat trials conducted over a period of 1 hour. Details of the kinematic and kinetic systems have been reported previously[4,12,14,16] and have recently been summarized in a recent report on walking pattern changes in the elderly.[17] For the foot-trajectory component of this study, a group of young adults (six men, five women), who ranged in age from 21 to 28 years ($\overline{X}$=24.9), were analyzed. Their average height was 1.73 m, and their average weight was 69.2 kg. Each subject walked at his or her natural cadence on a level walkway a minimum of 10 times; repeat trials were conducted over a period of 1 hour (one trial every 5 or 6 minutes). For the analysis of the energetic factors that affect step length, data were taken from analyses performed over the past 10 years using 55 young subjects averaging 22.6 years of age. Their average height was 1.75 m, and their average weight was 71.2 kg. The data-collection protocol of this analysis was identical to that of the foot-trajectory analysis, except each subject underwent only one walking trial at his or her natural cadence, at a fast cadence (defined as the subject's natural cadence+20 steps/min), or at a slow cadence (defined as the subject's natural cadence−20 steps/min). A total of 19 subjects were analyzed at slow and natural cadences, and 17 subjects were analyzed at fast cadences. Each subject provided informed consent before participation in the study.

Data Analysis

The trajectories of the heel and toe markers were plotted over the stride period, which was normalized to 100%, with HC at 0% and 100%. These heel and toe profiles were then averaged over the 10 repeat walking trials to assess intrasubject variability. Each intrasubject average was then ensemble-averaged to produce an intersubject average. Based on the variability measurements recorded at minimum toe clearance, each critical degree of freedom in the link chain was varied independently to demonstrate the sensitivity of the toe trajectory to small angular variations at each joint in the chain. In this way, the fine control necessary at each of the joints was documented. In a similar manner, the velocities of the heel in the vertical and horizontal directions were calculated in order to assess the rapid reduction in velocity of the heel during the latter half of swing and after HC. A similar sensitivity analysis on the angular velocities

Basic Research

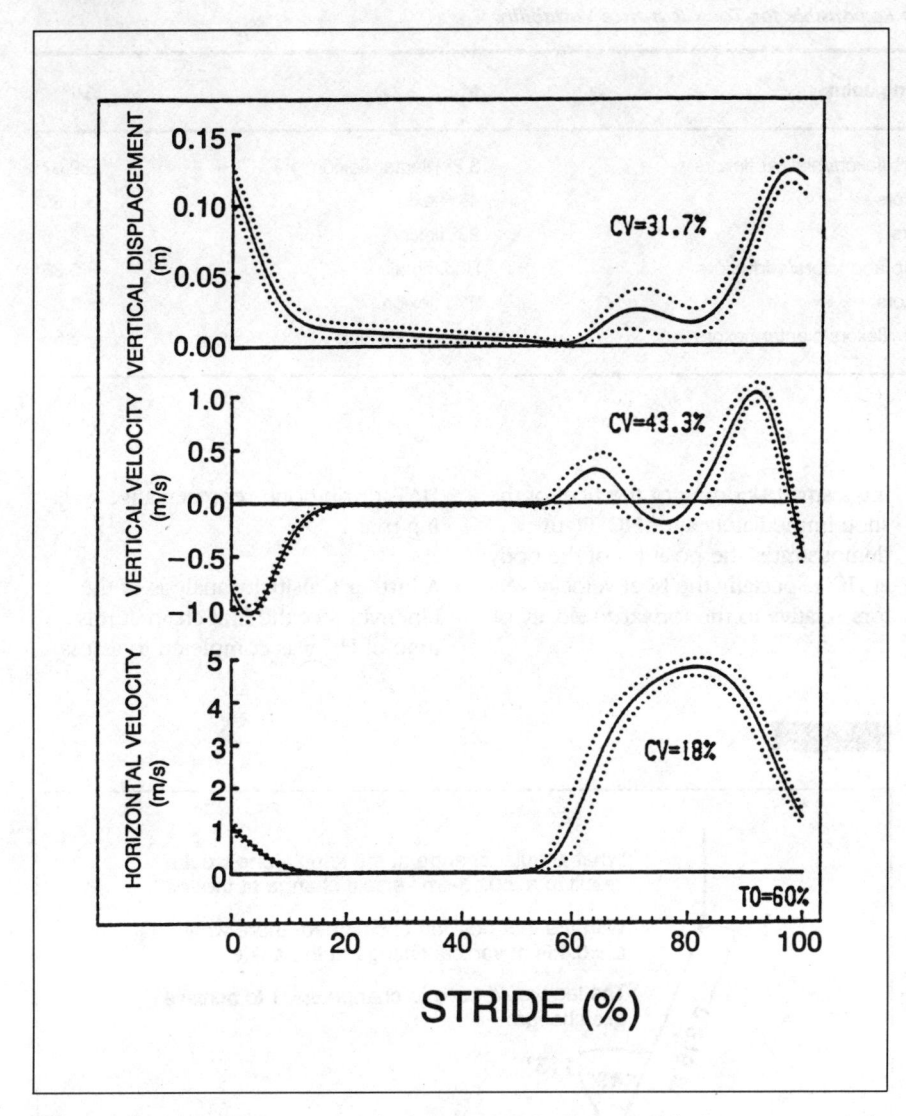

Figure 2. *Ensemble-averaged displacement and velocities of the toe over one stride of 11 subjects walking at their natural cadence. Heel contact was at 0% and 100% of stride, and toe-off (TO) was at 60% of stride. Minimum toe vertical displacement for each subject was set at zero at the minimum reached as the toe pressed downward into the floor immediately before TO. (CV=coefficient of variation.)*

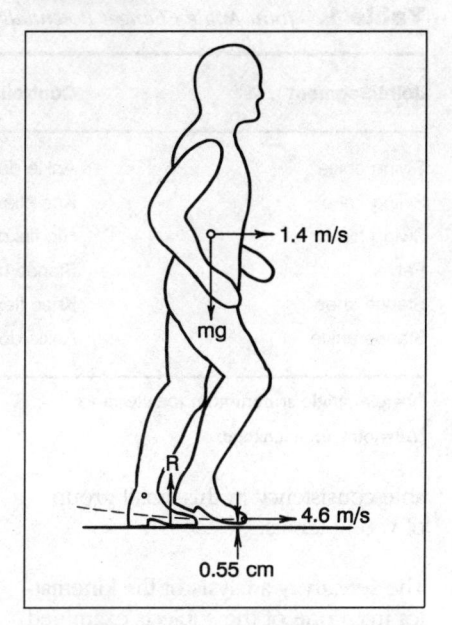

Figure 3. *Position of body at the instant of minimum toe clearance for one representative walking trial showing the high forward toe velocity (4.6 m/s) and center of gravity of the head, arms, and trunk located ahead of the stance foot. (R represents the ground-reaction-force vector, and mg represents the body's center-of-gravity vector.)*

of all segments in the link chain were examined at HC to determine their individual contributions to the slowing down of the heel at this potentially dangerous impact time. Finally, the joint mechanical power patterns immediately prior to and during swing were assessed[4] to determine how they changed as cadence and step length increased.

Results

Figure 2 plots the average vertical trajectory and both horizontal and vertical velocities of the toe for 11 subjects over the stride period. The toe trajectory showed the toe to reach its lowest point at about 56% of stride as the toe pushed downward during the final phase of push-off. This minimum on each trial was considered to be zero toe clearance for the purpose of plotting this displacement profile. After TO, the toe reached a height of a few centimeters. During mid-swing, the toe dropped to its minimum clearance; for these subjects, this mean clearance averaged 1.29 cm. During the latter half of swing, the toe rose to its maximum of about 15 cm just prior to HC. The mean intrasubject variability for this minimum toe clearance was 0.45 cm. Figure 2 shows that this minimum clearance was achieved when the forward velocity of the toe was at its maximum (ie, about 4.6 m/s). Figure 3 demonstrates the position of the stance and swing limbs and the upper body at this potentially dangerous tripping time during one representative walking trial. The forward velocity of the body was 1.4 m/s at this time, and the center of gravity of the HAT was just forward of the stance foot. The combination of this center-of-gravity location and the body's forward momentum means that, if a trip occurs, there is no possibility that the support limb can recover to return the body's center of gravity within the safe borders of the foot. The only possible safe recovery is by a safe placement of the swing limb itself. It is noted that the coefficients of variation (CVs) of these intersubject ensemble averages (Fig. 2) are quite low and indicate consider-

Table 1. *Joint Angle Changes Potentially Responsible for Toe Clearance Variability*

Joint/Segment	Controlling Joint	θ^a	$\Delta\theta^b$
Swing ankle	Ankle dorsiflexors/plantar flexors	3.2° plantar flexion	±2.07°
Swing knee	Knee flexors	49° flexion	±1.35°
Swing hip	Hip flexors	23° flexion	±2.16°
Pelvis	Stance hip abductors/adductors	Horizontal	±0.86°
Stance knee	Knee flexors	9.4° flexion	±3.3°
Stance ankle	Ankle dorsiflexors/plantar flexors	4.6° dorsiflexion	±2.6°

$^a\theta$=joint angle at minimum toe clearance.
$^b\Delta\theta$=joint angle change.

able consistency in this small group of young adults.

The sensitivity analysis of the kinematics from one of the subjects examined all joints in the link segment that had a potential for influencing the toe trajectory at the time of minimum toe clearance: swing ankle, swing knee, swing hip, stance hip abductor (pelvic list), stance knee, and stance ankle. The sensitivity analysis calculated the angular changes that, at each joint by itself, would cause the ±0.45-cm toe clearance variability. These results are reported in Table 1, and one typical calculation is presented in Figure 4. According to this interpretation of the results, if all the remaining joints remained unchanged, a change of ±0.86 degree at this time in stance hip abduction alone could be responsible for all of the variability seen in toe clearance.

Figure 5 plots the average vertical trajectory and both horizontal and vertical velocities of the heel for these same subjects over the stride period. The heel began rising in mid-stance at heel-off and reached a maximum of about 25 cm just after TO, then decreased rapidly, reaching about 1 cm above the ground at 90% of the stride period. During the last 10% of the stride prior to HC, the trajectory was almost horizontal; the horizontal velocity also decreased drastically from 4 m/s, reaching about 0.87 m/s at HC. This forward velocity decreased to zero at about 4% of the stride, indicating a small skidding of the heel of the shoe immediately after HC. Figure 6 demonstrates the position of the body at HC, especially the heel velocity vectors relative to the forward velocity of HAT, during one representative walking trial.

A further sensitivity analysis of the kinematics of the link chain at this time of HC was completed to assess

What angular change at the knee alone would result in a ±0.45-cm vertical change at the toe?

With the foot position unchanged, there would be a ±0.45-cm vertical change at the ankle.

The leg would have to change ±Δθ to achieve this change.

Vertical distance from knee to ankle = .425 sin 64°
= 0.382 m

∴ .425 sin (64 ±Δθ) = .382 ± .0045

sin (64 ±Δθ) = .9094 ∴ Δθ = 1.4°

sin (64 ±Δθ) = .8882 ∴ Δθ = 1.4°

∴ a ±1.4° change in knee angle by itself would cause the ±0.45-cm change in toe clearance.

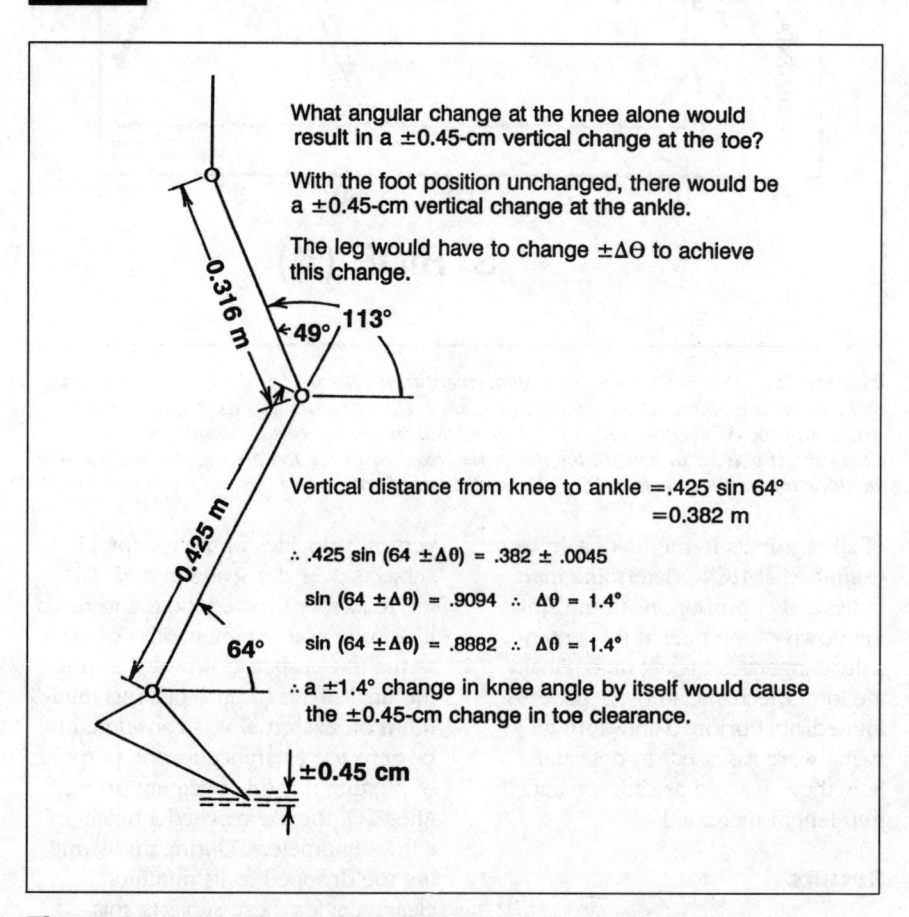

Figure 4. *Example of sensitivity calculation to determine the angular change (±Δθ) necessary at the knee alone to cause the ±0.45-cm displacement variability seen at the toe at the instant of minimum toe clearance.*

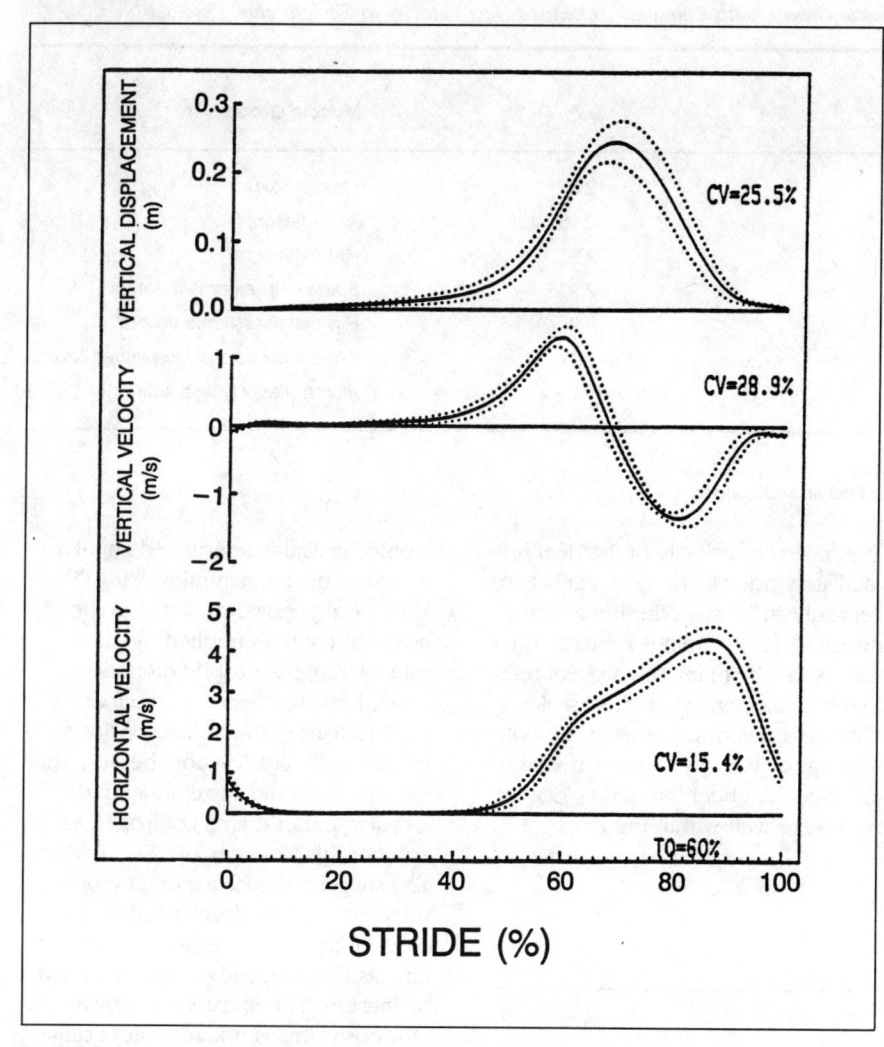

Figure 5. *Ensemble-averaged displacement and velocities of the heel of the same 11 subjects as represented in Fig. 2 over one stride, from heel contact (HC) to HC. Horizontal heel velocity reached a peak in mid-swing and decreased to virtually zero in the vertical direction and to a low value horizontally at HC. (CV=coefficient of variation; TO=toe-off.)*

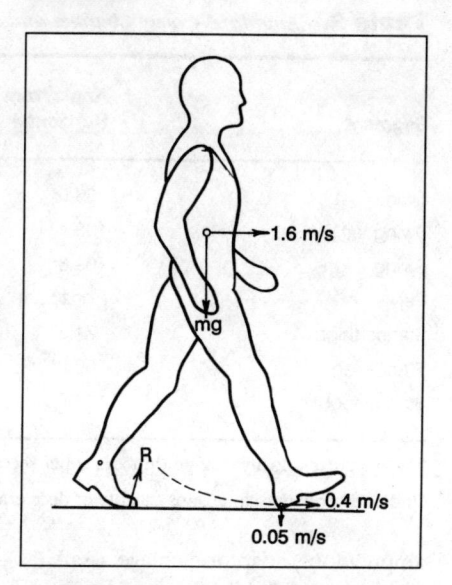

Figure 6. *Position of body at heel contact for one representative walking trial showing the low heel velocities relative to the forward velocity of the body's center of mass. (R represents the ground-reaction-force vector, and mg represents the body's center-of-gravity vector.)*

the angular velocity changes that, by themselves, would be necessary to reduce the forward heel velocity by 0.87 m/s, thus reducing it to exactly zero at HC. The potential angular velocities to which heel velocity is sensitive are swing foot, swing leg, swing thigh, pelvic horizontal velocity (controlled by hip rotators), stance thigh, stance leg, and stance foot. The necessary angular velocity changes are summarized in Table 2 with an indication of what muscle group would be responsible in each case (remembering that during stance the muscles at either the proximal or distal end of each segment can control). One typical calculation of the velocity sensitivity is presented in Figure 7.

The variability of the heel trajectories, as demonstrated by the CVs in the ensemble averages presented in Figure 5, is quite low. Again, this low variability is indicative of consistency in this small group of young adults.

Figures 8 through 10 present mechanical power profiles drawn from the database from subjects walking at three different cadences and at different step lengths. The 19 natural-cadence walkers had a mean cadence of 105.3 steps/min and a mean step length of 1.51 m (walking velocity=1.33 m/s). The 19 slow walkers had a cadence of 86.8 steps/min and a step length of 1.38 m (walking velocity=1.00 m/s), and the 17 fast walkers had a cadence of 123.1 steps/min and a step length of 1.64 m (walking velocity=1.68 m/s).

Discussion

Toe clearance has been considered to be a major responsibility of the swing leg dorsiflexors, and, as expected, it is quite sensitive to small angular changes ($\pm 2.07°$) of the swing ankle. The sensitivity analysis results (Tab. 1), however, show that the endpoint toe trajectory is also very sensitive to small angular changes at five other joints in the total link-segment chain. Toe clearance is sensitive to even smaller angular changes at the knee ($\pm 1.35°$) and during stance hip abduction and adduction ($\pm 0.86°$). Clinically, it is important to observe each walking patient and note any clearance problems and at which joint compensations are taking place. Thus, it is not surprising that certain patients, such as those with below-knee

Table 2. *Angular Velocity Changes and Muscle Groups with Potential to Reduce Heel Velocity to Zero at Heel Contact*

Segment	Angle from Horizontal	$\Delta\omega$[a]	Muscle Group
Swing foot	69.6°	12.3	Plantar flexors
Swing leg	108.4°	2.15	Knee flexors[a]
Swing thigh	109.6°	2.9	Hip extensors[a]
Pelvis	Horizontal	2.96	Stance hip external rotators[a]
Stance thigh	71.3°	2.9	Hip extensors/knee flexors
Stance leg	56.2°	2.47	Knee extensors/ankle plantar flexors
Stance foot	119.6°	5.5	Plantar flexors (less activity)

[a] $\Delta\omega$ = angular velocity change (in radians per second).
[b] Indicates only muscle groups capable of decelerating heel at heel contact.

amputations, adapt to achieve a safe foot clearance with increased knee flexion and "hip hiking" (increased stance hip abduction). Circumduction is also a common adaptation, but, because of the low sensitivity of the swing hip abductors, an appreciable angular change is required to make a significant change in toe clearance.

The trajectory velocity of the heel immediately prior to HC is virtually zero vertically and low in the horizontal direction; such findings raise the question as to why many researchers refer to this initial contact as "heel-strike." With the exception of the swing foot, the angular velocity changes necessary to reduce the heel forward velocity to zero were well within the range of biomechanically determined angular velocities during natural walking.[18] Functionally, however, some of the potential controls implied by the results of Table 2 must be discarded. A rapid plantar flexion of the foot (12.3 radians/s) immediately prior to HC is not a valid solution, because this movement would result in a rapid foot-slap rather than a controlled lowering of the foot after HC. The analysis also suggests the stance thigh's forward velocity could be decelerated by increased knee flexor activity at the same time as the stance leg was decelerated by increased knee extensor activity. Obviously, this is not an anatomically possible combination. Similarly, the tabulated results suggest that the stance ankle plantar flexors would have to increase their activity to decelerate the forward-rotating leg at the same time as they decreased activity to decrease foot plantar flexion. Again, this is not a compatible solution. Another possibility is hip extensor control of the stance thigh, but such control is not likely, because the stance hip extensors are not reported to be active at this time.[19,20]

Thus, the knee flexors, hip extensors, and stance hip external rotators are the only muscle groups that have the potential for decelerating the heel immediately prior to HC (Tab. 2). The most compatible combination of those three muscle groups are the knee flexors and the hip extensors,

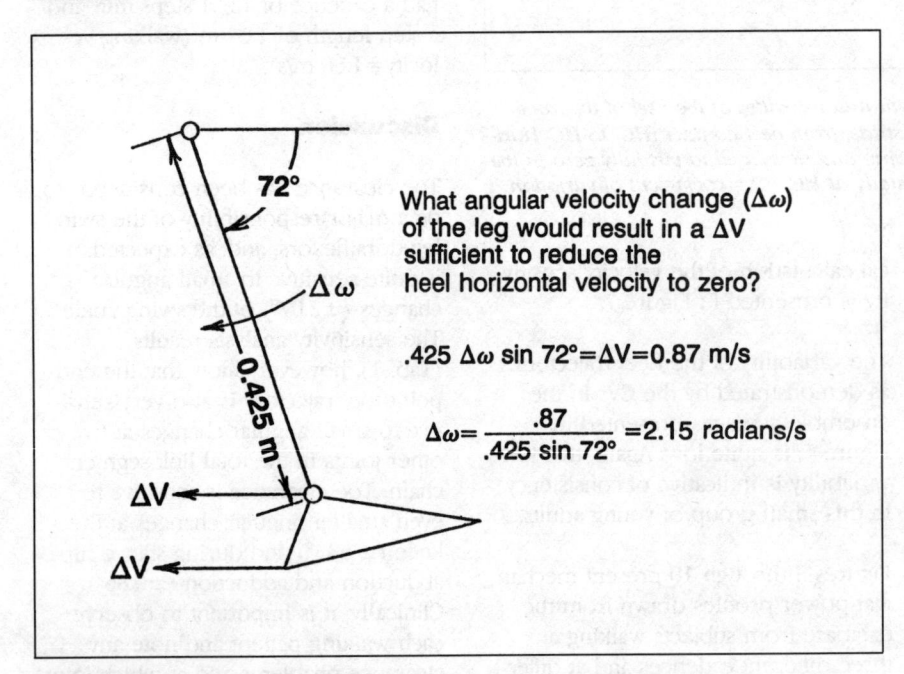

Figure 7. *Example of calculation to determine the sensitivity of the heel contact (HC) velocity to the velocity of the individual segment. ($\Delta\omega$ = angular velocity change that, by itself, could reduce the horizontal velocity of the heel at HC from its average value [0.87 m/s] to zero; ΔV = change in velocity.)*

What angular velocity change ($\Delta\omega$) of the leg would result in a ΔV sufficient to reduce the heel horizontal velocity to zero?

$$.425 \Delta\omega \sin 72° = \Delta V = 0.87 \text{ m/s}$$

$$\Delta\omega = \frac{.87}{.425 \sin 72°} = 2.15 \text{ radians/s}$$

Basic Research

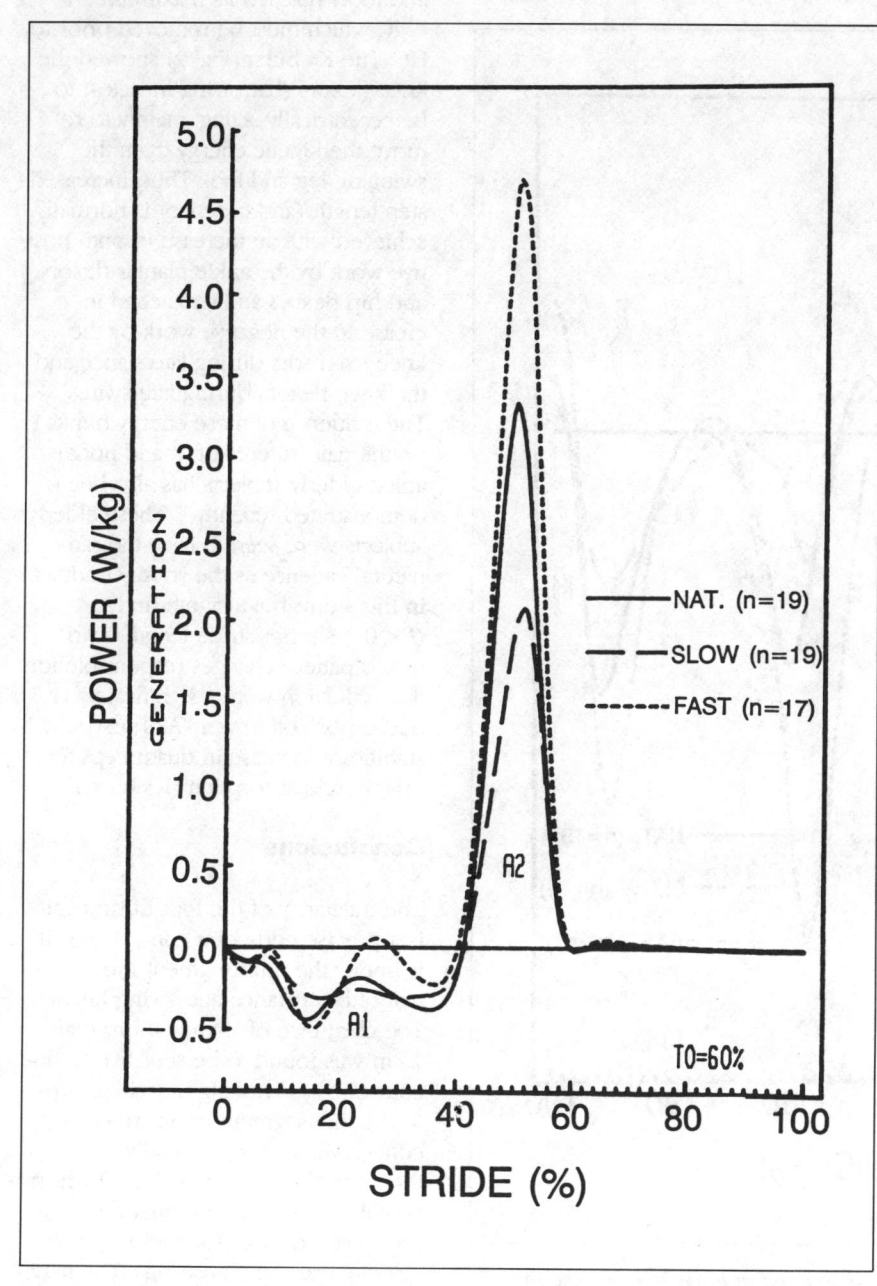

Figure 8. *Mechanical power generation and absorption profiles at the ankle for three walking-speed cadences: natural, slow, and fast. The push-off power (A2 burst) by the plantar flexors drastically increased from slow to fast walking cadences and represents over 75% of all energy generated in the stride period. The A1 power phase was the absorption of energy as the plantar flexors lengthen as the leg rotates forward over the foot. (TO=toe-off.)*

which means that the biarticulate hamstring muscles would be predicted to decelerate both the swing thigh and leg and therefore are the major decelerators of the foot. Electromyographic profiles show the hamstring muscles to be active in late swing.[19,20] Mechanical power analyses have also shown this to be true in both walking[4] and running,[21] during which the eccentric work done at the knee during the latter half of swing was dominant. In running,[21] a small, short-duration burst of positive power immediately followed this K4 negative work and was due to a concentric contraction as these same hamstring muscles momentarily accelerated the leg backward. This finding does not mean that the foot was traveling backward at this time. Rather, the body had a forward velocity of about 3 m/s, and, to reduce the foot velocity to near-zero, the foot would need a momentary backward velocity of about 3 m/s relative to the center of mass of the body. The central nervous system obviously recognizes the energetics of this fine control. The third possible muscle group noted in Table 2 that could control the swing limb's forward velocity are the stance hip external rotators. Because the angular rotation and velocity of the pelvis in the transverse plane were quite small, these rotators would have only minimal potential for control.

The clinical significance of this HC velocity analysis relates to the potential for a patient to slip at this critical phase of the gait cycle. Heel contact usually involves weight bearing on a small surface area of the heel, and, if the ground contact area is wet or slippery, there is an increased probability of slipping. In a study on fit and nondisabled elderly subjects, we have documented that their HC velocity was 1.15 m/s, which is significantly higher ($P<.01$) than for the younger adults in this study. Thus, these elderly individuals are at a greater risk for slipping, even though their walking velocity was significantly lower than that of the younger adults in this study (1.29 versus 1.43 m/s, respectively). To date, we have not documented the HC velocity for patients who are prone to fall; such studies are currently ongoing.

Four of the power bursts (ie, A2, K3, K4, and H3) shown in Figures 8 through 10 demonstrated drastic changes during push-off and swing that could influence step length. The ankle push-off burst (A2 in Fig. 8) showed a dramatic increase as the subjects accelerated their lower limb prior to TO to achieve a longer step length. Almost simultaneous to this push-off impulse was an increasing absorption of energy at the knee (K3 in Fig. 9) by the eccentrically act-

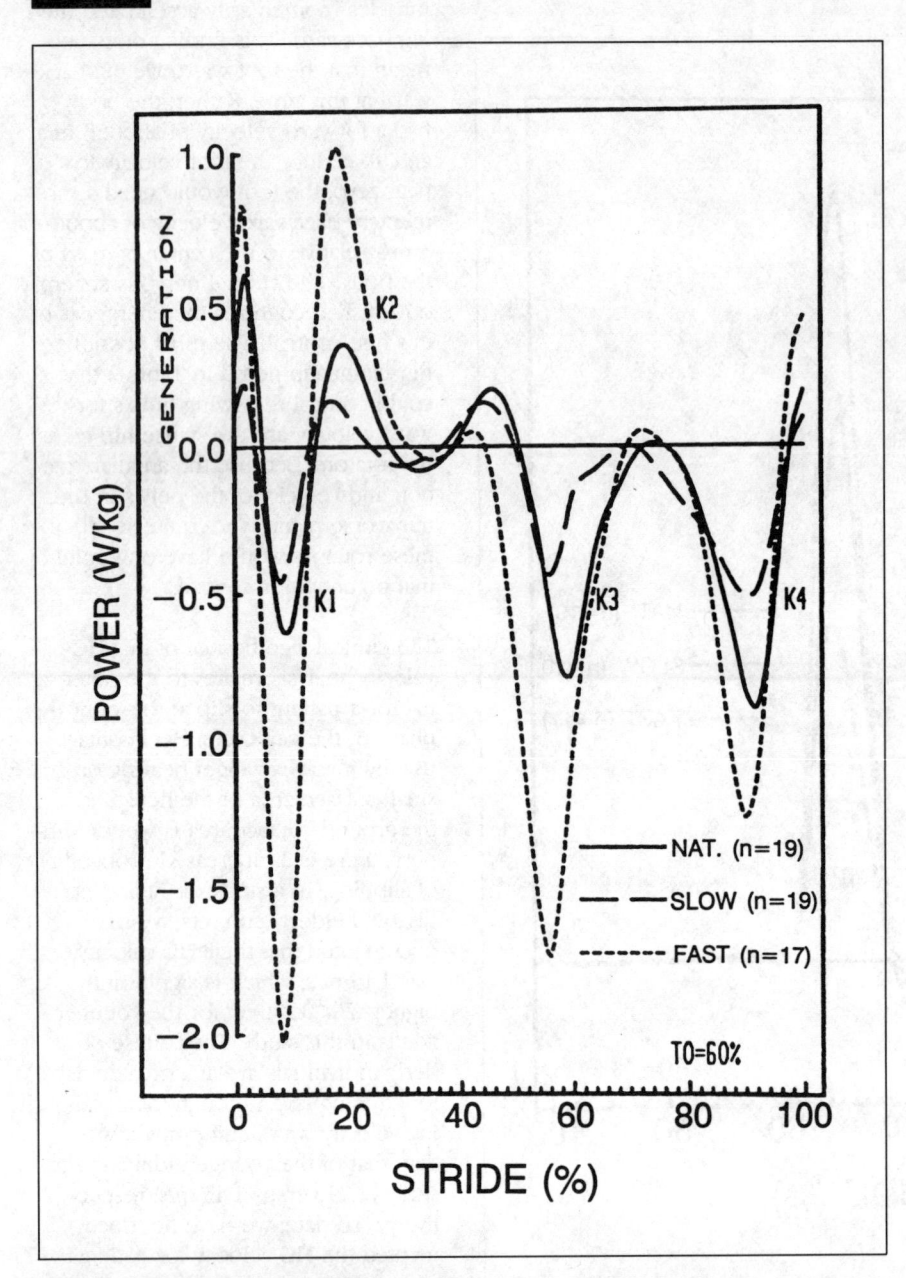

Figure 9. *Mechanical power absorption and generation at the knee for the same three cadence groups as represented in Fig. 8. The K3 burst was the power associated with the eccentrically contracting quadriceps femoris muscle necessary to control knee flexion caused by the "piston-like" push-off by the ankle in late stance. The K4 burst was due to the eccentrically contracting hamstring muscles decelerating the swinging leg prior to heel contact. Both K3 and K4 increased as cadence and stride length increased. The K1 burst was the absorption by the knee extensors as they lengthen when the knee flexes. The K2 burst was the generation by the same knee extensors as the knee extends during mid-stance. (TO=toe-off.)*

ing quadriceps femoris muscle. This absorption represents a necessary loss of energy to prevent too rapid a knee flexion prior to TO (60% of stride) resulting from the forceful upward acceleration of the leg caused by A2. At mid-double support (50% of stride), the hip flexors contracted concentrically to commence a pull-off of the lower limb (H3 in Fig. 10), which continued past TO until midswing. This impulse of pull-off energy also increased dramatically with increased cadence and step length. In mid-swing, the swinging lower limb (mainly leg and foot) reached its maximum energy, which must be removed prior to HC. The K4 burst (Fig. 9) showed the knee flexors (hamstring muscles) to be eccentrically acting, mainly to remove the kinetic energy from the swinging leg and foot. Thus, increased step length (and cadence) is normally achieved with an increase in both positive work by the ankle plantar flexors and hip flexors and a matched increase in the negative work by the knee extensors during late stance and the knee flexors during late swing. The influence of these energy bursts on the gait patterns of fit and nondisabled elderly subjects has also been demonstrated recently.[17] These elderly subjects were seen to have the same natural cadence as the younger adults in this study, but a significantly ($P<.01$) shorter stride length. Two motor pattern changes responsible for this reduction were a significantly reduced push-off power (A2 burst) and a significant increase in quadriceps femoris muscle absorption (K3 burst).

Conclusions

The trajectory of the foot during gait is a precise end-point control task. It is under the multisegment motor control of both stance and swing limbs. Toe clearance of slightly more than 1 cm was found to be sensitive to fine control by at least six muscle groups in the link-segment chain. Heel-contact velocity was virtually zero in the vertical direction, with a low horizontal velocity. The dominant muscle group responsible for reducing that velocity was the hamstrings. The magnitude of step length was found to be under the control of four concentric and eccentric motor patterns during late stance and swing. Step length and walking velocity were increased by increased plantar-flexor power during push-off and by increased hip-flexor power during "pull-off." Step length can be reduced by increased eccentric quadriceps femoris muscle activity during late stance and by increased eccentric hamstring muscle activity during late swing. In spite of the consistency in the foot trajectory profiles for this small group of young adults,

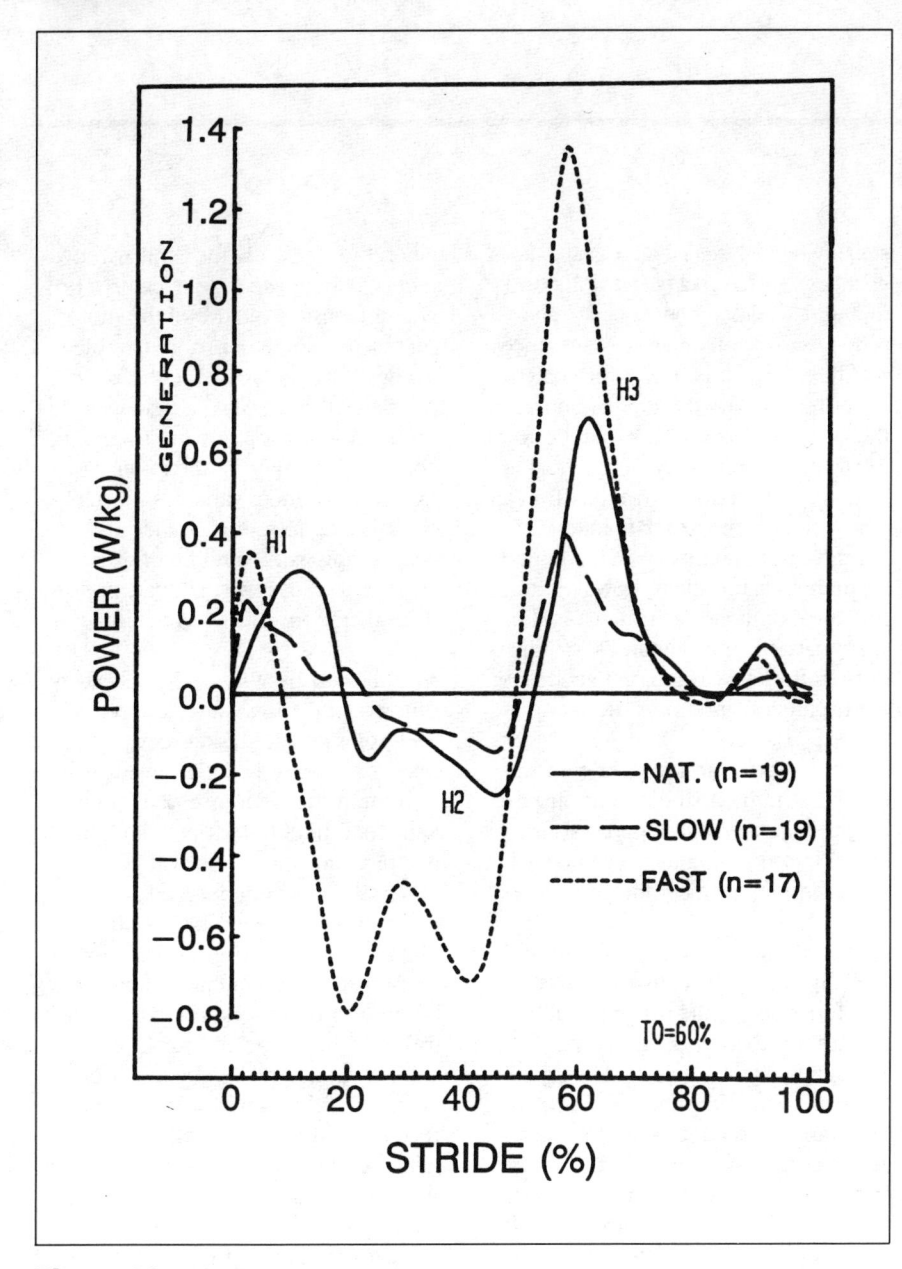

Figure 10. *Mechanical power generation and absorption at the hip for the same three cadence groups as represented in Fig. 8. The H3 burst represents the "pull-off" power generation by the hip flexors. This positive work began in late stance (50%), continued into mid-swing (80%), and increased drastically as cadence increased. The H1 power phase resulted from the hip extensors shortening immediately after heel contact. The H2 power burst resulted from the hip flexors; lengthening during mid-stance to decelerate the backward-rotating thigh. (TO=toe-off.)*

more research may be necessary to quantify any differences in larger groups of young adults and in other age groups.

Acknowledgment

I acknowledge the technical assistance of Mr Paul Guy.

References

1 Thorstensson A, Nilsson J, Carlson H, Zomlefer MR. Trunk movements in human normal walking. *Acta Physiol Scand.* 1984;121:9–22.

2 Patla AE, Frank JS, Winter DA. Assessment of balance control in the elderly: some issues. *Physiotherapy Canada.* 1990;42:89–98.

3 Winter DA. Overall principle of lower limb support during stance phase of gait. *J Biomech.* 1980;13:923–927.

4 Winter DA. Energy generation and absorption at the ankle and knee during fast, natural and slow cadences. *Clin Orthop.* 1983;197:147–154.

5 Finley FR, Karpovich PV. Electrogoniometric analysis of normal and pathological gaits. *Research Quarterly.* 1964;35:379–384.

6 Murray MP, Drought AB, Kory RC. Walking patterns of normal men. *J Bone Joint Surg [Am].* 1964;46:335–360.

7 Murray MP. Gait as a total pattern of movement. *Am J Phys Med.* 1967;46:290–333.

8 Murray MP. Walking patterns of normal women. *Arch Phys Med Rehabil.* 1967;51:637–650.

9 Johnston RC, Smidt GI. Measurement of hip-joint motion during walking. *J Bone Joint Surg [Am].* 1969;51:1083–1094.

10 Lamoreux LW. Kinematic measurements in the study of human walking. *Bulletin of Prosthetics Research.* Spring 1971:3–84.

11 Sutherland DH, Hagy JL. Measurement of gait movements from motion picture film. *J Bone Joint Surg [Am].* 1972;54:787–797.

12 Winter DA, Quanbury AO, Hobson DA, et al. Kinematics of normal locomotion: a statistical study based on TV data. *J Biomech.* 1974;1:479–486.

13 Murray MP, Clarkson BH. The vertical pathways of the foot during level walking, II: clinical examples of distorted pathways. *Phys Ther.* 1966;46:590–599.

14 Winter DA. Concerning scientific basis for the diagnosis of pathological gait and for rehabilitation protocols. *Physiotherapy Canada.* 1985;37:245–252.

15 Ralston HJ, Lukin L. Energy levels of human body segments during level walking. *Ergonomics.* 1969;12:39–46.

16 Winter DA, Quanbury AO, Reimer GD. Analysis of instantaneous energy of normal gait. *J Biomech.* 1976;9:253–257.

17 Winter DA, Patla AE, Frank JS, Walt SE. Biomechanical walking pattern changes in the fit and healthy elderly. *Phys Ther.* 1990;70:340–347.

18 Winter DA. *Biomechanics and Motor Control of Human Gait.* Waterloo, Ontario, Canada: University of Waterloo Press; 1987:25–26.

19 Winter DA, Yack HJ. EMG profiles during normal human walking: stride-to-stride and inter-subject variability. *Electroencephalogr Clin Neurophysiol.* 1987;67:402–411.

20 Shiavi R. Electromyographic patterns in adult locomotion: a comprehensive review. *J Rehabil Res Dev.* 1985;22:85–97.

21 Winter DA. Moments of force and mechanical power in slow jogging. *J Biomech.* 1983;16:91–97.

Commentary

The ability to formulate and apply effective clinical protocols for the evaluation and treatment of human movement dysfunction requires a knowledge of the kinesiological and pathokinesiological factors underlying the execution of functional tasks. Such information should not be limited to quantitative descriptions of movement performance, which merely chart the progress of a patient, but must also address the motor control processes responsible for producing movement. In the complex case of human gait, it may be useful to consider the general requirements of locomotor behavior outlined by Grillner and Wallen[1] as they pertain to both clinical and experimental perspectives. First, the movements underlying the locomotor synergy that provide propulsion must be produced. Second, adequate control of equilibrium during movement must be sustained. Third, locomotor movements must be adapted to (1) the goals of the subject and (2) the environment (ie, to avoid obstacles), which involves anticipatory control (eg, visuomotor skills) and compensation for actual perturbations.

Professor Winter has provided important fundamental information concerning normal human gait and has raised several important issues that have an impact on each of the foregoing requirements of locomotion. In the mechanical model of the link-chain system of body segments, which potentially may influence the control of the trajectory of the foot relative to the ground, he has illustrated the highly complex and interactive nature of the control problem, involving multiple segments, joints, motions at these joints, and muscles, that confronts the movement control system. Decomposition of the system through sensitivity analyses of angular position changes at each *individual* joint and of angular velocity changes at each *individual* segment on toe clearance and heel impact is useful for analytic purposes and for drawing attention to the potential for multiple components to affect foot trajectory. If our goal is to understand and restore function to a moving system rather than to its constituent elements (isolated joints, segments, or muscles), however, then the question arises as to what is the appropriate level of analysis for understanding how multijoint actions are organized and controlled.

There has been an increasing awareness[2] that individual elements are not the appropriate units of analysis for examining multijoint, or even conceptually simpler single-joint,[3] movements. Instead, several elements (eg, joints, muscles, movements) may be spatially and temporally organized into functional units that generally constrain the component elements to act together with respect to the goal of the system. Although such a form of organization should not be viewed as being rigidly fixed, it likely involves certain rules of combination that provide structure yet allow flexibility of performance.[4] The challenge then, for both the clinician and the motor control experimentalist, is to determine the equations of constraint or organizing principles that normally unite multiple system elements. In doing so, the consequences of pathological changes in movement function may be better predicted, prevented, evaluated, and rehabilitated.

One approach to solving this problem is to identify possible variant and invariant characteristics of movement. This line of analysis is drawn from the concept of motor programming, whereby, especially for familiar movements, certain repeatable features provide the basic structure of the motor output profile. Other more flexible features function to shape the basic profile to the specific task goal. Identification of such variant and invariant characteristics among the multisegmental movement variables normally contributing to foot clearance and impact during gait would be of great assistance to those who evaluate and treat alterations in these functions.

As indicated by Winter, the trajectory of the foot during gait is a precise end-point control task. Clearly, the results showing a modest average minimum toe clearance of 1.29 cm and small heel trajectory velocities just prior to floor contact suggest a generally underemphasized refinement of human locomotor output. The extent of the overall precision of foot trajectory is obscured, however, by the use of a measurement of central tendency as an index of intrasubject variability (ie, the mean of average heel and toe profiles over 10 repeat walking trials) rather than specific measurements of dispersion (eg, standard deviations, standard errors). Although it is noted that the coefficients of variation of intersubject ensemble averages are quite low and indicate considerable consistency, a quantitative determination of well-defined, low parametric limits of intertrial and intersubject response variability would strengthen the argument for precision of end-point control. It would also be of importance to determine whether the toe and heel profiles are unique to level overground walking in the gait laboratory or are retained during locomotion over a variety of terrains and in different environmental situations.

As Winter has previously emphasized,[5] a major problem exists when trying to determine the individual muscle forces that produce the net torque or mechanical power at a joint during locomotion. This situation, called *indeterminacy*, is due to the large number of muscles that can affect these variables. Consequently, there are innumerable possible solutions to producing the output as a joint. Similarly, as in the stance phase of gait, a specific, constant foot, leg, and thigh trajectory profile could be generated by a large number of combinations of lower-limb resultant joint torques in conjunction with an equally variable number of combinations of individual muscle forces. Therefore, despite a logical coalescence of mechanical and electromyographic data, the velocity of the heel at floor contact may equally be influenced by combinations of muscle activations or deactivations other than those of the hamstring muscles of the swing limb. This situation, in part, reflects the complex actions of multiarticular muscles, the lack of equivalent relative contributions of joint agonist muscles, and the influence of coactivation relationships among anatomically antagonistic muscles on resultant joint torques or mechanical power profiles. Such possible combinations are effectively countless and should be examined during locomotion in a variety of environments in which the same task goal could potentially be achieved by very different muscle-activation solutions. Such observations raise caution in an unqualified acceptance of the conclusion that the dominant muscle group responsible for controlling heel-contact velocity is the hamstrings. Similar reservations may also apply to the specific muscles implicated in producing the four mechanical power phases of gait.

Finally, Winter focuses on the influence of swing limb ankle, knee, and hip joint power profiles in altering step length and cadence. A fundamental question that arises from these and related[6] findings concerns what mechanical output of the neuromusculoskeletal system is being controlled. Are joint powers and the trajectory of the foot independently controlled? Winter's conclusion that the trajectory of the foot during gait is a precise end-point control task tends to support this viewpoint. Yet, might not the foot-to-ground trajectory be an emergent consequence of appropriate propulsion-related combinations of joint powers or torques that are determined by the experience of the individual with given surroundings and that may be adapted to changing goals of the system or to the environment? This issue is not only important from a motor-control perspective, but is critical to translating the results into effective strategies for evaluating and treating movement dysfunction. What are the critical factors of motor control for a goal-directed task? What are the characteristics of pathological motor profiles? What are the appropriate clinical variables that must be used to effect changes in these profiles? Though approaching the answers to such questions is indeed a challenging endeavor, Winter's article represents an important contribution to our understanding of the propulsive requirement of human gait and brings us closer to uniting our knowledge of cause and effect in the science of physical therapy practice.

Mark W Rogers, PhD, PT
Assistant Professor
Programs in Physical Therapy
Northwestern University
   Medical School
345 E Superior St, Rm 1323
Chicago, IL 60611

References

1 Grillner S, Wallen P. Central pattern generators for locomotion, with special reference to vertebrates. *Annu Rev Neurosci.* 1985;8:233–261.
2 Bernstein NA. *The Coordination and Regulation of Movements.* London, England: Pergamon Press; 1967.
3 Stein RB. What muscle variable(s) does the nervous system control in limb movements? *Behav Brain Sci.* 1982;5:535–577.
4 Macpherson JM. How flexible are muscle synergies? In: Humphrey DR, Freund HJ, eds. *Motor Control: Concepts and Issues.* Chichester, England: John Wiley & Sons Ltd; 1991:33–47.
5 Winter DA. Concerning the scientific basis for the diagnosis of pathological gait and for rehabilitation protocols. *Physiotherapy Canada.* 1985;37:245–252.
6 Winter DA, McFadyen, Dickey JP. Adaptability of the CNS in human walking. In: Patla AE, ed. *Adaptability of Human Gait.* Amsterdam, the Netherlands: Elsevier Science Publishers BV; 1991:127–144.

Author Response

I would like to thank Dr Rogers for his comments on my article. He has made some observations that reinforce my message and has gone on to pose some new questions that arise out of the analysis that was presented. I would like to comment on two of the points he raised. The first related to my use of ensemble averaging to obtain intrasubject and intersubject profiles. Such an approach is now routinely used and is the only way that I know to determine the mean pattern of any variable (in this case, heel and toe trajectories) and its variability over the stride period. It is nothing more than a calculation of the mean and standard deviation at each 2% interval over the stride. Thus, it is a descriptive process and as such gives the standard deviation at 50 points over the walking cycle. In this article, standard errors were not calculated except when significant differences were found, such as the difference in heel-contact velocity between young and elderly adults.

In his second point, Rogers has emphasized the problem of *indeterminacy* of the human neuromuscular system, and I have certainly focused on the necessity of identifying this problem in all of my publications over the past 8 years. This article is no exception, so I must question his doubts about my conclusion that the swing limb's hamstring muscles are the only reasonable decelerator of the foot immediately prior to heel contact. It was precisely because of that indeterminacy that I presented Table 2, which documented all potential motor patterns in the total link-segment chain that had the potential to reduce heel-contact velocity. By the process of elimination, I then rejected all but two muscle groups (ie, swing limb hamstring muscles and stance hip external rotators), and the hamstring muscles were, by far, the dominant muscle group.

David A Winter, PhD, PEng

Basic Research

Research Report

The Dual-Task Methodology and Assessing the Attentional Demands of Ambulation with Walking Devices

The purposes of this article are (1) to provide a preliminary examination of the attentional demands of ambulating with two commonly prescribed walking aids (a standard walker and a rolling walker) and (2) to introduce the dual-task methodology to the physical therapy community. Five subjects familiar with the appropriate use of the walkers and five subjects uninformed as to the correct use of the walkers participated in the study. Each subject completed the three phases of the experiment: (1) performing the reaction time (RT) task only; (2) performing each of the walking tasks only; and (3) performing each of the walking tasks in conjunction with the RT task, which constituted the dual-task conditions. The findings indicated that walking aided by either the rolling walker or the standard walker was highly attention demanding. More importantly, it appears that greater attentional demand was required when ambulating with the standard walker. These results are discussed with respect to the gait modifications and accuracy demands required when using these walkers. The usefulness of the dual-task methodology as a research tool for addressing clinically oriented questions is emphasized, and some potential applications of this methodology for the therapist within the clinic are discussed. [Wright DL, Kemp TL. The dual-task methodology and assessing the attentional demands of ambulation with walking devices. Phys Ther. 1992;72:306–315.]

Key Words: *Ambulation, Attention, dual-task methodology, Rolling walker, Standard walker.*

David L Wright
Tammy L Kemp

Patients with impaired physical status are often provided with assistive devices to ease the task of ambulation. Common devices include canes, standard walkers, and rolling walkers. A number of studies have addressed the metabolic and physiological demands associated with using these devices.[1,2] Little experimental rigor, however, has been devoted to other factors that may mediate the usefulness of prescribing a specific type of walking aid.

One factor that might influence a patient's ability to function with a walking device is the cognitive demands associated with using the device appropriately. Because many of the individuals prescribed a walking aid are using it for the first time, the attentional requirements of efficiently manipulating the device might be large enough to compromise the patient's stability. It would seem important, therefore, that the attentional demands associated with using the traditional walking aids be examined. Data from such an undertaking might offer insight into the underlying mechanisms contributing to possible

---

DL Wright, PhD, is Assistant Professor, Department of Health and Kinesiology, Elouise Beard Smith Human Performance Laboratories, Texas A&M University, College Station, TX 77843-4243 (USA). Address all correspondence to Dr Wright.

TL Kemp, PT, is Physical Therapist, Allied Physical Therapists, 2551 Texas Ave S, Ste J, College Station, TX 77840, and is currently completing her master's degree at Texas A&M University. She was Physical Therapist, Bryan Independent School District, Bryan, TX 77801, when this study was conducted.

The study was approved by the Texas A&M University Research Protocol Review Committee.

This article was submitted June 3, 1991, and was accepted October 11, 1991.

safety complications associated with ambulating with an assistive device.

The study of attention, or attentional capacity, has been a focus of the psychological literature for some time. Extensive investigations of the attentional process have been conducted on both verbal and motor skills in laboratory settings.[3-5] Only a few studies, however, have examined this process in applied settings such as rifle shooting,[6,7] diving,[8] and industrial and military settings typically involving mental work load (ie, cognitive effort required to meet the demand imposed by the task being performed).[9] Unfortunately, little evidence is available to address the attentional demands associated with using a device to aid in rehabilitation of a fundamental activity such as walking.

One method that has been used to determine the attentional demands of a particular task is called the *dual-task paradigm*.[10,11] As little, if any, evidence is available with respect to the attentional demand of tasks or procedures used in the clinic, it can be assumed that this methodology has yet to be used in the rehabilitation setting. The dual-task methodology requires an individual to perform a task that is being evaluated in terms of its attentional demand (primary task), while simultaneously performing an alternative task (commonly termed a "secondary probe task"). Vocal or manual reaction-time (RT) tasks are frequently used secondary probe tasks. These tasks consist of presenting a stimulus (usually visual or auditory) that the subject responds to as rapidly as possible, typically with a vocal or manual response. Secondary probe RT task performance in a dual-task condition is assumed to be inversely proportional to the attentional demands of the primary task. More specifically, as the primary task requires more attention to maintain a criterion level of performance, a coinciding detriment in secondary probe RT task performance should be observed, because less attention can be dedicated to performing the secondary task. In contrast, as the attentional demands of the primary task decrease, a corresponding improvement in secondary probe RT task performance would be expected.

Although the dual-task paradigm has been used extensively in addressing a variety of attention-related issues,[11] some researchers[3,12,13] have raised questions as to the validity of the dual-task paradigm. Primary concerns during dual-task conditions focus both on a subject's opportunity to anticipate the presentation of the secondary probe RT task stimulus and on trading off performance on the primary task for successful performance on the secondary probe RT task. These concerns can be reduced, however, by meeting a number of established criteria.[10]

First, it is important to reduce the opportunity for the subject to anticipate the temporal occurrence of the secondary-probe RT task stimulus during dual-task performance. Salmoni et al[14] suggest this problem can be counteracted by incorporating catch trials during at least 33% of the trials during dual-task conditions. *Catch trials* are those trials during the dual-task conditions in which the secondary probe RT task stimulus is not presented. We adhered to this suggested criterion by randomly presenting catch trials during 42% of the trials during the dual-task conditions.

Second, during dual-task performance, it is imperative that the subject not switch attention between the primary and secondary probe RT tasks. More specifically, a deterioration in performance of the primary task in order to improve secondary probe RT task performance cannot be tolerated. If this occurs, meaningful interpretation of these data becomes difficult, if not impossible. To evaluate whether this strategy was used by subjects in our experiment, a comparison was made between the performance of the primary task when performed in the absence of the secondary probe RT task and primary task performance when executed simultaneously with the secondary probe RT task. If a subject allocated attention to the secondary probe RT task during the dual-task condition, one would expect to see a change in the performance of the primary task, as compared with the subject's performance on the primary task alone.

Finally, consideration needs to be given to the type of secondary probe RT task used. Abernethy[10] suggests that when investigating residual attentional capacity, as in our study, an attempt should be made to reduce the possibility of structural interference between primary and secondary probe RT tasks. Structural interference is considered to occur when two tasks compete for a common attentional resource. This interference would be apparent, for example, when two tasks both rely on the auditory system to provide critical task information. Abernethy suggests that structural interference is more indicative of peripheral overload as opposed to revealing a limitation on the subject's residual attentional capacity. Structural interference was minimized in our experiment by using a secondary probe RT task that included an auditory stimulus and a vocal response paired with primary tasks that primarily consisted of visual input and a motor response.

The purposes of this study were (1) to conduct a preliminary assessment of the attentional demands of ambulating with two commonly prescribed assistive devices (ie, standard and rolling walkers) and (2) to provide an initial demonstration that the dual-task paradigm is a potentially useful methodology in the rehabilitative environment.

Method

Subjects

Ten subjects (5 male, 5 female) volunteered to participate in the study. The ages of the subjects ranged from 22 to 49 years ($\bar{X}=30.8$, $SD=8.1$). Prior to the study, no subject had used the assistive ambulatory devices for rehabilitative purposes. Five of the subjects (mean age=29.4 years, SD=10.0), however, were employed by the Brazos Valley Rehabilitation Center and were familiar with the appropriate use of each of the devices. The remaining 5 subjects (mean age=32.2 years, SD=4.0) were graduate students attending Texas A&M University (College Station, Tex). An informed consent form was signed by each subject prior to participation in the study.

Apparatus

Each subject was required to ambulate a distance of 6.1 m (20 ft) in three different walking conditions (primary tasks), namely, walking alone with no device, walking with a standard walker (Guardian model 30755 P*), and walking with a rolling walker (Lumex model 6050 with glide brakes[†]). The time necessary to walk 6.1 m for each of these tasks was recorded with a hand-held stopwatch. In addition, the experimenter recorded the total number of steps taken by the subject to ambulate 6.1 m. A *step* was defined as heel contact of right foot to heel contact of left foot or vice versa. The time and number of steps were subsequently used to calculate the cadence for each trial.

A secondary probe RT task consisted of making a vocal response as rapidly as possible to the presentation of a 1,500-Hz tone. In order to accomplish this task, each subject was required to wear a voice-activated response mechanism (Lafayette model 63040[‡]). This mechanism consisted of a throat microphone that was placed over the subject's larynx and fastened around the neck via a plastic collar. The throat microphone was connected to a control device and a millisecond timer (Lafayette model 54417-A[‡]). The control device allowed the experimenter to adjust the sensitivity of the throat microphone. The sensitivity of the throat microphone was set such that extraneous background noise, if present during a trial, would not cause a false triggering of the microphone. This procedure appeared to be very successful, as no trial had to be readministered because of this problem. The timer was initiated by the presentation of a 1,500-Hz tone and terminated by a vocal response that triggered the throat microphone worn by the subject. The time to make the vocal response after the presentation of the tone was termed *voice response time* (VRT).

Procedure

The procedure followed by each subject consisted of three phases administered in the following order: (1) performing the secondary probe RT task only, (2) performing each of the walking tasks only, and (3) performing each of the walking tasks in conjunction with the secondary probe RT task. The last phase constituted the dual-task conditions.

When the subjects entered the testing area, they were provided with instructions for performing each of the walking tasks. They were informed they would be required to walk a specified distance, which was indicated by a white tape line on the walkway. To ensure the subjects walked the 6.1-m distance in a straight line, they were told they should attempt to place one foot on either side of the taped line or their right foot on the taped line as they proceeded along the walkway. The subjects were then told that, when ambulating along the walkway, they should attempt to produce a comfortable cadence that could be reproduced from trial to trial. The subjects followed the same procedure for all walking tasks (ie, walking with a standard walker, walking with a rolling walker, and walking without an assistive device). To familiarize themselves with the task requirements and to demonstrate that they clearly understood the experimenters' instructions, the subjects then performed three trials for each of the three walking tasks.

The throat microphone was then placed on the subjects. They were asked to vocalize the letter "b" to allow the experimenter to adjust the sensitivity of the throat microphone. Pilot data indicated that vocalizing the letter "b" consistently triggered the VRT device, which led to its selection as the secondary task response. The subjects were instructed to respond by vocalizing the letter "b" whenever the tone was presented during the experiment. The initial phase of the experiment consisted of the subjects performing 10 trials of the secondary RT task only while standing at the start of the walkway. In each of these trials, the 1,500-Hz tone was presented, which activated the millisecond timer. The subjects vocalized the letter "b" as rapidly as possible in response to the presentation of this tone in order to terminate the timer. Voice response time was recorded for each of these trials.

In the next phase of the experiment, baseline data were obtained for each of the walking tasks when performed in the absence of the secondary probe RT task (ie, cadence during walking tasks only) (Tab. 1). These data are important in evaluating whether subjects trade off attention between the walking tasks and the secondary probe RT task in the third phase of the experiment (the dual-task conditions). In order to establish the baseline data, each subject performed a block of three trials for each walking task (ie, walking with the rolling walker, walking with the standard walker, and walking without an assistive device). The subject was informed at the beginning of each trial to produce a consistent cadence. The experimenter recorded the time it took the subject to ambulate the designated 6.1-m distance and the number of steps needed to accomplish this task. This procedure was the same for each of the walking tasks, with the order of presentation of these tasks counterbalanced across subjects.

---

*Guardian, PO Box 549, Simi Valley, CA 93062.

[†]Lumex, Div of Lumex Inc, 100 Spence St, Bay Shore, NY 11706-2290.

[‡]Lafayette Instrument Co, 3700 Sagamore Pkwy N, PO Box 5729, Lafayette, IN 47903.

Table 1. *Mean Cadence (in Steps per Minute) for Each Primary Task in Both Primary-Task-Only and Dual-Task Conditions for Each Subject*

	Primary-Task-Only Condition						Dual-Task Condition					
	No Device		Standard Walker		Rolling Walker		No Device		Standard Walker		Rolling Walker	
Subject	$\bar{X}$	SD	$\bar{X}$	SD	$\bar{X}$	SD	$\bar{X}$	SD	$\bar{X}$	SD	$\bar{X}$	SD
1[a]	113.83	2.70	96.03	3.00	76.70	1.74	110.37	4.49	116.30	6.57	73.97	3.68
2[a]	127.40	6.04	47.28	1.50	82.67	1.04	124.96	3.56	48.16	2.14	84.84	2.10
3[a]	108.46	4.06	67.13	1.25	84.38	6.70	103.34	9.37	66.44	3.87	80.11	5.46
4[a]	111.93	14.98	85.06	8.63	78.13	5.22	107.09	4.42	81.95	3.84	79.56	5.25
5[a]	114.82	4.22	43.12	3.47	74.94	8.34	117.61	11.15	47.60	2.24	77.13	4.23
6	109.28	4.44	77.77	5.50	51.04	4.56	112.52	6.40	77.35	7.35	54.02	3.81
7	133.61	2.62	33.71	1.65	36.45	0.89	121.12	3.35	35.70	1.96	33.95	2.95
8	123.82	4.49	59.91	7.65	88.24	4.17	131.25	5.65	61.56	2.94	89.52	2.38
9	105.50	3.09	54.45	0.85	61.66	5.95	108.32	2.01	54.96	1.37	63.39	2.94
10	110.11	3.02	53.03	0.71	66.15	1.88	115.39	3.39	52.68	1.02	66.71	2.64

[a]Subjects who were familiar with use of the walking device.

The final phase of the experiment consisted of the dual-task conditions. Twelve trials were administered to each subject for each of the walking tasks. The subject was informed that during some of these trials the tone would be presented. When this occurred, the subject was required to vocalize the letter "b" as rapidly as possible. The experimenter randomly presented the tone on 7 of the 12 trials. Therefore, 5 of the 12 trials for any particular walking task constituted catch trials. The tone was always presented only while the subject ambulated over the middle 4.3-m (14-ft) segment of the 6.1-m-long walkway. This procedure was used to eliminate additional attentional demands associated with starting or stopping the walking task. Prior to each trial, each subject was reminded to attempt to produce a consistent cadence from trial to trial. These procedures were followed for each of the three walking tasks, with the order of task presentation counterbalanced across subjects.

Data Analysis

Descriptive statistics were calculated for each subject on each dependent measure (cadence and VRT) for the appropriate experimental phases. Mean cadence (in steps per minute) was calculated for each walking task during experimental phases 2 and 3. Each of the three trials performed in experimental phase 2 was used to calculate cadence for each of the walking tasks. During experimental phase 3 (dual-task conditions), the seven trials that were accompanied by the presentation of the secondary probe RT task were used to calculate cadence for probed trials (Tab. 1). In addition, cadence was calculated for the five catch trials during the dual-task condition (Tab. 1). The cadence data were analyzed with a 3×3 (walking task × dual-task condition) analysis of variance (ANOVA) for repeated measures on both factors.

Mean VRT was calculated for trials on which the secondary probe RT task only was performed (experimental phase 1). In experimental phase 1, the VRTs for the first five trials were considered practice trials, because pilot work had indicated that the subjects needed to perform some trials to become familiar with the requirements of this task. Trials 6 through 10 were subsequently used to establish baseline VRTs for use in the analysis of secondary probe RT task data. An intraclass correlation (Type-I approach using ANOVA repeated-measures design) indicated the reliability of the VRTs for trials 6 through 10 (P_1=.89).[15] In addition, mean VRTs were calculated for the seven trials of experimental phase 3 in which the secondary probe RT task was presented. Analysis of the VRT data consisted of a one-way ANOVA. This analysis included VRT data from the following conditions: performing the secondary probe RT task only, walking only, walking with the standard walker, and walking with the rolling walker. The locus of any significant effects was identified by a Tukey *post hoc* test. The region of rejection was $P<.05$ for all analyses.

Results

Walking Task Performance

Mean cadences for performance of walking tasks only (experimental phase 2) and for the probe and catch trials in the dual-task conditions (experimental phase 3) for each subject are presented in Table 1. An important assumption of dual-task methodology is that primary task performance in a dual-task condition should not differ from that demonstrated in a primary-task-only condition.[10,16] Figure 1 depicts mean cadence and standard errors for each of the three walking tasks in experimental phases

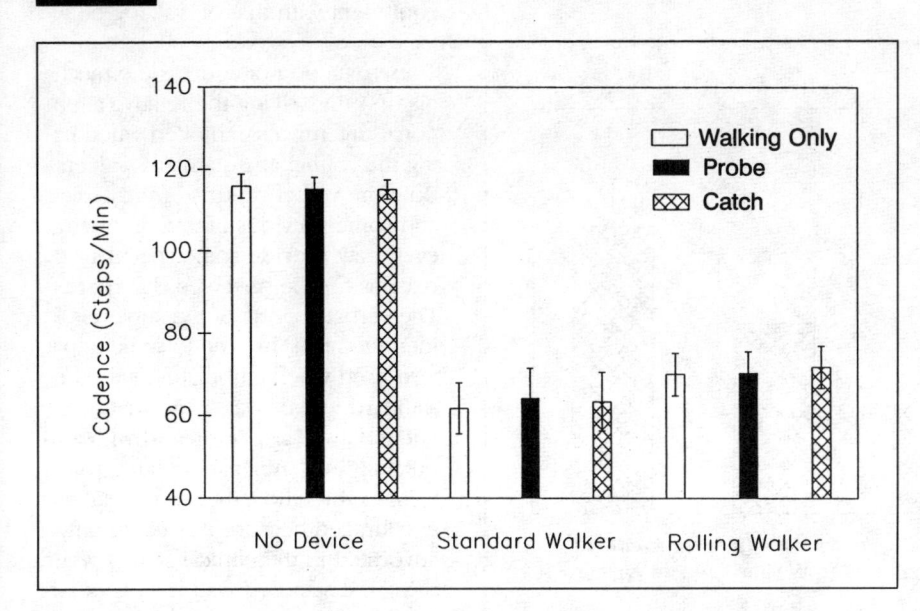

Figure 1. Mean cadence (in steps per minute) and standard errors for each primary task in primary-task-only and dual-task conditions collapsed across subjects.

2 and 3 collapsed across subjects. This figure shows that performance on each of the walking tasks was very similar. The ANOVA conducted on the cadence data revealed a main effect of walking task ($F=30.94$; $df=2,18$; $P<.01$). The effects of condition and the walking task×condition interaction were not significant ($F<1$).

Subsequent post hoc analysis indicated that the mean cadence when walking without an assistance device was significantly greater than the mean cadence when using the standard and rolling walkers (Fig. 1). The mean cadence did not differ significantly between the standard walker and rolling walker tasks. The lack of a walking task×condition interaction clearly indicates that the performance on each of the walking tasks was not compromised when performed in conjunction with the secondary probe RT task. This finding indicates that the subjects did not trade attention between the walking tasks and secondary probe RT task during the dual-task condition. It appears that subjects were attending to the walking task as instructed (Tab. 1).

Secondary Probe Reaction-Time Task Voice Response Times

Mean VRTs for each subject during experimental phases 1 and 3 are presented in Table 2. Mean VRT in both the secondary probe RT task only and with each of the walking tasks collapsed across subjects is depicted in Figure 2. This figure shows that the VRT was greater when walking with a rolling walker and standard walker than when walking with no assistive device or when performing the secondary task only. The ANOVA revealed that this effect was significant ($F=16.41$; $df=3,15$; $P<.01$). Subsequent post hoc analysis revealed that the VRT demonstrated when performing the walking tasks only did not significantly differ from the VRT demonstrated when performing the sec-

Table 2. Mean Voice Response Time (in Seconds) for Each Subject on the Secondary Probe Reaction-Time (RT) Task Only and When Performed in Conjunction with Each of the Walking Tasks

| | Secondary Probe RT Task-Only Condition | | Dual-Task Condition | | | | | |
| | | | No Device | | Standard Walker | | Rolling Walker | |
Subject	$\bar{X}$	SD	$\bar{X}$	SD	$\bar{X}$	SD	$\bar{X}$	SD
1[a]	0.312	0.031	0.402	0.052	0.627	0.248	0.581	0.123
2[a]	0.320	0.019	0.411	0.099	0.545	0.099	0.374	0.077
3[a]	0.342	0.019	0.353	0.039	0.493	0.067	0.417	0.024
4[a]	0.418	0.066	0.470	0.056	0.574	0.061	0.495	0.067
5[a]	0.338	0.022	0.400	0.083	0.520	0.079	0.397	0.028
6	0.216	0.034	0.263	0.035	0.350	0.038	0.290	0.039
7	0.430	0.072	0.423	0.054	0.636	0.103	0.523	0.074
8	0.415	0.076	0.627	0.138	0.854	0.222	0.717	0.091
9	0.387	0.076	0.431	0.024	0.985	0.458	0.552	0.229
10	0.424	0.042	0.429	0.041	0.460	0.043	0.484	0.067

[a]Subjects who were familiar with use of the walking devices.

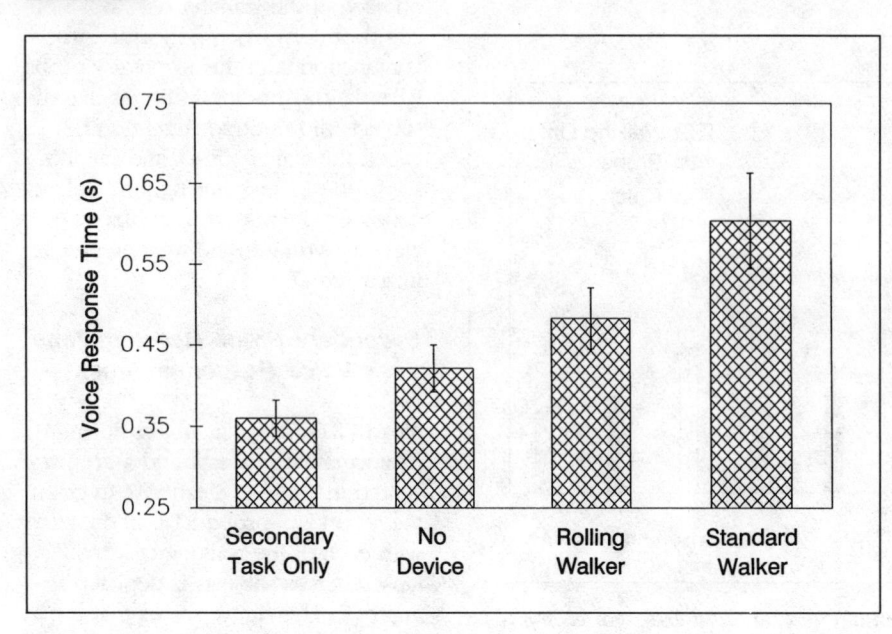

Figure 2. *Mean voice response time (VRT) (in seconds) and standard errors for secondary-task-only and dual-task conditions collapsed across subjects.*

ondary probe RT task only. Furthermore, the VRT demonstrated when walking with a rolling walker did not differ significantly from the VRT demonstrated when performing the walking tasks only, but was significantly slower than the VRT demonstrated for experimental phase 1. Finally, the VRT demonstrated when walking with the standard walker was significantly slower than the VRT demonstrated for all other walking tasks in the dual-task condition (Fig. 2).

Discussion

This study demonstrated that using either of the commonly prescribed walking aids, the standard walker or the rolling walker, is considerably attention demanding. This is clearly evident from the finding that the VRT obtained while walking with these aids was substantially increased beyond the VRT displayed when the secondary probe RT task only was performed (Fig. 2). From the standpoint of assessing the relative attentional demands of each of these walkers, the results revealed that the attentional requirements of using the rolling and standard walkers were not equivalent. It appears that ambulation assisted by a standard walker requires relatively greater attention than a similar activity aided by a rolling walker. Because a small, nondisabled sample was used in our study, generalizing this conclusion to patient populations may appear a large extrapolation. It is important to point out, however, that all subjects exhibited remarkably similar trends throughout the experiment, which suggests that the findings are relatively reliable (Tab. 2).

It should also be noted that five subjects had had considerable experience using the devices examined in this study (Tabs. 1 and 2). One might predict that these individuals would require relatively less attention to ambulate with the devices than would less experienced subjects. Each subject's performance, however, was very similar to that of those individuals who had no previous experience with the walking aids. This finding suggests that attentional demands of using these assistive devices may remain high even after relatively extensive experience using or teaching how to use the devices. One goal of future research, therefore, might be to identify the extent of practice that is necessary with these devices to reduce the attentional demands to a level equivalent with that of walking only.

These data do not address the mechanism(s) underlying the relative attentional differences exhibited when using the rolling and standard walkers. Both the verbal reports of the subjects and some previous literature, however, may provide some speculative reasons for the observed differences. The verbal reports of the subjects identified that the step-to walking pattern used when ambulating with the standard walker was somewhat different and awkward compared with ambulating in a "normal" walking pattern. In this study, the step-to gait was encouraged because it is commonly advocated in the clinical setting when using the standard walker. It is possible, however, that changing an automated step pattern (the step-through gait) used by an individual for many years may introduce a new source of attentional demand. The VRT obtained when performing the walking tasks only was found to be not significantly different from the VRT displayed while ambulating with the rolling walker. This finding supports the suggestion that offering the opportunity to engage a relatively automated gait may reduce the attentional demands when using an assistive device.

A second feature that may have contributed to the finding that additional attention is necessary when using the standard walker is the accuracy requirements inherent in correctly using this device. In this study, appropriate instructions concerning the use of the standard walker were given to the subject prior to using the device. These instructions were consistent with those that are provided by a therapist prior to rehabilitation using the standard walker. They stated that the walker should be placed on the floor with all four legs contacting the floor simultaneously.

Posner and Keele[17] demonstrated that as the accuracy demands of a movement increase, so do the attentional demands. One might speculate that because the standard walker needs to be lifted and appropriately placed during ambulation, this process requires relatively greater accuracy than merely pushing the rolling walker forward. If this is the case, the precise control required to manipulate the standard walker may also contribute to the attention necessary to efficiently function with this device. Obviously, further experimentation is necessary to examine the plausibility of these explanations, and the dual-task procedure used in this study would appear to be an ideal methodology to pursue these questions.

The dual-task methodology has a number of potential uses in clinical practice. For example, the procedures used in this study would be relatively simple to administer by a therapist as an additional evaluation tool to help examine the status of a patient's progress during a course of rehabilitation. More specifically, the dual-task methodology might be used at critical points throughout therapy to assess whether a patient has reduced the attentional demands associated with effecting a movement being rehabilitated. Furthermore, the rehabilitation procedures might also be evaluated using the dual-task methodology. By examining the rate at which a particular therapeutic approach reduces the attentional demands of a rehabilitated movement, therapists may discern the effectiveness of a therapeutic protocol. This methodology also has the potential to be used, with the development of appropriate criteria, to assist in the assessment of a patient's readiness for discharge or unsupervised use of an ambulatory device (in the case of a nursing home).

Conclusions

The dual-task methodology was used in this study to provide a preliminary assessment of the attentional demands of ambulating with two commonly prescribed walking aids. Because a population with no apparent physical or cognitive dysfunction was studied, extrapolation of these findings to clinical populations should be made with caution. An important contribution of this work, however, beyond the initial investigation of task attentional demands, is the introduction of a new methodology not currently used in the physical therapy community. It would appear that the dual-task methodology may potentially function as both a research and a clinical tool for this community. From a research standpoint, important attentional demand questions can be addressed for a wide variety of clinical populations (eg, geriatric, orthopedic, neurologic) as well as for a diverse set of activities. In the clinic, this methodology may provide a low-cost procedure that can be incorporated into a therapist's test battery to provide an additional means of assessing the progress a patient is making as a result of the program of rehabilitation.

Acknowledgments

We would like to extend our appreciation to the staff of the Brazos Valley Rehabilitation Center for their participation in the study. We would also like to thank Joel D Cowley and Clayton Gable for their assistance with data collection and analysis.

References

1 Fisher SV, Gullickson G. Energy costs of ambulation in health and disability: a literature review. *Arch Phys Med Rehabil.* 1978;59: 124–133.

2 Corcoran RJ, Brengelmann GL. Oxygen uptake in normal and handicapped subjects in relation to speed of walking beside velocity-controlled cart. *Arch Phys Med Rehabil.* 1970;51:78–87.

3 McLeod PA. Does probe RT measure central processing demand? *Q J Exp Psychol.* 1978;30:83–89.

4 Ells JG. Analysis of the temporal and attentional aspects of movement control. *J Exp Psychol.* 1973;99:10–21.

5 Posner MI, Boies SJ. Components of attention. *Psychol Rev.* 1971;78:391–408.

6 Rose DJ, Christina RW. Attention demands of precision pistol-shooting as a function of skill level. *Research Quarterly for Exercise and Sport.* 1990;61:111–113.

7 Landers DM, Wang MQ, Courtet P. Peripheral narrowing among experienced and inexperienced rifle shooters under low- and high-stress conditions. *Research Quarterly for Exercise and Sport.* 1985;56:122–130.

8 Weltman G, Egstrom GH. Perceptual narrowing in novice divers. *Hum Factors.* 1966;8:499–505.

9 Crosby JV, Parkinson S. A dual-task investigation of pilot's skill level. *Ergonomics.* 1976; 22:1301–1313.

10 Abernethy B. Dual-task methodology and motor skills research: some methodological constraints. *Journal of Human Movement Studies.* 1988;14:101–132.

11 Wickens CD. *Engineering Psychology and Human Performance.* Columbus, Ohio: Merrill Publishing Co; 1984.

12 McLeod PA. A dual-task response modality effect: support for multiprocessor models of attention. *Q J Exp Psychol.* 1977;29:651–667.

13 McLeod PA. What can probe RT tell us about the attentional demands of movements? In: Stelmach GE, Requin J, eds. *Tutorials in Motor Behavior.* Amsterdam, the Netherlands: Elsevier Science Publishers BV; 1980:579–589.

14 Salmoni AW, Sullivan JJ, Starkes JL. The attentional demands of movement: a critique of the probe technique. *Journal of Motor Behavior.* 1976;8:161–169.

15 Safrit MJ, Atwater AE, Baumgartner TA, West C. *Reliability Theory.* Washington, DC: American Alliance for Health, Physical Education, and Recreation; 1976.

16 Fisk AD, Derrick WL, Schneider W. A methodological assessment of dual-task paradigms. *Current Psychological Research and Reviews.* 1986;5:315–327.

17 Posner MI, Keele SW. Attention demands of movements. In: *Proceedings of the 16th Congress of Applied Psychology.* Amsterdam, the Netherlands: Swets en Zeitlinger BV; 1969.

Commentary

Wright and Kemp are to be commended for their "attention" to the cognitive factors accompanying the physical demands of a relatively routine and important task of daily life. Cognitive factors are routinely considered in physical therapy practice with respect to orientation to person and place, declarative memory, and social appropriateness. Little consideration, however, has been given to the information-processing demands inherent in the performance of physical activities of daily life. This is most likely because of the fact that such issues and their method of study draw heavily from the cognitive psychology literature, a literature that has only recently been recognized as providing potential theoretical perspectives and methodologies relevant to the development of the science and practice of physical therapy.[1] It is truly refreshing to see these concerns in our literature.

The authors' stated purposes were (1) to provide preliminary data regarding the attentional demands of ambulating with two different walking aids and (2) to introduce the dual-task methodology to the physical therapy community. My comments are directed at the "methodology" issue. In most branches of science, the methods used to address relevant questions can be thought of as operational extensions of theoretical or conceptual perspectives. Thus, to more fully appreciate the usefulness of the dual-task method and the assumptions on which it is based, we must briefly examine the various theories of attention.

As the authors note, the study of attention has been the focus of much psychological literature for some time—in fact, for over 100 years.[2] The authors assume that the reader knows what attention is, but there are at least six different definitions in the literature.[3,4] Some theories assume that attention is a fixed, undifferentiated capacity for processing information.[5–7] If this capacity or resource is exceeded by task requirements, then performance would deteriorate. In contrast to the fixed-capacity theories, Kahneman[8] argues for a flexible allocation of attention. In other words, the attentional capacity can change as the difficulty of two simultaneously performed tasks increases. For the present point, the notion of a flexible capacity creates difficulties for certain dual-task methods in which capacity is assumed to be fixed. More recent theories stress multiple resources in which attention is not a single resource, but a set of resource pools, each with its own capacity and specificity for information processes.[9]

In general, theories or perspectives about attention argue that information-processing requirements (ie, "cognitive demands") utilize some kind of capacity (eg, fixed, flexible). When several activities compete for this capacity, decrements in performance result.

Recently, Neumann[10] has opposed the "resource" theories in which the fundamental "direction of processing" is bottom-up, passive, and independent of intentions. He suggests that with the resource views, interference between two simultaneous tasks (eg, walking with a walker and responding to an auditory signal) occurs because attention (as a resource) is needed to perform the various processes. In contrast, he argues from a motor control perspective in which the level of control determines in a top-down fashion which processes are prevented from occurring. In other words, as a result of the selection of an action (eg, picking up the walker), certain processes are prevented from occurring or are delayed (eg, responding to the auditory tone). From this perspective, interference between two simultaneously performed tasks occurs *not* because of attentional resources, but because an action has been selected and the other processes are either completely or partially blocked. Presumably, this perspective has some ecological significance. For tasks that are important (for preservation of balance or prevention of injury) and hence selected, such an arrangement ensures that the selected action is completed and that other actions are prevented from interfering.

The purpose of this theoretical digression was to illustrate a potential problem with the dual-task methodology. First, and most important, the method assumes the acceptance of a fixed, unitary-capacity view of attention, and its interpretation depends on such a perspective. In particular, if the capacity required for walking with a rolling walker is low, then that remaining for processing the auditory signal will be relatively large, and the voice response time (VRT) will be fast. If, on the other hand, walking with a pick-up walker is very demanding, then little of the spare capacity will remain for the auditory signal, and its processing will be slow (ie, slow VRT). Thus, with this dual-task method, it is assumed that the duration of the secondary probe reaction time can provide an indication of the attentional requirements of the various walking tasks. Such an interpretation is incompatible with Kahneman's[8] flexible-capacity view, as well as that of others[9] incorporating multiple resource pools, and most certainly Neumann's[10] level-of-control view.

Another problem with the interpretation of dual-task data is that there is evidence that the probe reaction time does not actually measure some general residual capacity.[11,12] Rather, the changes in probe reaction time are now thought to reflect something about the specific pattern of interference between the processes required to perform the main task and those particular to the probe task. In other words, Wright and Kemp's results may reflect a specific form of interference between the processes involved in walking and talking tasks and not the attentional requirements of walking with assistive devices.

The authors do a fine job of addressing three important limitations of the dual-task methodology concerning structural interference, anticipation, and attentional trade-offs. They discuss how their study was modified in keeping with these methodological constraints.[13] I am still confused, however, as to how a "vocal" response is significantly different from a "motor" response. From my perspective, walking and talking are both sensorimotor responses, but the motor program for each may be formulated and transmitted by two distinct neural substrates.

In summary, this study and its important findings should be considered in light of the major assumptions underlying the use of the dual-task methodology. Let us not fall into the trap of presenting the "how to" without the "why."[14] If we are to borrow methodologies from other disciplines, we must make explicit the implicit assumptions and theoretical perspectives from which these methods evolved. Only then can we expect to create new knowledge and promote the development of our own theoretical frameworks.

Carolee J Winstein, PT, PhD
Assistant Professor
Department of Biokinesiology and Physical Therapy
University of Southern California
2250 Alcazar St, CSA 208
Los Angeles, CA 90033

References

1 Winstein CJ, Knecht HG. Movement science and its relevance to physical therapy. *Phys Ther*. 1990;70:759–762.
2 James W. *The Principles of Psychology*. New York, NY: Henry Holt & Co Inc; 1890.
3 Norman DA. *Memory and Attention*. New York, NY: John Wiley & Sons Inc; 1976.
4 Moray N. *Attention: Selective Processes in Vision and Hearing*. New York, NY: Academic Press Inc; 1970.
5 Broadbent DE. *Perception and Communication*. London, England: Pergamon Press; 1958.
6 Deutsch JD, Deutsch D. Attention: some theoretical considerations. *Psychol Rev*. 1963;70:80–90.
7 Keele SW. *Attention and Human Performance*. Pacific Palisades, Calif: Goodyear; 1973.
8 Kahneman D. *Attention and Effort*. Englewood Cliffs, NJ: Prentice-Hall; 1973.
9 Navon D, Gopher D. On the economy of the human processing system. *Psychol Rev*. 1979;86:214–255.
10 Neumann O. Beyond capacity: a functional view of attention. In: Heuer H, Sanders AF, eds. *Perspectives on Perception and Action*. Hillsdale, NJ: Lawrence Erlbaum Associates Inc; 1987:361–394.
11 McLeod PA. A dual task response modality effect: support for multiprocessor models of attention. *J Exp Psychol*. 1977;29:651–668.
12 McLeod PA. What can RT tell us about the attentional demands of movement? In: Stelmach GE, Requin J, eds. *Tutorials in Motor Behavior*. Amsterdam, the Netherlands: Elsevier Science Publishers BV; 1980;579–589.
13 Abernethy B. Dual-task methodology and motor skills research: some methodological constraints. *Journal of Human Movement Studies*. 1988;14:101–132.
14 Tammivaara J, Shepard KF. Theory: the guide to clinical practice and research. *Phys Ther*. 1990;70:578–582.

Author Response

Winstein provides an insightful overview of the current theoretical perspectives addressing attention. This overview consists of a brief, but precise, historical summary of the early attempts to understand the locus and structure of the bottleneck in human information processing via single-channel models,[1] limited processing resources,[2-4] and more recent functional frameworks championed by Allport[5] and Neumann.[6] We concur with Winstein that contemporary experimental methodologies are usually outgrowths of the theoretical wisdom of the time, thus necessitating an understanding of current theorizing in an area of research one wishes to examine. The synopsis provided by Winstein is therefore a timely and appropriate addition to our article.

Winstein appears to accept that an examination of the information-processing demands associated with the daily activities encountered by physical therapists is worthy of experimental "attention." She expressed some concerns, however, with the usefulness of the dual-task methodology to address such issues. Winstein identified two fundamental problems with this methodology. First, she states that using the dual-task methodology to study the attention process necessitates the acceptance of a nonflexible, unitary resource model. Second, findings from dual-task manipulations may not reflect a residual attention reservoir but may merely highlight specific interference patterns between particular activities.[7,8] More specifically, in our experiment, the walking task interfered with the talking task, or vice versa. The remainder of this thesis will attempt to address these "methodological" concerns.

The initial problem identified by Winstein, that the design of an experiment based on the dual-task methodology is inherently dependent on the theoretical position adopted by the researcher, is an important one. In our experiment, we assumed that attentional capacity is limited by a fixed, undifferentiated resource supply that can be allocated to the tasks being performed. Because we were attempting to examine the limitation on information processes attributable to what Kahneman[2] has labeled "capacity interference," we purposefully tried to avoid structural interference. This dictated the selection of the primary task-secondary task combination subsequently used in the study.

Had we chosen to approach this question from a multiple-resource perspective, the dual-task methodology would still be an appropriate protocol. In contrast to our study, however, a more suitable goal of secondary task selection adopting this viewpoint would have been to seek or create structural interference. This is because the secondary task must be sensitive to the primary task resource demand (for an example, see Fisk et al[9]). This procedure has been used rather extensively to detail the nature of the independent resource pools, which are central to this position.[4,5] As Abernethy states,

> Within these models, structural interference is the essence of the dual-task decrement, and interference is predicted to occur on any occasion in which the demands of the concurrent tasks exceed the available processing space within a specific resource pool.[10(p107)]

Therefore, the extent to which two simultaneously performed tasks will be executed successfully is dependent on the overlap in common resource pools utilized by each of the tasks. If there is a large overlap, interference would be predicted. Conversely, if little or no overlap in resource utilization occurs, no interference would be expected.

The purpose of this comparison between the single, undifferentiated resource and multiple resource positions is to point out the central role the dual-task "methodology" can play within both of these theoretical frameworks. It seems that a more fundamental problem in using the methodology is the fact that no single currently proposed theoretical position offers a compelling account of the available literature. The researcher is therefore faced with the dilemma of selecting the position that he or she believes is most appropriate and using the dual-task method accordingly. Winstein discusses a promising alternative to the "resource" models that was recently proposed by Neumann.[6] This perspective may prove to be particularly attractive to the clinical community because of its close alignment with the neurophysiology of attention. Although this perspective offers a "new" look at the limitations of human information processing, it should be noted that it has yet to withstand the experimental rigors that have characterized the resource models. At this stage, one might heed the caution expressed by Neumann that "the approach as it is presented here is far from being a theory of attention."[6(p375)]

Winstein's second concern stemmed from earlier criticism of the probe reaction task procedure by McLeod.[7,8] She suggests that changes in probe reaction time may reflect a specific source of interference between particular task combinations as opposed to a limitation on the attention process. The nature of the data obtained from our study, however, does not appear to support this contention. More specifically, if a source of interference occurring as a function of the processes subserving both walking and talking tasks accounted for the increase in the secondary task reaction time, this increase should also have occurred when walking without an assistive device was performed in combination with the secondary task. This was not the case. The mean secondary task reaction time in this condition did not differ reliably from the mean reaction time for the secondary task performed alone. Therefore, it appears that this walking task-talking task combination could be performed relatively interference free. Furthermore, if Winstein's speculation is correct, it would also be unlikely that there would be a differential increase in secondary task reaction time in the rolling and standard walker dual-task conditions.

In summary, Winstein's commentary functions as an important addition to our article in order to understand the intricacies of using the dual-task methodology. It is clear that the underlying processes subserving "attention" will continue to attract substantial research efforts in attempting to further delineate contemporary theoretical perspectives. We contend that this methodology has contributed much to the development of our understanding of the attention process to date and will continue to serve as a viable and important research protocol.

David L Wright, PhD
Tammy L Kemp, PT
Yuhua Li

References

1 Welford AT. Single-channel operation in the brain. *Acta Psychol (Amst)*. 1967;27:5–22.
2 Kahneman D. *Attention and Effort*. Englewood Cliffs, NJ: Prentice Hall; 1973.
3 Wickens CD. The structure of additional resources. In: Nickerson R, ed. *Attention and Performance VIII*. Hillsdale, NJ: Lawrence Erlbaum Associates Inc; 1980:239–257.
4 Wickens CD. Processing resources in attention. In: Parasuraman R, Davies R, eds. *Varieties in Attention*. New York, NY: Academic Press Inc; 1984:63–101.
5 Allport A. Attention and performance. In: Claxton G, ed. *Cognitive Psychology*. London, England: Routledge & Kegan Paul Ltd; 1980:112–153.
6 Neumann O. Beyond capacity: a functional view of attention. In: Heuer H, Sanders AF, eds. *Perspectives on Perception and Action*. Hillsdale, NJ: Lawrence Erlbaum Associates Inc; 1987:361–394.
7 McLeod PA. A dual-task response modality effect: support for multiprocessor models of attention. *Q J Exp Psychol*. 1977;29:651–668.
8 McLeod PA. What can probe RT tell us about the attentional demands of movements? In: Stelmach GE, Requin J, eds. *Tutorials in Motor Behavior*. Amsterdam, the Netherlands: Elsevier Science Publishers BV; 1980:579–589.
9 Fisk AD, Derrick WL, Schneider W. A methodological assessment of dual-task paradigms. *Current Psychological Research and Reviews*. 1986;5:315–327.
10 Abernethy B. Dual-task methodology and motor skills research: some methodological constraints. *Journal of Human Movement Studies*. 1988;14:101–132.

Basic Research

Research Report

Trunk Kinematics During Locomotor Activities

We investigated upper-body (ie, trunk) angular kinematics (motions) during gait, stair climbing and descending, and rising from a chair in two reference frames—relative to the pelvis and to room coordinates. Bilateral kinematic data were collected from 11 healthy subjects (6 female, 5 male), who were 27 to 88 years of age ($\overline{X}=58.9$, $SD=17.9$). During stair climbing, maximum trunk flexion relative to the room was at least double that during stair descending and gait. Arising from a chair required the most trunk flexion/extension range of motion (ROM) but the least abduction/adduction and medial/lateral (internal/external) rotation. Trunk ROM during gait was small ($\overline{X} \leq 12°$) and consistent with previous literature. Trunk range of motion relative to the room during stair climbing and descending was greater than trunk ROM during gait in all planes. The pelvis and trunk rotate in the transverse plane in greater synchrony during stair descending ($\overline{X}=8.1°$, $SD=5.6°$) than during gait ($\overline{X}=12.0°$, $SD=4.2°$). For all activities, trunk frontal and sagittal ROM relative to the pelvis was greater than that relative to the room coordinates. This finding suggests that trunk/pelvis coordination may be used to reduce potentially destabilizing anti-gravity trunk motions during daily activities. We conclude that upper-body kinematics relative to both pelvis and gravity during daily activities are important to locomotor control and should be considered in future studies of patients with locomotor disabilities. [Krebs DE, Wong D, Jevsevar D, et al. Trunk kinematics during locomotor activities. Phys Ther. 1992;72:505–514.]

Key Words: *Gait, Kinematics, Locomotion, Spinal mobility impairment.*

David E Krebs
Dennis Wong
David Jevsevar
Patrick O Riley
W Andrew Hodge

Trunk kinematics are critically important to the maintenance of body equilibrium and should be examined as a component of locomotion analysis.[1,2]

DE Krebs, PhD, PT, is Associate Professor, MGH Institute of Health Professions, 15 River St, Boston, MA 02108-3402 (USA). Address all correspondence to Dr Krebs.

D Wong, MD, PT, is Resident in Rehabilitation Medicine, Kingsbrook Jewish Medical Center, 585 Schenectady Ave, Brooklyn, NY 11203. He was an NIDRR Advanced Rehabilitation Fellow at the MGH/MIT Rehab Engineering Center when this work was conducted.

D Jevsevar, MD, is Resident in Orthopaedics, Tufts New England Medical Center, 750 Washington St, Boston, MA 02111. He was an NIDRR Advanced Rehabilitation Fellow at the MGH/MIT Rehab Engineering Center when this work was conducted.

PO Riley, PhD, is Technical Director, MGH Biomotion Laboratory, Massachusetts General Hospital, Boston, MA 02114.

WA Hodge, MD, is Assistant in Orthopaedics, Massachusetts General Hospital.

This study was approved by the Massachusetts General Hospital Institutional Review Board.

This work was supported in part by Grants H133P90005 and H133G00025 from the National Institute of Disability and Rehabilitation Research.

This article was submitted June 12, 1991, and was accepted March 18, 1992.

Three-dimensional, normal trunk kinematic data obtained during stair and chair locomotor daily activities, as well as during gait, enable therapists to make comparisons with pathological locomotion characteristics such as "gluteus medius" limp following hip surgery, trunk abduction lurch associated with knee arthroplasty or above-knee amputation, and spinal fusion. To date, however, most locomotor studies have only described lower-extremity kinematics.[3–8] Comparatively little information exists on the kinematics of upper-body movement during gait, and no three-dimensional kinematic data have been reported for other locomotor activities of daily living such as stair and chair activities.[9]

Table 1. Subject Characteristics and Locomotor Activities in Which Subjects' Data Are Included[a]

Subject No.	Sex	Age (y)	Height (cm)	Weight (kg)	Locomotor Activity				
					Free Gait (n=8)	Paced Gait (n=8)	Stair Climbing (n=11)	Stair Descending (n=10)	Rising from a Chair (n=11)
1	F	66	157.4	59.0	X	X	X	X	X
2	M	80	168.9	63.1	X	X	X	X	X
3	M	37	175.2	75.0	X	X	X	X	X
4	M	64	167.6	70.4	X	X	X	X	X
5	F	60	167.6	71.8			X	X	X
6	F	59	157.4	59.0			X	X	X
7	F	27	157.4	54.5			X		X
8	M	88	179.0	83.1	X	X	X	X	X
9	F	70	162.5	53.6	X	X	X	X	X
10	M	36	175.2	75.0	X	X	X	X	X
11	F	61	152.4	50.0	X	X	X	X	X
$\bar{X}$		58.9	165.5	65.0					
SD		17.9	8.4	10.2					
Range		27.0–88.0	152.4–179.0	50.0–83.1					

[a]Data from subjects 5, 6, and 7 are not included in some analyses because of technical limitations (eg, incomplete visibility while the subject was in the viewing volume).

In the 1960s, Murray and colleagues[7,8] described the transverse (medial/lateral [internal/external]) rotation of the trunk for free-speed walking of 60 healthy men (20–65 years of age) to be 6.9±1.9 degrees ($\bar{X}$±SD). Recently, Opila-Correia[10] reported three-dimensional trunk kinematics observed during the free-speed gait of 14 women, 21 to 54 years of age ($\bar{X}$=35.0, SD=10.4). In gait trials with subjects wearing low-heeled footwear, trunk flexion/extension, abduction/adduction, and medial/lateral total angular excursions relative to the pelvis averaged 11.1, 12.6, and 17.5 degrees, respectively. These same displacements relative to room coordinates were 9.2, 5.2, and 11.2 degrees, respectively. Treadmill gait was studied by Thorstensson et al[11] and Stokes et al.[12] Although Thorstensson et al did not report transverse-plane rotation, they did report net frontal-plane trunk range of motion (ROM) in 7 healthy 18- to 34-year-old subjects to be 2 to 9 degrees and net flexion/extension ROM to be 2 to 12 degrees in walking. Stokes and colleagues analyzed trunk movement of 3 female and 5 male subjects during treadmill walking (with shoes on) and reported "small" flexion/extension amplitudes, with a mean trunk abduction/adduction of 4.9±1.8 degrees and a mean transverse-plane trunk rotation of 4.7±1.6 degrees. It is clear that free walking differs from treadmill gait[13] and that room-referenced trunk kinematics differ from pelvis-referenced kinematics. Because the locomotor control system, using sensory input from the vestibular system, is aware of global, gravity-referenced coordinates, we examined both self-referenced (pelvis-referenced) and gravity-referenced (room-referenced) trunk motions during locomotor activities of daily living.

The purposes of our study were (1) to determine a sample of trunk ROM and maximum angular orientation in flexion/extension, abduction/adduction, and medial/lateral rotation in healthy subjects during daily activities, including free-speed and paced gait, during stair climbing and descending, and while rising from a chair and (2) to examine these kinematic findings in two frames of reference—the trunk relative to the pelvis and the trunk relative to room (gravity) coordinates.

Method

Subjects

Eleven volunteers who were free of musculoskeletal and neurological disease participated in the study. Subjects were recruited from the community and the laboratory staff. All subjects, to be eligible for the study, must have been (1) able to walk at least 1.6 km (1 mile) without stopping and to climb and descend stairs and arise from a chair without personal or upper-extremity assistance and (2) free of neuro-musculo-skeletal pathology and deformity as determined by history and physical examination. All subjects provided written informed consent. Participants' ages, sex, heights, and weights are presented in Table 1.

Basic Research

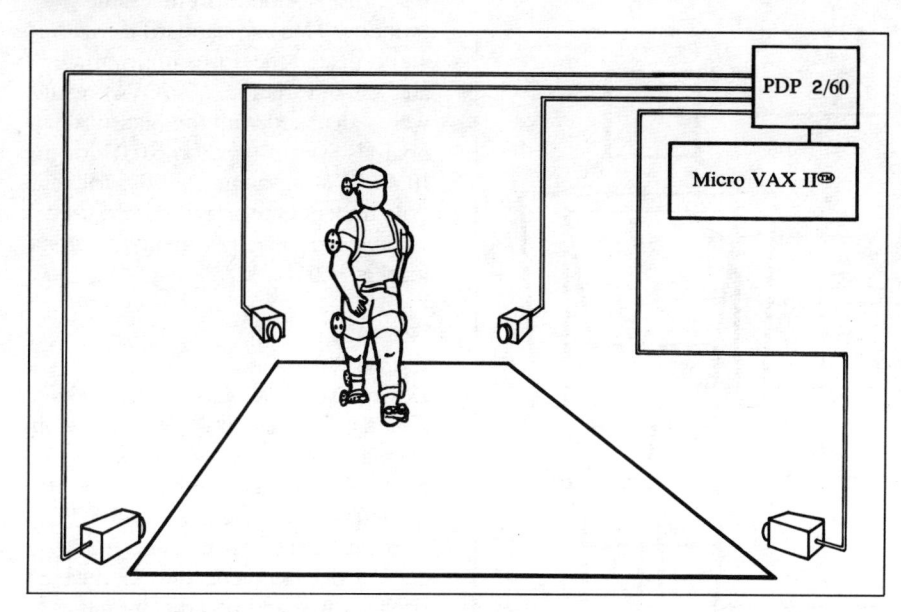

Figure 1. *Schematic depiction of laboratory kinematic data-acquisition apparatus. The PDP 11/60 and MicroVAX II® computers are used for data processing as well as acquisition.*

Instrumentation

Gait trials were conducted on a 10-m walkway. Stair trials involved the use of a weighted modular staircase with four stairs. The first of the four steps was 2.5 cm tall and was provided merely to help initiate steady-state stepping activity prior to data collection on three 18-×28-cm stairs. Chair trials involved the use of an armless and backless chair with an adjustable-height rigid seat.

Four Selspot II optoelectric cameras,* in addition to PDP 11/60† and MicroVAX II®† computers, were used to acquire bilateral kinematic data from the subjects. The Selspot system's infrared light-emitting diodes (LEDs) are tracked by an infrared detector within each camera; the system accuracy is <3 mm.[14] Camera placement resulted in a viewing volume of 1.8 m per side (Fig. 1). Kinematic data were acquired for 3 seconds at 153 Hz.

The LEDs were mounted in rigid arrays secured to 11 body segments: head, trunk, pelvis, thighs, shanks, feet, and upper arms (Fig. 2). Each segment was modeled as a rigid body having 6 degrees of freedom (three translations and three rotations), with its kinematics determined in part using TRACK© software‡[14] and the technique described by Riley et al.[15] The orientation of each body segment in space and the associated joint angles were calculated using 3-1-2 Cardan angles.[16,17]

Procedure

General. Barefoot subjects performed all activities in a single session. To prevent the preferred cadence from being influenced by the paced cadence, free (preferred-speed) gait was followed by paced gait. Stair climbing, then stair descending, then arising from a chair were performed with at least 1.5 minutes' rest between trials.

To approximate easily reproduced natural cadences and to prevent velocity-dependent kinematic changes from confounding the within- and between-subject comparisons, the paced cadence chosen for each activity was determined from pilot studies and previous literature.[7,8] All activities, except rising from a chair, were performed with unrestricted arm-swing.

Free gait. Subjects walked along a 10-m walkway at a comfortable speed; that is, subjects were instructed to "walk the way you usually do, as if you were taking a brisk stroll in the park." At least three strides were completed before the subjects entered the data-collection portion of the walkway. Several acclimation trials were completed before two trials of data were collected.

Paced gait. Subjects walked using an identical procedure to that of free gait, but each foot-strike was synchronized with a metronome set at 120 beats per minute (bpm). Subjects practiced this paced gait cadence until synchrony was comfortably achieved for each step, as determined by the heel-strike occurring at each metronome beat, without awkward-appearing "marching."

Stair climbing. Subjects climbed stairs to the beat of the metronome set at 80 bpm, with the left foot first striking the small step and the right foot striking the next full riser. Subjects stopped when they reached the top platform, which was the fourth step. Several practice trials were performed before the two data-collection trials to ensure smoothness and cadence synchronization with the metronome.

Descending stairs. Subjects started on the fourth (top) step and descended in time with the beat of a metronome set at 80 bpm and with the left foot striking the third step. Several practice trials preceded two data-collection trials, again to ensure smoothness and synchronization with the metronome.

---

*Selspot AB, Flöjelbergsgatan 14, S-431 37 Mölndal, Sweden, and Selspot System Ltd, Troy, MI 48093.

†Digital Equipment Corp, 146 Main St, Maynard, MA 01754.

‡Developed at the Massachusetts Institute of Technology, Cambridge, Mass.

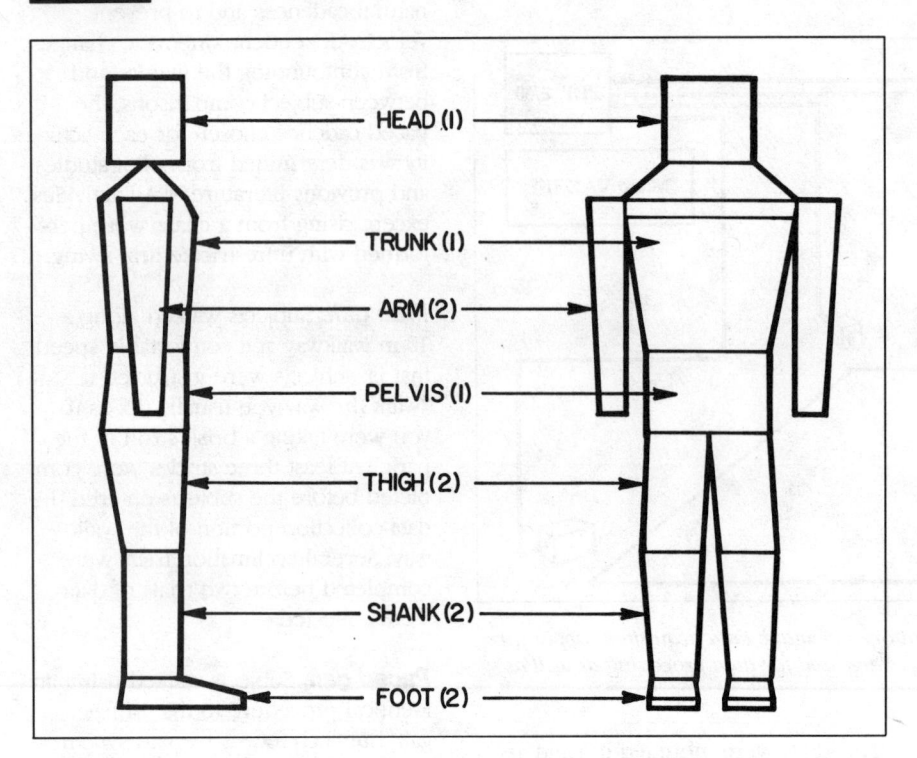

Figure 2. *Eleven-segment whole-body kinematic model. Note left and right views are rotated 90 degrees relative to room coordinates, but no trunk rotation relative to the pelvis frame of reference has occurred. Figure 1, by contrast, reveals trunk rotation relative to the pelvis as well as to room coordinates.*

Rising from a chair. Subjects were seated such that the chair edge was approximately 4 cm distal to their greater trochanters. Foot distance from the chair was determined by setting the angle between the floor and the tibia to 72 degrees (18° of dorsiflexion). The subjects' feet were equidistant from the chair and 10 cm apart. Chair height was adjusted to the height from the floor to the lateral knee joint line when the tibia was perpendicular to the floor (Fig. 3).[9] Subjects rose in time with the beat of a metronome set at 52 bpm, with arms folded to minimize the upper extremities' involvement in the movement. Subjects began rising at 1 beat and came to a full stand 1.2 seconds later at the next beat. Several practice trials were performed until subjects could perform the trials smoothly within the 1.2-second time limit. Data from two trials were collected for each subject.

Data Analysis

The trunk's peak (maximum or minimum) angle and total range of angular displacement were determined for both the trunk segment relative to room orientation and the trunk segment relative to the pelvic segment. Using a computer program that displayed maximum, minimum, and total ROM in the sagittal, frontal, and transverse planes (flexion/extension, abduction/adduction, and medial/lateral rotation, respectively), the angular values of each trial were obtained from a "window" corresponding to a complete cycle (right heel-strike to right heel-strike for gait and stair activities) or from the complete chair-rising activity. Velocity and cycle time were determined from center-of-mass (CM) displacements in this same "window." Means, standard deviations, and repeated-measures multivariate analysis of variance (MANOVA) results were calculated with the Statistical Analysis System (version 6.03)[§] for the IBM PC.[‖] Multivariate statistics for multiple dependent variables were used to preserve the experimentwise alpha level at <.05.

Results

Descriptive linear velocity and cycle-time data are shown in Table 2. Results from a typical subject are presented in Figure 4. Angular displacement results are illustrated in Figures 5 and 6. Velocity, pattern of motion, and trunk angular displacements did not differ between free and paced gait trials; hence, the remainder of the "Results" section and the "Discussion" section focus on paced gait trials, as compared with paced stair and paced chair activities of daily living. Between-trial reliability within subjects was very high ($\bar{X}<3°$ difference between trials for any activity/motion combination and trial-to-trial Pearson $r \geq .88$), but moderate between-subject variability is apparent in the standard deviations in Figures 5 and 6, being greater in peak motions than in ROM (peak-to-peak amplitude). No age-related peak motion or ROM differences were found in any plane or coordinate system ($F<1.2, P>.10$) in any activity.

Trunk ROM during rising from a chair in all planes relative to either the pelvis or room coordinates was significantly different from stair and gait ROM (Hotelling-Lawley $F>70.1$, $P<.01$) with the exception of chair versus gait abduction/adduction ($F=.18, P=.68$). Gait trunk ROM was similar ($F<5.6, P>.05$) to stair-descending trunk ROM in all planes, but gait ROM differed significantly from stair-climbing ROM in all planes ($F>10.7, P<.01$). Trunk ROM did not differ significantly between stair climbing and descending, except in medial/lateral rotation ($F=11.5$, $P=.01$).

[§]SAS Institute Inc, PO Box 8000, Cary, NC 27511.

[‖]International Business Machines Corp, PO Box 1328, Boca Raton, FL 33432.

Basic Research

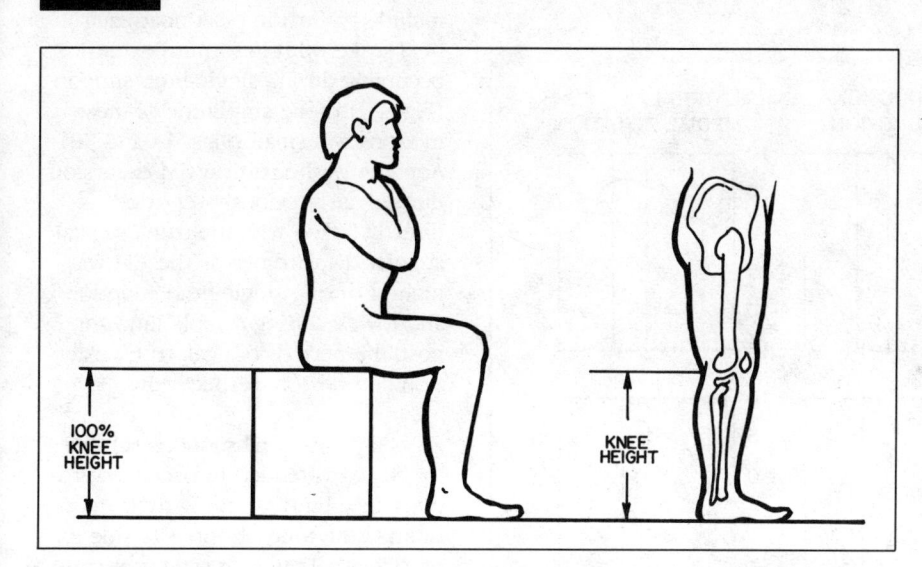

Figure 3. *Initial position for rising from a chair (left) and knee height determination landmarks (right).*

The greatest trunk ROM relative to the pelvis occurred for flexion/extension during rising from a chair (22.9°±9.5°) and was quadruple that of gait sagittal ROM (Fig. 5A). Room-referenced flexion/extension during rising from a chair (36.2°±7.7°) exceeded by >50% ($F=19.8$, $P<.001$) that relative to the pelvis (Fig. 5B). Nonsagittal trunk ROM relative to the pelvis was greatest in stair climbing (abduction/adduction, 14.7°±5.7°) and least in rising from a chair (medial/lateral rotation, 3.2°±1.1°) (Fig. 5A). Indeed, trunk ROM relative to the pelvis was greater than that relative to the room coordinates across all activities and motions ($F>5.3$, $P<.05$), except in chair and stair medial/lateral rotation (Figs. 5A and 5B).

During gait, trunk maximum sagittal orientation was about 3 degrees more toward extension relative to the pelvis (Fig. 6A) than relative to the room coordinates (Fig. 6B); this difference was not significant. Maximum trunk flexion relative to the room coordinates (Fig. 6B) during stair climbing was double that recorded during stair descending and sextuple that recorded during gait ($F>15.3$, $P<.01$). Relatively small, and statistically insignificant, differences existed in peak trunk lateral and transverse maximal positions among the five activities.

Discussion

We examined the kinematics of the trunk referenced to room coordinates and to the pelvis during common locomotor activities of daily living (ie, gait, stair, and chair activities). We chose a sample that was heterogeneous in height, weight, age, and gender to enhance the generalizability of the findings from this study. Because of the small sample size, however, firm conclusions await future studies. We found no differences in trunk kinematics with respect to age. We stipulated rather high functional locomotor levels, however, in the sample inclusion criteria for all subjects, including our elderly subjects. These high locomotor levels may have attenuated the possibly disease-induced differences that Murray and colleagues[7,8] observed in their elderly subjects compared with younger healthy subjects. As in the lower extremities, trunk kinematic patterns are very repeatable within subjects, but each subject used slightly idiosyncratic movement patterns and amplitudes, particularly in stair climbing and descending (note the standard deviation magnitudes in Figs. 5 and 6). In general, the standard deviations of peak motions exceeded those of ROM, because maximum angular displacements depend on each person's posture (segment anatomical orientations), which varies across subjects, and on movement amplitude. Range of motion depends only on movement peak-to-peak amplitude. For example, subjects could attain 36 degrees of trunk flexion ROM during rising from a chair by an excursion from 10 degrees of extension to 26 degrees of flexion, or 5 degrees of flexion to 41 degrees of flexion.

Kinematic Patterns

Trunk kinematic patterns are defined by Stokes et al[12] as the relative invariance of the number and cycle location of curve inflections and reversals. For example, in the chair-rising activity depicted in Figure 4, flexion relative to the room coordinates has a major reversal at about 55% of the cycle (abscissa) and an inflection (mathematically, where the second derivative of the curve changes sign) at about 35% of the cycle. The following discussion highlights only those movement patterns that were typical of the subjects investigated according to the criteria of Stokes et al. Future studies should consider larger representative samples, to delineate the extent of variations from the typical patterns determined in our study.

Control of the trunk is important for posture and balance, in part because of the upper body's relatively large mass (about two thirds of the body mass lies superior to the waist). It is difficult to depict three-dimensional

Table 2. *Average Forward Velocities and Cycle Times (±Standard Deviation) for Each Event*

Activity	Velocity (m/s)	Cycle Time (s)
Free gait	1.12±0.13	1.08±0.07
Paced gait	1.18±0.12	1.06±0.06
Stair climbing	0.39±0.05	1.60±0.29
Stair descending	0.36±0.07	1.58±0.25
Rising from a chair	. . .	1.32±0.12

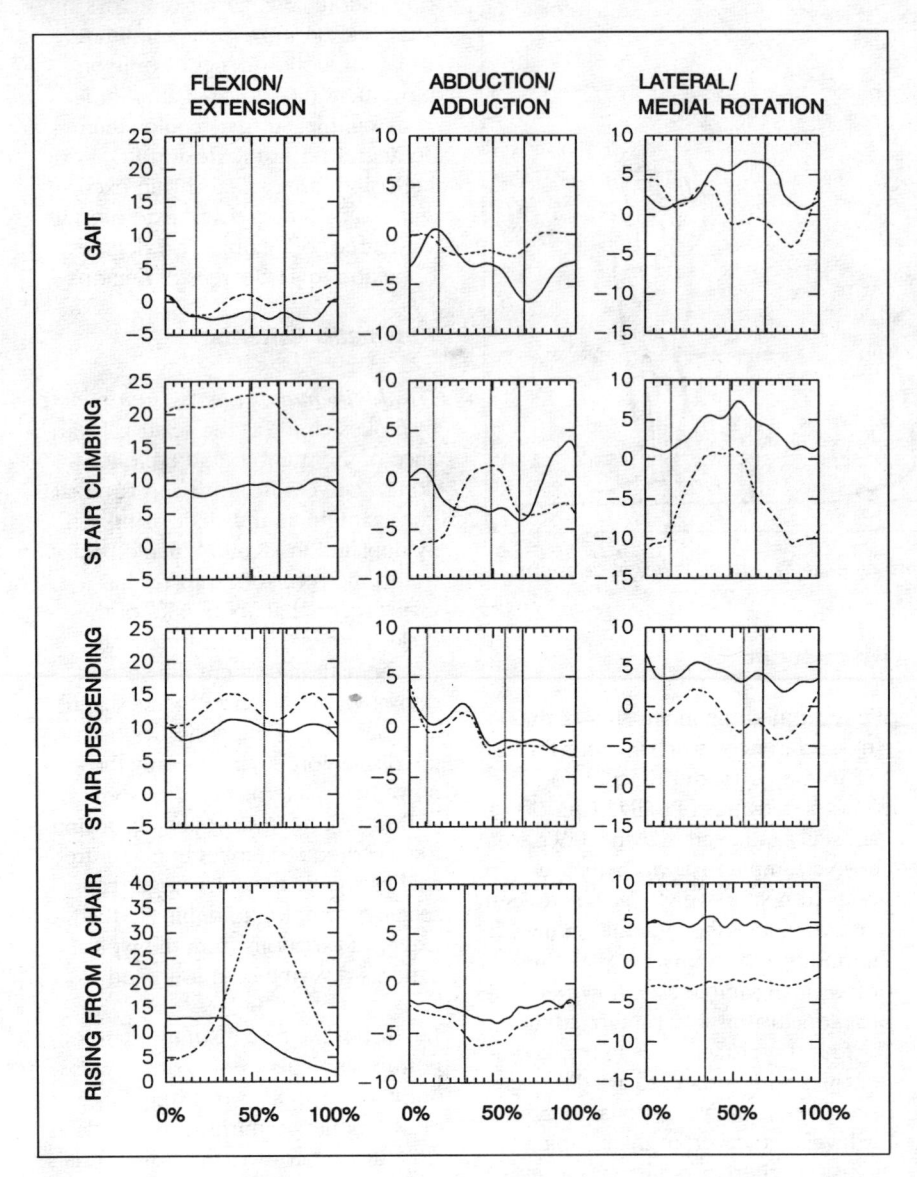

Figure 4. Trunk kinematics of a typical subject (subject 9, Tab. 1) during right-limb gait and stair cycles and during rising from a chair. Ordinate values are degrees of displacement: Note that the chair flexion/extension scale's values differ from those of the stair and gait scales to accommodate the subject's substantially greater flexion during the chair activity; positive values denote flexion, abduction to the right (right shoulder leaning downward), or lateral rotation to the right (clockwise rotation as viewed from above). Abscissa: 0% denotes start of a cycle (right foot-strike for gait and stairs, or first center-of-mass vertical displacement during rising from a chair), and 100% denotes next right foot-strike or, in rising from a chair, the point of maximum vertical center-of-mass displacement. Vertical dotted lines in gait and stair activities denote end of first double-support (left toe-off) phase, then beginning of second double-support (left foot-strike) phase, and finally end of right stance phase; in rising from a chair, the single vertical dotted line denotes loss of seat contact. In all graphs, the dashed curve represents trunk motion relative to room coordinates; the solid line denotes trunk motion relative to pelvis motion.

real-life patterns in two-dimensional written reports such as this; however, we include a comparative kinematic depiction to illustrate selected aspects of how trunk kinematic patterns may differ in subjects with known balance and locomotor control dysfunction (Fig. 7).

Gait. Sagittal-plane patterns typically included a flexion peak near each heel-strike, with maximum extension occurring during single-limb support (Fig. 4), but the amplitude of these motions was small (Figs. 5A and 5B). Apparently, the vertical CM excursion during gait previously reported[8] is directly linked with the trunk sagittal angular displacements. The CM was highest during single-limb support and lowest during double-limb support, the periods of peak trunk extension and flexion, respectively.

Frontal-plane trunk motions relative to the pelvis tended to occur toward the stance limb, reaching their maximum at the time of opposite side toe-off (Fig. 4). That is, at right foot contact, the trunk was midway in its movement from left to right leaning, and this motion continued until left toe-off, at which time a reversal occurred and the trunk began to lean toward the left side. Relative to room vertical, the abduction/adduction amplitude was lower and the curve had fewer inflection points than did pelvis-referenced motions; in general, we observed pelvis-referenced trunk abduction toward the soon-to-be stance limb, which reversed about 5% of a cycle following ipsilateral foot contact. The greater trunk-to-pelvis ROM was due to independent pelvis motions moving out of phase with the trunk.

Transverse trunk rotation relative to room coordinates was also 180 degrees out of phase with the pelvis and achieved maxima about 10% of a cycle after each heel-strike, rotating so that the ipsilateral shoulder was posterior to the heel-strike limb, nearly directly over the foot at mid-stance, and maximally anterior to the stance limb near toe-off (Fig. 4). A slight timing difference was apparent in the pattern relative to room coordinates in that curve reversals were achieved directly coincident with each heel-strike. That is, the trunk-pelvis transverse-plane maxima that occurred after each heel-strike were apparently caused by contravening pelvic rotation continuing after the trunk's intrinsic rotation period was

Basic Research

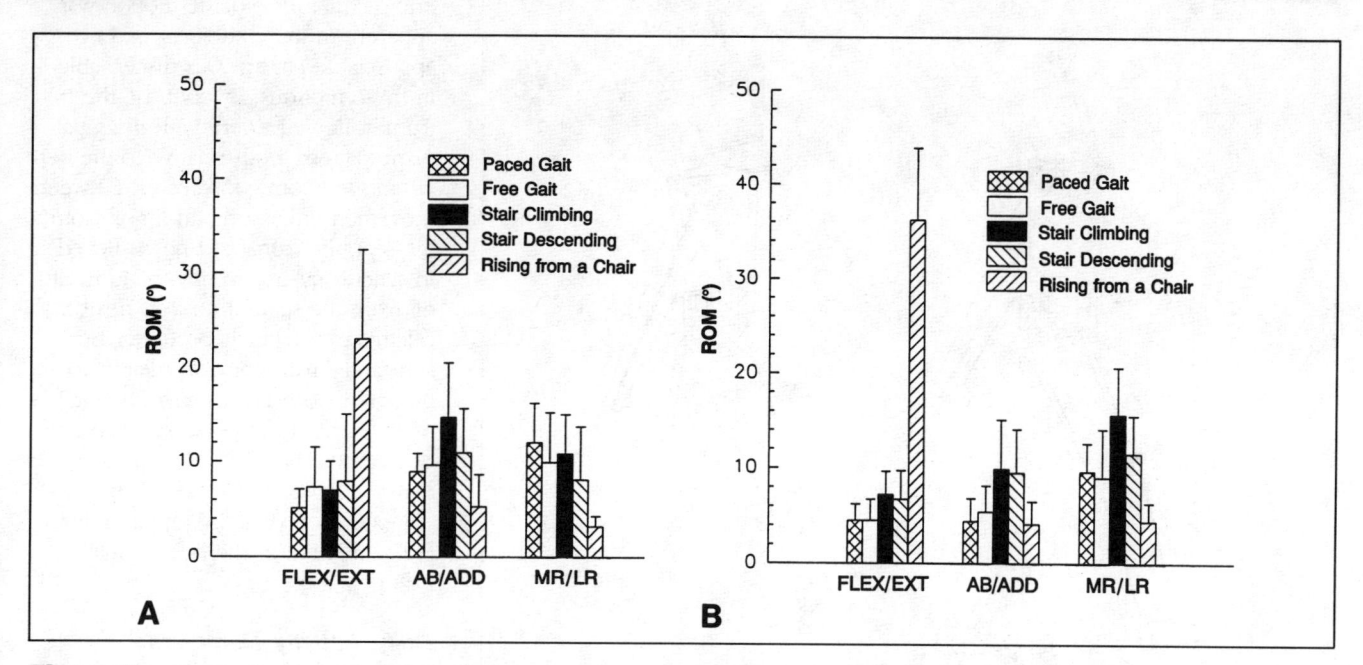

Figure 5. *Means (bars) and standard deviations (t-bars) for trunk range of motion (ROM) relative to the pelvis (A) and relative to room coordinates (B). (FLEX/EXT=flexion/extension, AB/ADD=abduction/adduction, MR/LR=medial rotation/lateral rotation.)*

complete. Thurston[18] reported virtually identical kinematic patterns for the trunk relative to the spine during gait, but did not consider room-referenced trunk motions.

Stair locomotion. Although sagittal- and transverse-plane ROM during stair locomotion was comparable to that observed during gait, trunk abduc-

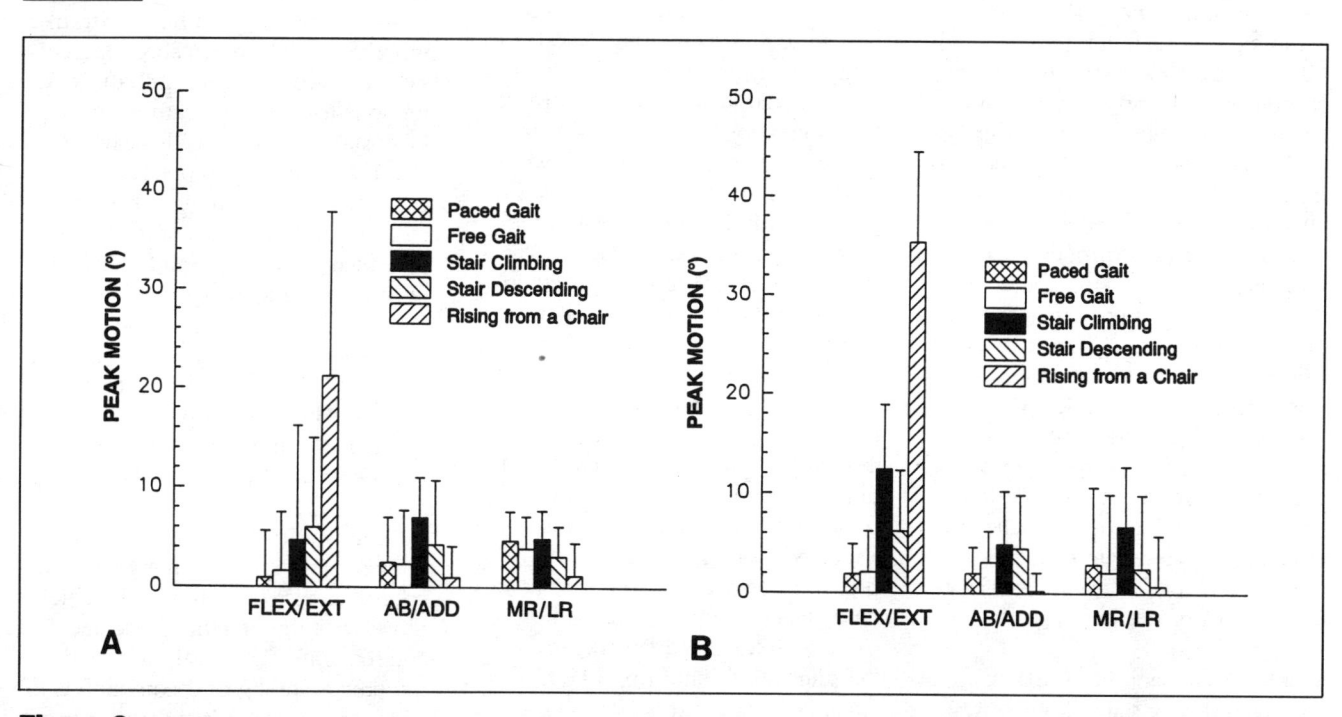

Figure 6. *Maximum trunk angular orientation, or peak motion, relative to the pelvis (A) and relative to room coordinates (B). Bars represent means, t-bars represent standard deviations. (FLEX/EXT=flexion/extension, AB/ADD=abduction/adduction, MR/LR=medial rotation/lateral rotation.)*

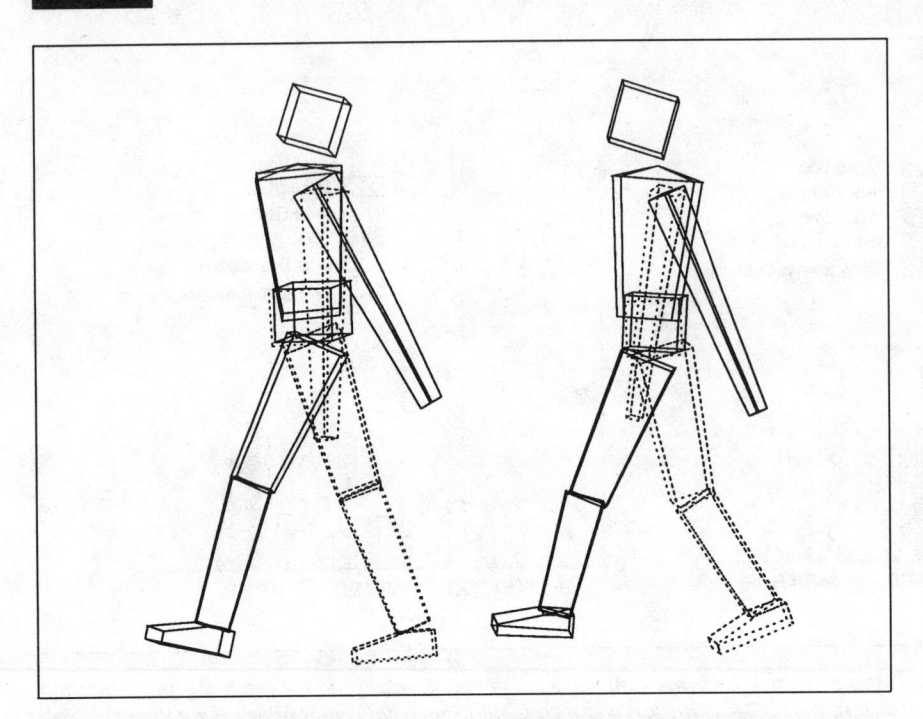

Figure 7. *Full-body kinematic model of a subject with bilateral vestibulopathy (left) and an age- and sex-matched healthy control subject (right) during identical gait cycle periods. Note both subjects have virtually identical head, pelvis, and lower-extremity orientations, but they differ in trunk position; the subject with vestibulopathy has more trunk flexion and less transverse rotation. Excessive trunk flexion during gait and decreased transverse rotation may enhance locomotor stability by positioning the center of gravity more anteriorly over the base of support and by decreasing the transverse oscillations associated with normal walking.*

tion/adduction ROM during stair climbing or descending (Figs. 5A and 5B) and peak flexion relative to room coordinates (Fig. 6B) substantially exceeded those values obtained during gait. Peak flexion values obtained during stair climbing reflect a more inclined posture assumed as subjects oriented the trunk to roughly parallel the stairs' 33-degree slope. After assuming this flexed posture, subjects demonstrated a trunk flexion/extension ROM relative to the pelvis that was similar to their gait ROM (Fig. 5A). Subjects descended the stairs with much less maximum trunk flexion relative to the room coordinates (Fig. 6B) than during stair climbing, apparently to maintain stability by shifting the trunk's mass away from the stairs' declension.

Trunk frontal-plane ROM relative to the pelvis during stair climbing was greater than that of gait, apparently to help clear the swing foot over the step and thus minimize lower-limb flexion requirements (Fig. 5A). These data also suggest that greater trunk abduction/adduction ROM relative to the room coordinates during stair locomotion, as compared with gait (Fig. 5B), stems from the relatively greater vertical displacement of the body's mass required to clear each stair.[8] The lesser transverse rotation relative to the pelvis during stair locomotion, as compared with gait, is apparently due to the shorter step and stride lengths required during stair climbing and descending. Motion patterns relative to the room coordinates during stair climbing were essentially opposite those of gait: the trunk was abducted away from the limb approaching stance, apparently to assist limb clearance over the step; the trunk then traversed laterally toward the stance limb until opposite footstrike occurred (Fig. 4). Trunk-to-pelvis frontal-plane patterns, however, were similar to gait patterns.

Rising from a chair. Very little lateral or transverse-plane motion was apparent in these subjects, and no appreciable pattern was discernable in these motions. By contrast, the trunk followed a very typical sagittal-plane pattern, both relative to the pelvis and to room coordinates. Between movement initiation and lift-off from the seat, the trunk and pelvis flexed in synchrony, apparently to aid in lift-off from the seat. Almost no flexion relative to the pelvis occurred, but substantial trunk flexion relative to the room coordinates was evident. After lift-off, trunk-pelvis synchrony decreased. The pelvis flexed more than the trunk, inducing relative spinal extension (Fig. 5A), apparently to aid attainment of the body's final vertical position.[9]

Daily Activity Demands on Spinal Mobility

These data suggest the extent of spinal mobility required by healthy subjects to perform activities of daily living. Trunk movements result from forces induced by trunk and limb muscles, or forces induced by ground reactions and inertia from the moving segments.[11] Patients with spinal fusion or muscle dysfunction limiting trunk mobility would presumably compensate with greater lower-limb or neck motions to maintain postural and head stability during daily activities. Combined spinal and pelvifemoral restriction could significantly disrupt all locomotor activities of daily living, particularly given the importance of pelvic kinematics suggested by these data. Trunk motion impairments would be expected to impede sagittal gait kinematics slightly, but stair and chair activities would probably be substantially impaired. This assertion deserves future study.

Normal whole-body kinematics apparently depend on smooth spinopelvic coordination. For example, the chief function of the erector spinae and anterolateral abdominal muscles during erect standing or locomotion is to resist gravity and to stabilize the trunk relative to gravity.[19] Unconstrained trunk flexion would be expected to occur during locomotion if muscle and hip movement restraints were absent, as in quadriplegia or if ligamentous integrity were surgically compromised. One function of the greater trunk flexion posture found during stair activities than during gait (Figs. 6A and 6B) may thus be that in stair locomotion, greater trunk inclination helps to project the whole-body CM forward along the moving base of support, providing a source of stable anteriorly directed momentum. Greater trunk inclination would also reduce lower-limb stress by damping the body's vertical CM oscillations. This hypothesis, although consistent with these data, should be investigated in future studies.

Trunk abduction/adduction ROM relative to the pelvis (Fig. 5A) was generally greater than that observed relative to room coordinates (Fig. 5B). These data support the assertions of Saunders et al[3] that lateral and transverse-plane pelvic rotations are particularly important for decreasing trunk-to-room oscillations that could excessively displace and thus destabilize the subject during locomotion. These data suggest that gravity's potentially destabilizing effects may be decreased by permitting relatively larger trunk-to-pelvis motions than trunk-to-room oscillations (Figs. 4 and 5B). For example, during the stance phase of normal gait, the pelvis and trunk shift toward the stance limb, and may approximate (abduct) on the swing-limb side, to assist foot clearance without necessitating excessive trunk and CM displacements. In gait, relatively little trunk abduction/adduction is required to effectively transfer weight and assist the swing leg. In stair locomotion, however, the greater lateral displacements may be needed to ensure that the trunk's substantial mass is securely over the supporting (stance) limb. Transverse-plane data support this contention. Medial/lateral rotation ROM relative to the room coordinates was greatest in stair negotiation (Fig. 5B), but transverse rotation ROM relative to the pelvis was relatively small because the pelvis moved in greater synchrony with the trunk in stair locomotion (Fig. 5A).

Table 3. *Comparison of Previously Reported Trunk Range of Motion (in Degrees) During Gait[a]*

	Murray[8]	Thorstensson et al[11b]	Stokes et al[12b,c]	Opila-Correia[10]	Thurston[18]
Sagittal plane	...	2–12	...	11.1	5.2
Frontal plane	...	2–9	4.9	12.6	6.8
Transverse plane	6.9	...	4.7	17.5	8.8
No. of subjects	60	7	8	14	10

[a]Trunk displacements are assumed to be relative to the pelvis unless otherwise indicated in the work cited.
[b]Treadmill (not overland) gait, with restricted arm-swing.
[c]Referenced to room vertical.

Because the chief task during rising from a chair is to transport the CM vertically, and only a small horizontal (forward) CM displacement is required,[4] it is not surprising that our subjects displayed little transverse- and frontal-plane trunk motions but large sagittal-plane displacements.

Comparison with Previous Results

Murray[8] found the range of thoracic transverse rotation for free-speed walking to be 6.9 degrees, compared with 9.0 degrees for our subjects. Stokes et al[12] reported the mean trunk range of abduction/adduction ROM in eight subjects during free-speed walking was 4.9 degrees, compared with 5.4 degrees for our subjects in the same movement. Stokes et al, however, provided no descriptive data other than gender for their subjects, and those data were collected during treadmill walking only. Thorstensson et al[11] reported the amplitude of apparently room-referenced trunk flexion/extension movements during treadmill gait of seven young subjects to be 2 to 12 degrees. Flexion/extension ROM in gait varied from 2 to 8 degrees relative to room coordinates and from 2.5 to 12.6 degrees relative to the pelvis in our study. Movement in all three planes of trunk ROM in Opila-Correia's study[10] exceeded that demonstrated by our subjects. Prior reports of trunk ROM during gait are presented in Table 3.

The data obtained in this study are within the range of the previously published results for gait, but no comparable data for stair or chair activities have been published, to our knowledge.

Further studies are needed (1) to investigate whether widely reported lower-limb phenomena such as increasing amplitude ROM with increasing locomotor velocity are also present in the trunk and (2) to investigate the effects of various diseases and impairments on trunk kinematics. Data on the effects of these angular rotations on translatory positions would complement the findings of previous studies and would be likely to further the understanding of whole-body stability (balance) control.

Clinical Considerations

We believe the primary utility of the data reported is in using these data to identify the relative demands of gait and activities of daily living on trunk mobility. Familiarity with trunk kinematics of healthy subjects should assist clinicians in designing therapeutic programs for patients with trunk and lower-limb impairments, such as those that occur after back surgery or following spinal cord injury with trunk weakness. Restoration of trunk mobility relative to room coordinates can be attained to some extent even in the absence of spinal mobility, through compensatory strategies. Restoration of normal trunk-to-pelvis kinematics, however, demands thoracolumbar spinal mobility. Hence, these data indicate that although lower-extremity gait kinematic studies will continue to be useful, full-body kinematic analyses during gait, stair climbing and descending, and arising from a chair are feasible and contribute unique information to locomotor assessments that solely lower-extremity studies cannot provide.

Summary

Because the trunk possesses the largest mass of any body segment and because we believe that computing resources continue to increase, whole-body assessments should be routinely included in "gait lab" analyses. Arising from a chair required the greatest trunk ROM in flexion/extension and the least trunk ROM in abduction/adduction and medial/lateral rotation. During gait and stair locomotion, trunk abduction/adduction ROM relative to the pelvis was significantly greater than that relative to room orientation, probably because of the downward tilting of the stance-side pelvis. Flexion/extension ROM was four times greater in rising from a chair than in stair activities, which in turn exceeded gait ROM. Because motions relative to the pelvis differ from room-referenced kinematics, observational locomotor assessments must carefully distinguish the two frames of reference. Each reference frame provides distinct information. The overall goals of locomotion probably include transporting the body with the least energy and the greatest stability, which are probably accomplished by optimizing the motions of the trunk relative to room coordinates and relative to the pelvis simultaneously.

These trunk ROM results during gait are consistent with previous findings, but they are the first three-dimensional kinematics to be reported on stair and chair activities. The functional trunk motions required by healthy subjects to perform activities of daily living should be considered in treating patients with locomotor disability, and especially when surgical procedures limit ROM of any of the body's segments. These findings should be considered tentative, however, until verified in larger, representative population studies.

References

1 Thorstensson A, Nilsson J, Carson H, Zomlefer MR. Trunk movements in human locomotion. *Acta Physiol Scand*. 1984;121:9–22.

2 Inman VT. Human locomotion. *Can Med Assoc J*. 1966;94:1047–1054.

3 Saunders M, Inman VT, Eberhart HD. The major determinants in normal and pathological gait. *J Bone Joint Surg [Am]*. 1953;35:543–558.

4 Pai Y-C, Rogers MW. Segmental contributions to total body momentum in sit-to-stand. *Med Sci Sport Exerc*. 1991;23:225–230.

5 Andriacchi TP, Andersson GBJ, Fermier RW, et al. A study of lower-limb mechanics during stair climbing. *J Bone Joint Surg [Am]*. 1980;62:749–757.

6 Craik RL. Gait and aging. In: Woollacott M, Shumway-Cook A, eds. *Posture and Gait Across the Life Span*. Columbia, SC: University of South Carolina Press; 1989:176–201.

7 Murray MP, Drought AB, Kory RC. Walking patterns of normal men. *J Bone Joint Surg [Am]*. 1964;46:335–360.

8 Murray MP. Gait as a total pattern of movement. *Am J Phys Med*. 1967;46:290–333.

9 Ikeda ER, Schenkman ML, Riley PO, Hodge WA. Influence of age on dynamics of rising from a chair. *Phys Ther*. 1991;71:473–481.

10 Opila-Correia KA. Kinematics of high-heeled gait. *Arch Phys Med Rehabil*. 1990;71:304–309.

11 Thorstensson A, Carlson H, Zomlefer MR, Nilsson J. Lumbar back muscle activity in relation to trunk movements during locomotion in man. *Acta Physiol Scand*. 1982;116:13–20.

12 Stokes VP, Andersson C, Forssberg H. Rotational and translational movement features of the pelvis and thorax during adult human locomotion. *J Biomech*. 1989;22:43–50.

13 Strathy GM, Chao EY, Laughman RK. Changes in knee function associated with treadmill ambulation. *J Biomech*. 1983;16:517–522.

14 Antonsson EK, Mann RW. Automatic 6-D.O.F. kinematic trajectory acquisition and analysis. *Journal of Dynamic Systems, Measurements, and Control*. 1989;111:31–39.

15 Riley PO, Fijan RS, Hodge WA, Mann RW. Determination of joint centers for posture studies. In: Stein JL, ed. *The Biomechanics of Normal and Prosthetic Gait: BED.ASME4*. New York, NY: American Society of Mechanical Engineers; 1987:131–136.

16 Tupling SJ, Peirrynowki MR. Use of Cardan angles to locate rigid bodies in three dimensional space. *Med Biol Eng Comput*. 1987;25:527–532.

17 Riley PO, Mann RW, Hodge WA. Modelling of the biomechanics of posture and balance. *J Biomech*. 1990;23:503–506.

18 Thurston AJ. Spinal and pelvic kinematics in osteoarthrosis of the hip joint. *Spine*. 1985;10:467–471.

19 Basmajian JV, DeLuca CJ. *Muscles Alive: Their Functions Revealed by Electromyography*. Baltimore, Md: Williams & Wilkins Co; 1985:358, 384.

Basic Research

Methods of Studying Gait

GARY L. SMIDT, Ph.D.

As one of the most frequently performed motor acts, walking has been studied from numerous approaches. In this paper, these methods of studying gait have been classified within four groups: observation, holistic, segmental, and comprehensive. Key contributions from a historical perspective, recent developments in the study of gait, and selected results from formal studies are presented for each category. Observation is a useful clinical tool and methods used in the laboratory have revealed selected characteristics of normal and abnormal walking and, to a limited extent, have provided information on the effectiveness of treatment for gait disorders. Methods which offer objective data have not been translated to widespread clinical use, but the need for this transition remains paramount if the health practitioner is to know whether various forms of treatment are beneficial to patients who walk in some abnormal fashion.

Walking is one of the most frequently performed motor acts and, as a result, is probably the most automatic activity accomplished by the human body.[1] To move the body synchronously, a complex synergy of nervous system conductor pathways and the muscular system are involved, including 636 muscles, 206 bones, dozens of organs, hundreds of sensing structures, thousands of communications circuits, and gallons of body fluid. The study of gait has been of interest to scientists and clinicians for several centuries. The progress made in gait analysis during this time is a fascinating evolutionary track to consider and reveals that many of the basic approaches to the problem have remained unchanged for the past fifty years. With advancing technology and apparent increased acceptance of the importance of gait study, however, the number of scientific investigations has begun to proliferate. Interestingly enough, at this time, no clear-cut definition of an ideal walking pattern exists. Depiction of an ideal gait pattern should evolve as inquiry regarding human walking continues for three reasons: 1) to measure the degree and extent of departure from normal, 2) to determine progressive changes resulting from therapeutic procedures, and 3) to evaluate the end result obtained as compared to initial disability.

The intent of this paper is to categorize the various approaches available for studying gait and to include key contributions on the topic of walking from a historical perspective, recent developments in the study of gait, and selected results from these studies.

CODIFICATION OF METHODS OF STUDYING GAIT

Observation

The first major category in identifying or describing a gait abnormality is observation. The first scientific observations of walking were probably done in 400 to 300 B.C. by Hippocrates and Aristotle. As a clinician observes a patient walking each day, the trained eye selects certain characteristics and makes an appropriate assessment. Some phrases and words which are used to describe how a patient walks are slow, fast, quick steps, limp, long

---

Dr. Smidt is Associate Professor and Director of Physical Therapy Programs, College of Medicine, University of Iowa, Iowa City, IA 52242.

steps, good, poor, smooth, the distance walked, recurvatum at knee during stance, one crutch, cane, walker, gluteus medius limp, antalgic gait, equinus gait, spastic calcaneal gait, hemiparetic gait, and drop foot gait. Observation is one approach to use in determining gait abnormality. Brunnstrom has used a standardized form to rate patients by assigning numerical values to indicate the quality of movement patterns for specific anatomical sites on the body.[2]

To enumerate on a potentially endless list of anatomical focuses for visual detection of atypical gait would be a fruitless undertaking. The examiner of gait might better center his attention on departures from symmetry of movement when viewing the patient from the front, back, or side. Asymmetry or asynchronous movement of body parts during walking may be manifested in terms of reduced or excessive displacements and speed of movement during particular points of the gait cycle. For example, a patient with an inflamed right hip may inadvertently spend a short period of time in right stance phase and rapidly execute the complete stride on the left side to minimize the loading period at the joint. This asynchronous movement of the lower extremities can be observed best from a side view. In an effort to reduce the pain during stance phase, the same patient may displace or thrust his trunk excessively over the right hip during stance phase, and, obviously, this asymmetric maneuver can best be seen from the front or rear view of the patient. Asynchronous segment movement for the sagittal plane, therefore, is usually detected when observing the patient from the side, and asymmetry in the coronal plane is best observed from the front or rear of the patient. An ideal location for viewing abnormalities in the transverse plane is not so distinct. When patients exhibit symmetry but have excessive or reduced movement simultaneously, the examiner should rely on the "trained eye" and knowledge of results of experimental studies which report information on parameters of gait for normals.

To increase the probability of benefit to the patient, the physical therapist must, in addition to differentiating between normal and abnormal gait, demonstrate the capability of differentiating among specific movement patterns which contribute to the total dysfunction during the act of walking. To accomplish this end, he must watch for displacement patterns of the head, arms, trunk, pelvis, thighs, shanks, and feet, plus angular movement, particularly of the major joints of the lower extremity.

Holistic Approach

Another major category or approach to the study of gait is a description of output in objective terms related to movement of the whole body during walking. Energy expenditure and the study of movement of the center of gravity during walking fall in this category. In studies of energy expenditure, expired air of the subject is usually collected in a bag and analysis of the gas reveals the amount of oxygen consumed by the body per unit of time. Bard and Ralston were among the first to apply this method to the study of gait.[3] Corcoran and associates demonstrated with a sample of hemiparetic patients that a 57 to 67 percent increase in energy expenditure above normal was required for walking at similar rates.[4] Based on his observation of the kinematics of the pelvis, Inman described the pathway of the center of gravity for walking with its lowest point occurring during double stance and its highest point during single limb balance.[5]

In 1928, Schwartz and Vaeth, who were also interested in the study of the movement of the center of gravity, developed a *basograph* which was a recorder attached to the back of a subject.[6] When the subject walked, the fully movable pen provided tracings which were generally indicative of vertical accelerations of the body. Recently, at the University of Iowa, three accelerometers were placed near the center of gravity and three signals illustrative of the fore-aft, vertical, and medial-lateral accelerations were obtained.[7] Harmonic analyses were applied to the acceleration curves and quantitative differences were reported among patients with a variety of disorders and among individuals with induced gait abnormalities. Amputees have also been studied by the accelerographic method.[8-10]

Segmental Method

Another major approach to the study of gait is a segmental description which yields objective measurements. In this approach, the investigator's attention is focused on local movement of the body parts in space or concomitant isolated activity. Photography has been a major aid in the study of the displacement of the body segments because the events are occurring too rapidly to be perceived by the human eye. In the nineteenth century, Marey was instrumental in using photography to study walking. He was also interested in studying many other animals, including birds.[11] He adorned his subjects with black with white strips of adhesive along the axis of the segments of interest. With his photographic method, he created, at twenty frames per second, stick figures depicting walking which are remarkably similar to those shown in contemporary reports.[12] At the turn of the century, Braune and Fischer also studied walking with photography.[13] Fischer was a mathematician who also did some work to determine the forces associated with the inertial effects of the movement of the limbs. Photography is generally used to obtain kinematic information during walking such as displacement, velocity, and acceleration of parts of the trunk and limbs and angular movement about the joints.

Electrogoniometry has also been used as a segmental approach to the study of gait. Karpovich and collaborators used a single axis goniometer for measuring joint motion in the sagittal plane at the knee and ankle.[14-16] In the late 1960s, investigators at the University of Iowa developed an electrogoniometer which measured joint motion in three planes. In patients with degenerative joint disease of the hip and knee, movement patterns were grossly different from normal, magnitude of motion was reduced for each plane, and the walking velocity influenced significantly the requirements for joint motion.[17-20]

Another segmental approach to the study of gait is electromyography. Surface or indwelling electrodes are used to obtain action potentials emanating from selected muscle sites. This activity is then related to the point in the gait cycle relative to location of foot (e.g., early stance, late swing), and most of the interest has been directed to the lower extremity.[21-26] For normal walking, the greatest demand for muscle tension among the major muscle groups is located near the transition from swing to stance (heel strike) and stance to swing (toe off). This phenomenon seems to be true for the tibialis anterior, quadriceps, hamstrings, gluteus maximus, hip adductor muscles, and even the erector spinae muscle group. A notable exception is the gastrocnemius-soleus muscle mass which tends to be active from midstance to late stance.

Measurement of floor reaction force (force between foot and walking surface) is another area in the segmental category. In 1939, Elftman reported the development of a force platform which was a mechanical system.[27] Contemporary systems are electronic in nature, and floor reaction force is reported in terms of vertical, fore-aft, and medial-lateral components.[28-30] In normal walking, the vertical component is the largest, followed by fore-aft and then medial-lateral. The magnitude of the vertical force is 115 percent of body weight during the weight acceptance and push-off aspects of stance phase and reduces to near body weight during midstance. The total fore-aft and medial-lateral excursion of force is in the vicinity of 27 and 14 percent of body weight respectively. The aft and lateral maximum forces tend to be larger than the forces associated with their counterparts in the opposite direction. The floor reaction force deviates drastically from normal in patients affected with joint disease, including those patients who use assistive devices. In general, the magnitudes of force are reduced and the rate to acceptance and release of maximum load is decidedly reduced.

Another approach in the category of segmental description uses temporal and distance factors which include parameters such as cadence, stride length, velocity, and stance time. The Weber brothers (1836) were the first to study quantitatively the parameters in this category.[31] Carlet (1872) used a chamber under the shoe surface from which pressure changes indicative of stance and swing phase were recorded on a kymograph as the subject walked a circular path.[32] Carlet also had an ingenious way of displaying the foot placement pattern to illustrate symmetry or asymmetry from a time perspective.

More recently, footswitches, photography, mechanical devices, and grid patterns on walkways have been used to obtain information necessary to determine temporal factors (cadence, stance time, swing time, swing/stance ratio, stride time, and double-stance time) and distance factors (stride length and step length).[18,19,31-36] Studies of walking velocity obviously require both temporal and distance data. Generally accepted figures for the most commonly mentioned parameters during normal walking are 130 centimeters per second (3.02 mph) for average velocity, 110 steps per minute for cadence, 150 centimeters for stride length, and 1.2 seconds for stride time; the first three factors tend to be positively related for each person. Usually, persons with an abnormal gait manifest reduced measurements for the distance factors and increased magnitudes for the parameters involving time.

Comprehensive Method

In addition to the methods already discussed in the observation, holistic, and segmental categories, the comprehensive approach probably represents the ideal objective for gait study. This approach requires a combination of methods previously identified which have the capability of yielding kinematic, kinetic, floor reaction force, and temporal and distance measures. Three-dimensional dynamic and floor reaction force data, however, are necessary to provide joint contact forces and moments at the major joints of the lower extremity, so temporal and distance factors should be added for the sake of completeness. Comprehensive methods have not yet been extensively employed. Braune and Fischer's attempt has already been mentioned. Bresler and Frankel (1950)[37] were the first to report forces and moments at the hip joint during stance phase, and similar information has been generated for the entire cycle for the hip by Paul[38] and Rydell[39] and for the knee (tibial-femoral joint) by Morrison.[25,40] The resultant joint force at the hip was found to be approximately four times body weight and three times body weight at the knee. No information of this type was located for abnormal walking.

SUMMARY

Different approaches to the study of walking have been discussed and a codification of these methods has been presented. Included were some key contributions from a historical perspective and selected results which might be clinically useful. Most of the methods currently available to study walking are conceptually similar to those employed several decades ago. Methods used in the laboratory have quantitatively revealed selected characteristics of normal and abnormal walking and, in a limited sense, these methods have provided information on the effectiveness of treatment for gait disorders. Although methods which yield objective forms of data for essential descriptive parameters have not been translated to widespread clinical use, the need for this transition remains paramount if the health practitioner is to know whether various forms of treatment are helpful to patients who walk in some abnormal fashion.

REFERENCES

1. Bernstein N: The Coordination and Regulation of Movements. New York, Pergamon Press, Inc., 1967
2. Brunnstrom S: Recording gait patterns of adult hemiplegic patients. J Amer Phys Ther Ass 44:11-18, 1964
3. Bard G, Ralston HJ: Measurement of energy expenditure during ambulation with special reference to evaluation of assistive devices. Arch Phys Med Rehabil 40:415-420, 1959
4. Corcoran PJ, Jebson RH, Brengelman GL, et al: Effects of plastic and metal leg braces on speed and energy cost of hemiparetic ambulation. Arch Phys Med Rehabil 51:69-77, 1970
5. Inman VT: Human locomotion. Can Med Assoc J 94:1047-1054, 1966
6. Schwartz PR, Vaeth WA: A method of making graphic records of normal and pathological gaits. JAMA 90:86-89, 1928
7. Smidt GL, Arora JS, Johnston RC: Accelerographic analysis of several types of walking. Am J Phys Med 50:285-300, 1971
8. Robinson J: Relation of Accelerographic Analysis and Selected Temporal and Distance Factors in Gait of Below Knee Amputees. Unpublished Master's Thesis, University of Iowa, 1972
9. Gage H: Accelerographic Analysis of Human Gait. American Society for Mechanical Engineers, Paper No. 64-WA/HUF 8, Washington, D.C. 1964
10. Lettre C, Contini R: Accelerographic Analysis of Human Gait. Tech. Report No. 1368.01, Office of Vocational Rehabilitation, Department of Health, Education, and Welfare. Washington, D.C., November, 1967
11. Marey EJ: De la locomotion terrestre chez les bipedes et les quadripeds. J Amer Physiol 9:42-80, 1873
12. Murray MP, Drought AB, Kory RC: Walking patterns of normal men. J Bone Joint Surg 46A:335-60, 1964

13. Braune CW, Fischer D: Der gang des menschen teil versuche unbelastin und belaetin menschen abhandl. Math-Phys Cl Sach Gesellsch Wissensch 21:153-322, 1895
14. Karpovich PV, Wilklow LB: A goniometric study of the human foot in standing and walking. US Armed Forces Med J 10:885-903, 1959
15. Karpovich PV, Gerden EL, Asa MM: Electrogoniometric study of joints. US Armed Forces Med J 11:424-450, 1960
16. Tipton CM, Karpovich PV: Electrogoniometric records of knee and ankle movements in pathologic gaits. Arch Phys Med Rehabil 46:267-272, 1965
17. Johnston RC, Smidt GL: Measurement of joint motion during walking. J Bone Joint Surg 51A:1083-1094, 1969
18. Smidt GL: Hip motion and related factors in walking. Phys Ther 51:9-21, 1971
19. Wadsworth JB, Smidt GL, Johnston RC: Gait characteristics of subjects with hip disease. Phys Ther 52:829-838, 1972
20. Kettelkamp DB, Johnson RJ, Smidt GL, et al: An electrogoniometric study of knee motion in normal gait. J Bone Joint Surg 52A:775-790, 1970
21. Joseph J: Electromyographic studies on muscle tone and the erect posture in man. Br J Surg 51:616-621, 1964
22. Eberhart HD, Inman VT, Saunders JB, et al: Fundamental Studies on Human Locomotion and Other Information Relating to the Design of Artificial Limbs. Report to the National Research Council, Committee on Artificial Limbs, University of California, Berkeley, 1947
23. Close JR, Todd FN: The phasic activity of the muscles of the lower extremity and the effect of tendon transfer. J Bone Joint Surg 41A:189-208, 1959
24. Milner M, Quanbury AO: Facets of control in human walking. Nature 227:734-735, 1970
25. Morrison JB: The mechanics of muscle function in locomotion. J Biomech 3:431-451, 1970
26. Sutherland H, Schottstaedt ER, Larsen LJ, et al: Clinical and electromyographic study of seven spastic children with internal rotation gait. J Bone Joint Surg 51A:1070-1082, 1969
27. Elftman H: The measurement of the external force during walking. Science 88:152-153, 1938
28. Hirsch C, Goldie I: Walkway studies after intertrochanteric osteotomy for osteoarthritis of the hip. Acta Orthop Scand 40:334-345, 1969
29. Jacobs NA, Skorecki J, Charnley J: Analysis of the vertical component of force in normal and pathological gait. J Biomech 5:11-34, 1972
30. Smidt GL, Wadsworth JB: Floor reaction forces during gait: comparison of patients with hip disease and normal subjects. Phys Ther 53:1056-1062, 1973
31. Weber W, Weber EF: Mechanik der menschlichen gehwerkzeuge. Gottingen, Dietrich, 1836
32. Carlet MG: Essai experimental sur la locomotion humaine. Etude de la march. Ann Sci Nat 16:1-92, 1872
33. Drillis RJ: Objective recording and biomechanics of pathological gait. Ann NY Acad Sci 74:86-109, 1958
34. Finley FR, Cody KA: Locomotion characteristics of urban pedestrians. Arch Phys Med Rehabil 51:423-426, 1970
35. Murray MP: Gait as a total pattern of movement. Am J Phys Med 46:290-333, 1967
36. Smidt GL, Simpson CM: How fast are you walking? Phys Ther 51:412-413, 1971
37. Bresler B, Frankel JP: Forces and moments in the leg during level walking. Trans Am Soc Mech Eng, 1950, pp 27-36
38. Paul JP: Forces transmitted by joints in the human body. Proc Inst Mech Eng 181:8-15, 1967
39. Rydell N: Forces in the Hip Joint II, in Intravital Studies in Biomechanics and Related Bioengineering Topics, edited by Kenedi RM, London, Pergamon Press, Inc., 1965
40. Morrison JB: The function of the knee joint in various activities. Biomed Eng 4:473-480, 1969

THE AUTHOR

Gary L. Smidt, Ph.D., received a bachelor's degree from Kearney State College, Kearney, Nebraska, and a master's degree, certificate in physical therapy, and Ph.D. degree from the University of Iowa. He is currently an associate professor and director of physical therapy programs at the University of Iowa. Dr. Smidt is active in the APTA Section on Research and has served as a consultant to the National Conference on Graduate Education in Physical Therapy. His research interests lie in the application of mechanical principles to disorders of the neuromusculoskeletal system.

Basic Research

A Method of Measuring the Duration of Foot-Floor Contact During Walking

GENA M. GARDNER, B.S.
and M. PATRICIA MURRAY, Ph.D.

A new method for monitoring the durations of foot-floor contact during walking is described. The method uses a screen walkway and conducting paper on the soles of the shoes as parts of an electrical circuit. The materials used are identified and circuit diagrams are provided. Several clinical applications and advantages and disadvantages of the method are discussed.

For more than a century, investigators have been interested in studying the temporal components of gait, and, consequently, a variety of innovations for monitoring the durations of foot-floor contact during walking has been devised.[1-13] After extensive experimentation with one of the more recent popular devices, foot switches, our dissatisfaction led to the development of a new technique which is unique in terms of materials, flexibility, and ease of application. Because the method is simple and accurate and offers immediate data visualization at relatively low cost, its use is applicable for treatment, research, and orthotic-prosthetic facilities, particularly where engineering assistance and financial support may be limited.

METHOD

A screen walkway and conducting paper on the soles of the shoe served as parts of an electrical circuit. During walking, contact of the shoe with the walkway produced an electrical signal which was recorded on a polygraph.

A thin rubber hall runner 15.25 meters long and 1.25 meters wide was glued* to an asbestos tile floor, and a sheet of aluminum window screen of the same dimensions was glued† to the hall runner. In cementing the screen to the hall runner, we obtained our best results when a second coat of contact cement was applied to the top of the screen after it was glued to the rubber. The contact cement did not prevent transmission of electrical signals by the screen. To complete the walkway, all edges of the screen were secured to the floor with 4-centimeter-wide cloth tape. Securing the screen in this manner leveled the edges of the walkway surface with the adjacent floor so that there was no danger of tripping when the area was used for other purposes.

Before the recording session, several meters of common shelf-lining Con-Tact® paper were rolled out and painted on the nonadhesive side with a thin coat of a conducting solution.‡ A second thin coat was applied with brush strokes at right angles to those of the first coat, and the paper was allowed to dry. As needed, a piece slightly larger than the sole of the shoe was cut

---

Ms Gardner is a Research Physical Therapist, Kinesiology Research Laboratory, Veterans Administration Center, Wood, WI 53193.

Dr. Murray is Chief, Kinesiology Research Laboratory, Veterans Administration Center, Wood, WI 53193, and Associate Professor of Physical Therapy, The Medical College of Wisconsin, Milwaukee, WI 53233.

This investigation was supported in part by United States Public Health Service Research Grant No. 13854 from the National Institute of Arthritis and Metabolic Diseases.

---

* #155 Waterproof Cement, W.W. Henry Co., Huntington Park, CA 90255.
† Weldwood® Contact Cement, U.S. Plywood-Champion Papers, Inc., Kalamazoo, MI 49003.
‡ #4817 Conductive Silver Composition, E.I. Du-Pont-De Nemours & Co. (Inc.), Niagara Falls, NY 14302.

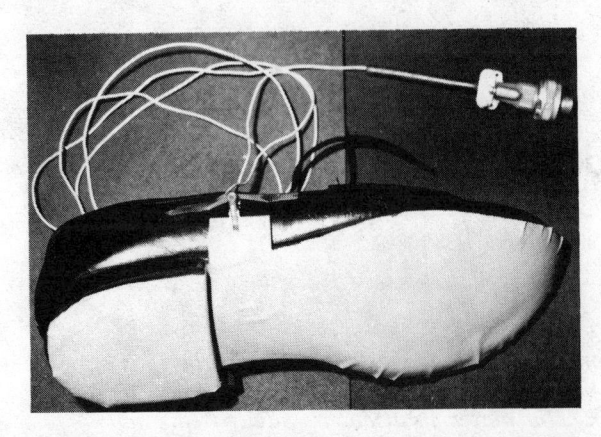

Fig. 1. Conducting paper and wire applied to a shoe.

from the roll and smoothed onto the bottom of each shoe in one continuous sheet. The paper was trimmed to the shape of the sole except for a lip of contact paper which projected from the lateral aspect of the sole in front of the heel for attachment of a wire with an alligator clip (Fig. 1). This wire was plugged into a small junction box strapped to the back of the subject's waist. Output wires from the junction box were carried to a polygraph by means of a pulley-type overhead traverse system. If a traverse system is not available, the length of wire from the patient to the polygraph can be carried manually behind the patient without interfering with his walking performance.

Discrete areas of each sole may be monitored by applying separate nonoverlapping pieces of contact paper to the desired areas of the sole and providing each piece with a tab for wire connection. When the paper is removed after the recording session, the sole of the shoe is rubbed with talcum powder to eliminate adhesive residue.

Diagrams of circuits for monitoring foot-floor contact times are shown in Figure 2. Circuits A and B are alike in that they permit recording of contact of both feet on a single recording channel, and they are designed to produce an electrical signal whenever any part of the undersurface of the shoe contacts the walkway. In contrast, circuit C is designed to monitor contact of two separate areas of each sole and requires one recording channel for each shoe monitored. Circuit A in Figure 2 employs a Grass Model 7 polygraph** while circuits B and C can be used with any type of strip chart recorder.

Figure 3 shows sample recordings of a normal man's foot-floor contact times obtained from circuits A through C in Figure 2, respectively. The polygraph strip from one of the shoes monitored with circuit C was omitted. The Grass polygraph has a remote up and down event marker which indicates foot-floor contact with an oscillating pen signal that provides records which are easy to measure and interpret (Fig. 3A).

For records B and C, the relative height of the pen signal identifies which contact sites are touching the screen at any given time. Precise timing measurements are made from instant to instant of abrupt changes in pen position which mark the initiation and termination of floor contact by the various contact sites.

CLINICAL APPLICATION

Circuit A was used to monitor the foot-floor contact of three subjects walking at free speed

---

** Grass Instrument Co., Quincy, MA 02169.

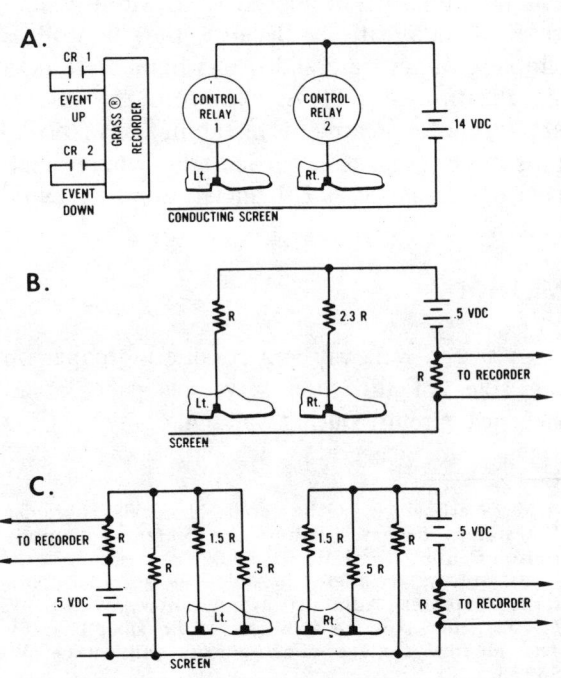

Fig. 2. Diagrams of circuits for monitoring foot-floor contact times (see text). R = resistor; VDC = volts direct current.

Basic Research

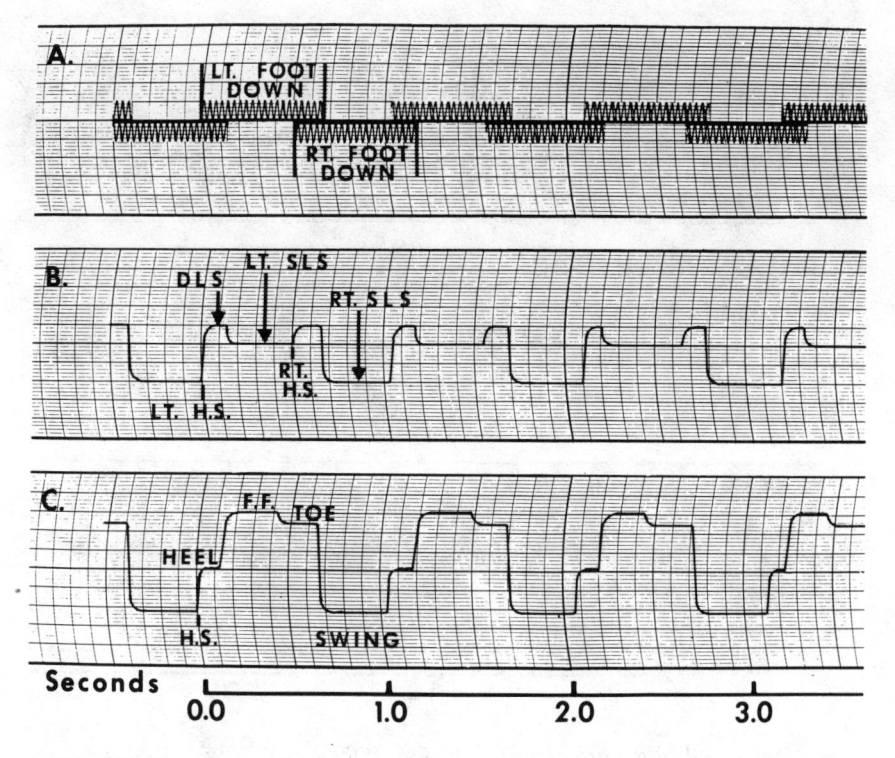

Fig. 3. Records of a normal man's foot-floor contact times made using circuits A-C of Figure 2, respectively. Records A and B were made with contact paper secured to the entire undersurface of both shoes; for C, separate nonoverlapping sheets of contact paper were applied to the heel and to the sole under the forepart of each shoe, but records for only one shoe are shown. HS = the instant of heel strike; DLS = double-limb support; SLS = single-limb support; FF = foot-flat; toe = the sole under the forepart of the shoe. For records B and C, the pen response was set at 15 cycles per second.

(Fig. 4). The percentages enclosed in the rectangles above and below the recorder pen oscillations indicate the proportion of time when each foot was in contact with the floor during each walking cycle (the walking cycle is the period of time between successive instants of initial foot-floor contact of the same foot and is inversely related to cadence). The unenclosed percentages represent the proportion of the cycle when one limb is in the swing phase and the opposite limb is in the single-limb support phase. The top record in Figure 4 is that of a normal man whose mean cycle duration was 1.06 seconds, which represents a cadence of 113 steps per minute. The feature most apparent is the symmetry of the right and left foot-floor contact and swing times.

In contrast, the asymmetry in timing of the man with unilateral hip pain is apparent and results mainly from an abnormally short swing phase of his sound limb, thus a shorter duration of single-limb support on his opposite painful limb. His walking cycles were abnormally long, and thus his cadence of 64 steps per minute was markedly slower than the free-speed cadence of normal men. The patient with Parkinson's disease walked with disproportionately long periods of foot-floor contact with respect to his swing phases (Fig. 4). The long periods of foot-floor contact no doubt resulted from his shuffling-type of gait, while the short periods of no contact related to his observed short step lengths.

Examples of the kind of information which can be obtained by monitoring signals from separate nonoverlapping pieces of contact paper applied to the heel and to the sole under the forepart of the shoe are shown in Figure 5. The

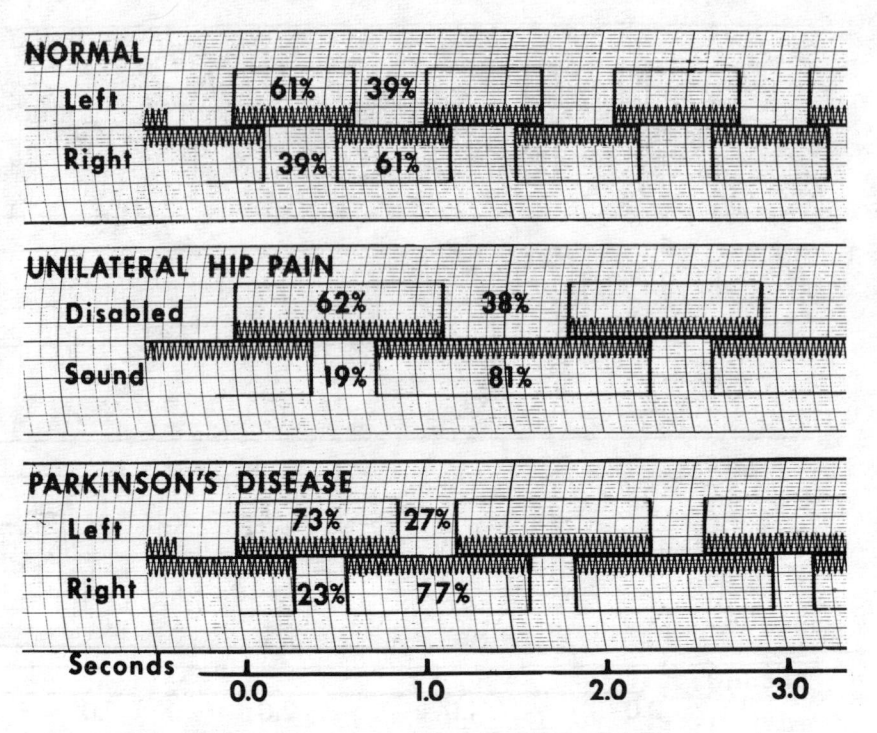

Fig. 4. Sample foot-floor contact records made with circuit A of Figure 2 during free-speed walking. The proportions of the walking cycle spent in foot-floor contact are indicated by recorder pen oscillations.

top timing bar depicts the floor-contact time of the foot of a normal man, and the lower four timing bars depict the abnormal sequence of contact times for the involved limb of disabled patients. The patient whose timing bar is shown second in Figure 5 had severe flexion deformities of the hip and knee of one limb, which gave rise to functional shortening of that limb. She compensated for the shortening by plantar flexing the ankle of that limb and walking on her toes during the entire weight-bearing phase.

The middle timing bar in Figure 5 illustrates the foot-floor contact time of a patient with hemiparesis with mild residual spasticity. Instead of the normal heel-toe sequence of floor contact, his initial floor contact was made with the full length of the lateral underside of his shoe, as observed visually.

The bottom timing bars in Figure 5 are those of a patient with syringomyelia with residual lower extremity muscle weakness which was more pronounced distally. When he walked without an orthosis, initial floor contact was made by the sole under the forepart of the shoe. This abnormality was corrected by the use of a short-leg orthosis with a dorsiflexion assist. When he was wearing the orthosis, however, his records suggested a need for adjustment of the tension of the spring assist, since 0.26 second elapsed before the forepart of his shoe made contact with the floor.

DISCUSSION

Although in some cases visual observation is sufficient to detect patterns of abnormal foot contact, even an experienced observer may sometimes have difficulty in detecting such abnormalities since they involve durations of only fractions of a second. When objective measurements are desired, the described method offers many advantages. The contact paper allows the subject to wear his own shoes, provides a nonslippery interface between the floor and the shoe, and is unobtrusive, thus eliminating awareness of apparatus under the shoe as a possible source of interference with

Basic Research

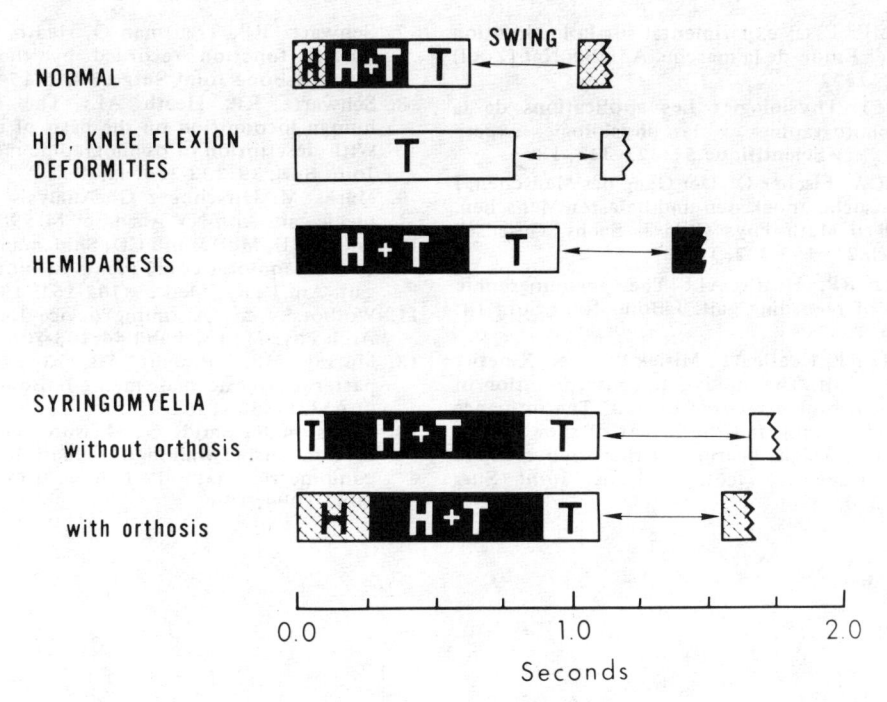

Fig. 5. Timing bars made from foot-floor contact records of free-speed walking of a normal man and of three patients who demonstrated abnormal sequences of contact by the heel (H) and sole under the forepart of the shoe (T).

the subject's usual pattern of walking. The described method also eliminates the need for foot placement restrictions, such as are imposed by methods in which the feet must fall on separate parallel walkways, which certainly would alter the gait of both normal and disabled subjects.

Whether one chooses to monitor contact by any part of the foot or contact of specific areas of the sole will probably depend upon the equipment available and the uses for the data. Our experience with placement of foot switches on discrete areas of the sole in subjects with neuromuscular or orthopedic disabilities has shown that it is possible for the foot to be in contact with the floor without detection by these switches. If the investigator or clinician intends to measure only foot-floor contact and swing times without respect to the area of the sole in contact with the floor, the described method has the advantage of being able to offer sensitivity of the entire sole of the shoe to contact with the floor. On the other hand, if the investigator intends to measure the duration of contact of specific areas of the sole, the described method offers a great deal of flexibility, since the size, shape, and site of placement of the contact paper can all be easily varied. The paper can be placed under the heel, great toe, first or fifth metatarsal head areas, or even under the medial versus lateral borders of the sole for use in subjects such as the patient with hemiparesis who inverted his foot and made initial contact with the lateral border.

The only limitation of the method which we have encountered is the wearing off of the conducting paint by patients who consistently scrape their feet, and, in these cases, the contact paper can be easily replaced.

Acknowledgments. We are grateful to Robert C. Scholz, M.S., and Ron M. Peterson, M.S., for their engineering assistance, and to Betty Clarkson, B.S., Susan Sepic, B.S., and Martin Schubring for their technical assistance.

REFERENCES

1. Weber W, Weber EF: Mechanik der menshlichen Gehwerkzenge. Gottingen, Dietrich, 1836

2. Carlet MF: Essai experimental sur la locomotion humaine: Etude de la marche. Ann Sci Nat (Zool) 16:1-92, 1872
3. Marey EJ: Physiologie: Les applications de la chronophotographie a la physiologie experimentale. Rev Scientifique 51:321-327, 1893
4. Braune CW, Fischer O: Der Gang des Menschen, I Teil. Versuche unbelasten und belasten Menschen. Abhandl d Math Phys Cl d k Sachs. Gesellsch Wissensch 21: 153-322, 1895
5. Schwartz RP, Heath AL: The pneumographic method of recording gait. J Bone Joint Surg 14: 783-794, 1932
6. Schwartz RP, Heath AL, Misiek W, et al: Kinetics of human gait. The making and interpretation of electrobasographic records of gait. The influence of rate of walking and the height of shoe heel on duration of weight bearing on the osseous tripod of the respective feet. J Bone Joint Surg 16:343-350, 1934
7. Schwartz RP, Trautman O, Heath AL: Gait and muscle function recorded by the electrobasograph. J Bone Joint Surg 18:445-454, 1936
8. Schwartz RP, Heath AL: The definition of human locomotion on the basis of measurement: With description of oscillographic method. J Bone Joint Surg 29:203-214, 1947
9. Marks M, Hirschberg G: Analysis of the hemiplegic gait. Ann NY Acad Sci 74:59-77, 1958
10. Smith KU, McDermid CD, Shideman FE: Analysis of the temporal components of motion in human gait. Am J Phys Med 39:142-151, 1960
11. Wolborsky M: A timing device for gait studies. Arch Phys Med Rehabil 44:105-108, 1963
12. Murray MP, Drought AB, Kory RC: Walking patterns of normal men. J Bone Joint Surg 46A:335-360, 1964
13. Johnston R, Smidt G: Measurement of hip joint motion during walking: Evaluation of an electrogoniometric method. J Bone Joint Surg 51A: 1083-1094, 1969

Evaluation of a Clinical Method of Gait Analysis

DARLENE D. BOENIG, BA

Data on step and stride length, step width, foot angle, and cadence were collected from a sample of 30 normal women. The gait patterns were recorded on white paper by a method which involved the application of moleskin and ink to the soles of the subjects' shoes. The purpose of the investigation was to demonstrate the validity, technical reliability, and clinical feasibility of this method of gait analysis. The values obtained compared favorably with the results of a similarly designed laboratory study cited in the literature. In addition, the method was easy to apply, was not time consuming, and yielded data which were reproducible at greater than a 95 percent level of confidence.

At the present time, the method of recording gait in most physical therapy departments is essentially subjective. Specialized instrumentation and other current methods available which offer objective data seem to be too time-consuming and complex in application to be of any practical use. The purpose of this investigation was to use a simple method of recording selected stride factors and to demonstrate the validity, technical reliability, and clinical feasibility of such a method.

REVIEW OF THE LITERATURE

A study by Murray and associates was designed to record various gait components of normal women.[1] The gait patterns of the subjects were recorded by a method of interrupted strobe-light photography. The subject sample consisted of 30 normal women with six subjects in each of five age groups ranging from 20 to 70 years. Murray's study of the gait of normal women was used as a major reference in the design and analysis of the data obtained in the investigation presented in this paper in an attempt to determine the validity of the method of gait analysis used. The method of measuring selected stride factors described in this investigation is similar to a technique developed in a pilot study in a Child Amputees Prosthetics project at the University of California Hospital in 1963.[2]

METHOD

Subject Sample

The test subjects were 30 women with no obvious gait abnormalities or symptomatic joint pathology. Six women between the ages of twenty and seventy years were selected in each of five age groups. All subjects wore shoes with heels 2.5 cm (1 in.) in height or less.

Data Collection

The following materials were used for data collection: precut triangle and square moleskin shapes, a roll of white paper (approximately ½ m wide), bottled inks of two different colors, masking tape, cotton swabs, and a stop watch. Permanent markers were placed on the floor in a specified test area six meters apart to facilitate measurement of the paper. A new strip of paper was taped to the floor with masking tape at the beginning of each trial. The subject was seated on a chair at one end of the paper. One triangle and one square of moleskin was placed

---

Ms. Boenig was a student of the University of Pennsylvania's School of Allied Medical Professions and was completing her internship in the Department of Physical Therapy at University Hospitals of Cleveland, Cleveland, Ohio, when this study was conducted. She is now Acting Chief in the Outpatient Division of the Department of Physical Therapy at University Hospitals of Cleveland, University Circle, Cleveland, OH 44106.

Adapted from a paper presented at the Fifty-second Annual Conference of the American Physical Therapy Association, New Orleans, LA, June 27–July 1, 1976.

shoe. The subject was then instructed to stand and was given the command to "walk as you normally do." The stop watch was used to record the time needed for the subject to walk the entire length of the paper. Figure 2 illustrates the ink imprints as they appeared on the paper strip following a trial.

Data Reduction

Five factors were measured in this study: stride length, step length, step width, foot angle, and cadence. Figure 3 illustrates the measurement of stride and step length from the paper. The midpoint of the heel square was used as a reference point for measurement. Stride length is the linear distance from heel-strike of one foot to heel-strike on the next successive step of the same foot. Step length is the distance from heel-strike of one foot to heel-strike on the next successive step of the opposite foot. Figure 4 describes the computa-

Fig. 1. (Left) Placement of the moleskin tapes on the soles of shoes. Fig. 2. (Right) Ink imprints as they appear on the paper strip following a walk by a subject.

on the approximated midline of the sole of each shoe on the toe and heel respectively (Fig. 1). Red ink applied with a cotton swab was used to saturate the moleskin on the right shoe; blue ink was applied to the moleskin on the left

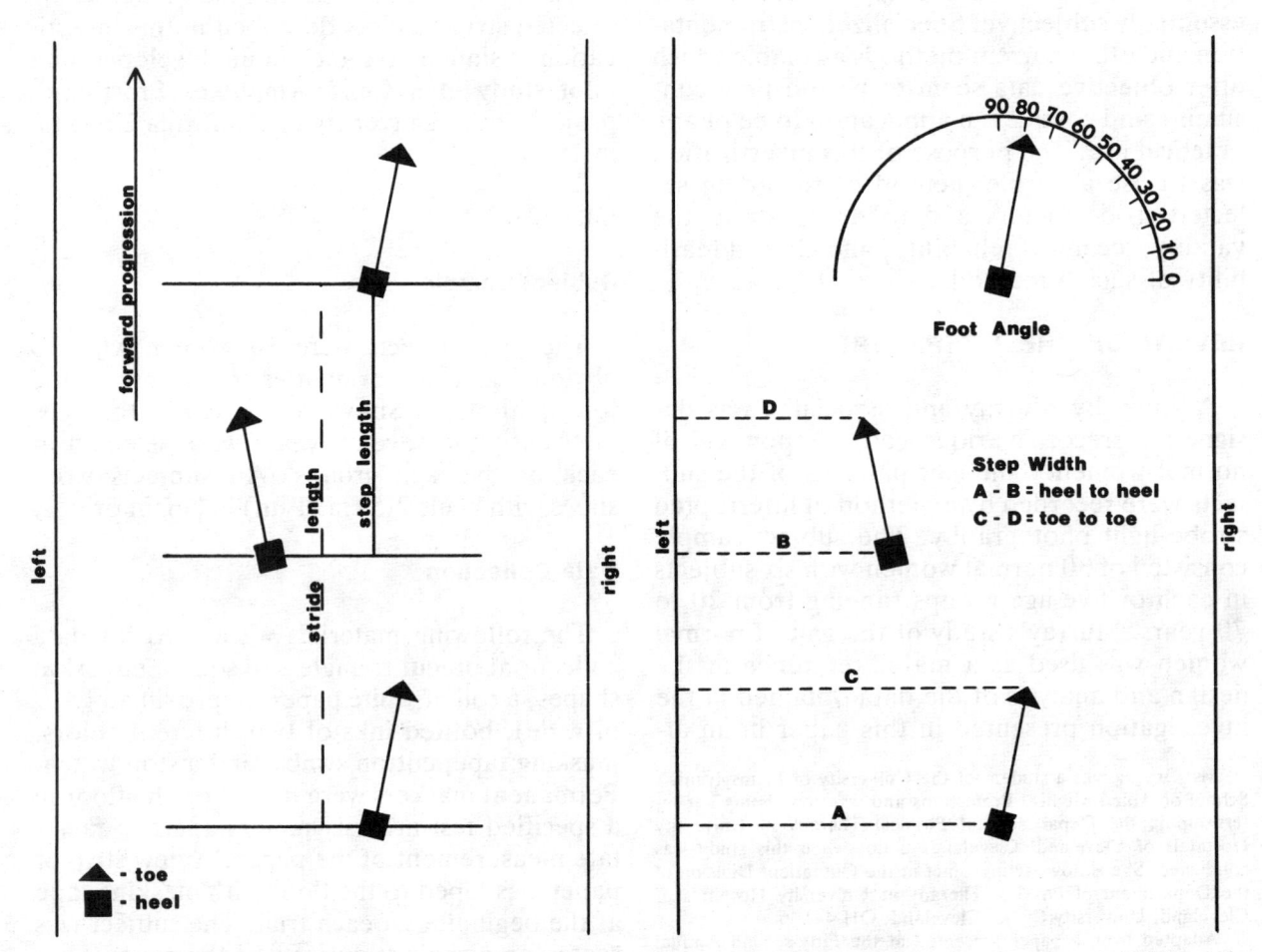

Fig. 3. Measurement of stride and step length from the ink imprints on the paper strip.

Fig. 4. Computation of foot angle and step width from the imprints on the paper strip.

TABLE 1
Comparison of the values obtained in this study with the results of the laboratory study by Murray.[1]

Stride Factor	Boenig		Murray	
	Mean	SD	Mean	SD
Stride length (cm)	124.0	14.0	133.0	9.0
Step width (cm)	6.8	1.4	6.9	2.9
Foot angle (degrees)				
left	9.0	4.1	6.4	7.2
right	8.0	3.3	5.1	5.7
Cadence (steps/min.)	90.0	—	117.0	—

tion of step-width and foot-angle factors. Step width is the transverse linear distance between points on two successive feet. Heel-to-heel step width is the calculated difference between distance A and distance B (refer to Figure 4) measured from the midpoints of two successive heel-squares of opposite feet to the edge of the paper. Toe-to-toe width was calculated in the same manner using the apices of the triangle squares as reference points for measurement (C-D). Foot angle refers to the amount of toe-out or toe-in of each foot. For each step, a long axis was drawn through the apex of the toe-triangle and the midpoint of the heel square. A line intersecting with the long axis was drawn perpendicular to the line of progression. A protractor placed on the perpendicular and intersecting with the long foot axis, was used to determine the angle of toe-out or toe-in. The angle was measured as the number of degrees the foot axis varied from the 90 degree mark on the protractor. The 90 degree mark was chosen as neutral because a line drawn parallel to this point is parallel to the line of progression. Cadence is a measurement of the number of steps taken per unit of time. This value was calculated by taking the amount of time needed to walk across the paper and dividing this number into the total number of steps taken on the paper.

RESULTS

Validity

Table 1 shows the mean values and standards of deviation for the stride parameters investigated in this study as compared to the values obtained in the similarly designed laboratory study by Murray.[1] On the average, the subjects in this study tended to have shorter stride lengths and took fewer steps per minute than the subjects in Murray's study. The mean values for step width, however, are very similar between the studies. Although the mean values for foot angle do differ between the studies, it is interesting to note the similar relationship of left to right in that the mean left foot angle is greater than the mean right foot angle in both studies.

Technical Reliability

Ten subjects were retested one week after their initial trial. New moleskin tapes were used but all other factors, ie, the time of day, location of testing, length of the paper strip, type of shoes used, were exactly the same. The data obtained for all stride factors in Trials I and II of the ten subjects were significantly correlated at greater than a 95 percent level of confidence using a Pearson's product test[3] (Table 2).

Clinical Feasibility

The calculated amount of time needed to conduct a trial was 5.4 minutes. The ink imprints on the paper dried in approximately 30 seconds. The paper could be easily rolled up, marked for identification, and stored for future reference. Detailed analysis of the data required an average of 20 minutes, but certain information could be obtained in an "on the spot" quick analysis using only a tape measure.

DISCUSSION

The mean values obtained in this study for the selected stride factors measured compare similarly with the values obtained in Murray's study (Table 1). A comparison of the two studies is appropriate for two reasons: first, the

TABLE 2
Pearson's product values for correlated data between Trials I and II of subjects retested (n = 10)

Stride Factor	Pearson's Product
Stride length	r = .925[a]
Step length	r = .972[a]
Step width	r = .782[a]
Foot angle	r = .694[b]
Cadence	r = .905[a]

[a] $p < .01$.
[b] $p < .05$.

design was similar in each study; and second, the subjects in each study were relatively free of restricting apparatus in testing. Murray found in a comparison of free and fast speed-walking trials that at faster speeds of walking, stride length increased while values for foot angle decreased.[4] A recent report of a study by Blessey and associates described similar findings in regard to the effect of speed of walking on stride length.[5] Blessey, et al, noted that the difference in stride length between two groups tended to increase with faster cadences. Based on this information, the differences noted in the values obtained in this study as compared to the values in Murray's study may be attributed to the fact that the subjects in Murray's study walked at a faster cadence.

Because the method did not involve the application of complex instrumentation or necessitate the use of a special walkway, it was readily adaptable to a clinical setting and testing occurred in less rigid surroundings than one might expect in a laboratory. It is likely that testing which is least offensive to the subject should provide more reliable information.

The advantages of obtaining an objective analysis in a clinical setting should be apparent to all in the field. The method described can provide the physical therapist with quantitative information in regard to the equality of successive step lengths, the amount of out-toeing or in-toeing of the lower extremity, and the base of support used by an individual in the gait cycle: pertinent items in a prosthetic or orthotic check-out. The following are several additional implications for the clinical use of this method: 1) the physical therapist can actually measure whether there has been improvement in the gait of patients as the result of an adjustment or procedure used in their treatment programs; 2) objective records can be provided of patients with progressive neuromuscular disorders or degenerative joint disease and of children with developmental disorders; and 3) the physical therapist is provided with a feasible means of conducting research on gait in a clinical setting.

CONCLUSIONS

The validity of the method of gait analysis used in this investigation was demonstrated by the fact that the results obtained compare favorably with the similarly designed laboratory study by Murray. The method was easy to apply, was not time-consuming, and yielded data which were reproducible: all factors which support the hypothesis that a method is clinically feasible and technically reliable. Implications for the clinical use of this method are discussed, but further investigation is necessary to determine the most appropriate areas of clinical applicability.

REFERENCES

1. Murray MP: Walking patterns of normal women. Arch Phys Med 51:637-650, 1970
2. Ogg LH: Measuring and evaluating the gait patterns of children. Phys Ther 43:717-720, 1963
3. Ferguson GA: Statistical Analysis in Psychology and Education. New York, McGraw-Hill, Inc, 1971
4. Murray MP: Gait as a total pattern of movement. Am J Phys Med 64:290-333, 1967
5. Blessey RL, Hislop HJ, Waters RL, et al: Metabolic energy cost of unrestrained walking. Phys Ther 56:1019-1024, 1976

Basic Research

A Light-Emitting Diode System for the Analysis of Gait
A Method and Selected Clinical Examples

**GARY L. SODERBERG, PhD,
and R.H. GABEL**

The purpose of this study was to design, develop, and test a comprehensive and potentially clinically useful system for the objective evaluation of sagittal-plane motion and temporal and distance factors during gait. A light-emitting diode and a 35 mm single-frame color slide photographic technique were utilized. The method of data collection and reduction produced values that are similar to reports in previous literature. High reliability was shown both within and across investigators testing normal subjects. Across-investigator comparisons when taking angular measurements demonstrated lower reliability. Specific clinical examples are presented to demonstrate the applicability of the light-emitting diode system.

Physical therapists use a variety of assessment procedures to assist with the determination of treatment goals and outcomes. Frequently, as a part of those techniques, the patient's gait is subjectively analyzed. Although simplified, more objective clinical analyses of gait have been advocated, no reported method satisfies the requirements of physical therapists.

Techniques that have been developed to evaluate objectively the parameters of gait are incomplete or have limitations. Stroboscopic photography,[1] electrogoniometric methods,[2] and television-computer analyses of the kinematics of human gait[3] all require excessive time and expensive equipment that may restrict the very movements intended to be evaluated. The system recently reported by Aptekar and associates used light patterns as a means of clinically assessing and recording gait, but the primary limitation of the Aptekar system is the inability to quantify the results.[4,5] A variety of other authors have confounded the limitations by combining techniques.[6-10] Minimization of these limitations and making provisions for a simple but objective method that can yield results in a relatively short period of time are the primary elements necessary for future usefulness of a technique in a clinical setting. The purpose of this study was, therefore, to design, develop, and test a comprehensive and potentially clinically useful system for the objective evaluation of uniplanar motion and the temporal and distance factors of gait.

METHOD

Data Collection

Experimentation, by the authors, with various techniques that provide objective data established the suitability of light-emitting diodes (LEDs) and a 35 mm single-frame color slide photographic technique. The LEDs* chosen for use were 3 mm in diameter and available in a number of colors. In addition, they were durable and capable of a wide range of pulse rates and illumination intensities. Orthoplast† fixtures fabricated in the laboratory were found to be suitable for the mounting of the LEDs.

Preparation for testing included the insertion of bilateral insoles‡ that were responsible for providing footswitch data. The fixtures (Fig. 1) were attached so that the designated LED of each array was located over the joint centers on the lateral aspect of the subject's lower extremity nearest the camera.

Dr. Soderberg is Assistant Professor and Associate Director, Physical Therapy Education, College of Medicine, S118 Westlawn, University of Iowa, Iowa City, IA 52242.

Mr. Gabel is an electronics engineer, Department of Orthopedics, University of Iowa Hospitals and Clinics, Iowa City, IA 52242.

Adapted from a paper presented at the Fifty-First Annual Conference of the American Physical Therapy Association in Anaheim, CA, June 1975.

The study was supported in part by Grant No. 2R01-AM-14486, US Department of Health, Education, and Welfare.

---

* Newark Electronics, 1019 First Ave SW, Cedar Rapids, IA 52404.
† Johnson & Johnson Co, 501 George St, New Brunswick, NJ 08901.
‡ Orthopaedics Biomechanics Lab, University of Iowa, Iowa City, IA 52242.

The greater trochanter of the hip, the lateral epicondyle at the knee, an area 1 cm proximal to the lateral malleolus, and the base of the fifth metatarsal were the specific landmarks used. An additional fixture was placed on the medial aspect of the contralateral foot at a location where the LED would be equivalent to the base of the fifth metatarsal. Total weight of the fixtures was 450 gms and each had been constructed so as to be applicable to any subject or patient. Wiring from each semirigid fixture and from the footswitches was connected to a 5 by 10 by 15 cm control unit mounted over the sacrum. This unit weighed 1.2 kg and contained batteries and all electronics necessary for LED control. A pushbutton, also attached to this unit, allowed the clinician on-off control of the LEDs preset rate of 25 flashes per second.

For the purpose of evaluation of distance, a thin board was placed on the floor parallel to the direction of the patient's walk. LEDs had been fixed to the board at 5-cm intervals and illuminated at low intensity via a small battery. A 35-mm camera** was then located at knee-level 360 cms from the walkway center with the EHB/135 high speed Ektachrome†† film plane parallel to the direction of walk. A time exposure at an *f* stop of two was used for all filming.

After a static exposure of the illuminated LEDs was taken, the subject walked through the field of view in a dimly illuminated room. Either a clinician or an assistant illuminated the LEDs after the subject was in the field of view of the camera. Several trials were then completed to ensure the likelihood that analyzable data frames were available. Data collection took approximately 20 minutes.

Because the LEDs appeared in proximity on the data frame, filming was done in color. This procedure significantly decreased the time and difficulty of data analysis. A black-and-white reproduction of an actual data frame from a normal subject is shown in Figure 2. The LED attached over the iliac crest, necessary only for the measurement of sagittal plane hip motion, was mounted directly above the LED located at the level of the center of the hip joint. Also shown in Figure 2 is the bilateral footswitch pattern provided by the four LEDs mounted on the fixture attached inferior to the knee. System circuitry automatically activated one of the four LEDs when the respective portion of the left or right foot, toe or heel, was in contact with the floor. Provided, therefore, were heel strike (HS), flat foot (FF), heel off (HO), and toe off (TO) events for both lower extremities. The LED pattern seen in Figure 2 provided an accurate and quantifiable means of identifying and relating portions of the gait cycles.

Also seen in Figure 2 is the location of the LED on the contralateral foot. Location of this single light at a point representing the fifth metatarsal was necessary for the determination of step distances and subsequent comparisons made for symmetry.

In addition, a green LED was placed adjacent to each red LED. This green bulb flashed at every 10th illumination of the red LED. Besides providing a means of synchronization, the green LED facilitated the determination of the temporal factors associated with any part of the gait cycle by allowing for counting by tens. A small current to the red bulbs during every other interval of 10 flashes provided an interval of "streaking" (Fig. 2) which also assisted the process of synchronization.

Data Reduction

Temporal and Distance Factors. Determination of stride time was accomplished by counting the

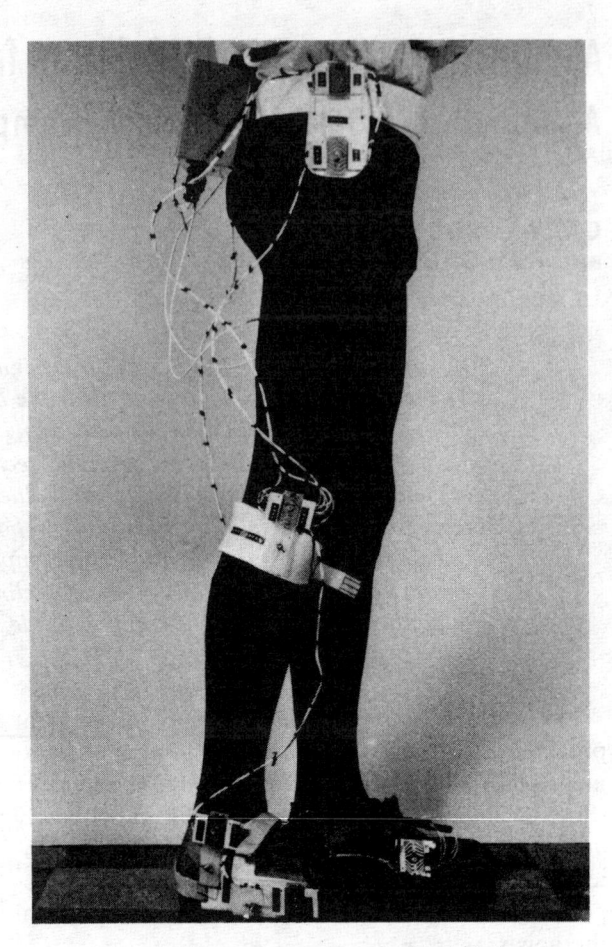

Fig. 1. Normal subject prepared for filming.

** Honeywell Pentax, Ashai Optical, 15 East 25th St, New York, NY 10010.
†† Eastman Kodak Co, 343 State St, Rochester, NY 14650.

Basic Research

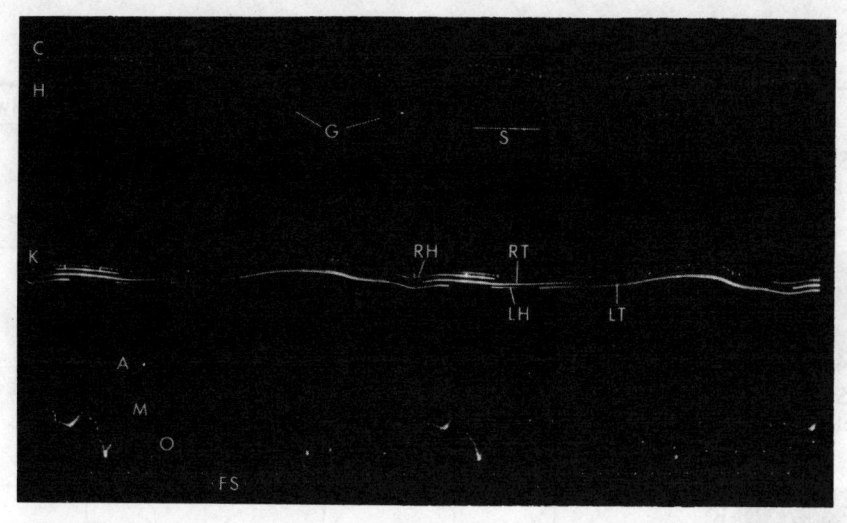

Fig. 2. Black-and-white data frame from a normal subject walking from left to right. Designation for LED locations or functions are: C = iliac crest, H = hip center, K = knee center, A = ankle center, M = forefoot center (head of the fifth metatarsal), O = green light attached to forefoot center of extremity distal from camera, G = green synchronizing lights, RH = right heel contact, RT = right toe contact, LH = left heel contact, LT = left toe contact, S = streak interval, FS = floor scale.

number of red LED flashes from the first to the second right heel strike (onset of RH, Fig. 2). This number was counted from the knee array and recorded on the form shown in Table 1. Similarly, the number of LED flashes from right HS to right TO (stance phase), from right TO to right HS (swing phase) and from right HS to left HS (R-L step time) was recorded. Division by pulse rate, 25 flashes per second, yielded results in seconds. Similar counting of LED flashes was completed to determine left stride, stance, swing, and left-to-right step times.

The lights over the metatarsals were used to determine distance factors because they were closest to the array positioned on the walkway surface. Right stride length was attained by counting the number of floor scale lights between two consecutive cycles of right FF (M, Fig. 2). Likewise, step distance was determined by counting the number of lights from the right metatarsal to the equivalent location on the metatarsal of the contralateral foot. These and other entries were made on the form shown in Table 2. Once entered they were multiplied by the 5-cm distance between each of the floor

TABLE 1
Data Sheet for Temporal Factors

Leg	Event	Formula	# Flashes	Results (sec)
Right	Stride	$HS_{Rt\ 1st}$ to $HS_{Rt\ 2nd}$	38	1.52
	Stance	$HS_{Rt\ 1st}$ to $TO_{Rt\ 1st}$	22	0.88
	Swing	$TO_{Rt\ 1st}$ to $HS_{Rt\ 2nd}$	16	0.64
	Step (Rt-Lt)	$HS_{Rt\ 1st}$ to $HS_{Lt\ 1st}$	18	0.72

(÷ 25)

TABLE 2
Data Sheet for Distance Factors

Leg	Event	Formula	# Flashes	Results (cm)
Right	Stride	$M_{Rt\ 1st}$ to $M_{Rt\ 2nd}$	28	140
	Step (Rt-Lt)	$M_{Rt\ 1st}$ to $M_{Lt\ 1st}$	14	70
Left	Stride	$M_{Lt\ 1st}$ to $M_{Lt\ 2nd}$	28	140
	Step (Rt-Lt)	$M_{Lt\ 1st}$ to $M_{Rt\ 2nd}$	14	70

(× 5)

TABLE 3
Angular Measurement Form

			HS	FF	MSt	HO	TO	MSw	HS
HIP	Static (Reference) −6°	F I L M	29°						
		R E A L	35°						
KNEE	Static (Reference) °	F I L M							
		R E A L							
ANKLE	Static (Reference) °	F I L M							
		R E A L							

scale LEDs yielding results in centimeters.

Velocity and Cadence. From the above calculations, velocity can be determined by dividing stride length by stride time. Cadence can also be calculated by substituting into the expression:

$$\frac{1 \text{ step}}{\text{step time}} \times 60 \text{ (secs)} = \text{steps/minute}$$

Angular Motion. For sagittal-plane joint motion determination the photograph was projected onto glass covered with plain white paper. Angles from the static exposure were measured with a clear plastic goniometer. This procedure provided a reference, or correction value, needed to produce the real joint angle.

Use of the synchronizing mechanisms made it possible to readily select LED flashes associated with specific gait-cycle events. Included in the analysis were HS, FF, midstance (MSt), HO, TO, midswing (MSw), and the next HS. Midstance and midswing points were selected on the basis of the previously determined temporal factors, whereas all other points were determined from the footswitch lights. Angular measurements were recorded from the film for sagittal plane motion for the hip, knee, and ankle joint for each event in the cycle. One sample measurement of 29 degrees has been entered in Table 3 for the hip at heel strike. To account for the position of the LEDs, the static value of six degrees was added to determine the real value of 35 degrees. All other joint angles were determined in like manner. When one is familiar with the process, all data can be reduced in approximately 30 minutes.

Testing

To evaluate reliability of the data-reduction methods, exposures were taken of four normal subjects, aged 27 to 35 years, walking at a free velocity. Film frames from two trials of each of the four subjects were analyzed by three physical therapists and one laboratory technician. In addition, four of the eight frames available were selected for reduction on a second occasion. All reduction of retest data was completed without knowledge of results from the initial analysis. Twelve temporal and distance factors were recorded by each evaluator for each frame, but only 11 of the 12 were subjected to analysis because the step time from left to right was unavailable on a number of frames.

Patients with hemiplegia, amputations, cerebral palsy, and total joint replacements were also evaluated in order to test the applicability of the system

Basic Research

TABLE 4
Temporal and Distance Reliability Coefficients (significant at .01)

Variable		Within-investigators (n = 4) test-retest for four frames	Across-investigators (n = 4) for eight frames
Time			
Right	−Stride	.99	.97
	Stance	.98	.91
	Swing	.98	.97
	Step (R-L)	.99	.94
Left	−Stride	.99	.92
	Stance	.93[a]	.81[a]
	Swing	.99	.95
	Step (L-R)	.98	.92
Distance			
Right	−Stride	.99	.99
	Step (R-L)	.99	.99
Left	−Step (L-R)	.99	.97

[a] significant at .05.

to clinical disorders. Only data collected on two patients are reported in this study, one a hemiplegic and the other a right above-knee amputee.

RESULTS

Reliability

Table 4 shows the temporal and distance factor reliability coefficients for the four investigators. Using the confidence interval for interclass correlations, all coefficients for the within-investigator test-retest, with the exception of left stance time, were significant at the .01 level. The same pattern of significance was demonstrated when investigators were compared across the two frames analyzed for each subject.

Reliability coefficients for the angular measurements, both within and across investigators, are presented in Table 5. Statistical significance was shown at the .01 level with the exception of three within- and eight across-investigator coefficients. These three were significant at the .05 level.

Normal Subjects

Data for a randomly selected gait cycle of a normal subject are presented in Table 6. The data frame from the same subject was used in the determination of the sagittal plane angular values shown in Table 7.

Selected Patients

The data for the hemiplegic patient evaluated are presented in Figure 3. Representative temporal and distance data for the two patients are shown in Table 8.

DISCUSSION

The high values of both the within- and across-investigator reliability coefficients for the temporal and distance factors (Tab. 4) indicated that these factors can be repeatedly measured with a high degree of accuracy. The lowest coefficient of .81 for left stance time was due to inaccurate measurement by one therapist. Evaluation of the raw data indicated that in the analysis of one frame a count of 10 had been dropped, and a value of 14 LED flashes had been entered instead of the correct value of 24.

The coefficients for the angular measurement (Tab. 5) are also generally acceptable. The nonsignificant within-investigator coefficients occur for the knee at the first HS and at HO. Review of the raw data indicated that in several instances errors were made because the reference frame value was added to instead of subtracted from the value determined

TABLE 5
Angular Measurement Reliability Coefficients (significant at .01)

Variable	Within-investigators (n = 4) test-retest for four frames			Across-investigators (n = 4) for eight frames		
	Hip	Knee	Ankle	Hip	Knee	Ankle
HS	.99	.54[b]	.99	.97	.54[b]	.97
FF	.99	.99	.99	.96	.99	.16[b]
MSt	.99	.98	.93[a]	.99	.95	.66[a]
HO	.99	.94[b]	.95	.98	.94	.91[a]
TO	.98	.95	.99	.80[b]	.46[b]	.64[b]
MSw	.99	.99	.98	.98	.81	.97
HS	.99	.98	.96	.99	.92[b]	—

[a] significant at .05.
[b] non-significant.

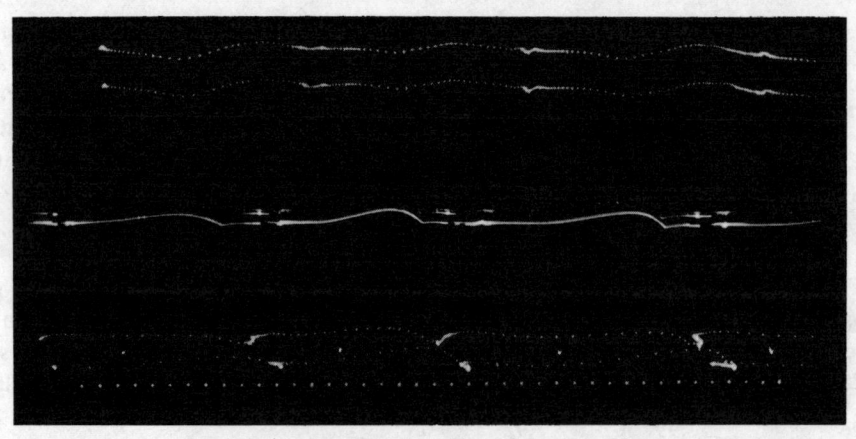

Fig. 3. Right hemiplegic walking from left to right.

from the film frame. Because the knee is near zero degrees at the point of HS, a small change in the static or reference angle produced a large change in the magnitude of the knee angle relative to the other reported values. Subsequently, the HS coefficient was lowered to the reported value of .54.

The across-investigator coefficients are low for several points on the gait cycle. The coefficient of .54 for the knee at HS reflects the difficulties that lowered the within-investigator coefficient for the same joint and event. The coefficient of .46 for the knee at TO appears, from the evaluation of the raw data, to be due to unexplainable reduction difficulties arising from the analysis of one subject. In addition, the laboratory technician data for knee angle at TO appeared to be consistently high by approximately 10 degrees. The nonsignificant coefficient for the hip at TO, when considered with the knee and ankle at TO, indicates that adequate identification of this point in the gait cycle was difficult. The lower coefficients for the ankle at FF, MSt, and TO appear to be due to confusion that arises from the bunching up of LED flashes while the foot is on the ground. Measurements for the FF ankle angle varied from 6 to 14 degrees; for MSt, from 5 to 6 degrees; and for TO, from 6 to 11 degrees. The temporal, distance, and sagittal-plane motion values reported in this paper and those recorded for all normal subjects are in general agreement with data reported by Bajd and associates,[7] Kettelkamp and associates,[8] Richards and Knutsson,[9] Sutherland and Hagy,[10] Smidt,[11] and others. The symmetry of the data reported in Table 6 apparently represents the ability of the LED system to record accurately the parameters under investigation in this study.

TABLE 6
Temporal and Distance Factors Normal Subject

	Right	Left
Stride Time (secs)	1.40	1.36
Stance Time (secs)	0.88	0.88
Swing Time (secs)	0.52	0.48
Stride Distance (cms)	140.00	138.00
	Right to Left	Left to Right
Step Time (secs)	0.68	0.72
Step Distance (cms)	70.00	70.00

TABLE 7
Sagittal Plane Angular Measurements Normal Subject

	HS	FF	MSt	HO	TO	MSw	HS
Hip	36	32	11	3	3	31	29
Knee	5	16	10	4	41	62	7
Ankle[a]	-5	6	-4	-9	16	3	1

[a] Dorsiflexion of ankle is negative value.

TABLE 8
Temporal and Distance Factors

	Right Hemiplegic		Right AK Amputee	
	R	L	R	L
Stride Time (secs)	2.24	2.12	2.16	2.20
Stance Time (secs)	0.64	1.88	1.44	1.80
Swing Time (secs)	1.60	0.24	0.72	0.40
Stride Distance (cms)	55.00	60.00	57.50	55.00
	R to L	L to R	R to L	L to R
Step Time (secs)	0.64	1.60	1.16	1.20
Step Distance (cms)	17.50	37.50	30.00	27.50

Hemiplegic patient data in Table 8 showed remarkable gait asymmetry. Although stride times were approximately equal, the stance and swing times for the two extremities were dramatically different. Other gross differences are apparent in the step time and distance. On the other hand, the amputee demonstrated reasonable symmetry in all parameters. As perhaps could be expected, and as Robinson[12] has reported, less time was spent in the stance phase on the amputed extremity. The literature reveals little other available data for comparison with the results presented in this paper. The quantities presented are representative of what is clinically observable.

Other investigators have attempted to use similar techniques or to devise systems that are clinically applicable.[1] Gutewort reported on a system using pulsed lights, but he made no attempt to relate his technique to the clinical situation.[13] Although the chronocyclography technique of Bauman has some advantages, the sophisticated instrumentation and data reduction techniques required would preclude general clinical application.[14] The most recent reports of Aptekar and associates present systems that are useful but provide the clinician with only a series of comparative light traces.[4,5] A comprehensive discussion of the advantages and disadvantages of all systems available for motion analysis is beyond the scope of this paper, but Neuhauser[15] and Peat[16] provide helpful reviews of various techniques and their requirements.

The method of assessing gait parameters of clinical interest as reported in this paper has demonstrated the capability to produce a more complete and objective analysis than has been reported to date. The use of multiple frame exposures, as in motion pictures, and the lack of footswitch data in stroboscopic photographic techniques are obstacles that have been overcome. Although time and expense can be cited as limiting factors, a nominal investment of both is necessary in order to implement the system reported in this paper.

The LED system offers specific quantifiable data that can be used for purposes of patient feedback, sequential assessments, or documentation and evaluation of treatment techniques. The high reliability demonstrated for normal subjects offers promise for use of the method with pathological subjects. Additional clinical trials have, indeed, been performed on a variety of patients, and usefulness across the full spectrum of patients treated by physical therapists is ensured.

SUMMARY

This paper has presented a system that provides a more complete and objective method of assessing parameters of gait than has been reported to date. Reliability studies showed 1) the method can be used with a high degree of reliability within investigators, and 2) use of across-investigators comparisons of angular measurements cannot be done for all parts of the gait cycle at the ankle until higher reliability can be specifically demonstrated. When compared to the literature, all temporal and distance factors and sagittal-plane motions evaluated were accurately measured with the technique reported.

Acknowledgment. Statistical analyses were completed by Steven E. Nelson, Research Assistant, University of Iowa, Iowa City, IA.

REFERENCES

1. Murray MP, Drought AB, Kory RC: Walking patterns of normal men. J Bone Joint Surg [Am] 48:335-360, 1964
2. Johnston RC, Smidt GL: Measurement of hip-joint motion during walking. J Bone Joint Surg [Am] 51:1083-1094, 1969
3. Winter DA, Greenlaw RK, Hobson DA: Television-computer analysis of kinematics of human gait. Comput Biomed Res 5:498, 1972
4. Aptekar RG, Ford F, Bleck EE: Light patterns as a means of assessing and recording gait: I. Methods and results in normal children. Dev Med Child Neurol 18:31-36, 1976
5. Aptekar RG, Ford F, Bleck EE: Light patterns as a means of assessing and recording gait. II: Results in children with cerebral palsy. Dev Med Child Neurol 18:37-40, 1976
6. Bajd TM, Kljajic M, Trnkoczy A, et al: Electro-goniometric measurement of step length. Scand J Rehabil Med 6:78-80, 1974
7. Bajd T, Stanic U, Tomsic M, et al: Computer measurement and analysis of gait parameters. Proceedings of the 9th Yugoslav International Symposium on Information Processing, October, 1974
8. Kettelkamp DB, Johnson RC, Smidt GL, et al: An electrogoniometric study of knee motion in normal gait. J Bone Joint Surg [Am] 52:775-790, 1970
9. Richard C, Knutsson E: Evaluation of abnormal gait patterns by intermittent-light photography and electromyograph. Scand J Rehabil Med [Suppl] 3:61-68, 1974
10. Sutherland DH, Hagy JL: Measurement of gait movements from motion picture film. J Bone Joint Surg [Am] 54:787-797, 1972
11. Smidt GL: Hip motion and related factors in walking. Phys Ther 51:9-21, 1971
12. Robinson JL: Relation of Accelerographic Analysis and Selected Temporal and Distance Factors in Gait of Below-Knee Amputees. Thesis. Iowa City, IA, University of Iowa, 1972
13. Gutewort W: The numerical presentation of the kinematics of human body motions. In Vredenbregt J, Wartenweiler J (eds): Biomechanics II. Baltimore, University Park Press, 1971, pp 290-298
14. Bauman W: New chronophotographic methods for three-dimensional movement analysis. In Nelson RC, Morehouse CA (eds): Biomechanics IV. Baltimore, University Park Press, 1974, pp 463-468
15. Neuhauser G: Methods of assessing and recording motor skills and movement patterns. Dev Med Child Neurol 17:369-386, 1975
16. Peat M: Gait analysis instrumentation in the clinical facility. In Proceedings World Confederation of Physical Therapy, Seventh International Congress, Montreal, Canada, June, 1974

Basic Research

Footprint Analysis in Gait Documentation

An Instructional Sheet Format

MARION SHORES, MS

We have developed a method of gait documentation using footprints. Written in the format of an instructional sheet, it serves as a useful training aid for new staff, students, and volunteers. The sheet has illustrations next to the instructions, and the wording is intentionally simplified. It enables performance of the test with little or no further instruction. Many of our staff members have used the original instruction sheet, which has been revised according to their feedback during the past year.

Footprint analysis can be performed quickly; is economical; is quantitative and objective; allows visualization of uneven weight-bearing distribution or pressure areas, of toe drag, and of asymmetry of

---

Ms. Shores is Physical Therapist, Department of Physical Therapy, Daniel Freeman Hospital, Inglewood, CA 90301 (USA).

This article was submitted December 18, 1979, and accepted April 15, 1980.

	DATE					
VELOCITY (112.0 cm/sec)*						
CADENCE (110–120 steps/min)*						
BASE OF SUPPORT (5–10 cm)*						
STRIDE LENGTH (132.0 cm)*	Right					
	Left					
STEP LENGTH (66.0 cm)*	Right					
	Left					
FOOT ANGLE (°)	Right					
	Left					
TOE DRAG (+cm)	Right					
	Left					
ASYMMETRY (Describe)						
* Numbers in parentheses are normal values.[2]						

Fig. 1. *Gait flow sheet.*

anatomical structures; and serves as a permanent record for later comparison and for motivating the patient to walk more effectively.

Cost for equipment is two to three dollars per month for the paper and paint. The test takes two to three minutes to complete. Computation and recording may take another 10 minutes and may be completed at a more convenient time. The entire procedure may be performed by supportive personnel.

The gait flow sheet (Fig. 1) enables the clinician to pinpoint changes in gait performance. The measurements from the footprints include: velocity, cadence, angle of the foot from the line of progression, base of support, stride and step lengths, toe drag, and asymmetry. Blank spaces are provided on the flow sheet so that other gait characteristics being monitored may be added to the same flow sheet.

The term "foot angle" indicates the amount of out-toeing or in-toeing between the long axis of each foot and the line of progression.[1]

Quantitative gait evaluation is useful for objective documentation of gait changes and the justification of gait training to physicians and to health care providers, for treatment or program evaluation, and as a clinical research tool.[2]

THE INSTRUCTIONAL SHEET

The following materials are required:
- Tray long enough to accommodate both feet
- Tempera paint to cover bottom of tray

Basic Research

- Paper 610 cm by 457 cm (20 ft x 15 ft) to place on floor
- One chair at each end of paper
- Water and towel for cleansing feet placed at far end
- Stopwatch
- Scissors
- Yardstick
- Goniometer

Part One: Test

1. Preliminaries:
- Have materials ready.
- Mark a line 15.2 cm (6 in) from the far end of paper.
- Dip patient's feet into tray of paint.
2. Instruct patient to walk at his regular speed across the paper looking straight ahead.
3. Set the timer on the stopwatch from the third heelstrike to the line drawn at the far end. This allows the patient time to get started and rules out slowing down at the end of the timed sequence. There should be at least three sets of footprints for the analysis.
4. Cut off the first two footprints up to the edge of the third heelstrike. Cut off the far end of the paper up to the drawn line (Fig. 2).

Part Two: Computation and Recording

1. Velocity. Divide the total walking distance timed by stopwatch by the elapsed time. Record in centimeters per second.
2. Cadence. Divide the number of steps taken during the timed sequence by the elapsed time. Transfer to steps per minute using this conversion formula:

$$\frac{steps}{seconds} \times \frac{60\ seconds}{1\ minute}$$

3. Foot Angle (see Fig. 3):
- Draw longitudinal lines AJ bisecting each footprint.
- Divide length of one footprint into equal thirds.
- Draw horizontal line CD through the posterior third of footprint perpendicular to line AJ. Use measurement from base of heel to line CD for drawing horizontal lines on the other footprints.
- Connect intersections CD and AJ of two ipsilateral feet for line FF. FF is the line of progression.

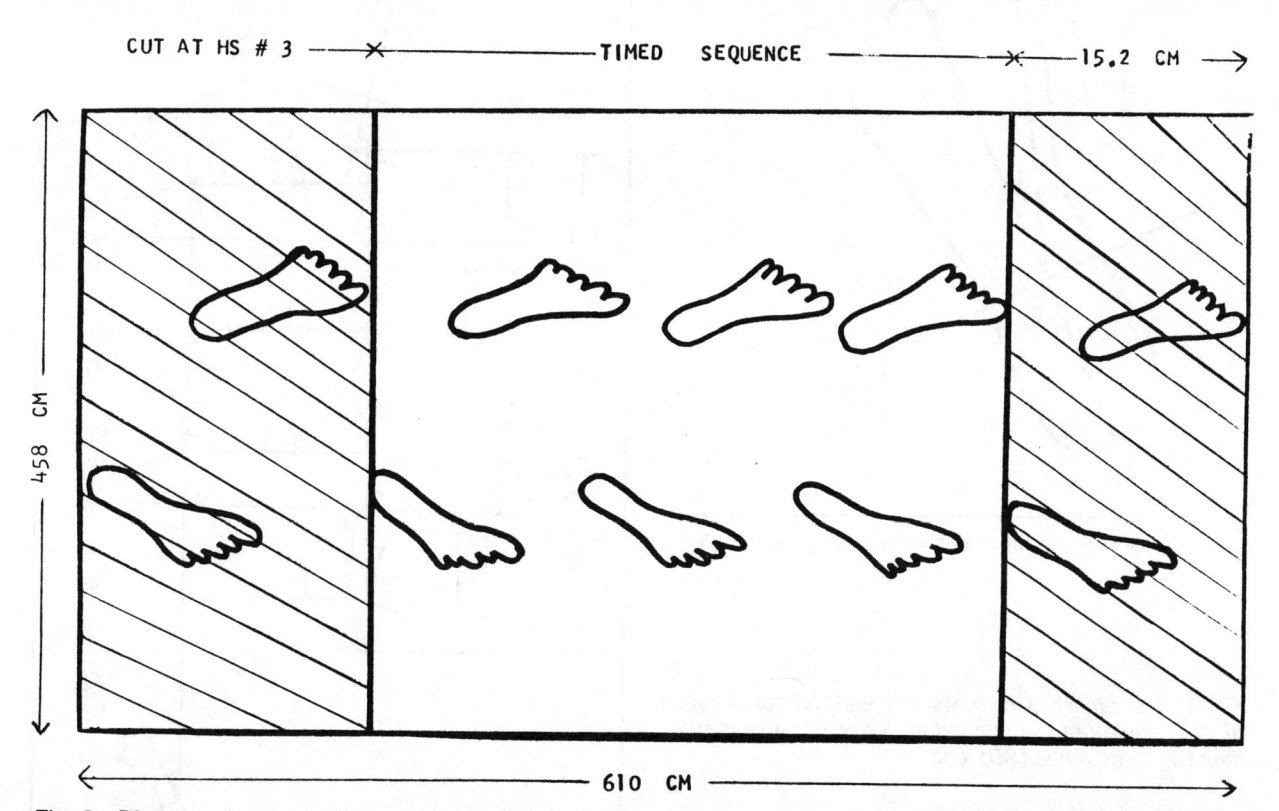

Fig. 2. Diagram showing paper measurements and footprints prior to computation. Shaded areas are to be cut off. HS: heelstrike.

Fig. 4. Determination of base of support (BS).

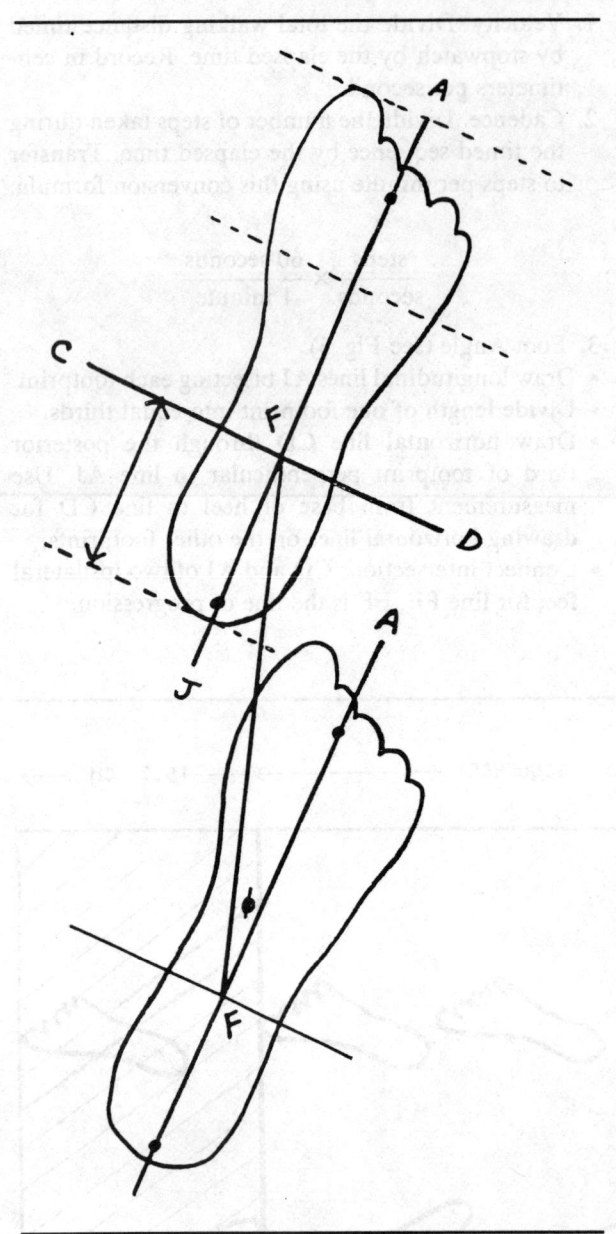

Fig. 3. Schematic method for computing foot angle.

Fig. 5. Diagram showing footprints ready for computation of: foot angle (angles shaded in), base of support (BS), step (ST), and stride (SD) lengths.

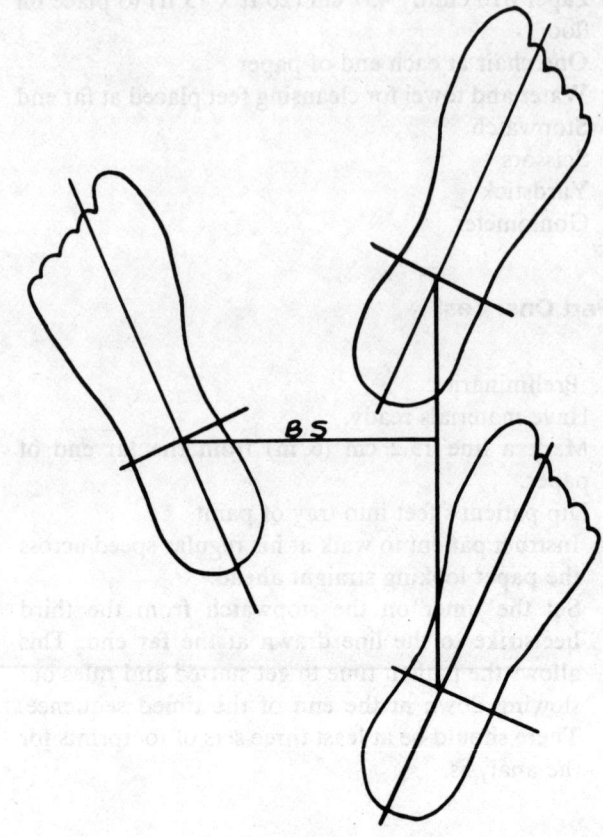

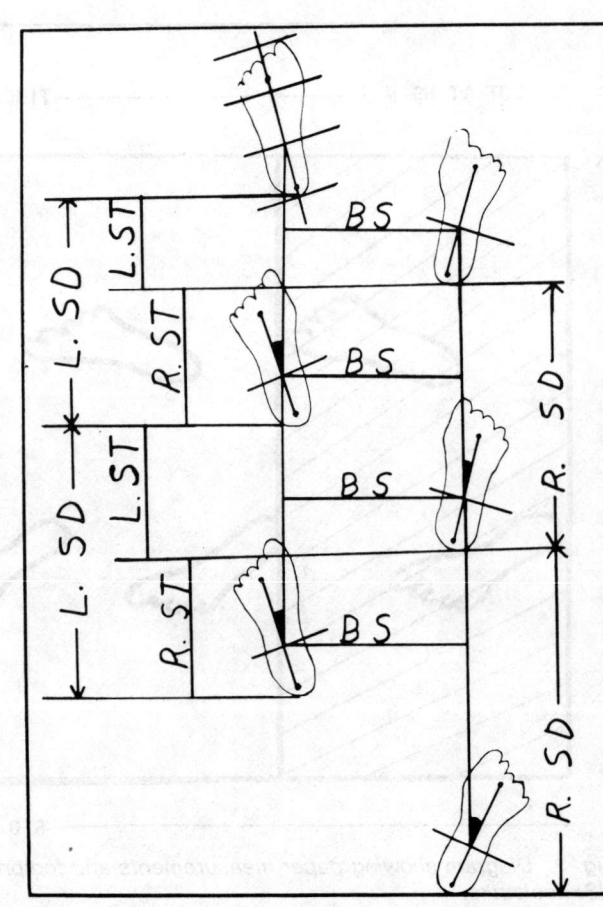

- Record resulting angle ∅ as index of foot angle of the lower foot; indicate whether foot is out-toeing or in-toeing.
4. Base of Support. Draw line BS from intersection CD and AJ of contralateral foot perpendicular to line FF (Fig. 4).
5. Stride and Step Length (see Fig. 5):
- Stride Length. Measure distance between two consecutive ipsilateral heelstrikes.
- Step Length. Measure distance between two consecutive contralateral heelstrikes.
6. Summary:
- Compute velocity and cadence.
- Take average of two sets of footprints for recording the foot angle, base of support, stride, and step lengths (Fig. 5).
- Observe and compare footprints of both feet. Look for toe drag or any asymmetry of pressure areas or weight-bearing distribution at arches, heels, and balls of feet.
- Record findings on the gait flow sheet (Fig. 1).

Acknowledgments. Ms. Linda Lindholm, former physical therapist at Daniel Freeman Hospital, supplied the information on computation of the foot angle.

REFERENCES

1. Murray MP, Kory RC: Walking patterns of normal women. Arch Phys Med Rehabil 51:637–650, 1970
2. Robinson JL: Quantitative gait evaluation in the clinic. Bulletin of the Orthopaedic Section, American Physical Therapy Association 2 (2):11–12, 1977

suggestions from the field

A Portable Feedback Gait Apparatus for Five Segments of Footfall

DEBRA WARSHAL, MS,
MIKE JACOBS, PhD,
WYNNE LEE, PhD,
and MIGUEL GARCIA, BS

Proper control of gait is a problem for some individuals who have had insults to the CNS. After attaining proper balance and hip control—requisites for walking—these persons may still have gait abnormalities. Failure to achieve a normal pattern of foot placement (that is, sequential placement of the heel, ball, and big toe) is one such abnormality. Typically, either the heel or big toe phase of foot placement is missed. These abnormalities can cause awkward gait patterns characterized by resultant muscle imbalances or extraneous compensatory movements. Correction to a more normal foot placement pattern is, therefore, a desirable goal for therapeutic exercise programs designed for ambulatory, motor impaired individuals who have foot placement abnormalities. This article describes a portable device used to assist individuals in achieving proper foot placement through the use of augmented audio feedback (AFB) during walking.

Recently the Krusen Center for Research and Engineering, Philadelphia, PA, developed an auditory monitoring system similar to ours that recorded differential weight bearing through the use of a device implanted in a shoe insole (Personal communication, DR Taylor, Jr, PhD, May 1980). Our system differs from theirs by responding to the sequence of weight distribution characterized in gait patterning; theirs recorded only change in absolute limb load. Until this present design, most clinical apparatus designed to monitor foot placement patterning in gait through AFB has been cumbersome and immovable and has required the patient to practice gait patterns in a fixed location on a metal walkway.[1,2] Major advantages of the present equipment design are portability, light weight, and potential application in varied ambulatory and balance tasks. The AFB system described herein consists of a sequential contact system (SCS) developed by Windell and Kindig[2] plus an auditory tone generator. Output from the system provides signals for generating permanent records of the foot placement pattern through a physiograph or for generating AFB during ambulation. The rationale for using AFB as a corrective technique was based upon previously successful use of AFB in a number of clinical applications including foot drop.[1,3-5]

The SCS used to measure and provide feedback to subjects about the pattern of footfall during walking has four fundamental components: 1) a differential voltage circuit with output of three voltage levels; 2) wire contacts attached to the heel, ball, and large toe of one foot; 3) wire leads from the foot contacts to the differential circuit, plus a ground wire; and 4) a 28-ft by 2-ft* metal walkway that, with the ground and foot contact wires, completes the circuit when either the ball, heel, or big toe makes contact.[2]

Our apparatus is a three-way modification of the SCS for portable use. The first modification is the attachment of wire contacts to elastic straps that slip around the foot to keep the contacts in place. Secondly, an additional differential voltage circuit that operates a battery-powered, five-tone audio signal generator is included. This additional circuit provides

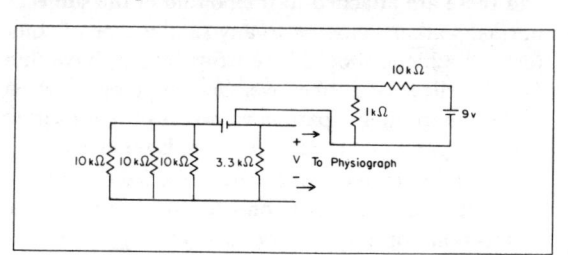

Fig. 1. Voltage divider to physiograph.

Ms. Warshal is a graduate student in the Department of Physical and Health Education, Neuromotor Control, College of Education, The University of Texas at Austin, Austin, TX 78712 (USA).

Dr. Jacobs is Director, Austin Developmental and Therapy Clinic, 4211 Medical Parkway, Suite A, Austin, TX 78756.

Dr. Lee is a postgraduate fellow, Department of Cognitive Science, University of California at San Diego, San Diego, CA 92111.

Mr. Garcia is an undergraduate student in the Department of Electrical Engineering, University of Texas at Austin, Austin, TX 78712.

This study was supported in part by a grant from the Bureau of Education for the Handicapped (Office of Education) #G007801892.

Requests for reprints should be made to Ms. Warshal at The University of Texas at Austin, Department of Physical and Health Education, Austin, TX 78712.

This article was submitted October 14, 1980, and accepted March 14, 1981.

* 1 foot = .305 meters.

AFB when the volume control is turned on. Third, the foot apparatus includes a contact wire, thus eliminating the need for a walkway. This contact wire permits use of the device in a number of activities not possible with the stationary design, such as treadmill walking or maneuvering in a changing environment.

The differential voltage output can be used for permanent recording of a subject's foot placement pattern, immediate auditory feedback, or both, simultaneously. For recording, the voltage signals from one circuit can be preamplified and fed through one channel of a Physiograph 6 amplifier/recorder system†; each voltage level (in other words, each separate contact location) is then recorded on paper with a different amplitude deflection of the pen writer. Slight voltage changes associated with simultaneous contact of two or more leads also occur, but are less distinct. The output of the tone generator circuit can simultaneously provide subjects with immediate auditory information about which part of their foot is in contact with the ground and the proper sequencing of gait pattern. The AFB device can be turned off when feedback is not desired.

Description

In order to construct a portable feedback system capable of discriminating the pattern of footfall during walking with both auditory and visual assessment, the following necessities were determined
1) a "sound box" that produces one different note for each of the five parts of the step (heel; heel and ball; heel, ball, and big toe; ball and big toe; big toe).
2) a circuit connecting the system to the "sound box" and to a physiograph for simultaneous auditory feedback and physiographic recordings.
3) an on-off switch to allow recording on the physiograph without auditory feedback, as well as the reverse, without disconnecting the foot apparatus.
4) a voltage divider in the physiograph able to provide five distinct voltage levels.
5) one set of connections for each foot providing contacts to operate the AFB system without the metal walkway. These connections should be light, thin, and comfortable enough to walk on.
6) a wiring system free of any excess wires that might become tangled and limit movement range and potential.
7) wiring that is flexible and strong.

The resistor placement that yielded the best results is shown in Figure 1. The voltage source comes from a 9-V battery attached to a voltage divider. The purpose of this low voltage is to keep the initial voltage in the physiograph small in comparison to the differences in the voltages generated in the gait task.

† Hewlett-Packard Corp, 171 Wyman St, Waltham, MA 02154.

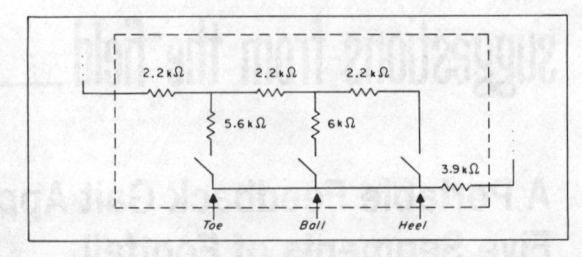

Fig. 2. Voltage divider in sound box.

The sound box (Fig. 2) is designed to give variable resistances in series, with 3.9 kΩ resulting in a new note for each of the five combinations of the step. With the heel down, a 6.6 kΩ resistor is in series with 3.9 kΩ; with the ball of the foot down as well as the heel, a 6.0 kΩ resistor is in series; and when all three critical parts of the foot touch the floor, a 4.5 kΩ resistor is in series. As the heel lifts off of the floor, a 5.5 kΩ resistor is in series, and with just the toe down, a 7.8 kΩ resistor is in series. As the resistance goes down, the pitch of the note goes up, and as the resistance increases, the audio notes go down. One problem was encountered when these values were used for the resistors: the resistance of the sound box was greater than that of the physiograph. This prevented current from passing into the sound box to produce the tones. Therefore, it was necessary to increase the total resistance from the resistors going into the physiograph to the values shown in Figure 2.

Strips of flexible plastic wrapped with copper tape for conduction purposes are used for connections, and these are attached to the bottom of the subject's normal walking shoe, or to any similar type of comfortable athletic shoe. These strips (Fig. 3) have thin (3-mm) slices of rubber weather stripping between them to keep them apart when no pressure is exerted and to separate them after pressure is removed.

An additional piece of plastic is applied to the floor side of the connections to ensure a contact area. The connection for the big toe is 4 cm by 1 cm, the connection for the ball of the foot is 9 cm by 1 cm, and the connection for the heel is 7 cm by 1 cm. In addition, an elastic strip provides a pocket for each "sandwich." This addition provides protection and a place to strap the connection to the shoe. Velcro is

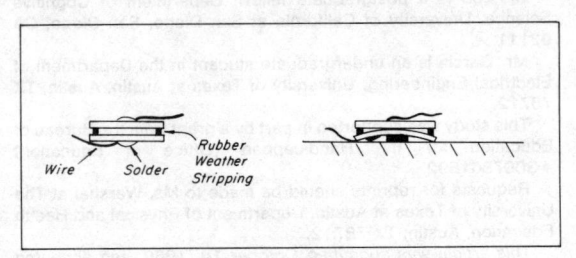

Fig. 3. Side view of connection strips.

Basic Research

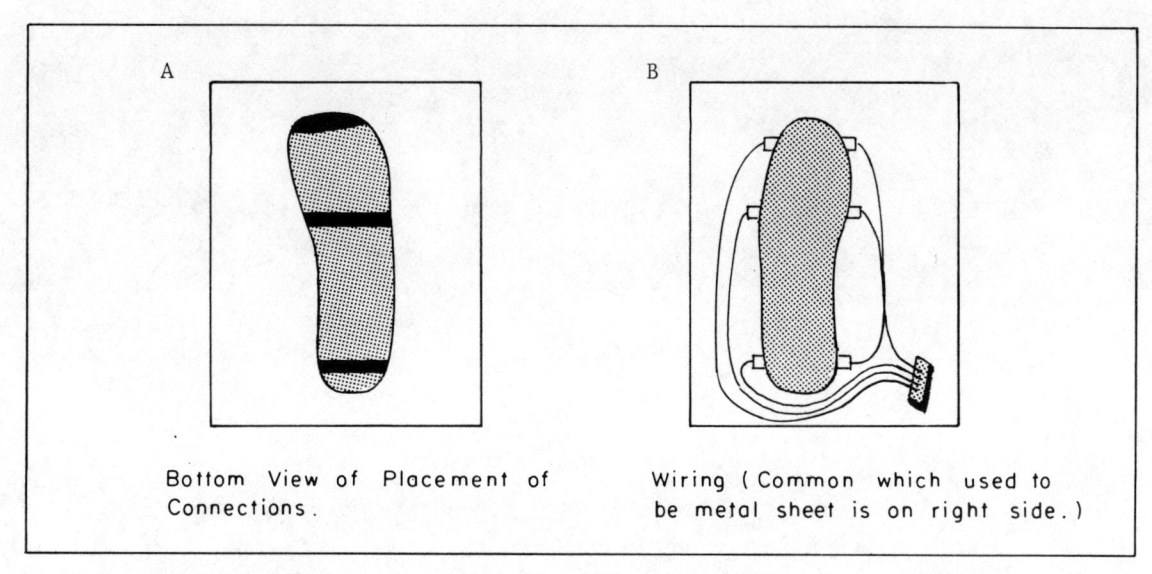

Fig. 4a. Bottom view of placement of connections. **Fig. 4b.** Wiring (common is on right side).

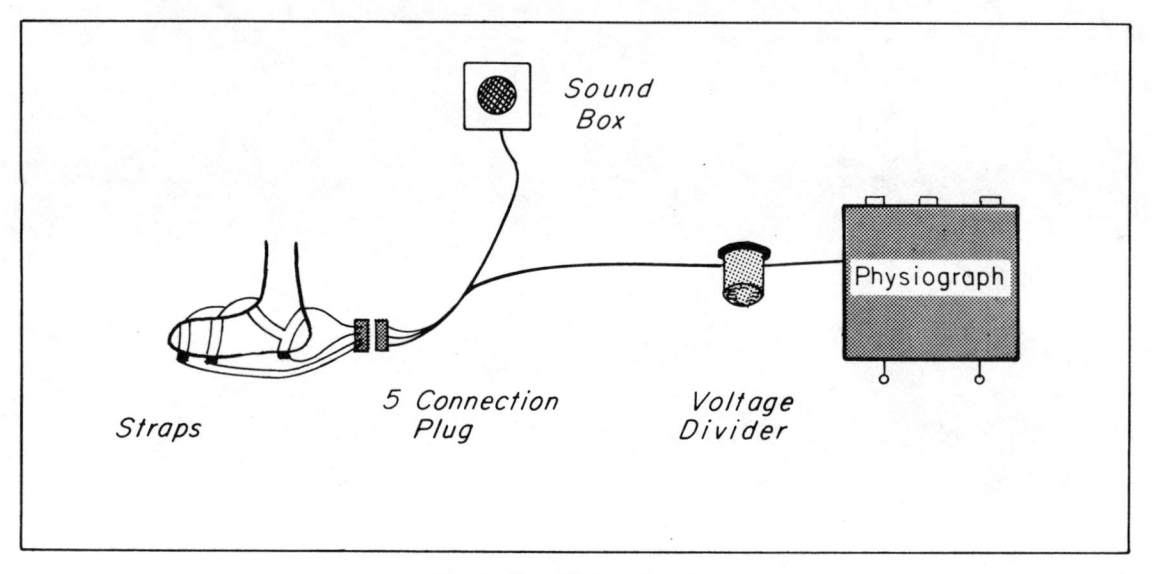

Fig. 5. Overall view of system.

used for the strapping material. The connections are placed at the bottom of the shoe as shown in Figure 4a. Figure 4b shows the manner in which the wire attaches and wraps around the foot.

When written records are desired, a telephone cord with four internal, insulated wires is attached from the physiograph to the sound box to permit freedom in walking direction and distance while eliminating excess wires. A five-connection plug onto the shoe device provides an easy way to switch assessment from one foot to the other. Figure 5 shows an overall view of the entire system.

REFERENCES

1. Gabel RN, Johnston RC, Crowninshield RD: A gait analyzer/trainer instrumentation system. J Biomech 12:543-549, 1979
2. Windell EJ, Kindig LE: A sequential contact system for determining time and events in movement analysis. Am Correct Ther J 29:134-137, 1975
3. Basmajian JV: Biofeedback in therapeutic exercise. In Basmajian JV (ed): Therapeutic Exercise, ed 3. Baltimore, MD, Williams & Wilkins Co, 1978, pp 220-227
4. Basmajian JV, Kukulka CG, Narayan MG, et al: Biofeedback treatment of foot drop after stroke compared with standard rehabilitation techniques: Effects on voluntary control and strength. Arch Phys Med Rehabil 56:231-236, 1975
5. Spearing D, Poppen R: The use of feedback in the reduction of foot dragging in a cerebral palsied client. J Nerv Ment Dis 159:148-151, 1974

Basic Research

Methods of Measurement in Soviet Gait Analysis Research, 1963-1974

LEOPOLD G. SELKER, MS

A description of several of the methods of measurement used by gait researchers in the Soviet Union from 1963 to 1974 is provided. The discussion encompasses seven major categories of measurement: 1) methods of measuring the rotatory position or angular displacements of the joints of the lower extremities, 2) the phase durations and their temporal relations, 3) support reactions, 4) the rotatory position of the pelvis and vertebral column, 5) step length, stride length, and other linear displacements of components of gait phase activity, 6) muscular activity, and 7) miscellaneous types of analysis. The incorporation of Soviet findings and methods of gait analysis into the design and operation of future investigations of human walking is recommended.

The increasing demand for more precise and objective methods of measurement in gait analysis has, within the last 25 years, been the source of incredible ingenuity and unique and obvious productivity which is demonstrated by the recent development of new techniques and instrumentation in gait analysis research. Researchers in the Soviet Union have been major contributors to the development and application of these methods of quantitative measurement.

From an overview of the gait research literature produced in the Union of Soviet Socialist Republics (USSR) in the decade 1963 to 1974, seven major categories of measurement have been identified for further discussion: 1) the rotatory position or angular displacements of the joints of the lower extremities, 2) the phase durations of the gait cycle and their temporal relationships, 3) support reactions or forces, 4) the rotatory position or angular displacements of the pelvis and vertebral column, 5) step length, stride length, and other linear displacements of components of gait phase activity, 6) muscular activity, and 7) miscellaneous types of analysis which include the use of mathematical constructs, for example, correlations or ratios and models.

ANGULAR DISPLACEMENTS OF LOWER EXTREMITY JOINTS

Soviet gait researchers have used a device developed by Cherskov, called the angulometer, to record the angular displacements of the joints of the lower extremity.[1] Examination of the various descriptions[2-4] of the angulometer and its use supports the notion that it is essentially not different from an elgon or electrogoniometer. A potentiometer or variable resistor is employed to register and translate changes in angular position into proportional changes in resistance. These changes in resistance are often displayed on a loop oscillograph. Examination of the Soviet literature on gait analysis research in the decade 1963–1974 revealed no elaboration of the basic uniplanar unit for two- and three-planar registration of angular displacement. One might characterize the potentiometric method of measurement of angular displacement in the Soviet Union during this period as roughly equivalent to those used in this country in the mid-sixties.

The angulographic method of measurement has proven utility, as demonstrated in a

---

Mr. Selker is director of physical therapy services, University Hospital, Texas A & M University, College Station, TX 77843.

number of practical applications. With this method, Borozdina and her associates[2,4] have assessed not only the compensatory mechanisms of patients with unilateral arthrosis deformans of the hip joint but also the effectiveness of the operative procedure of alloplasty of the hip joint for patients with rheumatoid arthritis.

PHASE DURATIONS OF GAIT CYCLE

A technique reported frequently to record the various temporal relations of the components of the gait cycle relies on an apparatus called the podograph or chronometer of step phase. This apparatus is not essentially different from the footswitch metal-walkway equipment employed in this country and was developed by Cherskov in 1952.[1] Since then, various elaborations of this basic method and apparatus have been devised,[4-6] the exact details of which have not been described clearly in the Soviet literature. Clearly, however, this method has developed to the point that now the various temporal components of the gait cycle not only are clearly displayed on either an oscillograph or a pen-recorder printout but also are adequately registered by using only one channel of these display instruments. The characteristic deflections from the horizontal axis, observed on a printout or oscilloscope, reflect the changes in in-circuit current that are produced by the definitive unit resistance values assigned to the foot switches on the heel and toe.

With this podographic method, the investigator Panfilov uncovered a quantitative dependence of time variables (duration of single support, swing, and double support) on cadence.[7] The subjects in Panfilov's study walked to the light-sound signal of a photo-phono stimulator while wearing over their shoes thin rubber galoshes with the mounted contact pickups of the podogram. A second oscillograph and a dynamometric platform were employed to check the period of double support.

In addition to this application to normal subjects, the podographic method has found wide application in investigations of pathologic gaits. Podography has been applied in determination of the temporal gait characteristics of patients with flat feet, destructive forms of tubercular coxitis, and various lower extremity pathological conditions.[2,4-6,8]

SUPPORT REACTIONS

Soviet gait researchers have used electrotensometers to measure the loading or support reactions on the heel and forefoot during walking. These force transducers have been placed on the subject's support surface, that is, the bottom of the shoe as tensometric pickups or, more often, on the walking surface mounted as force plates that are part of a shifting dynamometric or force platform. Not uncommonly, the recordings of such instrumentation are displayed on an oscillograph or pen-recorder printout. With the former method of display, Apshtein and Potikhanova have examined the distribution of weight bearing on the support surfaces of the foot in normal and flat-footed subjects.[9] Lenskii and Myakotina have employed an eight-channel pen recorder in their investigation of the distribution of weight bearing, duration of step phases, and hip and knee angles in patients with equinus, calcaneus, and drop-foot deformities.[10] In this study, the force platform type of electrotensometry was used in combination with podography and angulography to assess the effectiveness of the operation of posterior arthrorhysis of the ankle joint.

Engineers in the USSR have developed a silicon tensometric detector or force transducer that perhaps is the most highly sensitive transducer in existence. This transducer is of the crystal type and is called a nesistor.[11]

ANGULAR DISPLACEMENTS OF THE PELVIS AND VERTEBRAL COLUMN

Cyclogramometry, developed by Popova in 1940, has been described by Soviet researchers as the most objective and precise method of measuring variables in gait research.[3] This technique is essentially not considerably different from interrupted light photography. The linear and angular displacements of joints and body segments and their respective velocities and accelerations may be determined through an analysis of the successively displaced images of the subject on the scaled cyclogram picture or photograph. Bernshtein has used this chronocyclograph technique extensively to study the linear and angular displacements of various parts of the body, including the spine and pelvis, during locomotion.[12]

Kosilov introduced the use of a cine-apparatus with modifications to produce the resultant cinecyclogram of the walking cycle.[3] This method is capable of producing information on the linear and angular displacements of various body segments, their respective velocity and acceleration, and the durations of the various phases of the gait cycle. The various photometric techniques used in the Soviet Union appear to be not unlike those used by gait researchers in this country to measure the individual variables of gait phase activity.

With respect to nonphotometric techniques, the researcher Belenkii developed an apparatus which makes use of a free gyroscope, weighs only 800 grams, and is capable of measuring the pelvic or vertebral rotational displacements in three planes to within ± 0.5 degrees.[13] This mechanism is capable of recording the rotational displacements in one plane at a time, but the recording axis may be reoriented to allow recording in each of the other two planes. The proportional changes in in-line current that these rotations produce may be displayed on an oscillograph that is connected to the apparatus. The apparatus, which includes the gyroscope and its housing, a fixture mechanism, a gauge, and a step calibrator, is simply aligned and fastened to the subject's bony landmarks (iliac crests or transverse processes) by an elastic belt.

The utility of the gyroscopic apparatus is demonstrated in a number of applications. Belenkii has used the device to examine the pathologic gaits of patients with scoliosis, hip and ankle dislocations, and prostheses.[13] Clearly, the method may also be used to monitor the maintenance of proper pelvic posture during the ambulation of normals.

Approximately one year after his original description of this method, Belenkii described a further elaboration of the basic apparatus for the recording of the angular displacements of the spine and pelvis during walking.[12] This newer technique also employs a gyroscope in addition to a special instrument containing tensometric pickups or force transducers to record the angular displacement of the vertebrae T2 and T11 in relation to the pelvis. The sophistication of this nonphotometric technique is apparent. The equipment not only registers the rotations of the vertebrae and the pelvis separately and the rotation of the vertebrae in relation to the pelvis but also allows the recording of fine angular displacements of individual vertebrae.

STEP LENGTH AND OTHER LINEAR DISPLACEMENTS

A number of methods of measurement, already discussed, may be used to determine the step and stride length during walking. These lengths may be determined from the footswitch-produced deflections on an oscilloscope screen or they may be measured directly from the photograph of activity produced by various photometric techniques. The linear displacements of the head and other body segments may be determined by the latter method. Another method of recording step and stride length data is that of measuring the displacement of ink markings produced on a plain paper walkway by ink pads attached to the soles of the feet.

A technique called *ikhnography* has found increasingly wide applications in gait analysis research. The ikhnogram is produced on a 50- by 600-cm paper pathway by India ink from a plastic, balloonlike, syringe self-recorder fastened to each leg.[14] The syringe sends ink to a thin open-ended tube that leads to the relative center of support, just anterior to the heel. The perimeter of the sole and heel of each shoe is soaked with ink mixed with sand. This method allows recording to occur not only during support phase but also during swing phase. The horizontal plane trajectory of the swing extremity, therefore, may be examined objectively, as well as the step length, stride length, and the frontal plane component of these lengths. The last named value is the transverse distance from the midline, in the general direction of walking and passing through the relative center of support of the stance footprint, to the swing marking of the same foot. The uncomplicated technique of inkhnography has been shown to be a valuable adjunct in the investigation of prosthetic problems and in the improvement of prosthetic fitting.[14]

MUSCULAR ACTIVITY

An examination of the use of electromyography (EMG) showed the tendency to use EMG in conjunction with other methods of variable measurement. One group of Soviet gait researchers studied various muscles of the lower extremity by EMG in conjunction with podography of the phases of the step cycle, angulography of the joints of the lower extremity, and cinecyclography of the entire walking cycle.[3]

Slavutskii and Borozdina, in a study of gait with above-knee and below-knee prostheses, simultaneously registered the recordings of the electrical activity of muscles through EMG, the data of the angulogram of the angular displacements of the joints of the lower extremity, and the podogram of the temporal relations of phases of the step cycle, as well as the vertical and longitudinal components of the support reactions via force transducers.[15] A detailed description of the method of quantifying EMG activity is available elsewhere.[16]

MISCELLANEOUS TYPES OF ANALYSIS

Soviet gait researchers have made use of mathematical models for the determination, explanation, and prediction of changes in the force moments and energy expenditure of ambulation with prostheses of different constructions.[17] Such modeling has resulted in calculation of the ratio of the means and the mean square deviation values (and their dispersions) of the angular displacements of the joints of the lower extremities.[18] This ratio allows evaluation of the angular motion in the lower extremity joints. In this elaborate work, the angular displacements of the hip, knee, and ankle were recorded on an oscillogram which was processed by a computer for calculation of the mathematical functions. Another gait researcher has expressed mathematically the angular displacements of the hip, knee, and ankle joints during gait as a function of time.[3]

A number of mathematical constructs appear to be used quite frequently in Soviet gait research. The coefficient of rhythm is the ratio of the swing duration of one extremity to the swing duration of the other.[4, 5, 15] The coefficient of similarity is the ratio of the energy expenditure during prosthetic ambulation to the energy expenditure of normal ambulation for a specified joint and its force moments.[17] The coefficient of asymmetry, the ratio of the support reaction or force on the affected extremity to the sound extremity,[19] and the coefficient of support, the ratio of support duration of one extremity to the other,[8] have been applied in investigations of the gait of subjects with spondylolisthesis and congenital hip dislocation, respectively. These constructs provide the investigator with a quantitative index of the degree of disturbance of asymmetrical and symmetrical processes of gait phase activity.

CONCLUSION

This review of Soviet gait research literature shows that the various methods of measurement are extensively and practically applied. The methods are often used to evaluate and justify various forms of surgical intervention. These measurement techniques are frequently helpful in improving the quality of prosthetic fitting. Their use in the evaluation and study of pathological conditions of the walking mechanism has been demonstrated repeatedly.

A wealth of information regarding methods of measurement is available to those who examine the gait research produced in the USSR in the last decade. Gait researchers may best serve their purpose by incorporating information from such a rich resource into the formulation, design, and operation of their investigations of human walking.

REFERENCES

Editor's note. Abstracts of most of the articles cited in this presentation appear in the "Abstracts of Current Literature" section of this issue.

1. Cherskov MI: 2nd scientific session of the central scientific-research design and planning institute of prosthetic and orthotic appliances. Moscow, p.49, 1952
2. Borozdina AA, Panova MI: Biomechanical indices of walking prior to and following alloplasty of the hip joints in patients with rheumatoid arthritis. Ortop Travmatol Protez 33:9-12, January 1972
3. Bukreeva DP, Kosilov SA, Tambieva AP: On age related characteristics of the biodynamics and control of gait. Fiziol Zh SSSR Sechenov 55:77-86, January 1969
4. Borozdina AA, Travkin AA: Compensatory-adaptive mechanisms in patients with unilateral arthrosis deformans of the hip joint. Ortop Travmatol Protez 26:15-20, October 1965
5. Chenskikh EP, Nikolaenko NK: Biomechanical characteristics of gait in destructive forms of tubercular coxitis. Ortop Travmatol Protez 32:38-41, January 1971
6. Lepkhina LP: Walking peculiarities of patients with flat foot according to podography data. Ortop Travmatol Protez 29:66-69, February 1968
7. Panfilov VY: Investigation of the biomechanical parameters of human walking: 1. Time structure of the step. Biofizika 15 (5):927-931, 1970
8. Polonskii MN, Tikhonenkov ES: Peculiarities of the statics and kinematics of children with congenital subluxation and luxation of the hip before and after treatment. Ortop Travmatol Protez 30:49-55, August 1969
9. Apshtein ZV, Potikhanova GG: A study of the supportive function of the anterior part of the foot during walking. Ortop Travmatol Protez 25:60, April 1964
10. Lenskii VM, Myakotina LI: Some biomechanical indices of the efficacy of operation of posterior arthrorhysis of the talo-crural joint. Ortop Travmatol Protez 27:41-44, January 1966
11. Gordetskii AF, Shadrin US, Khazan AD: "Nesistor" silicon tensometric detectors highly sensitive transducers. Biomed Eng (Russ) 3:33-35, May-June 1967
12. Belenkii VE: Study of the movements of the spine and pelvis during walking. Ortop Travmatol Protez 32:37-43, July 1971
13. Belenkii VE: A method of recording the angular pelvic displacement during walking. Ortop Travmatol Protez 31:67-69, May 1970
14. Shuliak IP, Voltskaya II: Ikhnography with self-recorders in the study of walking in normal individuals and after prosthesis. Ortop Travmatol Protez 32:67-69, January 1971
15. Slavutskii IL, Borozdina AA: A complex quantitative study of the muscle electric activity and elements of kinematics and dynamics of walking: 2. Study of gait on leg and thigh prostheses. Ortop Travmatol Protez 27:52-60, October 1966
16. Slavutskii IL, Borozdina AA: A complex quantitative study of the muscle electric activity and elements of kinematics and dynamics of walking: 1. Study of Normal Walking. Ortop Travmatol Protez 27:1-51, October 1966
17. Naroditzkaya RE: Correlation of the force moments determining the movements of the hip joint of the truncated limb and the knee joint part of the prosthesis. Ortop Travmatol Protez 32:18-20, July 1971
18. Lialin VS, Matveev AP: Correlation analysis of the kinematic-dynamic structure of human gait. Fiziol Zh SSSR Sechenov 57:218-222, February 1971
19. Glazirin DI, Myakotina LI: Objective evaluation of some features of spondylolisthesis while standing and walking. Ortop Travmatol Protez 30:66-69, April 1969

Basic Research

Clinical Determination of Energy Cost and Walking Velocity via Stopwatch or Speedometer Cane and Conversion Graphs

DAVID H. NIELSEN,
DAVID G. GERLEMAN,
LOUIS R. AMUNDSEN,
and DAVID A. HOEPER

Work rate and relative exercise intensity are basic considerations in developing optimal exercise training prescriptions in cardiopulmonary rehabilitation. Velocity and energy cost of walking are directly related to work rate and indirectly related to exercise intensity. This article describes two methods of measuring walking velocity and estimating oxygen uptake. The validity and accuracy of the methods are discussed. We believe both methods are clinically useful. Because one method uses a specially instrumented speedometer cane, technical information concerning its design and construction is also included.

Key Words: Energy metabolism, Gait, Cardiovascular system, Exercise test.

Patients hospitalized with ischemic heart disease, chronic respiratory disease, or any condition that requires prolonged bed rest need supervised physical reconditioning. Various clinical studies have shown that mobilization exercises benefit these types of patients because this form of activity prevents venous thrombosis, pulmonary embolism, muscular catabolism, and other complications such as orthostatic intolerance.[1,2] Mobilization exercises usually include breathing maneuvers, assisted and active range-of-motion activities, and a gradual progression from lying to sitting, standing, and graded walking.

Relative exercise intensity and work rate are basic considerations in the rehabilitation process. Heart rate is the standard indicator of total body exercise intensity.[3] Heart rate is usually monitored by surface electrodes and radiotelemetry or by manual palpation of the carotid or radial arteries. Commercially available devices, such as the Exersentry* and the Insta-Pulse† heart rate monitors, are also used.

Dr. Nielsen is Associate Professor, Physical Therapy Education, College of Medicine, The University of Iowa, Iowa City, IA 52242 (USA).

Mr. Gerleman is Engineer, Physical Therapy Education Staff, College of Medicine, The University of Iowa, Iowa City, IA 52242.

Dr. Amundsen is Director, Graduate Programs in Physical Therapy, Department of Physical Medicine, University of Minnesota, Minneapolis, MN 55455.

Mr. Hoeper is Director, Cardiology Service, Memorial Hospital of South Bend, 615 N Michigan St, South Bend, IN 46601.

This article was submitted October 15, 1980, and accepted September 11, 1981.

* Respironics, 650 Seco Rd, Monroeville, PA 15146.

† Biosig, Inc, PO Box 651, NDG Montreal, Quebec, Canada H4A 391.

Estimation of energy cost, in terms of the oxygen uptake of physical activities, by using published tables and charts[3] is one customary method of determining work rate.[4] The accuracy of such estimates is related to the degree of standardization of the activity. In the absence of pathological gait deviations, walking is a reasonably standardized activity and energy cost determinations are fairly reproducible.[5]

In the past, energy cost data were reported for continuous walking that was typically conducted on large circular or oval courses.[5-9] Only recently has information been available for segmental (back-and-forth) walking that does not require facilities uncommon to the clinical setting (first author, DHN, unpublished study).

The main purpose of this article is to describe two clinical methods of measuring velocity and estimating energy cost of walking. The first method requires a stopwatch and a segmental walkway; the second incorporates an electromechanically instrumented speedometer cane for monitoring continuous walking velocity. Because this cane is not commercially available, this article also provides information for the construction and use of this cane.

METHOD

Segmental Walking

A simple and practical method of determining segmental walking velocity uses a stopwatch and a segmental walkway. A hallway or clinic gymnasium

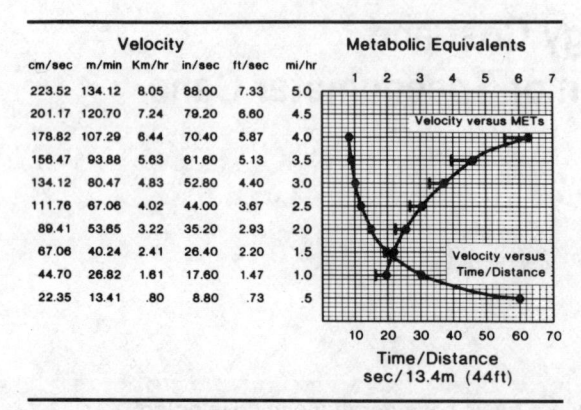

Velocity						Metabolic Equivalents
cm/sec	m/min	Km/hr	in/sec	ft/sec	mi/hr	
223.52	134.12	8.05	88.00	7.33	5.0	
201.17	120.70	7.24	79.20	6.60	4.5	
178.82	107.29	6.44	70.40	5.87	4.0	
156.47	93.88	5.63	61.60	5.13	3.5	
134.12	80.47	4.83	52.80	4.40	3.0	
111.76	67.06	4.02	44.00	3.67	2.5	
89.41	53.65	3.22	35.20	2.93	2.0	
67.06	40.24	2.41	26.40	2.20	1.5	
44.70	26.82	1.61	17.60	1.47	1.0	
22.35	13.41	.80	8.80	.73	.5	

Fig. 1. Time, velocity, energy cost conversion graph for segmental walking (1 MET = 3.5 ml O_2/kg·min).

Velocity						
cm/sec	m/min	Km/hr	in/sec	ft/sec	mi/hr	
223.52	134.12	8.05	88.00	7.33	5.0	
201.17	120.70	7.24	79.20	6.60	4.5	
178.82	107.29	6.44	70.40	5.87	4.0	
156.47	93.88	5.63	61.60	5.13	3.5	
134.12	80.47	4.83	52.80	4.40	3.0	
111.76	67.06	4.02	44.00	3.67	2.5	
89.41	53.65	3.22	35.20	2.93	2.0	
67.06	40.24	2.41	26.40	2.20	1.5	
44.70	26.82	1.61	17.60	1.47	1.0	
22.35	13.41	.80	8.80	.73	.5	

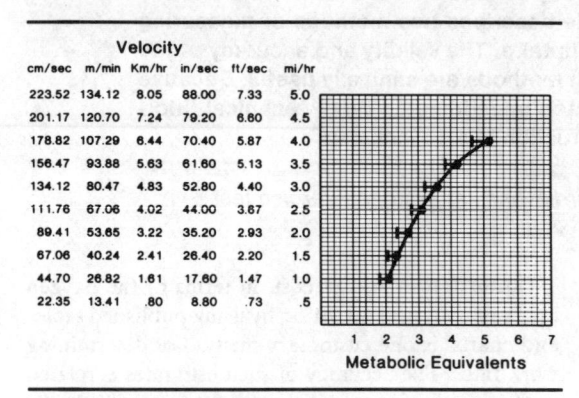

Fig. 2. Energy cost versus velocity conversion graph for continuous walking (1 MET = 3.5 ml O_2/kg·min).

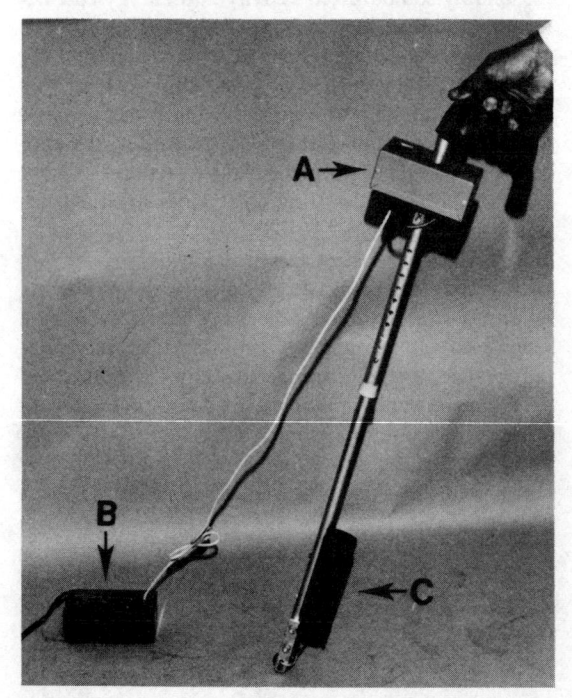

Fig. 3. Speedometer Cane: A—electronic counter, B—battery recharger, C—rechargeable battery.

is sufficient for this method. The floor should be flat and have a smooth, low-resistance surface (hardwood or tile). The walkway course should include a segment 13.41 m (44 ft) long designated with masking tape fixed to the floor. The walking distance should extend 2 to 3 m at each end to eliminate measurement error associated with velocity changes resulting from turning (DHN, unpublished study).

The procedure for determining walking velocity involves timing the patient while he walks back and forth over the measured walkway. We have found that three to five measurements provide reliable results. If heart rate is measured, the patient should walk for a minimum of three to five minutes to ensure physiological steady state conditions. Stable heart rate measurements in a previous study have verified steady state conditions for this type of activity (DHN, unpublished study). Walking velocity, based on the mean of the time measurements, can be graphically determined from the velocity versus time/distance curve presented in Figure 1. (Because various measurement units are referred to in the literature, a measurement unit conversion table was included in the figure.) Walking velocity might also be calculated using the following equation: walking velocity (cm/sec) = 134 cm/mean time (sec). For instance, if the observed mean of three time measurements were 30 seconds, the walking velocity, according to the graph, would be 44.70 cm/sec (1 mi/hr). Mathematically determined, walking velocity would be 134 cm/30 sec = 44.70 cm/sec. Once the walking velocity is known, energy cost can be estimated from the energy cost versus velocity conversion graph presented in Figure 1 or it can be predicted from the following regression equation: E (segmental walking) = .001362 v^2 + 5.22, where E equals energy cost in ml O_2/kg·min and v equals velocity in m/min (DHN, unpublished study). To express the energy cost in terms of metabolic equivalent units (METs), divide by 3.5 ml O_2/kg·min. The corresponding energy cost as determined from the graph for the preceding example (v = 44.70 cm/sec) is 2 METs.

Continuous Walking with Speedometer Cane

If continuous walking is employed in the reconditioning program, determining walking velocity by the same stopwatch method as above has certain drawbacks because the repeated distances must be measured and visibly designated on the floor. In addition, patient freedom is restricted because the patient is confined to a specific walking course. A mobile velocity measuring device, such as the speedometer cane described below, may be more appropriate. The cane could be held by the patient or by the therapist who paces the patient. With this method, accurate, immediate, and continuous feedback of the direct measurement of walking velocity would be provided.

Energy cost measurements for continuous versus segmental walking have been shown to vary significantly (DHN, unpublished study). This invalidates the previously referred to conversion graph and prediction equation for application with the continuous mode of walking. The observed differences appeared to be related to walking speed. The velocities investigated were between 44.70 cm/sec (1.0 mi/hr) and 178.82 cm/sec (4.0 mi/hr). The energy cost for segmental walking was significantly greater than that for continuous walking when walking velocities were equal to and above 111.76 cm/sec (2.5 mi/hr). The differences were attributed to increased physical activity associated with stopping, starting, and turning movements inherent in back-and-forth segmental walking.

The appropriate graph for estimating energy cost from continuous walking velocity measurements is presented in Figure 2. The corresponding prediction equation is E (continuous walking) = $.00094 v^2 + 6.10$, where E equals energy cost expressed in ml O_2/kg·min and v equals velocity in m/min (DHN, unpublished study).

Speedometer Cane

Because the speedometer cane is not available commercially and has not been described elsewhere, it is described below. Necessary technical information is discussed, including a mechanical drawing, an electrical schematic, and photographs for construction of the device. Instrument calibration and validation are also discussed.

Description. A standard, adjustable aluminum cane was adapted (Fig. 3). A revolving brass wheel was attached to the foot of the cane, and an electronic digital rate counter was attached to the handle. A rechargeable battery was mounted on the distal portion of the shaft of the cane. The wheel revolved in the sagittal plane and activated the counter as the cane was pushed along the floor.

A small rubber O-ring was placed around the perimeter of the revolving wheel. The circumference of the wheel, including the O-ring, was 10 cm. Ten equally spaced holes were drilled around the perimeter of the wheel. A precision bearing was pressed into a hole drilled in the center of the wheel, and a small metal axle shaft was inserted through the bearing. Two opposing metal plates, bolted to the foot of the cane, served as mounting struts for the axle shaft. The ends of the axle shaft were inserted into small holes drilled in the mounting plates. Two metal sleeves were placed concentrically on the axle shaft, one on each side of the wheel, to maintain the wheel

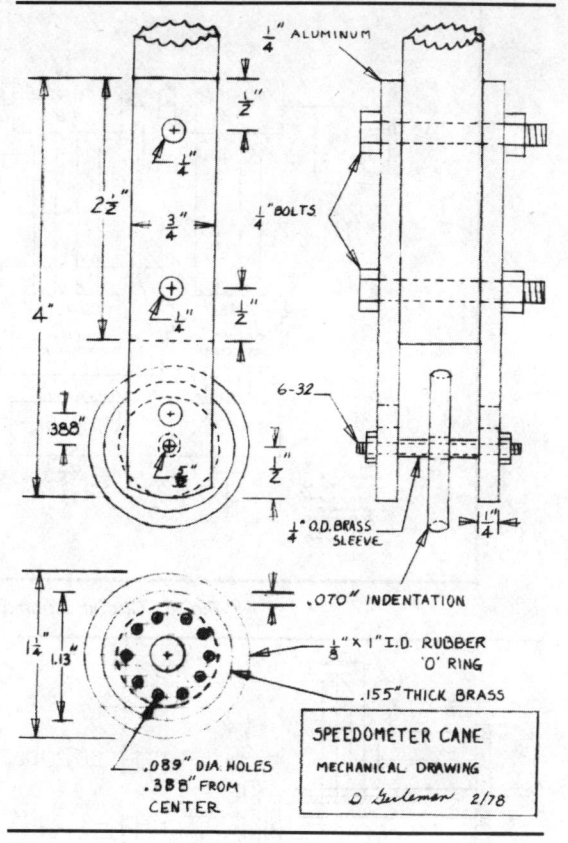

Fig. 4. Physical specifications, foot of speedometer cane.

in the center position between the mounting struts (Figs. 4, 5). A miniature ultraviolet light source was mounted on the inner side of one of the metal plates. The light source was positioned in line with the holes

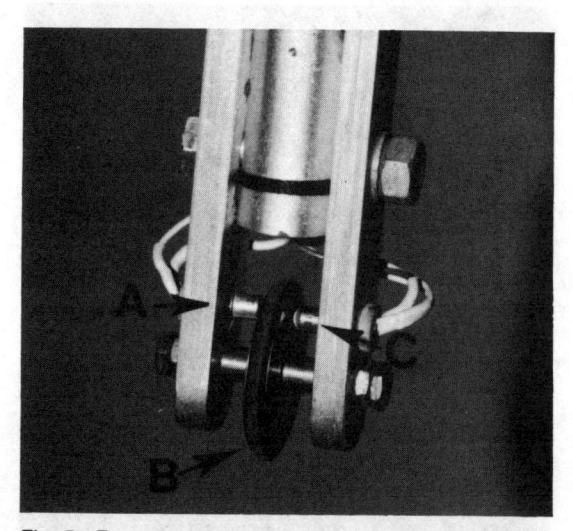

Fig. 5. Foot of speedometer cane: A—miniature ultraviolet light, B—revolving wheel, C—miniature electrophoto cell.

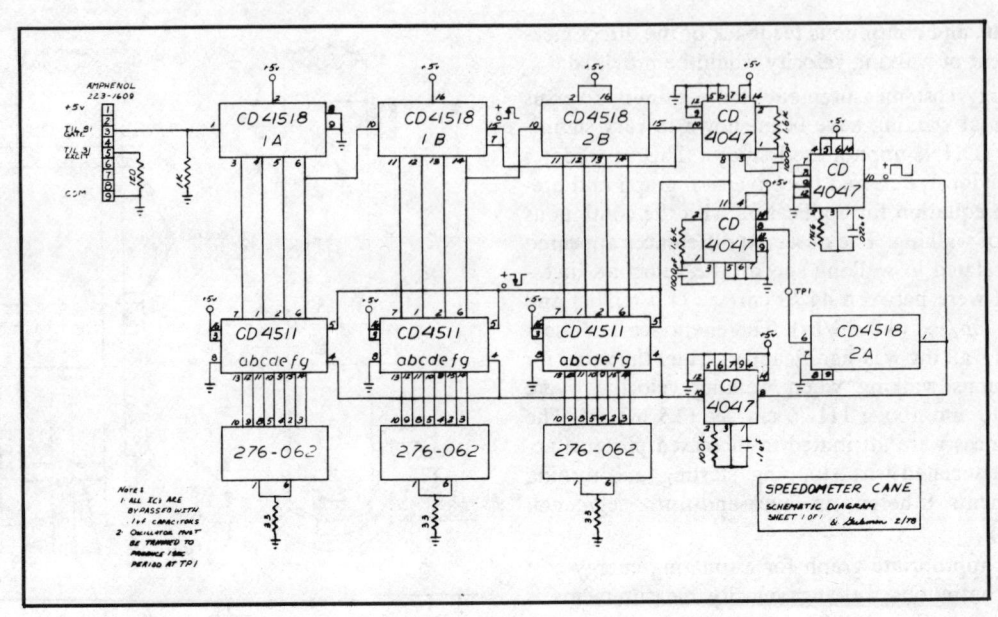

Fig. 6. Circuit schematic of speedometer cane.

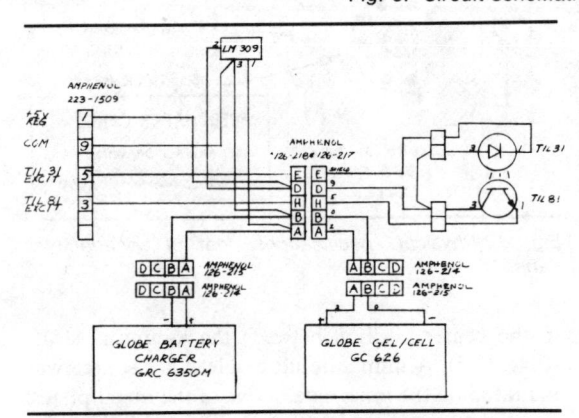

Fig. 7. External wiring diagram of speedometer cane.

Fig. 8. Digital display of electronic counter.

drilled around the perimeter of the revolving wheel. A miniature photocell was mounted on the opposing metal plate in line with the light source. With this arrangement, wheel revolutions caused interruptions in the light beam (10 interruptions each revolution) and alterations in the electrical output (electrical pulses) of the photocell, which activated an electronic counter (Figs. 6, 7). Each second, the count was shifted to a digital display (Fig. 8) and the counter was reset to zero. The display represented the linear velocity of the cane in centimeters per second. The theoretical measurement precision was ± 1 cm/sec (.02 m/hr). A 6-V rechargeable battery powered the system. The functional capacity of the fully charged battery was about 15 hours. The battery charger had fast and standby recharge modes.

Calibration and Validation. The speedometer cane was calibrated on a motor-driven treadmill at four velocities (Tab. 1). The cane was held in a steady position, the wheel of the cane riding on the revolving belt of the treadmill. The velocity as measured with the speedometer cane was compared to the velocity of the treadmill belt, which was simultaneously determined using an electronic timing system. The timing system measured the time intervals between the belt revolutions of the treadmill. The mean time intervals of three trials of 25 belt revolutions were determined at each of four velocities. The treadmill belt velocity was calculated from the mean time measurements and the measured circumference of the belt. The maximum percent error between the speedometer cane measured velocity and treadmill determined velocity was less than 1.5 percent.

The ability of the speedometer cane to be used to regulate walking velocity during repeated trials was analyzed to determine instrument validity. One investigator, controlling his walking velocity by using the cane, walked repeatedly over a 20-m (65.62-ft) distance in a hallway at seven different target velocities (Tab. 2). The time required to walk that distance was measured with an electronic timing system. Five walking trials were performed at each velocity. Walking velocities were calculated by dividing the mean of each of the five time measurements into the 20-m distance. (This distance was expressed in centimeters, ie, 2,000 cm/mean time.) A comparison of the target velocities with the calculated velocities produced a maximal error of less than 2 percent.

DISCUSSION

The accuracy of the two methods of measuring walking velocity and estimating energy cost described above is dependent on the measurement error of the velocity determinations and the intersubject variability inherent in oxygen uptake assessment. The relative velocity measurement error (velocity error/velocity) for the stopwatch method, because of its derivation, increased with walking speed and, in general, was greater than that observed for the speedometer cane. As an illustration, if one assumes 0.5 second as a reasonable stopwatch time error, the relative percent errors in calculating velocity for three general speeds—slow, 89.41 cm/sec (2.0 mph); medium, 134.12 cm/sec (3.0 mi/hr, equal to free pace or self-selected walking velocity); and fast, 178.82 cm/sec (4.0 mi/hr)—would be ± 2.3 percent, ± 3.4 percent, and ± 5.0 percent, respectively. Stopwatch velocity measurement errors of these magnitudes would produce a maximal error in estimating energy cost of ± 7 percent, which would equal ± 0.4 METs at the fast speed and less at the slower speed. In contrast, the observed 2 percent velocity measurement error for the speedometer cane would result in a ± 3 percent error in estimating energy cost. This means a maximal error of ± 0.2 METs for the fast velocity and less for slower speeds.

The magnitude of intersubject variability inherent in oxygen uptake determinations for walking is reflected by the standard deviations presented in Figures 1 and 2. The energy cost for segmental walking at higher speeds varied more than that for continuous walking (s = 0.29–0.74 METs, mean s = 0.44 METs vs s = 0.27–0.49 METs, mean s = 0.34 METs). Because these values were larger than the maximal error values attributed to the velocity measurement errors, they were the primary determinants of the accuracy of energy cost estimates for the two methods described here. Accordingly, the accuracy for estimating energy cost based on the mean standard deviations for the segmental walking/stopwatch method was ± 0.4 METs and ± 0.3 METs for the continuous walking/speedometer cane method. In the worst case, based on the maximum standard deviations, the accuracy of each method was 0.8 METs and 0.5 METs, respectively. We find either pair of accuracy values acceptable for clinical application.

The use of energy cost predictions based on walking velocity requires that the patient have a normal gait. Gait abnormalities alter walking efficiency and oxygen uptake values. The two methods of estimating energy cost described above are thus useful only for patients with normal gait performance. The reader is reminded that the energy costs for segmental and continuous walking are different; therefore, conversion graphs and prediction equations for these two modes of walking are not interchangeable.

Close supervision and regulation of exercise in reconditioning programs are mandatory to ensure patient safety and optimal progress. Guidelines for establishing appropriate activity and training levels are available.[10] The two methods of measuring walking velocity and estimating energy cost described here should facilitate implementation of these programs. Both methods have been used successfully in rehabilitation programs at the University of Iowa Hospitals. These methods enable systematic regulation of exercise, and the objective data provided expedites the documentation of patient progress and staff communication. The speedometer cane has been well-received by both clinicians and patients. Audio biofeedback for programing exercise walking velocity and a heart rate monitor are being considered as possible adaptations for the cane.

TABLE 1
Speedometer Cane Calibration Data

Treadmill Velocity		Speedometer Cane Velocity		% Error
cm/sec	mi/hr	cm/sec	mi/hr	
40.84	0.91	41.0	0.92	0.39
87.47	1.96	88.5	1.98	1.18
136.57	3.06	137.5	3.08	0.68
184.97	4.14	186.0	4.16	0.56

TABLE 2
Speedometer Cane Validation Data

Target Velocity		Calculated Velocity		% Error
cm/sec	mi/hr	cm/sec	mi/hr	
44.70	1.00	44.97	1.01	0.60
67.06	1.50	66.91	1.50	0.22
89.41	2.00	87.87	1.97	1.72
111.76	2.50	110.07	2.46	1.51
134.11	3.00	131.49	2.94	1.95
156.46	3.50	155.52	3.48	0.60
178.82	4.00	175.75	3.92	1.72

REFERENCES

1. Kellerman JJ: Rehabilitation of patients with coronary heart disease. In Sonnenblick EH, Lesch M (eds): Exercise and Heart Disease. New York, NY, Grune & Stratton Inc, 1977, p 184
2. Wenger NK: Early ambulation after myocardial infarction: Rationale, program components and results. In Wenger NK, Hellerstein HK (eds): Rehabilitation of the Coronary Patient. New York, NY, John Wiley & Sons Inc, 1979, pp 55-56
3. Naughton JP, Hellerstein HK (eds): Exercise Testing and the Training of Apparently Healthy Individuals: A Handbook for Physicians. New York, NY, The Committee on Exercise, American Heart Association, 1972
4. Zohman LR: Early ambulation of post-MI patients: Montefiore Hospital. In Naughton JP, Hellerstein HK, Mohler IC (eds): Exercise Testing and Training in Coronary Heart Disease. New York, NY, Academic Press Inc, 1973, pp 330-331
5. Blessey RL, Hislop HJ, Waters RL, et al: Metabolic energy cost of unrestrained walking. Phys Ther 56:1019-1024, 1976
6. Corcoran PJ, Brengelmann GL: Oxygen uptake in normal and handicapped subjects in relation to speed of walking beside velocity controlled cart. Arch Phys Med Rehabil 51:78-87, 1970
7. Ralston HJ: Energy-speed relation and optimal speed during level walking. Int Z Angew Physiol 17:277-283, 1958
8. Grimby G, Soderholm B: Energy expenditure of men in different age groups during level walking and bicycle ergometry. Scand J Clin Lab Invest 14:321-328, 1962
9. Jankowski LW, Ferguson RJ, Langelier M, et al: Accuracy of methods for estimating O_2 cost of walking in coronary patients. J Appl Physiol 33:672-673, 1972
10. Amundsen LR: Establishing activity and training levels for patients with ischemic heart disease. Phys Ther 59:754-758, 1979

Basic Research

Use of the Krusen Limb Load Monitor to Quantify Temporal and Loading Measurements of Gait

STEVEN L. WOLF and
STUART A. BINDER-MACLEOD

The purposes of this investigation were to determine whether the temporal and force measurements from the Krusen Limb Load Monitor produced clinically reliable data and to begin identifying the factors that determine the monitor's reliability. Temporal and loading measurements were made from the output of the Krusen Limb Load Monitor and compared to values obtained from a calibrated force platform. Such comparisons were made for 30 steps taken by two subjects on three separate occasions and from the same two subjects plus a third subject for 100 consecutive steps. For most measures, mean values from the limb-load monitor were significantly different from those recorded from the force platform. From a clinical perspective, however, the range of measures was narrow for the 95 percent confidence level of the observed differences for the temporal components of stance between the limb-load monitor and force platform, with the narrowest range of measures related to the appropriateness of "fit" of the limb-load monitor force plate within the shoe. The loading components of stance showed a relatively wide 95 percent confidence interval that appeared unrelated to fit. Thus, given a "good fitting" force plate insert, the therapist can make clinically meaningful measurements of the temporal components of the stance phase of gait using the limb-load monitor.

Key Words: *Equipment, Gait.*

Demonstration of a therapeutic intervention's effectiveness in improving gait is ultimately dependent upon some quantitative assessment of appropriate loading and temporal changes in the gait pattern. In 1974, Smidt clearly delineated methods to analyze qualitative aspects of gait.[1] Since that time, numerous techniques based upon cinematography,[2-4] computer interface,[5-8] and telemetry[9] have been advocated as mechanisms for more accurately examining gait characteristics. These methods are costly, require unique skills to perform, and are excessively time-consuming.

Dr. Wolf is Coordinator, Biofeedback Programs, Emory University Regional Rehabilitation Research and Training Center, Emory University School of Medicine, 1441 Clifton Rd NE, Atlanta, GA 30322 (USA).

Mr. Binder-Macleod is Instructor, Division of Physical Therapy, Department of Community Health, Emory University, and is Research Associate, Emory University Regional Rehabilitation Research and Training Center.

This work was supported by Grant No. 16-P-56808/4-16 from the Institute for Handicapped Research, Department of Health and Human Services, Washington, DC 20005, and was presented in part at the annual conference of the American Physical Therapy Association, Washington, DC, on July 1, 1981.

This article was submitted September 10, 1981, and accepted November 23, 1981.

Therefore, at present, such sophisticated techniques are not of practical value to the clinician who has limited time and resources. Nelson and Tucker have attempted to remedy these concerns by developing a functional ambulation profile,[10] but the scoring system used in this procedure is still time-consuming and requires several calculations. Perry et al have developed a device for temporal components of gait[11]; however, this device is expensive.

The Krusen Limb Load Monitor (LLM) was designed to provide patients with auditory feedback as weight exceeding a predetermined magnitude is placed on a force plate inserted within a shoe.[12] This device, and the modifications that we have added,[13] are quite effective in helping patients increase unilateral weight bearing or walk with more equal bilateral weight distributions and stance times.

The LLM enables the clinician to examine the analog (voltage) signal generated by weight placed upon the force plate. This signal can be displayed on an oscilloscope or, more appropriately for permanent recording, on a strip-chart recorder. By comparing the force exerted through the plate to a known weight and by knowing the speed of paper movement through the strip-chart recorder, one can measure the loading and temporal elements of the stance phase, respectively. The recorded temporal and force components of the LLM are depicted in Figure 1. Because the LLM (model 101*) and the accompanying shoe inserts are comparatively inexpensive, and because the output from the LLM is easily connected to any strip-chart recorder (such as a single-channel EEG machine usually accessible in most hospitals), the clinician can have an inexpensive and permanent record of unilateral gait or, with two LLMs and a dual-channel strip-chart recorder, a record of bilateral gait activities. Total cost for two LLMs with three different sizes of shoe inserts and a quality two-channel strip-chart recorder is approximately $2,500. These recordings are easy to make and take less than 15 minutes.[14] Analysis of recordings requires an additional 15 minutes and, when repeated at predetermined intervals, provides a documentation of gait changes that can demonstrate evidence for efficacy of treatment.

Before investing in such an effort, however, one must determine whether this procedure yields reliable information. The purposes of this investigation were to determine whether the temporal and force measurements from the LLM produced clinically reliable data and to begin identifying the factors that determine the LLM's reliability.

METHOD

Evaluation of the LLM was performed by comparing the output of the LLM to the output of a standardized measuring device (force platform). The force platform is a stationary pressure-sensitive plate† approximately 40 cm by 59 cm (15.7 in x 23.2 in) positioned within a 3-m (9.8-ft) walkway. From our experience and from the information provided by the manufacturer, the force platform represents an extremely reliable and accurate measuring tool. The LLM consists of a pressure-sensitive force plate worn within the subject's right shoe and a small direct-current amplifier attached to the subject's belt. The subjects were trained so that only the right foot would strike the force platform. Subjects attempted to maintain a consistent speed as they walked down the 3-m runway onto the force platform. Therefore, one step that contained the "simultaneous" output of the LLM and force platform could be recorded for each trial.

The outputs of the LLM and force platform were recorded on two channels of a Gould brush 200 pen recorder.‡ The temporal and loading measures (Fig.

---

* Krusen Research Center, Moss Rehabilitation Hospital, 12th and Tabor Rd, Philadelphia, PA 19141.
† Kristal Instrument Corp, 2475 Grand Island Blvd, Grand Island, NY 14072.
‡ Gould Inc, 3631 Perkins Ave, Cleveland, OH 44114.

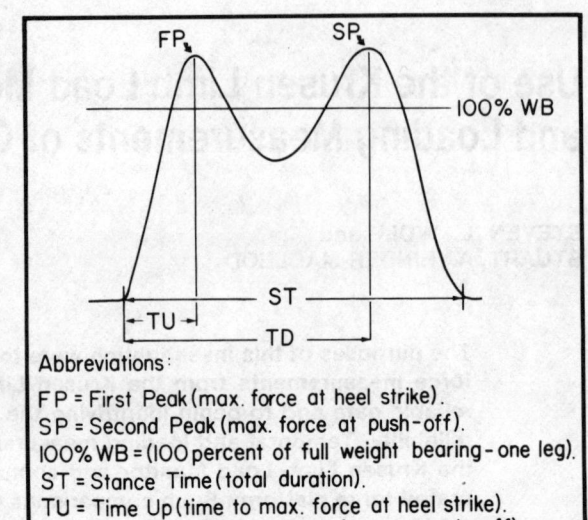

Fig. 1. Analog signal generated from output of LLM or force platform through the stance phase of gait. The line depicting 100 percent weight bearing (WB) is superimposed and determined during unilateral full weight bearing. FP and SP values are expressed as a percentage of 100 percent WB.

Abbreviations:
FP = First Peak (max. force at heel strike).
SP = Second Peak (max. force at push-off).
100% WB = (100 percent of full weight bearing - one leg).
ST = Stance Time (total duration).
TU = Time Up (time to max. force at heel strike).
TD = Time Down (time to max. force at push-off).

1) of each step were analyzed from this pen recording. A paper speed of 50 mm/sec (20 msec/mm) was used throughout the study. Force calibrations for the LLM and the force platform were performed for each subject before each recording session by determining the subject's 100 percent body weight (Fig. 1). All force measurements were calculated as a percentage of the subject's body weight. Analyses were performed on the differences between the data from the force platform and the LLM. Differences between each loading and temporal measure were computed by subtracting the force-platform value from the LLM value for each step. All data were measured by hand from the strip-chart recordings. These data points were then entered into a Univac 90/80 computer** for subsequent analyses using a Biomedical Computer Programs P-Series statistical package.[15]

Data were obtained from two men and one woman, all of whom had no previous lower extremity pathological condition. Right shoe inserts attached to the LLM were such that the woman (Subject 3) had the best fit of the foot plate insert and Subject 1 had the worst (loosest) fit. The appropriateness of the fit was purely subjective, and no attempt was made to quantify the measure.

The study was performed in two phases. In Phase 1, each male subject performed 10 trials on three different days. At each session the subjects wore the same shoes. For Phase 2, each of the three subjects walked for 100 recorded steps at one session.

** Sperry Univac, PO Box 500, Blue Bill, PA 19424.

TABLE 1
Differences Between Shoe-Insert and Force-Platform Values Computed for all Temporal and Limb-Loading Measures From 30 Steps Taken by Each Male Subject

Gait Measure	Subject	X̄	s	Paired t test			Range of 95% Confidence Interval	
				T	p	df		
Stance time	1	45.4 msec	47.0	5.19	<.0001	29	−48.6 to	139.4 msec
	2	32.1 msec	10.3	17.06	<.0001	29	11.5 to	52.7 msec
Time up	1	13.9 msec	34.9	2.15	<.04	29	−55.9 to	83.7 msec
	2	−9.7 msec	14.3	−3.71	<.001	29	−38.3 to	18.9 msec
Time down	1	−0.7 msec	41.2	−0.09	.93	29	−83.0 to	81.9 msec
	2	4.4 msec	21.7	1.11	.28	29	−39.0 to	47.8 msec
First peak	1	−06%	10	−3.18	<.004	29	−14 to	26%
	2	−03%	12	−1.59	.12	29	−27 to	21%
Second peak	1	−33%	07	−24.37	<.0001	29	−51 to	−23%
	2	−07%	09	−3.97	<.0001	29	−24 to	12%

Data Analysis

In each phase of the study, data for all steps taken by a subject were combined and paired t tests were performed for each of the variables. Results from the paired t tests reflect the ability of the LLM to produce a similar value on the average compared to the force-platform value and thus reflect the accuracy of the LLM. The calculations were made by subtracting the force-platform value from the LLM value. In the tables that follow, therefore, a positive value for the mean difference denotes that on the average the LLM value was larger than the force-platform value. Small p-values indicate a low probability that the mean difference would be as large as that observed if there is no difference between the "true" mean values of the LLM and the force platform. Although a small p-value means there is little likelihood the result occurred by chance, the magnitude of the difference must be considered in assessing the accuracy of the LLM compared to the force platform. Although statistically significant, a small mean difference may not be clinically important.

While the accuracy of the LLM on the average is important, of perhaps greater interest to the clinician is the magnitude of the difference that is likely to occur on any given step. For this reason, the 95 percent confidence intervals of the differences for each subject were examined. Assuming a normal distribution for the differences, the 95 percent confidence interval is calculated by mean ± 1.96 standard deviations. As the variability of the difference decreases, the narrower the 95 percent confidence limit becomes. The 95 percent confidence interval provides a range into which about 95 percent of future observations will fall. If the range of values in the 95 percent confidence interval are all within clinically acceptable limits, the clinician can be reasonably confident that a future LLM value will be close to the value that would have been gotten from using the force platform. The 95 percent confidence interval describes the reliability of the LLM.

RESULTS

Phase 1

Data analyses for each of the two male subjects were essentially identical, whether examined as three separate trials of 10 steps or grouped as a total of 30 steps. Table 1 shows the mean differences between shoe-insert and force-platform values, the p-values for paired t tests, and 95 percent confidence intervals for each loading and temporal measure from the 30 steps taken by each of the male subjects. With the exception of the time-down measurement, mean differences for all temporal and loading measures for Subject 1 were greater than for Subject 2. This indicates generally greater accuracy of the LLM for Subject 2 compared to the accuracy for Subject 1. Statistically insignificant differences between LLM and platform are noted for the time-down measurements of Subjects 1 and 2 and for the first peak of Subject 2. All other p-values were significant at $p \leq .05$.

The 95 percent confidence intervals for differences between shoe-insert and force-platform values are also depicted in Table 1. The 95 percent confidence interval is much more restricted among temporal values (stance time, time up, time down) for Subject 2 compared to Subject 1. In contrast, Subject 2 demonstrates a slightly wider range for the 95 percent confidence intervals for the loading measures than Subject 1.

Phase 2

Data analyses for the 100 steps taken by each subject are shown in Table 2. With the exception of the first-peak measurements, progressively smaller mean differences are observed for all measurements as one proceeds from Subject 1 to Subject 3. Examination of p-values now indicates that all mean differences between the shoe-insert and force-platform values, except time-down measurements for Subjects 2 and 3, are statistically different from zero and hence the LLM and force platform are not measuring the same values for each temporal and force characteristic of stance.

Comparing the 95 percent confidence intervals from Table 2 (100 steps) to Table 1 (30 steps) for Subjects 1 and 2 reveals that among temporal values, the range is expanded for Subject 1 and reduced for Subject 2. The first subject shows a narrower range for loading (first peak) and wider range for unloading (second peak) after taking 100 steps compared to taking 30 steps. The second subject, however, shows a narrower absolute range for both loading and unloading for taking 100 steps compared to taking 30 steps.

DISCUSSION

Following a review of the data, several questions arise. What is the source of the error between the shoe-insert and force-platform measures and why did this error change across subjects? In addition, What is the clinical relevance of the observed differences between these two measuring devices? (ie, How accurate and sensitive is the LLM?)

The sources of error between the shoe-insert and force-platform measures are difficult to explain. If some sort of regular bias was operating, one would, at the very least, expect the sign of the differences between the means (column 2 of Tabs. 1 and 2) to remain the same; that is, the insert values would consistently be either greater than (a positive number) or less than (a negative number) the force-platform values for all of the temporal or force measures. This did not occur. For each individual measure, both positive and negative values were observed. Similarly, excluding loading measures for Subject 2, inspection of collective temporal or loading measures (temporal measures grouped together and loading measures grouped together) for each subject fails to reveal a consistent bias. Thus, the data analysis suggests that some sort of consistent bias is not operating either within or across subjects. The source(s) of error must therefore be due to some variable factor(s).

One such source of variable error may be related to the fit of shoe insert within the shoe. Because a better fit would result in 1) less sliding of the insert within the shoe and 2) a larger percentage of the forces of the foot being translated through the shoe insert (less force transmitted directly through the shoe without passing through the transducer), we postulated the better the fit the more accurate the recording. This hypothesis was borne out by the results. Subject 3, who had the best fit, showed the smallest mean differences between shoe-insert and force-platform values and the narrowest confidence intervals for all measures. Thus the fit may, in part, be responsible for the accuracy and variability in the measures observed.

Another possible source of error may be explained within the data analysis procedure. During the data analysis, the results were read from a strip-chart recording with the finest gradation being 1 mm. With a paper speed of 50 mm/sec, each millimeter represents 20 msec in real time. Similarly, the amplitudes of the subjects' body weights were such that each millimeter represented up to 6.7 percent of the subject's body weight. Thus, a recording error of 20 msec or 6.7 percent of body weight could easily be explained by random variability in the evaluator's approximation of the recording to the nearest millimeter. Certainly, other possible sources of error exist.

TABLE 2
Differences Between Shoe-Insert and Force-Platform Values Computed for all Temporal and Limb-Loading Measures From 100 Steps Taken by Subjects 1 and 2 (men) and 3 (woman)

Gait Measure	Subject	$\bar{X}$	s	Paired t test			Range of 95% Confidence Interval	
				T	p	df		
Stance time	1	76.0 msec	50.5	15.04	<.0001	99	−25.0 to	177.0 msec
	2	19.8 msec	13.8	14.37	<.0001	99	−7.8 to	47.4 msec
	3	−3.6 msec	8.8	−4.08	<.0001	99	−21.2 to	14.0 msec
Time up	1	37.2 msec	44.8	8.31	<.0001	99	−52.4 to	126.8 msec
	2	−24.2 msec	12.8	−18.90	<.0001	99	−49.8 to	1.5 msec
	3	−14.0 msec	8.5	−16.42	<.0001	99	−31.0 to	3.0 msec
Time down	1	35.6 msec	49.7	7.16	<.0001	99	−63.8 to	135.0 msec
	2	−2.4 msec	14.8	−1.62	.109	99	−32.0 to	27.2 msec
	3	0.8 msec	9.6	0.83	.407	99	−18.4 to	20.0 msec
First peak	1	08%	08	10.17	<.0001	99	−8 to	24%
	2	−26%	10	−24.53	<.0001	99	−46 to	6%
	3	13%	05	25.98	<.0001	99	3 to	23%
Second peak	1	36%	08	−41.41	<.0001	99	−52 to	−20%
	2	−28%	07	−41.19	<.0001	99	−42 to	−14%
	3	−19%	07	−26.33	<.0001	99	−33 to	−5%

Basic Research

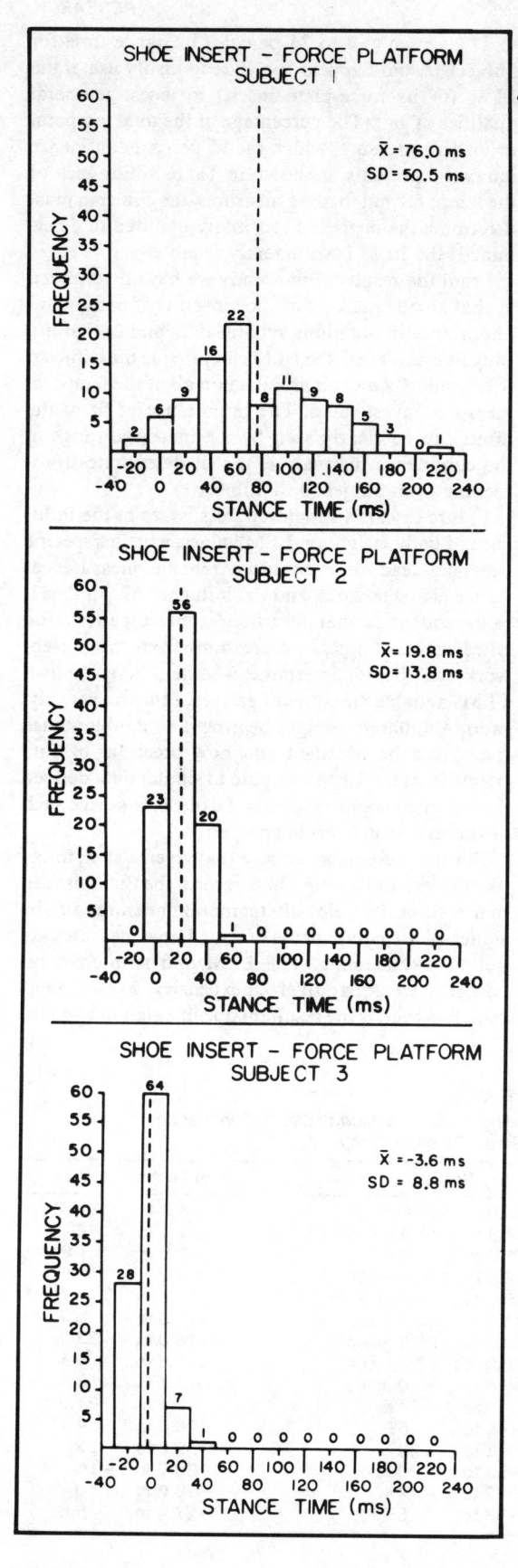

Fig. 2. Mean difference of the stance times between LLM and force-platform values for 100 steps. The "fit" of the insert was progressively better from Subject 1 to Subject 3. Dashed vertical line represents mean ($\bar{X}$), number of observations within ranges are depicted above bars, and SD represents standard deviation.

Uncontrolled variables, including the weight of the subjects and the cadence during testing, may also account for part of the error.

The clinical significance of the observed differences between the LLM and force-platform data must be addressed. In so doing, several questions are raised. Even if small mean differences are noted between the LLM and force platform, how sensitive is the LLM in detecting differences in patient performance? Is the LLM more accurate for one subject or one measure than for another subject or measure?

In addressing the clinical relevance of the differences between the means calculated for the LLM versus the force platform, one must recall that for the second phase of the study, all but one measure (time down) showed statistically significant mean differences for all of the subjects. Does statistical difference mean that the majority of the measures are so inaccurate as to be clinically useless? Obviously, careful inspection and analysis of the data are necessary. Though a statistically significant difference between the means may exist, the magnitude of this difference is what is important.

The difference between statistical significance and clinical relevance is most clearly highlighted when, for example, stance-time data for Subject 3 (Fig. 2, bottom) are closely scrutinized. The mean difference between total stance-time measures for the LLM and the force platform was −3.6 msec; that is, on the average, the measurements from the force platform were slightly greater than those from the LLM. Yet the paired t-test analysis indicates that such differences are significantly different from zero (the expected difference) and hence one would conclude that stance-time data from the LLM and the force platform are not the same. In order to determine the clinical relevance of this difference, the magnitude of the error needs to be compared to the entire stance time. Murray and co-workers[16] have shown that the stance duration among 60 subjects walking at a cadence of 112 steps/min ranges from 610 to 650 msec depending upon the subject's height. Typically, our stance durations were slightly longer probably because of the reluctance of the subjects to increase walking velocity in light of the short ramp distance. The mean stance time for Subject 3 is approximately 670 msec, which is comparable to data from Murray and colleagues.[16,17] Therefore, even using the median of Murray and co-workers' observations (630 msec), the percentage error that 3.6 msec represents is only 0.57 percent of the stance time. Such a difference is not clinically relevant. A similar argument could be made for all of Subject 3's temporal measures.

In examining the loading measures during stance, again by way of example, the first peak of Subject 3 will be scrutinized. Table 2, column 2, shows a 13 percent mean difference between the LLM and force-platform values. Because the actual force generated during the first peak for Subject 3 was approximately 120 percent of the subject's body weight, the percentage error is actually only 11 percent. The clinician must decide if this difference is clinically relevant. We believe that the loading measures do show a clinically significant mean difference for each subject.

Even, as in the case of the temporal measures for Subject 3, if the difference between means for a measure was not clinically significant, the sensitivity of the LLM as a clinical tool is uncertain. To determine the sensitivity of the LLM the 95 percent confidence intervals must be studied. The 95 percent confidence interval describes the range of the variability of the LLM. From a clinical perspective, the variability or possible recording error of the LLM must be smaller than the smallest changes in measured patient performance; that is, if the range of the confidence interval is wider than a clinically significant change in a patient's performance, then the LLM would not be a sufficiently sensitive measuring device.

For the stance time of Subject 3, a confidence interval of −21.2 to 14.0 msec is noted (Tab. 2, column 6). This confidence interval represents a range of −2.9 to 1.9 percent of the entire stance time. Thus, for any one observation, the difference between the stance-time values for the LLM versus the force platform will range from −2.9 to 1.9 percent of the stance time with 95 percent certainty. This difference appears to be of a sufficiently small magnitude to permit the use of the LLM to evaluate stance time within most clinical situations. In contrast, a range of −25.0 to 177.0 msec (−4 to 24 percent) in stance time for Subject 1 would appear too wide to justify use of the LLM (or its force-plate insert) to assess temporal qualities of gait. The percentage of the total temporal or loading measure, which the 95 percent confidence interval represents, is shown in Table 3. For each of the temporal and loading measures, the clinician must determine the degree of sensitivity required to determine if the LLM is sufficiently reliable.

From the results of our study we have determined 1) that an adequate fit of shoe insert is of paramount importance in obtaining reliable data and 2) that if a snug fit is achieved, the LLM can provide the clinician with clinically meaningful data on all of the temporal measures investigated. The importance of fit of the insert can be clearly seen by the increased range in the confidence intervals as the fit deteriorates (proceeding from Subject 3 to Subject 1).

Future research should address issues of the influence of body weight on LLM values, whether specific cadences lead to even more reliable measures of stance characteristics, and the influence of gait deviations and abnormal foot-floor contact patterns on LLM values. Logical progressions from the present work will be to determine whether, in using two LLMs, reliable measures of gait (eg, cadence, velocity, ratio of unilateral weight bearing, and double stance times) can be obtained and how precision of data points from the LLM compare to similar data derived from a more sophisticated and expensive device such as the foot-switch stride analyzer.[11]

For the present, we believe that given a snug force-plate insert and a strip-chart record, the therapist can make clinically relevant temporal measures of the stance-phase components of gait. These measures are not time-consuming. The LLM and strip-chart recorder are comparatively inexpensive. Most important, however, is the resultant documentation of quan-

TABLE 3

Percentage of the Total Temporal and Loading Measure Which the 95% Confidence Represents for the 100 Steps Taken by Subjects

Gait Measure	Subject	Range 95% Confidence Interval	% of Total Measure
Stance time	1	−25.0 to 177.0 msec	−4.0 to 24
	2	−7.8 to 47.4 msec	−1.0 to 7
	3	−21.2 to 14.0 msec	−2.9 to 1.9
Time up	1	−52.4 to 126.8 msec	−24.0 to 60
	2	−49.8 to 1.5 msec	−25.2 to .7
	3	−31.0 to 3.0 msec	−13.0 to .2
Time down	1	−63.8 to 135.0 msec	−16.0 to 22
	2	−32.0 to 27.2 msec	−5.6 to 6.4
	3	−18.4 to 20.0 msec	−3.1 to 3.3
First peak	1	−8 to 24%	−7.1 to 21.3
	2	−46 to 6%	−26.0 to 4.4
	3	3 to 23%	2.0 to 22
Second peak	1	−52 to −20%	−38.0 to −16
	2	−42 to −14%	−26.0 to −10
	3	−33 to −5%	−27.4 to −5.6

tifiable aspects of gait against which the efficacy of therapeutic interventions can be measured.

CONCLUSION

From a statistical perspective, the LLM and a standardized force platform do not appear to measure the same aspects (stance time, time up, time down, first peak, second peak) of the stance phase of gait. This observation is true whether data are obtained on separate occasions or at one session. Ninety-five percent confidence interval determinations, however, indicate that for temporal measures, the difference between the LLM and the force platform become substantially reduced as the fit of the force-plate insert within the shoe improves. Under this improved condition, data from the LLM can be used to quantify the temporal components of the stance characteristics of gait in an effort to document effectiveness of clinical treatment. Loading measurements showed sufficiently wide 95 percent confidence intervals, thereby suggesting that the accuracy of LLM may be insufficient to quantify the loading measures of gait.

Acknowledgments. We wish to express our appreciation to Mike Lynn for assistance with statistical analyses and preparation of this manuscript; James Perry for help in preparing tables and illustrations; Gloria Bassett for typing; and Ray Burdette, PhD, for help in the initial phase of data collection.

REFERENCES

1. Smidt GL: Methods of studying gait. Phys Ther 54:13–17, 1974
2. Holt KS, Jones RB, Wilson R: Gait analysis by means of a multiple sequence exposure camera. Dev Med Child Neurol 16:742–745, 1974
3. Aptekar RG, Ford F, Bleck EE: Light patterns as a means of assessing and recording gait. I: Methods and results in normal children. Dev Med Child Neurol 18:31–36, 1976
4. Grieve DW, Leggett D, Wetherstone B: The analysis of normal stepping movements as a possible basis for locomotion assessment of the lower limbs. J Anat 127:515–532, 1978
5. Cappozzo A, Figura F, Marchetti M: The interplay of muscular and external forces in human ambulation. J Biomech 9:35–43, 1976
6. Wall JC, Charteris J, Hoare JW: An automated on-line system for measuring the temporal patterns of foot/floor contact. J Med Eng Technol 2:187–190, 1978
7. Lesh MD, Mansour JM, Simon SR: A gait analysis subsystem for smoothing and differentiation of human motion data. J Biomech Eng 101:205–212, 1979
8. Gifford GE, Hutton WC: A microprocessor controlled system for evaluating treatments for disabilities affecting the lower limbs. J Biomech 2:38–41, 1980
9. Winter DA, Quanbury AO: Multichannel biotelemetry systems for use in EMG studies, particularly in locomotion. Am J Phys Med 54:142–147, 1975
10. Nelson AJ, Tucker J: Strategies for the improvement of hemiparetic locomotion, abstracted. Phys Ther 60:594, 1980
11. Perry J, Antonelli D, Barnes L, et al: Foot Switch Stride Analyzer. Veterans Administration Contract No. V101(134), Downey, CA, 1979, p 538
12. Wannstedt GT, Herman RM: Use of augmented sensory feedback to achieve symmetrical standing. Phys Ther 58:553–559, 1978
13. Wolf SL, Hudson JE: Feedback signal based upon force and time delay: Modification of the Krusen Limb Load Monitor. Phys Ther 60:1289–1290, 1980
14. Binder SA, Wolf SL: Use of a portable transducer to objectively analyze temporal and force components of the gait cycle, abstracted. Phys Ther 61:713, 1981
15. Dixon WJ, Brown MB: BMDP-79, Biomedical Computer Programs P-Series, Berkeley, CA, University of California Press, 1979
16. Murray MP, Drought B, Kory RC: Walking patterns of normal men. J Bone Joint Surg [AM] 46:335–360, 1964
17. Murray MP, Kory RC, Sepic SB: Walking patterns of normal women. Arch Phys Med Rehabil 51:637–650, 1970

Commentary

It seems to me that the authors undertook this study with a preconceived conviction that the Krusen Limb Load Monitor was a good gait measuring device and then proceeded to interpret their data in a manner which would justify that assumption.

As initially conceived, the study could have been good. It compared a new gait device with an established force platform. Reliability was tested by using multiple measurements on the same subjects. The appropriate statistical analysis technique was selected. Execution of the study, however, became a series of data exclusions until the desired outcome was found.

They began with only two subjects. A third was added and the last half of the study was repeated when deviations between the data from the two men's tests suggested a poor fit between LLM and foot size was a significant factor. Further data interpretations focused on the better results obtained with the third subject.

From the recordings with the two force units (LLM and platform), five factors were calculated. Even for the one preferred subject, three of these items (first and second peak forces and time-up) exhibited inconsistent deviation, with magnitudes of 13, 22, and 27 percent. These were omitted from further consideration. Time down also was excluded, possibly because the interval from floor contact to attainment of the second peak force has no clinical significance. As a result of these serial exclusions, the recommendation that the LLM was a useful gait measuring device was based on the one measurement (stance time) that provided consistent data with less than 3 percent deviation for the one subject. This leaves three unanswered questions.

First, Is this degree of stance time accuracy repeatable in other subjects with a similar goodness of fit?

Secondly, How accurately does the LLM have to fit to provide reliable data? (The authors even excluded from consideration the data of the second subject.) The corollary to this question is, How many LLM units are required to test an average adult population? For simple footswitches that have a 1-in length adjustment, the answer is four pairs. Each pair of LLM will cost at least $1,000 and their lengths are fixed. Does this mean eight pairs are needed ($8,000)?

Thirdly, What is the clinical significance of measuring just stance time? Is it worth an initial outlay of $2,500 to test the few appropriately sized patients or $9,500 to measure all adults? It commonly is assumed that stance time indicates the limb's weight-bearing capacity. At times it does, but not consistently. A study of 149 patients with degenerative hip disease in which the clinical descriptions of disability (limp, distance, support) and footswitch stride measurements were compared showed no statistical correlation nor significance for total stance. The most pertinent and significant values were single stance, velocity, stride length, and cadence. This study considered none of these questions, yet the device was declared clinically useful.

In recommending the LLM as an inexpensive gait measuring unit, the authors anticipated that the clinical therapist would spend about 15 minutes measuring the gait values from the printed record after taking a similar amount of time to gain the data. Our experience suggests this is unrealistic. A two-channel footswitch recording system was provided each of six clinical centers testing the NMA electronic brace for drop foot. Hand calculations were required, and virtually no one compiled these data. That is why the Rancho-VA Gait Analyzer has a precalculated display. We further learned that busy clinicians did not even want to write down displayed numbers. Nor would they settle for just two values (velocity and single stance). Hence, the latest version (now called the "Stride Analyzer") provides a printed record of all the gait values that can be measured from footswitches in both absolute and percent-of-normal terms. These are the reasons gait measuring devices cost more than $2,500. The Stride Analyzer now is being used regularly in five clinical centers.

One has to conclude that this study was an inhouse preliminary investigation. Either further study to determine appropriate calibration factor for providing dependable data with device sizing or selection of another system should have occurred.

JACQUELIN PERRY, MD
Chief, Pathokinesiology Service
Rancho Los Amigos Hospital
Downey, CA 90242
Professor of Orthopaedics,
University of Southern California

Authors' Response

We are most honored and flattered that Dr. Perry has taken time to critique our article and hope that the reader will find this work, her commentary, and our response of clinical benefit. In critiquing a manuscript, the reviewer has the awesome responsibility to objectively and constructively inform the reader and author(s) about strengths and weaknesses related to the content of the manuscript. This task requires careful scrutiny of the written word. We must express our disappointment in Dr. Perry's efforts to do this. We have chosen to respond to her concerns in the same sequence in which they were presented.

To suggest that we undertook this study with a "preconceived conviction" that the Krusen Limb Load Monitor (LLM) was a good gait measuring device and that we proceeded to justify this "assumption" is to misread the clearly stated purposes of the work, both of which dealt with ascertaining the reliability of the device before justifying its clinical efficacy.

Data exclusions were neither intended nor implicated. As stated, one of the purposes of this investigation was to attempt to identify relevant factors related to the LLM's reliability. We are not sure what data were omitted or to what specific "desired outcome" Dr. Perry is referring. We believe that our data are clearly laid out in tables provided.

The progression of this study started with a "purely subjective" fit of inserts and ultimately led to the determination of the "best fit" to yield data that on repeated step analyses produced clinically meaningful, tight values at 95 percent confidence intervals. To make such a determination does not require multiple subjects, but, rather, few subjects undergoing multiple steps. Undoubtedly, the study could have been strengthened if data generated by Subject 3 could be replicated with other subjects wearing a snug fitting insert. Thus the intent of this work was to find if a "desired outcome," or, more appropriately, clinically relevant data could be obtained from the LLM.

There are five identified measures (three temporal and two loading) that we calculated from the LLM. Expressed as an absolute range within 95 percent confidence intervals, these percentages of the total measure—to which we presume Dr. Perry is referring to as "deviations"—for first peak, second peak, and time-up were 20 percent, 21.8 percent, and 13.2 percent, respectively, for Subject 3 and not 13, 22, and 27 percent as noted in the critique. We are not sure why Dr. Perry states that these three values and the time-down were omitted from further consideration. We merely chose to use the stance time as an illustration and stated, based upon all of the data presented, that "the clinician must determine the degree of sensitivity required to determine if the LLM is sufficiently reliable."

Contrary to the finances for LLMs indicated by Dr. Perry, a pair of LLM devices would not cost $1,000 but rather $650, and only additional inserts would need to be purchased. Each pair of inserts costs about $25. Although in the case of degenerative hip disease, stance times may not be significant, certainly in other gait disorders, such as those resulting from virtually any CNS insult, this measure is quite important. Furthermore, we are confident Dr. Perry realizes that important measures of cadence, single stance and double stance time, and velocity can be clearly ascertained from LLM outputs.

We believe that the Rancho-VA Gait Analyzer is an excellent device. Contrary to experiences shared by Dr. Perry, therapists at our facility required 20 minutes to make the 5 calculations for 10 steps, or a total of 50 measures. Therefore, time to make gait measurements by clinical physical therapists is not excessively overwhelming and certainly within the realm of possibility. Ultimately, time, expense, and the necessity to obtain quantitative measures about qualitative aspects of gait will help to dictate the types of measuring devices chosen by clinicians. We offer the LLM as an example of how, under proper circumstances, a comparatively cheaper instrument may be used to achieve this end.

STEVEN L. WOLF
STUART A. BINDER-MACLEOD

Basic Research

Absorbent Paper Method for Recording Foot Placement During Gait

Suggestion from the Field

BERTHA H. CLARKSON

PROBLEM

A simple inexpensive method was needed to record foot placement of the gait pattern of severely to profoundly retarded clients. The methods previously reported in the literature were impractical for a variety of reasons. Some were too expensive and too sophisticated for the basic information we needed to collect.[1-3] Others required the use of indelible substances applied to the feet or footwear,[3-5] and our subjects could not be depended upon to confine their walking to the prepared walkway. Our entire facility is carpeted, so this was a major concern. Furthermore, our subjects would not readily tolerate the application of pads or other material to their feet or shoes, nor could they understand the necessity for leaving these in place until gait recording was completed. Their interest in the materials on their feet precluded conducting natural walking trials.

METHOD

To circumvent these difficulties, a method was devised using absorbent paper overlaying water-soaked material placed on top of a protective floor covering. The entire walkway area which was 8.5 m (28 ft) long and 1.0 m (3.3 ft) wide was first covered with a protective sheet of heavy brown wrapping paper that was taped with masking tape to the carpet. The brown paper had sufficient finish to be somewhat water-repellent without being slippery. Thoroughly moistened absorbent material, such as paper toweling, lightweight turkish towels, terry cloth, or desk blotters, was placed over the brown paper to cover all but a 3-cm (1.2-in) border. Next, the recording layer of absorbent paper was placed over the moistened material and taped in place at each end of the walkway. The most convenient and inexpensive material to use for this top layer was paper tablecloth purchased by the roll from a party supply house, although rolls of regular paper towels taped together lengthwise provided equally satisfactory records.

As the subject walked, the pressure of his bodyweight caused the water from the middle layer to be absorbed into the dry top layer to form clearly defined wet footprints. Because the three layers placed on the thin carpet, which is cemented directly to a concrete floor, provided a suitably firm walking surface without excess compressibility, no distortion of the footprints occurred. Immediately after the walking trial was completed, the wet footprints on the top layer were outlined with a felt-tip pen and the marked paper was taken up, labeled, and laid aside to dry (Figure) while another layer was put in place on the walkway for recording the next trial. Unless unusual delays occurred, it was not necessary to remoisten the middle layer during a recording session of up to 10 walking trials. No change in the dimension of the

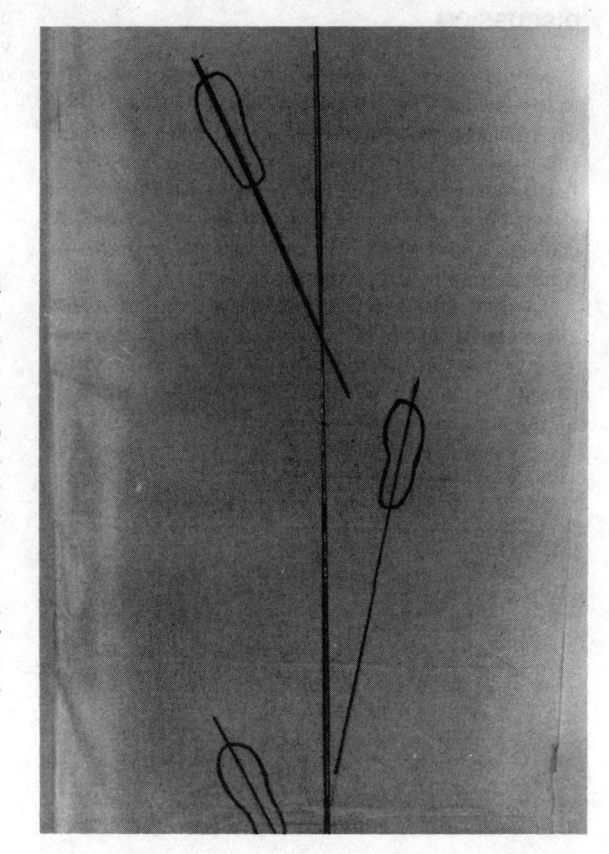

Figure. Section of the recording layer ready for measurement.

Mrs. Clarkson is Physical Therapy Program Advisor, Georgia Retardation Center, 4770 N Peachtree Rd, Atlanta, GA 30338 (USA).

This article was submitted March 12, 1982, and accepted September 10, 1982.

recording layer was found to occur from the wetting and subsequent drying, undoubtedly because of the relatively small area that became wet in a walking trial.

DISCUSSION

An unpublished study (Adler and associates, 1978) that recorded foot placement during walking of 25 normal children from two to five years of age showed this method to work for their subjects walking barefoot as well as it did for our retarded clients who wore shoes. The feel of the walking surface on the bare feet did not appear to cause any sign of withdrawal, hesitation, or other adverse reaction.

The advantages of the method are obvious. Accessible, inexpensive materials that are easy to work with give an accurate, permanent walking record under relatively natural walking conditions for clients wearing shoes or walking barefoot, without danger of stains, without distracting attachments, and without causing aversive tactile sensations. The resultant record allows easy measurement of step and stride lengths, stride width, and foot angles of in-toeing and out-toeing. If desired, additional data for computing velocity and cadence may easily be obtained by marking lines on the walkway and using a stopwatch.

REFERENCES

1. Murray MP, Drought AB, Kory RC: Walking patterns of normal men. J Bone Joint Surg [Am] 46:335-360, 1964
2. Statham L, Murray MP: Early walking patterns of normal children. Clin Orthop 79:8-24, 1971
3. Burnett CN, Johnson EW: Development of gait in children: 1. Method. Dev Med Child Neurol 13:196-206, 1971
4. Ogg HL: Measuring and evaluating the gait patterns of children. J Amer Phys Ther Assoc 43:717-720, 1963
5. Shores M: Footprint analysis in gait documentation: An instructional sheet format. Phys Ther 60:1163-1167, 1980

Basic Research

A Clinical Method of Quantitative Gait Analysis

Suggestion from the Field

KAY CERNY

Quantitative methods of gait analysis are needed for documenting patient progress and for doing clinical research. Abnormal measures of velocity, step length, step width, stride length, and cadence have been shown to be important indicators of gait dysfunction.[1-4] Improvements in function are paralleled by improvements in these measures.

Sophisticated electronic methods are the simplest and most accurate methods of gait analysis; however, they are not usually available to the clinician. These methods are expensive[2] and are found in only a few locomotion laboratories.[5]

This paper describes a clinical method of quantitative gait analysis, different from the clinical methods that have been described in the literature,[2,6] that can be used in any clinical setting. This method can be used to quantify stride length, step length, step width, cadence, and velocity of walking, and it requires less equipment and may be simpler to use than other clinical methods. Also, this method provides a way of teaching objective gait analysis to physical therapy students.

PROCEDURE

The procedure requires only a stopwatch, two felt-tip marking pens with washable ink, and a 16-m (53 ft) walkway that is premeasured and marked with masking tape at four points. A hallway, an outside cement area at a clinic, or patient's home, as well as a portion of a clinic floor can be used for the walkway. The walkway is marked to show a center area 6 m long and two 5-m areas on each end (Fig. 1). Measurements are made within the 6-m area only; the two 5-m areas allow for warming up to "normal" velocity before measurement and slowing down after measurement. Using these extensions of the measurement area of the walkway is intended to eliminate measurement errors.

Felt-tip marking pens are taped to the back of the patient's shoes so that the tip just reaches the floor

---

Ms. Cerny is Associate Professor, Department of Physical Therapy, California State University at Long Beach, Long Beach, CA 90840 (USA).

This article was submitted August 3, 1982; was with the author for revision 14 weeks; and was accepted for publication February 18, 1983.

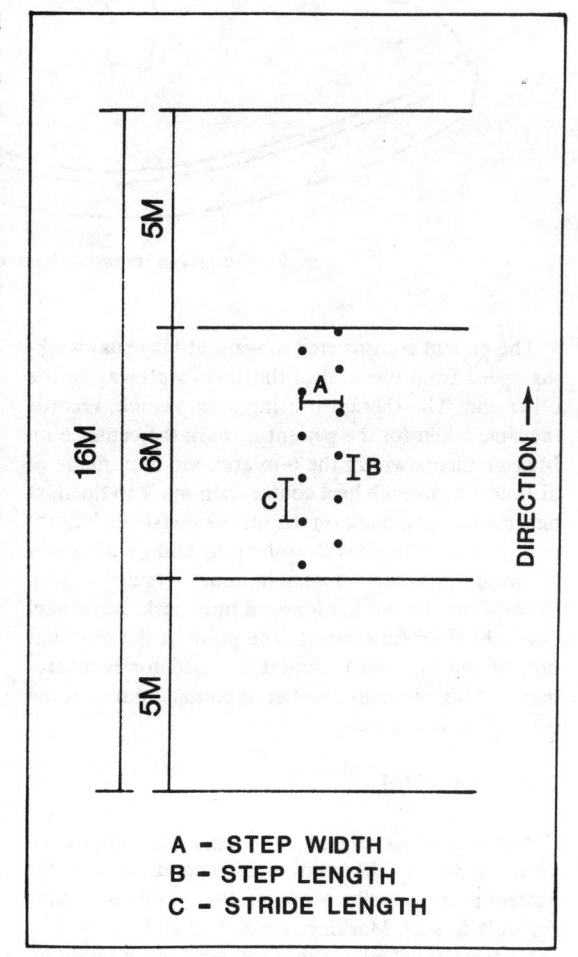

Fig. 1. Walkway with heel contact marks shown in center 6-m measurement area. Five-meter areas on each end of measurement area are used for warming up to normal velocity and slowing down after measurement.

when he is standing (Fig. 2). Before the procedure, the patient should take a few steps at the side of the walkway to ensure that the markers are correctly positioned to indicate heel contact. If several trials are done on the same walkway, marks must be erased after each trial. If several patients are to be tested at the same time, different colored pens can be used to eliminate the need to erase the marks after each patient's walk.

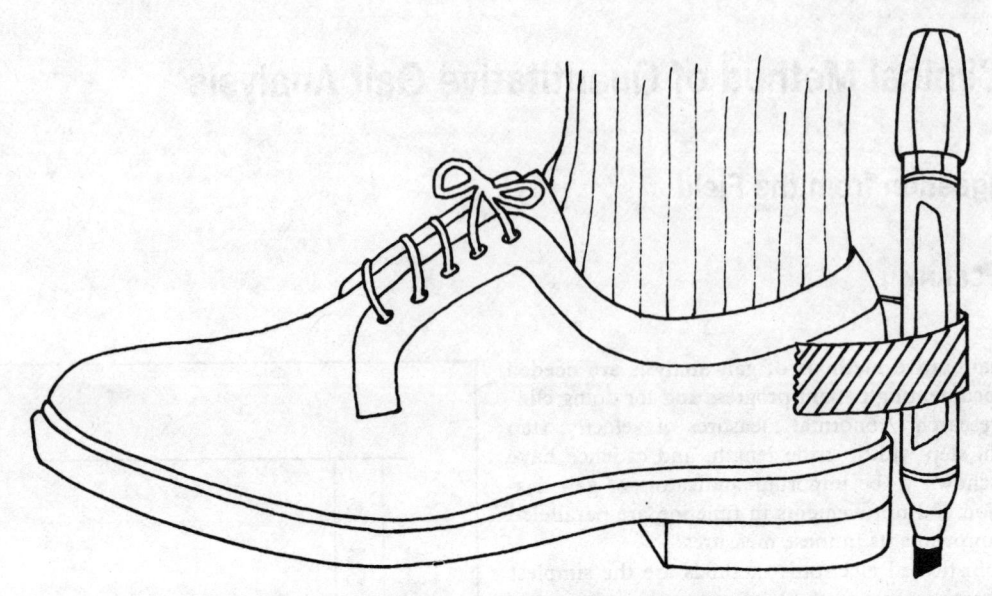

Fig. 2. Felt-tip pen taped to back of shoe for marking heel contact.

The patient is instructed to walk at his usual walking speed from one end of the 16-m walkway to the other end. The therapist, using a stopwatch, records the time taken for the patient to walk the center 6 m. Measurements within the 6-m area are then made of distances from each heel contact pen mark to the next heel contact pen mark on the same side (stride length) and on alternate sides (step length) and of distances of width between successive marks (step width). (Sometimes the marker leaves a line mark as the heel nears the floor for contact. The point at the termination of the line mark should be used for measurement.) Also, the total number of contact marks in the center 6 m is counted.

CALCULATIONS

Velocity is calculated in meters per minute by dividing 360 by the number of seconds it took the patient to traverse the 6-m area (6m × 60 sec ÷ time for walk in sec). Markings are not used for this.

Cadence is calculated in steps per minute by dividing the product of the number of marks in the center 6 m and 60 by the number of seconds it took to traverse the 6-m area (# marks × 60 ÷ time for walk in sec).

Stride length, the distance from heel contact mark to heel contact mark by the same foot, is calculated by averaging the middle three strides (Fig. 1).

The step length measurement is the average of the middle three steps on the right side by measurements in the line of progression from the left contact pen mark to the right contact pen mark and on the left side by measurements from the right contact mark to the left contact pen mark (Fig. 1).

Step width is the distance perpendicular to the line of progression from left contact mark to right and from right to left. The middle three steps are averaged for each side (Fig. 1) to obtain this measurement.

DISCUSSION

There are several advantages of this method of quantitative gait analysis. The costs are low; a good stopwatch can be purchased for approximately $30, and felt-tip marking pens and masking tape are readily available and inexpensive. Other advantages are that the method can be performed by one therapist in any clinical setting, that setup time and equipment requirements are minimal, and that the information gained can be used to document a patient's walking ability before and after treatment.

REFERENCES

1. Perry J, Antonelli DJ, Bontrager EL: VA-Rancho Gait Analyzer Final Project Report. Washington, DC, Veterans Administration (contract no. V101 (134) P-244, July 1, 1974-Sept 30, 1976), 1976
2. Robinson JL, Smidt GL: Quantitative gait evaluation in the clinic. Phys Ther 61:351-353, 1981
3. Gore DR, Murray MP, Sepic SB, et al: Walking patterns of men with unilateral surgical hip fusion. J Bone Joint Surg [Am] 57:759-765, 1975
4. Waters RL, Perry J, Antonelli DJ, et al: Energy cost of walking of amputees: The influence of level of amputation. J Bone Joint Surg [Am] 58:42-46, 1976
5. Winter DA: The locomotion laboratory as a clinical assessment system. Med Prog Technol 4:95-106, 1976
6. Boenig DD: Evaluation of a clinical method of gait analysis. Phys Ther 57:795-798, 1977

Basic Research

Assessing the Reliability of Measurements from the Krusen Limb Load Monitor to Analyze Temporal and Loading Characteristics of Normal Gait

PATRICIA B. CAREY,
STEVEN L. WOLF,
STUART A. BINDER-MACLEOD,
and RAYMOND L. BAIN

This study assessed the reliability of measurements made by four physical therapists on healthy subject gait data recorded from the Krusen limb load monitor. The five components of step (stance time, time up, time to second peak, and force at the first and second peaks) were analyzed. Six components contributing to gait (ambulation time; velocity; cadence; average swing phase duration, left lower extremity; average swing phase duration, right lower extremity; and ratio of unilateral weight bearing, right lower extremity to left lower extremity) were also analyzed. Intraclass correlation coefficients for the five step components and the gait measures of ambulation time, velocity, and cadence showed high measurement reliability. The other measures of gait showed low intraclass correlation coefficients. The limb load monitor can, therefore, be used by clinicians to measure the five step components and three gait measures (ambulation time, velocity, and cadence) with high measurement reliability.

Key Words: Gait, Physical therapy.

Documenting quantitative data on the temporal aspects of the stance phase of gait and on components such as cadence, velocity, and duration of swing phase provide a basis for the clinician to assess the effect of a therapeutic intervention on lower extremity function. Such objective documentation at preset intervals may also help to justify related medical costs. Highly technical methods of studying gait to derive quantitative data have been proposed. Photography,[1] force-sensitive walkways,[2] telemetry,[3] and computer interfaces[4] are a few examples of technical advances designed to measure components of gait. These methods have various limitations in terms of cost, type of data produced, space requirements, and professional time and training needed to use the assessment tool effectively.

Ms. Carey was a graduate physical therapy student in the Division of Physical Therapy, Emory University School of Medicine, Atlanta, GA, when this study was performed; she is now Assistant Director of Physical Therapy, Rehabilitation Institute of New Orleans, F. Edward Hebert Rehabilitation Hospital, New Orleans, LA 70114 (USA).

Dr. Wolf is a senior investigator at Emory University Regional Rehabilitation Research and Training Center, Emory University School of Medicine, 1441 Clifton Road NE, Atlanta, GA.

Mr. Binder-Macleod was a research associate at the Emory University Rehabilitation Research and Training Center, and an Instructor, Department of Community Health, Division of Physical Therapy, Emory University School of Medicine, when this study was performed. He is now a doctoral student, Department of Physiology, Medical College of Virginia, Virginia Commonwealth University, Richmond, VA.

Dr. Bain is Assistant Professor, Department of Biometry, Emory University School of Medicine, Atlanta, GA.

This study was completed in partial fulfillment of the requirements for Ms. Carey's Master of Medical Science degree, Emory University, and was partially supported by Grant G008003042, Department of Education, National Institute of Handicapped Research, Washington, DC, and Allied Health Advanced Training Grant No. 5 AO2 AH00724-03, Department of Health and Human Services.

This article was submitted October 28, 1982; was with the authors for revision 14 weeks; and was accepted September 2, 1983.

The Krusen limb load monitor (LLM) may remedy attempts to find a clinically feasible gait assessment tool by minimizing the potential drawbacks of present methods. The LLM was developed to provide patients with a proportional auditory signal correlated to the amount of weight placed on the lower extremity and can be used to assist amputees to shift increasing weight onto a prosthesis.[5] The LLM consists of a pressure transducer that may be worn within a shoe and a control box worn on the patient's waist. Output from the LLM provides auditory feedback as weight exceeding a predetermined threshold is sensed. The LLM has been used successfully to teach symmetrical standing to stroke patients and proper limb loading during gait training with orthopaedic, neurologic, and amputee patients.[6] An inexpensive, variable potentiometer can be added to the LLM to control the force and time delay components of the feedback tone separately, which supplies the patient with more precise feedback about the force and duration of stance.[7]

Besides the clinical use of the LLM for patient treatment, the LLM has recently gained some credence as an assessment tool. The LLM is equipped with an output connection to an oscilloscope or a strip-chart recorder (SCR) to analyze the analog (voltage) signal of the amount of weight exerted on the pressure transducer. When the most secure fitting shoe insert is used, the LLM output recorded on a SCR provides reliable and accurate measurements of the temporal components of a step compared with similar measurements derived from a force platform.[8]

The purpose of this study was to assess the LLM's potential as a clinical gait assessment tool by determining the interrater and intrarater reliability of four physical therapists' computations of 11 variables from LLM data on 10 healthy subjects.

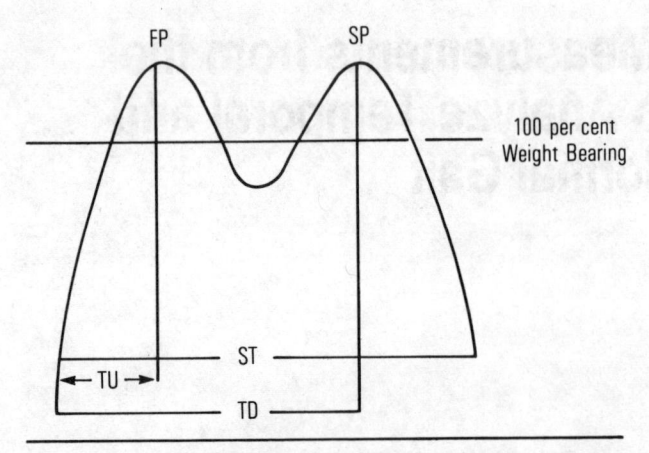

Fig. 1. Temporal and loading aspects of LLM step data: ST, TU, TD, 100 percent weight bearing superimposed and determined during unilateral full weight bearing; FP, SP. See Appendix for definitions.

METHOD

Subjects

Ten healthy subjects, 7 women and 3 men, between the ages of 21 to 28 years and without lower extremity pathology were used to generate data for this study. All subjects wore shoes with heel heights of 3.2 cm (1.25 in) or less. Written informed consent was obtained from all subjects before their participation in the study.

Ambulation Data

A dual channel LLM* (Model 101) was attached to the subject's waist. Each subject was fitted with the largest LLM shoe insert that could lie securely in each shoe. The LLM output was led to a Gould brush 200 pen writer† to record the LLM outputs from each lower extremity. Standard paper speed for recording each gait pattern was 50 mm per second (20 ms per mm). The LLM and the pen writer were calibrated before data collection for each subject. Unilateral weight-bearing measurements for each lower extremity were recorded on the SCR, first in a static position while the subject stood on one leg to determine 100 percent body weight and second, during ambulation. An event marker output, which was manually operated by an investigator, was led to a third channel of the SCR and used to signal when each subject passed over each distance marker.

Subjects performed three practice trials by ambulating at their natural cadences over a 9.1-m (30-ft) level walkway that contained distance markers at 0 m, 1.5 m, 7.6 m, and 9.1 m (0 ft, 5 ft, 25 ft, and 30 ft). The average time for each subject to ambulate from the 1.5-m (5-ft) marker to the 7.6-m (25-ft) marker on the walkway was computed from the SCR for three trials. The additional 1.5 m (5 ft) at the beginning and end of the walkway minimized any acceleration and deceleration stride changes. During actual data collection, subjects were required to complete one ambulation trial to within 0.5 second of their average practice ambulation times. Only data collected when the subjects were between the 1.5-m (5-ft) marker and the 7.6-m (25-ft) marker on the 9.1-m (30-ft) walkway were used for analysis.

---

* *Krusen Research Center, Moss Rehabilitation Hospital, 12th St & Tabor Rd, Philadelphia, PA 19141.*
† *Gould Inc, 3631 Perkins Ave, Cleveland, OH 44114.*

Raters

At the end of data collection, four physical therapists participated as raters in analyzing the LLM output components of both the individual steps and overall gait patterns of the 10 subjects. All of the raters were experienced physical therapists with no previous familiarity with the LLM gait computations.

The raters first attended a practice workshop to learn how to analyze individual steps. The raters received written and verbal instructions for computing three temporal and two loading (force) components of the step data. The analog signals for the LLM that generated such data are depicted in Figure 1. The three time measurements: time up (TU), time second peak (TD), and stance time (ST) represent the time from heel strike to the first peak, second peak, and toe off, respectively. The loading components consisted of the maximum forces recorded at the first peak (FP) and second (SP) peaks expressed as a percentage of the subject's total body weight (100% weight bearing). Unlike Figure 1, the SCR printout paper is divided into millimeter partitions. By counting the partitions between designated landmarks, the raters computed the temporal and loading aspects of each individual step. Raters were instructed to round off all values to the nearest millimeter.

Each rater then attended three subsequent sessions to analyze the step data. The raters computed the temporal and loading component from photocopies of one randomly drawn step from each of the 10 subjects. The 10 steps were presented to the raters in random order during the three analysis sessions.

Next, the raters attended a practice session to learn how to analyze six components of the overall gait pattern. Written and verbal instructions were again presented. The six gait measurements were 1) ambulation time (T_t); 2) velocity (V); 3) cadence (C); 4) average swing phase duration of the left lower extremity (SWP_l); 5) average swing phase duration of the right lower extremity (SWP_r); and 6) ratio of unilateral weight bearing, right lower extremity to left lower extremity ($R_{r:l}$).

To compute ambulation time (defined as the number of seconds the subject took to ambulate 6.1 m [20 ft]), the raters counted the number of millimeters on the SCR record between the starting point and the ending point of data collection. The formula used to compute T_t was

$$T_t = \frac{\text{number of mm from 1.5 m to 7.6 m}}{50 \text{ mm/sec (paper speed)}} \quad (1)$$

Velocity, defined as the total ambulation distance traversed divided by the time necessary to complete the distance, was computed by use of the following formula:

$$V = \frac{6.1 \text{ m (distance covered)}}{T_t \text{ (total time)}} \quad (2)$$

Cadence, defined as the number of steps taken each minute, was computed by use of the following formula:

$$C = \frac{\text{number of steps} \times 60 \text{ sec/min}}{T_t} \quad (3)$$

In the event that only a portion of a step was completed when the data collection period ended, raters were instructed to express the step as a fraction. The numerator represented the number of millimeters in the step actually included in the

Basic Research

TABLE 1
Means and Estimated Standard Deviations for Variance Components for Step Measurements

Component Measured[a] (mm)	$\bar{X}$	Variance Component (estimated s)			
		S	R	SR	T(SR)
ST	35.6	6.45[b]	.203[b]	.174[b]	.231[b]
TU	9.4	2.73[b]	.204[b]	.184[b]	.137
TD	27.3	5.81[b]	.067	.220[b]	.360
FP	21.6	35.3[b]	.003	.017	.479
SP	22.3	26.2[b]	.007	.021	.731

[a] See Appendix for abbreviation key.
[b] $p < .01$.

data collection. The denominator represented the number of millimeters in the entire step.

The average duration of swing phase on the left lower extremity and the right lower extremity was determined by counting the number of millimeters on the printout between toe off and initial contact at heel strike for each lower extremity. The SWP_l was computed by use of the following formula:

$$SWP_l = \frac{\text{total mm of all swing phases on the left LE}}{\text{total number of steps taken on the left LE}} \quad (4)$$

The same formula was used to compute SWP_r, using data pertaining to the right lower extremity.

The $R_{r:l}$, defined as the ratio between the number of seconds the right lower extremity is in unilateral weight bearing versus the number of seconds the left lower extremity is in unilateral weight bearing, was computed by use of the following formula:

$$R_{r:l} = \frac{SWP_r}{SWP_l} \quad (5)$$

The raters analyzed photocopies of all 10 SCR records in random order during three subsequent sessions. The SCR records were rounded to the nearest millimeter in reducing data. We used calculators to compute the six gait measurements from the formulas provided.

Data Analysis

The analysis of the triplicate LLM determinations obtained by the four raters on the 10 subjects is summarized for the five step components and six overall gait measurements. A two-factor (subjects and raters) analysis of variance model was used to analyze the measurements.[9] A variance component model was formulated in which the factors in the model corresponded to the sources of variability (variance components) to be estimated. The four variance components were 1) variability between subjects (S), 2) variability between raters (R), 3) variability of subject by rater combinations (SR), and 4) variability of the triplicate determinations within each subject by rater combination (T[SR]).

The intraclass correlation coefficient (ICC) was used as a measure of association between the average of the triplicate determinations of a rater obtained on a subject with the average of the triplicate determinations of a second rater on the same subject. The estimated ICC was obtained by taking the ratio of the estimated subject variance component to the sum of all the estimated variance components.[10] Reliable measurements, as suggested by Shrout and Fleiss, are those for which the subject variance component is dominant over the other variance components (ie, greater than 50%).[10]

The results are summarized in terms of the estimated variance components. Significance testing, using an F-ratio statistic, indicates which of the subject, rater, or subject by rater variance components were significantly different from zero. The abbreviations used for the five step measurements, the six gait measurements, and the four variance components are summarized in the Appendix.

TABLE 3
Intraclass Correlation Coefficients for LLM Step and Gait Variables[a]

Step Measurements	Correlation Coefficient (r)	Gait Measurement	Correlation Coefficient (r)
ST	.934	T_t	.984
TU	.861	V	.998
TD	.900	C	.831
FP	.995	SWP_l	.475
		SWP_r	.642
SP	.990	$R_{r:l}$	.726

[a] See Appendix for abbreviation key.

RESULTS

Tables 1 and 2 contain the summary analysis for the step and gait measurements, respectively. The overall mean of each measurement is presented along with the estimated standard deviation of the particular variance component.

Step Measurements

The summary analysis for the step component measurements is presented in Table 1. The estimated standard deviation for each of the subject variance components was significantly ($p < .01$) greater than zero, indicating significant subject (S) variability, for all five step measurements. Significant ($p < .01$) rater (R) variability occurred for the ST and TU measurements; ST, TU, and TD measurements had significant ($p < .01$) subject by rater (SR) variability.

The estimated ICC for the five step measurements (Tab. 3) ranged from $r = .861$ for TU to $r = .995$ for FP. The largest percentage of total variability was attributed to the subject variance component (Fig. 2). Therefore, the five step measurements were indicative of reliable measurements.

TABLE 2
Means and Estimated Standard Deviations for Variance Components for Gait Measurements

Measurement[a]	$\bar{X}$	Variance Component (estimated s)			
		S	R	SR	T(SR)
T_t (sec)	4.96	.542[b]	.002	—[c]	.033
V (ft/sec)	4.08	.365[b]	—	—	.003
C (steps/min)	110.6	43.6[b]	—	6.33[b]	8.60
SWP_r (mm)	18.03	2.02[b]	.483	1.39[b]	.626
SWP_l (mm)	17.03	2.60[b]	.520[b]	.737[b]	.358
$R_{r:l}$	1.08	.032[b]	—	.013[b]	.003

[a] See Appendix for abbreviation key.
[b] $p < .01$.
[c] — indicate s of $< .001$.

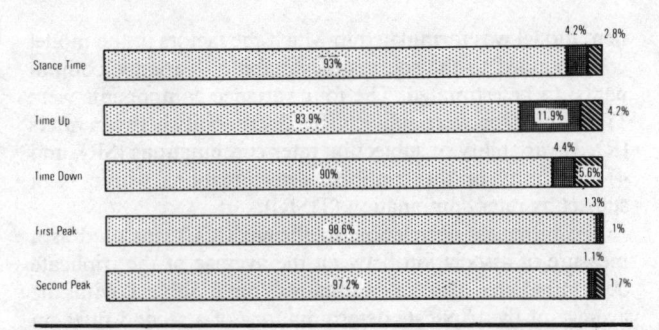

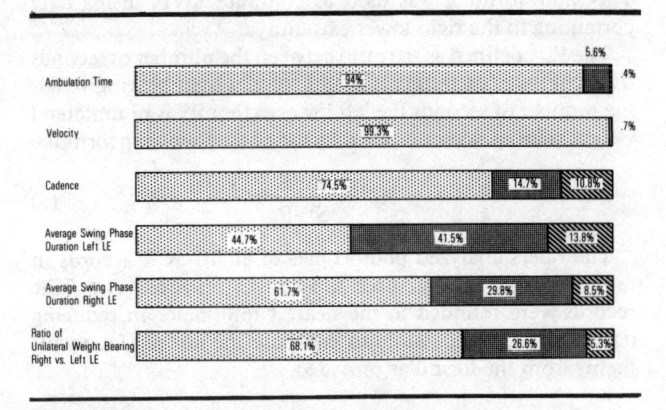

Fig. 2. Sources of variability in percentages for the LLM step variables. Subject variability, stipple; rater and rater by subject (interrater) variability, checks; triplicate (intrarater) variability, diagonal lines.

Gait Measurements

The summary analysis for the gait measurements is presented in Table 2. The estimated standard deviation for each of the subject variance components was significantly ($p < .01$) greater than zero, indicating significant subject variability for the six gait measurements. Rater variability was significant for the SWP_l measurement ($p < .01$); four of the six gait measurements (C, SWP_r, SWP_l, and $R_{r:l}$) had significant subject by rater variability ($p < .01$).

Table 3 presents the estimated ICC for the six gait measurements. The estimated ICC ranged from $r = .475$ for SWP_l to $r = .998$ for V. The estimated ICC for SWP_l ($r = .475$) indicates that it was not a reliable measurement because the subject variability was dominated by the other sources of variability. The percentage of total variability attributed to each variance component for each of the six gait measurements is shown in Figure 3.

DISCUSSION

The purpose of this study was to determine interrater and intrarater reliability for four physical therapists' computations of 11 variables from the LLM data to evaluate the LLM's potential as a clinical gait assessment tool. The results of this study unquestionably portray a high degree of interrater and intrarater reliability for all five of the LLM step variables and for T_t, V, and C from the LLM gait variables.

Statistical analysis of the LLM step and gait variables did not yield an exact ICC value to use as a guideline for classifying a variable as sufficiently reliable to serve as a clinical tool. But surely the step variables and the gait variables T_t and V (with the large percentages of between-subject variability and small percentages of between-rater [rater and subject by rater] and within-rater [triplicate] variability associated with these results) can be regarded as highly reliable. Clinicians may safely accept the ICC of $r = .831$ for C as a highly reliable estimation because approximately 75 percent of the variability associated with C was because of between-subject variability; the remaining 25 percent of the variability was almost equally split within-raters and between-raters. A possible source of rater variability appears from a review of one rater's measurements that suggested a misinterpretation of instructions about strip-chart measurements. Another source of nonagreement between-raters and within-raters may have arisen from computational errors. A *post hoc* review of the raters' worksheets confirmed that such errors occasionally occurred.

The gait variables with the smallest ICC and the greatest interrater variability (C, SWP_l, SWP_r, and $R_{r:l}$) are dependent on each other for accurate computation. Computing SWP_l and SWP_r requires using the number of steps taken on each lower extremity previously determined for the variable C. An error in counting the correct number of steps taken on each lower extremity could, therefore, negatively influence rater agreement for SWP_l and SWP_r. Interestingly, the results suggest that interrater and intrarater reliability are higher for the variable $R_{r:l}$ than SWP_l and SWP_r; these swing phase variables are the less reliable components of $R_{r:l}$. Apparently any discrepancy involved in determining the values of SWP_l and SWP_r to compute $R_{r:l}$ had a canceling effect on each other when used in conjunction to form a ratio of the two values.

Raters were more reliable within themselves than between themselves for determining SWP_l, SWP_r, and $R_{r:l}$. From a clinical perspective, use of the same physical therapist to compute the gait variables for a subject eliminates a major source of variability (rater and subject by rater) and thus ensures greater reliability of results.

Other possible explanations for greater variability associated with the results of SWP_l, SWP_r, and $R_{r:l}$ include the quality of photocopied strips presented to the raters or the misinterpretation of the directions and demonstrations provided for the raters. Portions of the strips photocopied had varying degrees of legibility and possibly obliterated the exact location of designated landmarks. Photocopying a patient's LLM data

Fig. 3. Sources of variability in percentages for the LLM gait variables. Subject variability, stipple; rater and rater by subject (interrater) variability, checks; triplicate (intrarater) variability, diagonal lines.

would be unnecessary in the clinic. Further investigation into more exacting, self-explanatory directions for determining SWP_l, SWP_r, and $R_{r:l}$ may be necessary to ensure greater accuracy of results.

The potential uses of the LLM as a clinical tool are untapped. To use one device for gait assessment and for gait training purposes could prove to be cost and time efficient.

Essential considerations for selecting the proper clinical gait recording technique have been outlined.[11] The LLM gait recordings require only careful personnel and take about 15 minutes to complete. Familiarity with the procedures for computing the LLM variables substantially reduces the time necessary to calculate the variables. In the present study, the clinicians initially averaged about 6 minutes to compute the LLM data for one step and 16 minutes to complete one LLM gait data record. By the third rating session, clinicians required an average of 2 minutes to compute the variables for one step and 13 minutes to compute the LLM gait variables.

The LLM apparatus is lightweight and the leads from the shoe inserts to the pen recorder do not appear to influence the subject's walking performance. Bilateral or unilateral permanent gait recordings can be made. The cost of the LLM and variously sized shoe inserts is relatively inexpensive compared with other gait assessment equipment providing similar gait information.[8] Equipment readily available in most hospitals, such as a single channel EKG machine, or a strip-chart recorder can be used to record the LLM data.

CONCLUSION

We believe that the LLM provides a rapid, easily operated, objective, qualitative impression of healthy subjects' gait patterns. Interrater and intrarater reliability is very dependable for individual step time (ST, TU, and TD), and force variables (FP and SP) and the overall gait variables (T_t, V, and C). Additional studies with various patient groups and pathological gaits need to be conducted to establish the clinical utility of this device.

Acknowledgements. The authors wish to express their appreciation to Karen Chambers, Barbara Jerry, Lisa Lovelace, and Toni Oliff for their participation as raters in this study and to Gloria Bassett and Janet Vogt for secretarial assistance.

REFERENCES

1. Holt KS, Jones RB, Wilson R: Gait analysis by means of a multiple sequence exposure camera. Dev Med Child Neurol 16:742-745, 1974
2. Charnley J, Pusso R: The recording and the analysis of gait in relation to the surgery of the hip joint. Clin Orthop 58:153-164, 1968
3. Winter DA, Quanbury AO: Multichannel biotelemetry systems for use in EMG studies, particularly in locomotion. Am J Phys Med 54:142-147, 1975
4. Gifford GE, Hutton WC: A microprocessor controlled system for evaluating treatments for disabilities affecting the lower limbs. J Biomed Eng 2:38-41, 1980
5. Wannstedt GT, Craik RL: Clinical evaluation of a sensory feedback device: The limb load monitor. Bull Prosthet Res 29:8-38, 1978
6. Wannstedt GT, Herman RM: Use of augmented sensory feedback to achieve symmetrical standing. Phys Ther 58:553-559, 1978
7. Wolf SL, Hudson JE: Feedback signal based upon force and time delay: Modification of the Krusen limb load monitor—Suggestion from the field. Phys Ther 60:1289-1290, 1980
8. Wolf SL, Binder-Macleod SA: Use of the Krusen limb load monitor to quantify temporal and loading measurements of gait. Phys Ther 62:976-982, 1982
9. Neter J, Wasserman W: Applied Linear Statistical Models. Homewood, IL, Richard D Irwin Inc, 1976, pp 616-627
10. Shrout PE, Fleiss JL: Intraclass correlations: Uses in assessing rater reliability. Psychol Bull 86:420-428, 1979
11. Ganguli S, Mukhejie P: A gait recording technique for clinical use. J Biomed Eng 2:38-41, 1980

APPENDIX

Abbreviations Used for the Step Component Measurements, Gait Measurements, and Variance Components

Step Component Measurements

ST	Stance time, from heel strike to toe off
TU	Time up, from heel strike to peak of vertical force at heel strike
TD	Time second peak, from heel strike to peak of vertical force at push off
FP	First peak, from baseline to peak at heel strike
SP	Second peak, from baseline to peak at push off

Gait Measurements

T_t	Ambulation time
V	Velocity
C	Cadence
SWP_l	Average swing phase, duration left LE
SWP_r	Average swing phase, duration right LE
$R_{r:l}$	Ratio of unilateral weight bearing: right LE to left LE

Variance Components

S	Subject variability
R	Rater variability
SR	Subject by rater variability
T(SR)	Triplicate variability

Gait Analysis Techniques
Rancho Los Amigos Hospital Gait Laboratory

JoANNE K. GRONLEY
and JACQUELIN PERRY

In the gait laboratory at Rancho Los Amigos Hospital, the emphasis is on patient testing to identify functional problems and determine the effectiveness of treatment programs. Footswitch stride analysis, dynamic EMG, energy-cost measurements, force plate, and instrumented motion analysis are the techniques most often used. Stride data define the temporal and distance factors of gait. We use this information to classify the patient's ability to walk and measure response to treatment programs. Inappropriate muscle action in the patient disabled by an upper motor neuron lesion is identified with dynamic EMG. Intramuscular wire electrodes are used to differentiate the action of adjacent muscles. We use the information to localize the source of abnormal function so that selection of treatment procedures is more precise. Force and motion data aid in determining the functional requirement and the muscular response necessary to meet the demand. Determining the optimum mode of locomotion and developing criteria for program planning have become more realistic with the aid of energy-cost measurements. Microprocessors and personal computer systems have made compact and reliable single-concept instrumentation available for basic gait analysis in the standard clinical environment at a modest cost. The more elaborate composite systems, however, still require custom instrumentation and engineering support.

Key Words: Biomechanics, Electrodiagnosis, Gait, Laboratories.

Walking is such a complex function that analysis of a patient's performance can follow any of several directions. To find out what the subject is doing, motion analysis is the appropriate technique to use. To answer the question why a particular gait pattern occurs, dynamic EMG and vector analysis are indicated. To determine the effectiveness of the person's gait, energy cost and stride analysis are required. Each approach involves specialized instrumentation and data processing capability and adequate testing space. Although the ideal testing situation is a facility that can do everything, few clinicians have their opportunity to work in such a facility. Consequently, gait laboratories differ in their organization. The direction taken generally is determined by the clinical environment, budget and space constraints, and interests of the investigators.

Rancho Los Amigos Hospital (RLAH) is a 400-bed inpatient rehabilitation program with a correspondingly large outpatient program. The disabling factor for most of our patients is paralysis from cerebral palsy, stroke, spinal cord injury, muscular dystrophy, and myelodysplasia. The hospital also has a large amputee service and an active arthritis program. The dominant clinical questions are "why" and "how effective." Consequently, the technical emphasis of the pathokinesiology service has been on dynamic EMG, footswitch stride analysis, and energy-cost measurements. Force plate and instrumented motion analysis are used to a more limited extent.

MOTION ANALYSIS

To meet daily clinical need, we developed a system of observational gait analysis.[1] This method is taught regularly to all incoming physical therapists and orthopedic residents. Because the staff can visually identify the patient's major gait deficits, an instrumented system becomes less important. The development of an observation system was fortunate because we could not meet the staffing load needed to reduce the film data from the three cameras that are used in comprehensive motion analysis. The necessity to transfer manually the marker locations from film to computer on a frame-by-frame basis generally takes an hour for each stride studied. The number of patients seen made this an overwhelming task.

The observational technique of gait analysis involves systematically assessing the motion pattern of each segment (foot, ankle, knee, hip, pelvis, and trunk). Notations of their occurrence and timing in the gait cycle are made on a chart that lists the 32 most common errors and indicates in which gait phases they are apt to occur.[1] From this work sheet, the sequence of composite limb postures is identified. By phasically relating the events at one joint to those occurring in adjacent segments, the observer can differentiate primary gait deficits from compensatory actions. Mechanisms that obstruct standing stability, inhibit progression, or increase energy cost are identified, and therapeutic plans are formulated.

For quantitated motion analysis, we most often use electrogoniometers.* We

---

Ms. Gronley is Research Coordinator and Research Physical Therapist, Pathokinesiology Service, Rancho Los Amigos Hospital, 7601 E Imperial Hwy, Downey, CA 90242 (USA).

Dr. Perry is Director, Pathokinesiology Service, Rancho Los Amigos Hospital, and Professor of Orthopaedic Surgery, University of Southern California, Los Angeles, CA.

* Pathokinesiology Laboratory, Rancho Los Amigos Hospital, 7601 E Imperial Hwy, Downey, CA 90242.

have appropriate sizes for the knee, ankle, and subtalar joints. The design is a single-axis parallelogram that accommodates a shifting joint axis and minor deviations in alignment. The electrogoniometers for the knee and ankle are designed to cross the joint anteriorly so that measurements can be made of subjects wearing orthoses (Fig. 1). Accuracy has been determined by roentgenogram comparison. At the knee, soft tissue movement under the base plates resulting from muscle contraction and skin motion introduces an 8-degree error with 90 degrees of flexion. The ankle goniometer tracks within 2 degrees. A hip goniometer is not used. Several designs have been tested but proved to be inadequate to record true hip-joint motion. For all but the thinnest patient, the soft tissues prevent a firm grasp of the pelvis when applying the device. Instead, the proximal band encircles the waist. Consequently, the resulting data represent an undifferentiated mixture of spine and hip mobility. Because gait related to hip dysfunction involves markedly different mixtures of trunk, pelvis, and hip motion, we prefer not to record a composite number. When needed, observational analysis is used to distinguish trunk, pelvis, and thigh motion.

To provide automated recordings of the entire limb, the advantages of a videodisk camera† and a microcomputer‡ have been combined into a single plane system. With passive markers identifying appropriate landmarks on the limb, the subject's performance is recorded on videocassette.§ A technique for automatic tracking of the anatomical markers was developed. The system's output can be either stick figures of the limb's motion pattern or graphs of the individual joint actions. Use of the technique is limited by the time required to process the data (one hour).

DYNAMIC ELECTROMYOGRAPHY

In a patient disabled by paralysis, the primary cause of the gait deficit is inappropriate muscle action. When the

---

† Rotary Shutter Camera RSC-1050 and Video Disc Recorder SVM-1010, AVIDD Electronics Co, 2210 Bellflower Blvd, Long Beach, CA 90815.
‡ Apple II, Gateway Computer Center, 11470 South St, Cerritos, CA 90701.
§ JVC Model CR 6060U Video Cassette Recording System, JVC Industries Company, 1011 W Artesia Blvd, Compton, CA 90220.

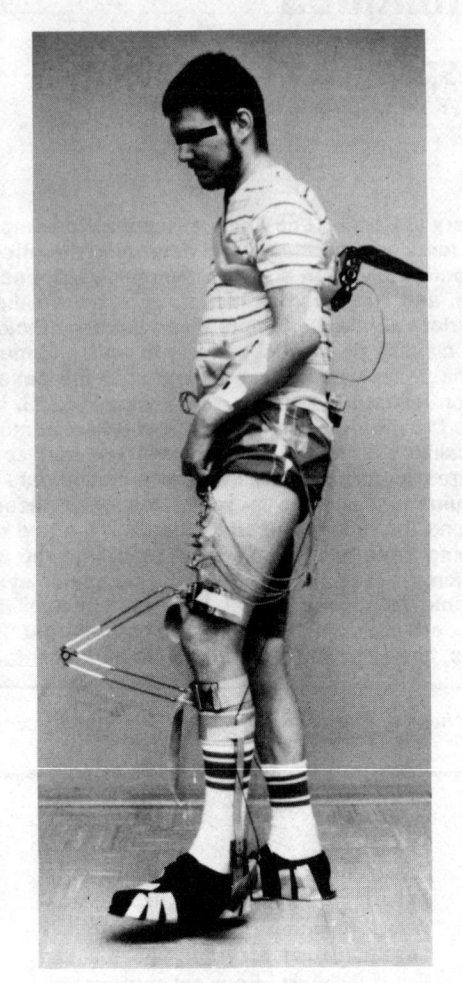

Fig. 1. Double parallelogram goniometer that provides a continuous record of knee motion as the patient walks.

cause is an upper motor neuron lesion, the response to manual tests of strength or the stretch reactions commonly are not equivalent to the muscle action generated by the more complex demands of walking. Thus, to understand the patient's problem, the clinician must use instrumentation that can record muscle function as it occurs rather than infer muscle action from clinical tests. For this purpose, we use dynamic EMG. The system is designed to identify timing and relative intensity of muscle action. Designation of muscle force has been approached only on a preliminary experimental basis.[2]

Myoelectric signals are generated as a neural stimulus spreads along the muscle fibers to initiate a contraction. Response is on an all-or-none basis, that is, each stimulus totally activates every muscle fiber in the motor unit. The resulting muscle force is proportional to the number of motor fibers involved. Recording of these activating signals constitutes the EMG. Milner-Brown and associates have confirmed at the motor-unit level that the all-or-none relationship between stimulation and force also applies to the EMG recorded.[3] This finding validates extending functional interpretations beyond simple timing. It also adds value to reading the raw recordings rather than reducing the data to a simple on and off line representing the muscle activity.

Surface electrodes are the most convenient means of recording muscle function because they are painless to apply and can be used by any clinician. Unfortunately, however, they do not differentiate the signals of adjacent muscles.[4] Consequently, only gross group activity can be determined, and even this may capture signals from adjacent groups. At RLAH, the emphasis is on differentiating the action of adjacent muscles by the use of intramuscular electrodes.

Reliance on wire electrodes is based on confidence that their recordings are representative of total muscle action. Early studies by Buchthal et al, however, raised some doubts.[5] By combining the finding that the signal of a single motor unit traveled a very short distance with the assumption that similar muscle fibers were in close proximity, they estimated the coaxial needle sampled only 25 motor units. Improved equipment increased the estimate to 75 motor units. Knowing that muscles contain several hundred motor units, kinesiologists assumed that the small sampling area of wire electrodes might not record significant muscle action. More recently, Burke et al have shown that the muscle fibers of one motor unit are widely dispersed both longitudinally and transversely within the muscle.[6,7] Although the area recorded by the wire electrode is small, the motor units included are representative of the action of the entire muscle. Consistency in our clinical data supports the anatomical finding of Burke et al.[6,7]

The electrodes we use consist of a pair of 50-μ, nylon-coated stainless steel wires‖ with the distal 2 mm stripped of insulation and bent to hook in the muscle for stability. Short circuiting is

---

‖ .002" Stablohm 800, California Fine Wire Co, PO Box 446, Grover City, CA 93433.

avoided by making the hooked ends of different length so that the noninsulated areas do not touch. Basmajian and Stecko's technique of simultaneously inserting the two wires with a single 25-gauge needle is used.[8] Accuracy of the insertion in the desired muscle is confirmed by mild electrical stimulation through the wires combined with palpation of the muscle belly or tendon.

Because relative motion between adjacent muscles or at the subcutaneous tissue interface can be considerable, drawing extra wire into the fascial planes is necessary. Carrying the joint through a full range of motion and having the muscle contract strongly before the test are the techniques we use to gain wire mobility. This technique allows vigorous activity without tethering. A free loop of wire is formed at the point of skin penetration. With these precautions, the electrodes maintain their position within the muscle for the duration of the clinical testing. Because the posttest data of our research findings are similar to those obtained on the initial tests, this method of handling the wire has apparently overcome the displacement problems identified by Jonsson and Komi.[9]

Usually the test involves studying eight muscles. The myoelectric signals are transmitted by FM-FM telemetry[#] to the recording instrumentation. Bandwidth of the system is 40 to 1,000 Hz with a roll-off at the low end beginning at 200 Hz. A 60-Hz notch filter excludes noise generated by the AC power. The overall system gain is 1,000. This signal processing eliminates most cross-talk (signal spread among muscles) so that the action of adjacent muscles can be clearly differentiated. The resulting data are stored on 7-channel analog tape.[**] A printout on light sensitive paper[††] is provided for immediate visual analysis.

Timing within the gait phase is the most common determination made from the EMG record. Using normal performance as the reference, patient's muscle control patterns are identified. Deviations from normal timing are classified as premature, prolonged, or continuous and out-of-phase action (Fig. 2).

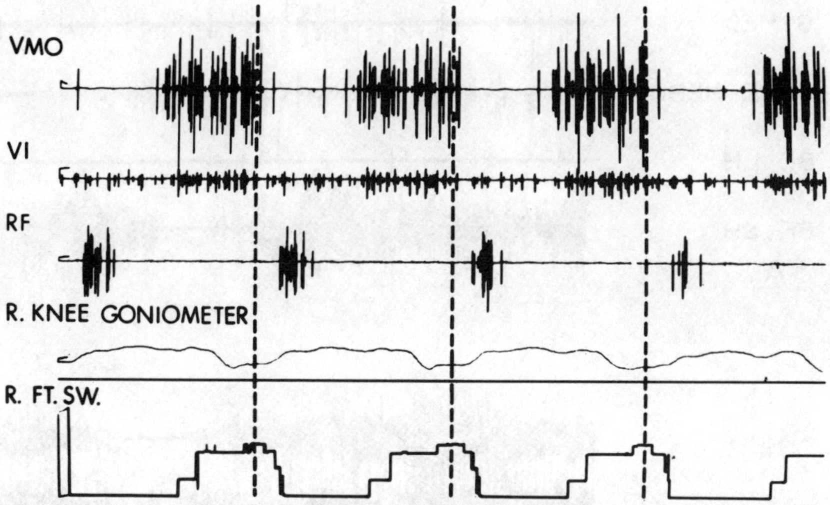

[#] Model 2600, Bio Sentry Telemetry Inc, 20720G Earl St, Torrance, CA 90503.

[**] Ampex FR 1300, Ampex Corp, 401 Broadway, Redwood City, CA 84063.

[††] Honeywell 2106 Visicorder, Honeywell, Inc, 7037 Havenhurst Ave, Van Nuys, CA 91407.

Fig. 2. An EMG study of a stroke patient with inadequate knee flexion in swing (30°). Causes are out of phase activity of the rectus femoris muscle (RF) and continuous activity of the vastus intermedius muscle (VI). Although timing of the vastus medialis oblique muscle (VMO) is prolonged, it is not impeding knee flexion as activity terminates at the time of contralateral floor contact (vertical dashed line). The right footswitch signal (R. FT. SW.) identifies the intervals of swing (baseline) and stance (irregular staircase). The support sequence is heel, heel and first metatarsal with a brief period of flat foot (heel plus first and fifth metatarsals).

The orthopedic surgeons use this information to differentiate normal from abnormal function within muscle groups for final surgical planning of tendon releases or transfers in patients with upper motor neuron lesions (stroke, cerebral palsy, head trauma, or incomplete spinal cord injury). This technique has led to far better clinical results because normal function is not sacrificed and abnormal action is more precisely defined.

The physical therapists use this information to gain a clearer understanding of the problems they are treating. Of particular significance to them are the preliminary tests used both to assess the patient's neurological control and the quality of the electrode insertions. Before the walking runs, EMG recordings are made during manual muscle strength (selective control) and quick stretch (spasticity) tests and resisted flexor and extensor patterns. These data provide a basis for comparing the clinical findings with the muscle activity during walking.

From the raw EMG records, changes in the relative intensity of muscle action during the various tests and within the gait stride can be appreciated. These EMG records extend understanding of the patient's control capability. For example, a muscle that responds poorly to the manual muscle test and mass patterns in the supine position may function adequately during walking; at that time, a primitive synergy was used (Fig. 3). This finding means standard resistive equipment will not be an effective strengthening technique.

Dynamic EMG also is used to test the therapeutic effectiveness of physical therapy procedures. Adler et al found that approximation applied to normal subjects stimulates hip and knee extensor action and inhibits the plantar flexor muscles (unpublished data, 1976).

Quantification of the EMG to provide information on the relative intensity of muscle action is reserved for tests involving normal subjects. This limitation is imposed because exact placement of the electrode within the muscle cannot be controlled. As a result, the number of motor units sampled is unknown. To circumvent the problem of an undeterminable sample size, all functional data are normalized against a quantitated test.[10] Most often, a maximum sustained effort (highest one of four seconds) is used and the data are reported as a ratio (percent maximum). If several test situations are involved, dependence on a single reference value can be avoided by using the sum of all the data obtained with that electrode as the normalizing base.

Through quantitated EMG, functional differences among muscles con-

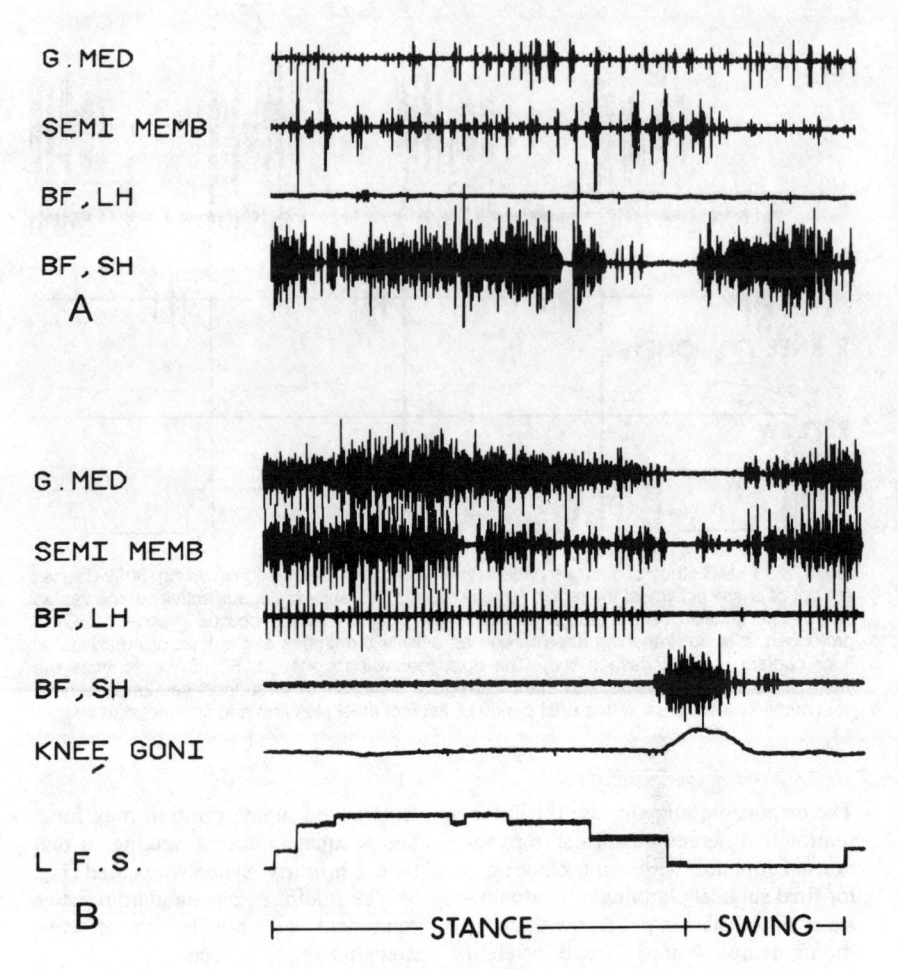

Fig. 3. Use of EMG to clarify the discrepancy in muscle action between the manual muscle test and walking. A. EMG during side-lying manual test of the gluteus medius muscle (G. Med); note attempted substitution of the knee flexors for inability to abduct the limb. B. Activity level of same muscles during walking: G. MED = gluteus medius; SEMI MEMB = semimembranosus; BF, LH = biceps femoris, long head; BF, SH = biceps femoris, short head; KNEE GONI = knee goniometer; and L F. S. = left footswitch.

sidered to have a common action have been identified. For example, within the hip extensor muscle group, the medial hamstring muscles limit their intense action to limb deceleration in terminal swing.[10] Persistence of biceps femoris muscle action through the loading response is consistent with the external tibial torsion that must be restrained. Activity of the gluteus maximus and adductor magnus muscles focuses on stabilizing the hip during the loading response when an added knee flexion torque by the hamstring muscles would be undesirable.

Extending the concept of relative intensity to correlations between muscle force and the dynamic EMG record, muscle quality (force per unit EMG) and mode of contraction must be considered. The maximum isometric force obtained in the optimum joint position is reduced by velocity, a concentric effort, and positional changes.[11,12] Therefore, motion measurements and the individual's basic strength are essential to making muscle-force determinations.

STRIDE ANALYSIS

The temporal and distance factors of gait are categorized by our laboratory as stride characteristics. To obtain this data, we insert appropriately sized insoles with compression-closing switches covering the areas of the heel, heads of the first and fifth metatarsals, and great toe into each shoe of the subject (or taped to the sole of the bare foot). As the subject walks a 10-m distance, timing of foot contact on the floor is recorded during the middle 6 m. This data interval, bracketed by photoelectric cells, represents the subject's steady state with the variations of starting and stopping excluded. The signals are transmitted by FM-FM telemetry to the recording equipment and subsequently presented in two forms. On the analog records, stance is represented by the duration of the footswitch activity and swing is the baseline interval. The foot-support pattern is displayed as steps in a staircase (Fig. 2). Steps of equal height designate the normal sequence of heel only (H), heel and fifth metatarsal (H,5), flat foot (H,1,5), and heel-off (1,5). Abnormal modes of foot support are differentiated by half steps (fifth or first metatarsal only or heel and first metatarsal). Toe contact is identified by an oscillating signal superimposed on the step recording. Currently, the analog footswitch record is used to define the stride timing of the other data such as EMG, electrogoniometers, or force measurements (Fig. 2).

Automatic calculation of the stride variables is attained with a microcomputer-based stride analyzer.§§ The quantitated factors are expressed both as absolute values and as a percentage of normal values. In addition, a display of the foot-support pattern is provided (Fig. 4).

From the footswitch signals, a clinician can calculate nine gait measurements: velocity, stride length, cadence, single stance, initial and terminal double stance, total stance, gait-cycle duration, and the foot-support pattern.

These data are used to summarize the patient's ability to walk. Gait velocity (distance covered per minute) is the basic measurement. Relative effectiveness of different therapeutic procedures and the value of orthoses and prosthetic alignment are defined by gait velocity. Single stance time identifies the weight-bearing capability of the limb. This time has proved more accurate than total stance time because the latter measurement is a mixture of double and single support times. Transfer of body weight during the double stance period is highly variable when there is limb impairment; hence, floor contact per se may not be representative of the limb's support capability. The therapists and orthopedic surgeons at RLAH follow the clinical course of their patients by using a portable microcomputer unit for stride

---

§§ Footswitch Stride Analyzer, B & L Engineering, 9618 Santa Fe Springs Rd #8, Santa Fe Springs, CA 90670.

analysis because the clinical areas are several blocks from the gait laboratory (Fig. 5).

ENERGY COST

The effort involved in walking and in propelling a wheelchair has been of particular concern to the RLAH clinical staff. Because most of our patients are severely disabled, choosing the optimum mode of locomotion for them is a daily challenge to the staff. These decisions have been facilitated by measuring the energy expended by the patients during these activities. The criteria for program planning are more realistic when they are based on energy measurements.

In determining the most effective way of measuring energy cost, our initial efforts with a treadmill demonstrated subjects with impaired limbs either cannot walk on a treadmill at all or must use a velocity that is slower than their customary speeds. Additionally, the walking must continue a sufficient time to assure an aerobic functional state. Thus, for representative data, the energy-cost measurements must be made on a circular, stationary walkway that allows uninterrupted travel for approximately five minutes. We use an outside, circular, concrete, 60-m track (Fig. 6). Performance data are collected with ECG electrodes for heart rate, a thermal sensor for respiratory rate, and a footswitch for gait velocity. The subject's expired air is collected in a plastic Douglas bag and then, after the test, is analyzed for volume and content of oxygen and carbon dioxide. Testing consists of an initial rest period for baseline data followed by the five-minute walk. The first three minutes are used to attain a steady aerobic state (heart and respiratory rates stabilize). Performance data and expired air are collected for the next two minutes. Comparison of heart rate during the first and last 30-second intervals of the data collection period are made to assure that a steady state was maintained. Data collected in other than a steady state underestimates the subject's energy requirements.

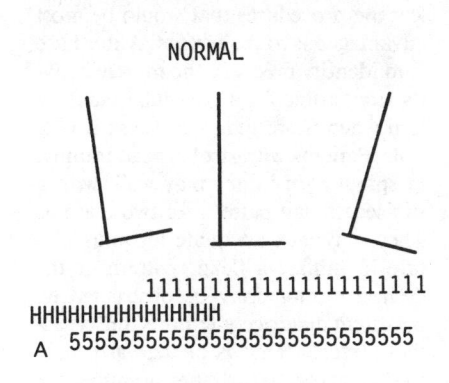

From such testing, we have learned that most patients accommodate to their disability by slowing their gait velocity to reduce the demand more than they increase their rate of energy use.[13] Only children and young adults with sound cardiopulmonary systems and adequate arm and trunk strength to substitute for disabled legs register notable increases in their rate of energy use. Dependence on crutches is particularly costly of energy. Propelling a wheelchair is equivalent to normal walking in both velocity and energy cost.[14,15]

Testing has largely focused on the various clinical groups to determine appropriate functional criteria. Individual assessments are done periodically to determine the effectiveness of a particular orthosis. Another purpose is to demonstrate to the patient, family, or finance source, the dual need for a wheelchair for long distances and an orthosis for indoor and irregular terrain travel.

FORCE MEASUREMENTS

A force plate[||] with piezoelectric crystal sensors initially was obtained as a research tool. The purpose was to measure the weight-bearing capacity of the hips of patients with arthritis. Re-

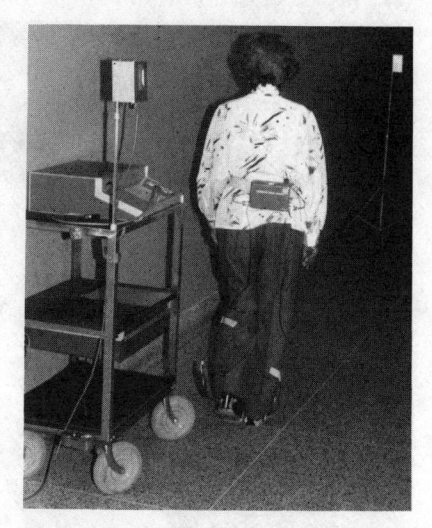

Fig. 5. Portable stride analysis system. Patient wearing footswitches and a memory unit on the belt. A 6-m distance is designated by the two light switches and a calculator is sitting on the cart.

sults of baseline studies, however, demonstrated the intensity of the ground reaction forces was primarily determined by the subject's gait velocity.[16] The customary, double peak pattern found in normal walking disappeared when subjects reduced their walking speed from 80 to 60 m a minute. As most patients walk at the slower speed, the only remaining variable was duration of maximum load. This is more easily obtained from the footswitch measurement of single limb support. Patients with unilateral pathology do show loading-rate differences, but they vary with the individual more than with the type of disability. A major limitation to using a force plate is the difficulty in obtaining reliable data. The entire foot and only that foot must contact the plate during the recording period. Yet, deliberate loading (targeting) will give false data. Setting the subject's starting point so that the desired foot will naturally strike the plate during the walking trial may require numerous repetitions. Arthritic or severely paralyzed patients commonly lack the endurance needed. The force plate, thus, was not considered clinically useful until the Moss visible vector system was developed.[17,18] After modifying the system so that the signals started with initial floor contact rather than later in the limb-loading period, this technique was adopted at RLAH to display normal and pathological weight-bearing patterns. It has proved to be an excellent teaching tool. A vector gener-

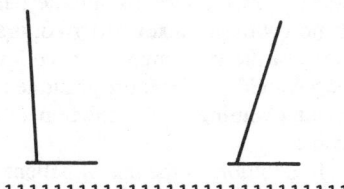

Fig. 4. Stride analyzer footswitch patterns. A. Normal sequence. Stick figures have been added to identify the pattern of floor contact. B. Type of foot-support pattern with weak calf. Note flat foot contact and lack of heel-off. H = heel, 1 = first metatarsal, 5 = fifth metatarsal.

[||] Measuring Platform, Kristal Instrument Corp, 2475 Grand Island Blvd, Grand Island, NY 14072.

Fig. 6. Child with cerebral palsy undergoing an energy-cost test.

ator transposes the vertical and horizontal (saggital or frontal) ground reaction forces into a resultant vector that is displayed on an oscilloscope (Fig. 7). This represents the line of body weight. The height of the vector is made proportional to the load intensity. A lens magnifies the oscilloscope line to the size of the subject viewed in the camera lens. Simultaneous photography of the subject and the oscilloscope is accomplished with a half-silvered mirror. By the laborious technique of manually measuring the perpendicular distance between the vector and joint center markers from each frame of film and by multiplying these moment arms by the vector magnitude, we can determine the flexion and extension (or frontal plane abduction and adduction) torques that the muscles must control. The same information can be obtained by combining instrumented motion analysis and force-plate data, but currently we lack half of that system. Also color photography that superimposes the oscilloscope representation of the body weight line (vector) on the walking subject shows the demands for muscle control more clearly (Fig. 8).

Clinically, the visible vector system has been used to differentiate appropriate and inappropriate muscle action. If the vector displays a flexion torque at the knee, activity of the quadriceps femoris muscle is appropriate. Conversely, the presence of quadriceps femoris action when the vector is on the extensor side of the joint shows faulty muscle activity. A second cause for this event must be sought. Such information on muscle action has assisted interpretation of prosthetic alignment, postural patterns of the arthritic patient, and surgical planning.

THERAPIST INVOLVEMENT

Gait laboratories provide therapists with an opportunity to obtain objective performance data on patients they are treating. In fact, in our clinical studies, the University of Southern California physical therapy students' participation is an established practice. The information necessary to document change or examine the problem depends on the clinical question to be answered but may be as simple as recording the foot-support pattern with different types of footwear or identifying the stride characteristics before and after changing foot or socket alignment in the amputee.

The clinical question may relate to a population of patients or to a particular treatment program as it is applied to a group of patients. Many questions could be asked. For instance, is constant passive motion after total knee replacement (or any other pathological category) more effective at regaining the rapid knee motion needed in gait than other postoperative treatment programs? One approach to answering this question might be by combining electrogoniometry and stride analysis.

Dynamic EMG opens a new avenue for therapists to confirm the type of muscle actions that result from the various therapeutic approaches designed to improve voluntary control by patients with neurological dysfunction. This information can be used in several ways. One use is organizing investigations to assess new procedures. A second use is assessing patient performance when it is contrary to what is customarily expected. The therapist can define the problems better and more accurately select the procedures that would be most advantageous to the patient. A third use is to identify precisely the muscle activity responsible for a particular gait pattern when more than one cause is possible. Patients with stroke, head trauma, or spinal cord injury may walk with a stiff-legged gait pattern for two reasons. They may lack adequate hip musculature to initiate a flexor pattern or the desired motion may be obstructed by persistent quadriceps femoris muscle activity (rectus femoris or the vasti muscles) in swing. In the latter situation, the therapist also would have the opportunity to learn whether the obstructive force is rectus femoris muscle participation with the flexor pattern or is spasticity of the vasti muscles. Such knowledge would permit more accurate treatment planning and prediction of outcome.

In addition to the use of patient testing laboratories, the physical therapist also is well-suited to contribute to the testing program. Because the value of dynamic EMG depends on the accurate anatomical placement of electrodes and the selection of muscles appropriate to the clinical question, the physical therapist is an integral member of the investigative team. Professional training of the physical therapist in anatomy, ki-

Basic Research

nesiology, and clinical characteristics of the different types of disability are the factors critical to effective, definitive testing. The only additional knowledge the therapist need gain is a three-dimensional awareness of muscular relationships and that of the major vessels and nerves.

Informative motion analysis requires accurate anatomical placement of the skin markers or electrogoniometer. Some of the landmarks are very subtle and pathology can introduce significant change. The physical therapist is best prepared to meet these challenges. Appreciation for the functional potential of patients and accommodating the testing procedure to their capability are facilitated by the knowledge that therapists have.

In addition to making major contributions to the testing process, the therapist also can be of great value in study design and data interpretation. So many types of data can be obtained and methods of processing the outcome can be used that producing meaningful information depends on including a functional perspective that a skilled therapist is able to provide.

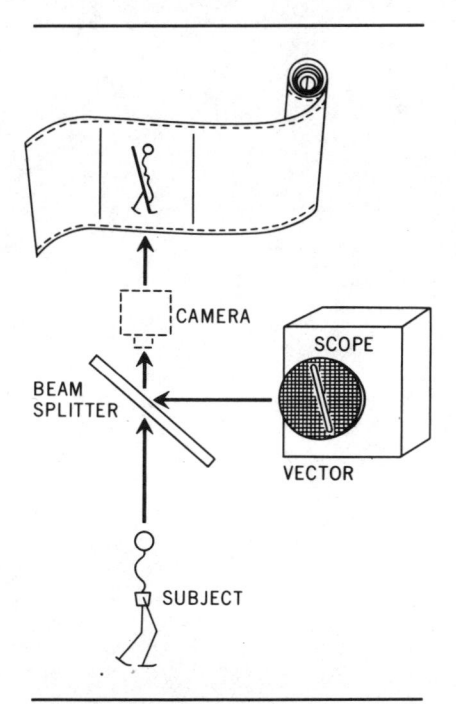

Fig. 7. Diagram of the visible vector system. Through a half-silvered mirror (beam splitter), the oscilloscope display (scope) of the instantaneous vector is simultaneously recorded with the photograph of the walking subject.

AVAILABILITY

Observational analysis is the basic technique available to all physical therapists. To profit from a systematic approach, the therapist needs training, regular practice, a pad of recording forms, and an unobstructed area within the clinic. Formal training through a short continuing education course is the most expedient way to start, but self-study also is possible. Expertise is developed by participating regularly in a consultative gait clinic where difficult problems are assessed in an organized fashion. Everyone in attendance evaluates the patient's walk and completes a gait analysis form. The findings are compared to assure all are considering the same events. Differences in interpretations are resolved, a consensus is attained, and therapeutic recommendations are formulated.

Instrumented laboratories can be developed at several levels. The determinants are training, space, funds, and engineering support. The present state of microprocessors and personal computer systems have made it possible to develop compact and reliable instrumentation. Currently, several commercial single function units are available that allow the therapists to select the types of measurements most appropriate for their clinical practice.

A system to define the patient's stride characteristics should be the basic instrument because stride identifies the patient's walking ability (eg, velocity, stride length, and single limb support). Recording the patient's foot support pattern also is very helpful. The therapist can document the patient's current ability, the progress gained by various therapeutic procedures, the effect of an orthosis, or a cardiopulmonary conditioning program. Stride measurements also define the gait variables during collection of other data types such as motion or force. If a 20-ft (6-m) hallway is available, no additional space is needed to use a system such as a stride analyzer. This type of instrumentation costs approximately $7,000.

The addition of an exoskeleton goniometer (about $3,000) makes it possible to measure knee (and hip) motion. Such a system would be of particular value in a clinic having numerous arthritic patients or others undergoing various types of knee reconstruction. As

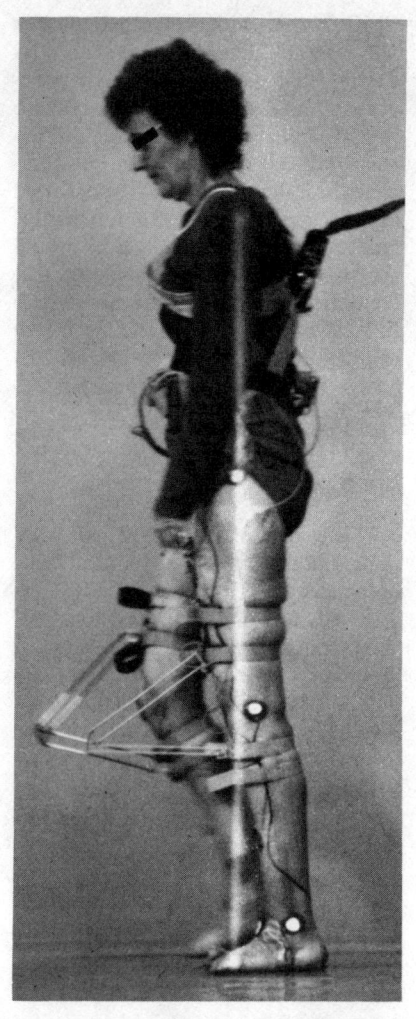

Fig. 8. Vector display of the demands of walking during late midstance after total knee arthroplasty. Location of the body weight line indicates a dorsiflexion torque at the ankle necessitating soleus-muscle restraint. The slight extensor torque at the knee relieves the quadriceps femoris muscle of any work.

yet, none of the commercial systems are designed for the ankle.

Availability of dynamic EMG is still difficult. Theoretically, one could develop a limited system for under $10,000 that would provide a quasi-accurate printout and rely on cables for data transmission. The output is too imprecise to carry clinical significance. The therapist must consider more sophisticated equipment for improved accuracy and precise coordination with other gait measurements such as footswitches or motion.

Additional laboratory capability requires a much larger financial investment and engineering assistance. An

EMG system with full capability not only would have amplifiers and recorders that could handle signals between 40 and 1,000 Hz but would also have filtering that could distinguish signals from adjacent muscles. For complete flexibility, the facility would need to add telemetry and a multichannel tape recorder. A system like this would approach $50,000. We anticipate that the difficulties in obtaining an effective EMG system will be overcome in the near future on completion of an EMG analyzer that is currently in progress.

SUMMARY

This review of the RLAH gait laboratory has emphasized our clinical focus on patient care. Research projects have followed two directions. Technical developments have related to developing the footswitch, energy cost, and dynamic EMG systems. Functional research has assessed normal performance to provide baselines for interpreting pathological activities.

REFERENCES

1. Normal and Pathological Gait Syllabus. Downey, CA, Professional Staff Association of Rancho Los Amigos Hospital Inc, 1981
2. Hsu AT, Perry J, Hislop HJ, et al: Analysis of quadriceps muscle force and myoelectric activity during flexed knee stance. Abstract. Phys Ther 63:765-766, 1983
3. Milner-Brown HS, Stein RB, Vemm R: The orderly recruitment of human motor units during voluntary isometric contractions. J Physiol (Lond) 230:359-370, 1973
4. Perry J, Easterday CS, Antonelli DJ: Surface versus intramuscular electrodes for electromyography of superficial and deep muscles. Phys Ther 61:7-15, 1981
5. Buchthal F, Guld C, Rosenfalck P: Multi-electrode study of the territory of a motor unit. Acta Physiol Scand 39:83-104, 1957
6. Burke RE, Tsairis P: Anatomy and innervation ratios in motor units of cat gastrocnemius. J Physiol (Lond) 234:749-765, 1973
7. Burke RE, Levine DN, Salcman M, et al: Motor units in cat soleus muscle: Physiological, histochemical and morphological characteristics. J Physiol (Lond) 238:503-514, 1974
8. Basmajian JV, Stecko GA: A new bipolar indwelling electrode for electromyography. J Appl Physiol 17:849, 1962
9. Jonsson B, Komi PV: Reproducibility problems when using wire electrodes in electromyographic kinesiology. In Desmedt JE (ed): New Developments in Electromyography and Clinical Neurophysiology. Basel, Switzerland, Karger, 1973, vol 1, pp 540-546
10. Lyons K, Perry J, Gronley JK, et al: Timing and relative intensity of hip extensor and abductor muscle action during level and stair ambulation: An EMG study. Phys Ther 63:1597-1605, 1983
11. Komi PV: Relationship between muscle tension, EMG and velocity of contraction under concentric and eccentric work. In Desmedt JE (ed): New Developments in Electromyography and Clinical Neurophysiology. Basel, Switzerland, Karger, 1973, vol 1, pp 596-606
12. Smidt GL: Biomechanical analysis of knee flexion and extension. J Biomech 6:79-82, 1973
13. Waters RL, Perry J, Antonelli DJ, et al: Energy cost of walking of amputees: The influence of level of amputation. J Bone Joint Surg [Am] 58:42-46, 1976
14. Waters RL, Perry J, Antonelli DJ, et al: Energetics: Application to the study and management of locomotor disabilities. Orthop Clin North Am 9:351-377, 1978
15. Waters RL, Campbell J, Thomas L, et al: Energy cost of walking in lower-extremity plaster casts. J Bone Joint Surg [Am] 64:896-899, 1982
16. Skinner SR: The correlation between gait velocity and rate of lower-extremity loading and unloading. Bulletin of Prosthetic Research 18(1):303-304, Spring, 1981
17. Cook TM, Cozzens BA, Kenosian H: Real-time force line visualization: Bioengineering applications. In: Advances in Bioengineering. New York, NY, American Society of Mechanical Engineers, 1977
18. Cook TM, Cozzens BA: The effects of heel height and ankle-foot-orthosis configuration on weight line location: A demonstration of principles. Orthotics and Prosthetics 30(4):43-46, 1976

Basic Research

Objective Clinical Evaluation of Function
Gait Analysis

R. KEITH LAUGHMAN,
LINDA J. ASKEW,
ROBERT R. BLEIMEYER,
and EDMUND Y. CHAO

Automated gait analysis allows us to document and quantify objectively normal gait, functional deficits, and patient response to therapeutic intervention. Instrumentation for this analysis at the Mayo Clinic Gait Laboratory includes three-dimensional electrogoniometers for measurement of relative joint rotation at the hip, knee, and ankle; footswitches that record foot-floor contact sequences; instrumented mats that measure step length and width; piezoelectric force plates for measurement of floor reaction forces; and two walkways that simulate a variety of ground conditions. We use a DEC-PDP 11/34 computer for acquisition, storage, and analysis of data and for generation of a gait report form that displays a patient's results relative to normal and previous evaluations. Applications of these techniques include assessment of function preoperative and postoperative total joint arthroplasty, quantification of gait faults, and documentation of effectiveness of exercise and gait training techniques. We have demonstrated the reliability of the techniques, accumulated a sizeable normal data bank, and developed a concise, effective data summary for communication with referring practitioners.

Key Words: *Biomechanics, Computers, Gait, Physical therapy.*

In physical therapy, as in all areas of medical care, a need exists for instrumentation and procedures to document objectively patient response to treatment. Because much of physical therapy is concerned with the restoration of function, the ability to quantify functional progress is a logical application of existing technology.

Objective gait analysis is relatively new in the clinic, but much progress has been made recently. Currently, a number of gait analysis laboratories exist with many different gait analysis systems.[1-7] Within the constraints of available instrumentation and expertise, each laboratory has been designed to meet the needs of specific patient populations.

The Mayo Clinic Gait Laboratory is part of the Orthopedic Biomechanics Laboratory, Mayo Clinic/Foundation. The staff with full-time assignment to evaluation includes two physical therapists, one laboratory assistant, and one secretary. Other laboratory personnel with expertise in mechanical and electrical engineering and computer science work part-time in the area of patient assessment.

According to a classification of gait analysis outlined by Quanbury,[8] our laboratory is currently evolving through the documentation stage. In sequential assessments, we can record and monitor most gait problems that present themselves as an abnormal motion pattern. This capability allows for more detailed monitoring of patients' progress and thus more effective follow-up.

In 1981, Brand and Crowninshield specified six general conditions that must be met for a laboratory to be useful in a clinical setting.[9] Their conditions centered around the accuracy, reliability, validity, and encumbrance of the instrumentation. These conditions are indisputable. In the following sections, we describe our gait analysis methods and illustrate how they meet or exceed these basic criteria.

METHOD
Equipment

Our gait laboratory has two walkways: the first is used in the measurement of level walking, and the second walkway is used in the measurement of walking on various ground conditions, such as stairs, ramps, and side-slopes (Fig. 1). The instrumentation and analysis methods have been described and justified previously in detail.[10,11] Briefly, they consist of three-dimensional electrogoniometers for measurement of relative joint rotation at the hip, knee, or ankle. Because of the attachment of our instrumentation to the soft tissue overlying the bony structures, we are monitoring primarily the osteokinematics at a particular joint rather than the detailed arthrokinematics. The electrogoniometers consist of three precision potentiometers oriented orthogonal to one another (Fig. 2). Changes in voltage output occur with motion and are calibrated to be read in degrees. The electrogoniometers are aligned so that the proximal coordinate system is fixed, whereas the distal coordinate system is movable. The flexion/extension axis is aligned relative to bony landmarks to ensure consistent placement. The patient's foot-floor contact

---

Mr. Laughman is Research Associate, Department of Orthopedics, and a member of the faculty, Mayo Program of Physical Therapy, Orthopedic Biomechanics Laboratory, Mayo Clinic/Foundation, Rochester, MN 55905 (USA).

Ms. Askew is Research Physical Therapist, Orthopedic Biomechanics Laboratory, Mayo Clinic/Foundation.

Mr. Bleimeyer is System Analyst/Programmer, Orthopedic Biomechanics Laboratory, Mayo Clinic/Foundation.

Dr. Chao is Director of the Orthopedic Biomechanics Laboratory, Mayo Clinic/Foundation.

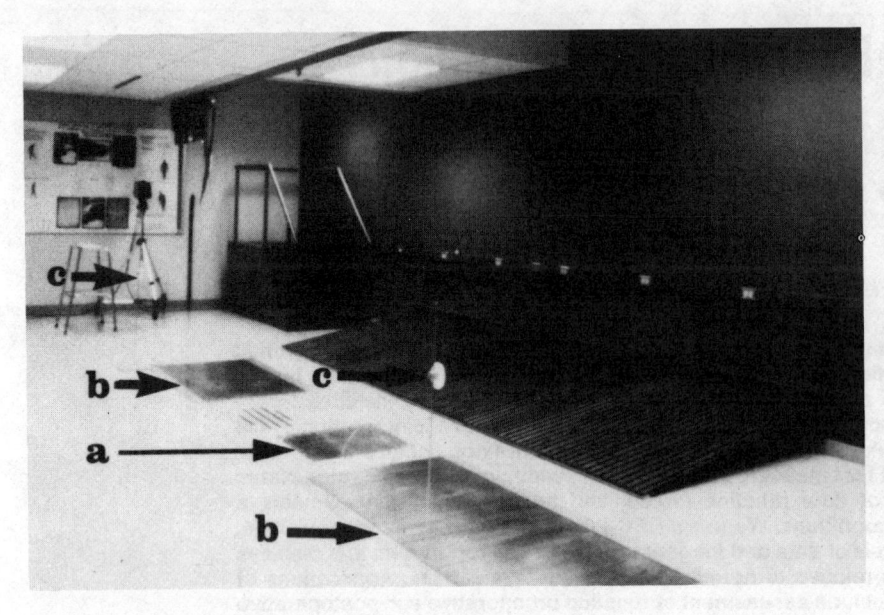

Fig. 1. Gait laboratory walkways (level and varied ground conditions): a = piezoelectric force plates, b = instrumented stainless steel mats for the measurement of step length and width, and c = infrared light switches used in the measurement of walking velocity and as triggers for computer sampling.

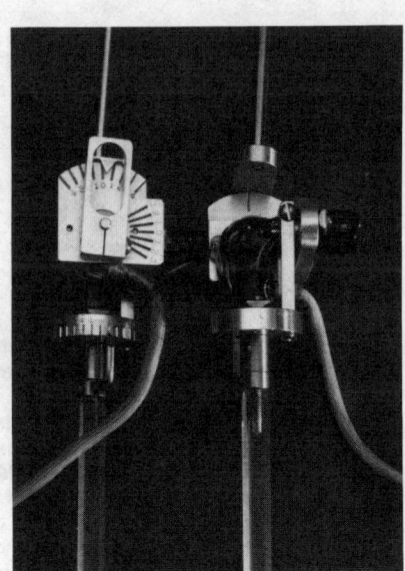

Fig. 2. Lateral (L) and medial (R) views of the three-dimensional electrogoniometer.

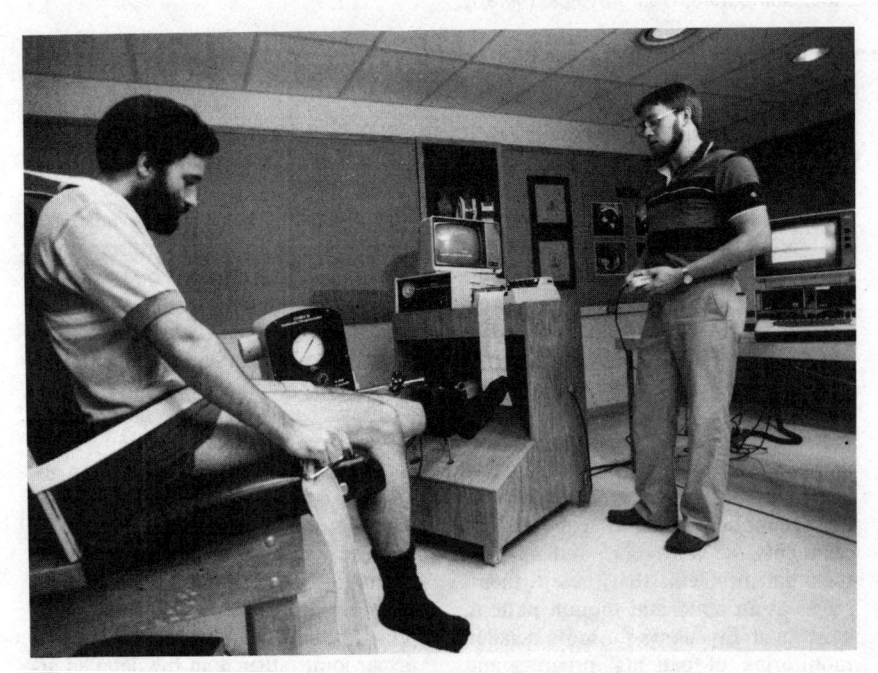

Fig. 3. Modified Cybex® II dynamometer with Apple II® computer automation.

sequence is measured with binary footswitches at the heel, fifth and first metatarsal heads, and great toe; step lengths and widths are measured using instrumented mats (Fig. 1b). The three component forces of the floor-reaction force vector are measured with two piezoelectric force plates (Fig. 1a). One force plate has been modified to measure the areas of pressure beneath the foot during walking. These gait data are subjected to automated data reduction and storage using a DEC-PDP 11/34 computer.* To supplement the motion and force data, we also measure the patient's lower extremity strength as part of our assessment. This assessment is accomplished using either an isometric torque cell or a modifed Cybex® II dynamometer† with Apple II® computer‡ automation (Fig. 3).

Analysis

Based on test-retest results, we have found the instrumentation and analysis methods to be reliable and reproducible when proper attention is paid to the attachment and alignment of the electrogoniometer. Figure 4 is the average knee motion pattern obtained from a group of five subjects with normal gait patterns who were in a test-retest investigation. The figure demonstrates the reproducibility of the curves.

The automated analysis methods involve the generation of average motion patterns from 15 to 20 gait cycles (heel strike to ipsilateral heel strike). Each curve is fitted with Fourier coefficients that facilitate curve averaging, storage, and analysis. We obtained the average motion patterns in Figure 4 with the aid of these Fourier coefficients. Peak-to-peak motion values are also measured at various intervals throughout the gait cycle.

* Digital Equipment Corp, 200 Baker Ave, W Concord, MA 01742.

† Cybex, Division of Lumex, Inc, 2100 Smithtown Ave, Ronkonkoma, NY 11779.

‡ Apple Computer, 10260 Bandley Dr, Cupertino, CA 95014.

In addition, we calculate composite indexes[11] to provide a general overview of the patient's function. We use the index approach for two reasons: 1) it facilitates data analysis and comparison and 2) it provides a better indication of the patient's overall function through the combined effects of selected gait characteristics.

The form used for reporting test results is a matter of considerable importance. If the data are not communicated in an effective format, they will be of little use in the clinic regardless of quality. The data presentation should be accurate, clear, and concise, because often this information is communicated to clinicians who may not be experts in gait analysis and, therefore, may not take the time to sort through pages of numbers and charts. In an attempt to meet this need, our patients' results are sent to the referring clinician by a computerized report form (Fig. 5). The individual characteristics represent a group of variables that are familiar to clinicians and that they have found helpful. In addition, the previously discussed indexes are also reported in the last two columns of the form. The data are presented in both graphic and numeric forms with the rectangles representing values ranging from the 10th to 90th percentile of a normal sample (N = 69). The numbers and symbols then represent the patient's individual values obtained during the evaluation. Sequential evaluations can be plotted on the same form so that we may observe a patient's progress relative to normal and relative to earlier assessments.

The assessment procedure itself requires about 45 minutes. This procedure includes a detailed explanation to the patient, a brief physical examination, applying the instruments to the patient, and data collection. The report form is printed once we obtain the patient's average results. The data analysis and summarization require about 30 to 45 minutes.

CLINICAL APPLICATIONS

We use the gait analysis method to document and assess virtually all of the classic gait faults that consist of abnormal motion patterns or temporal and distance abnormalities. Some examples of typical deviations that we have observed follow.

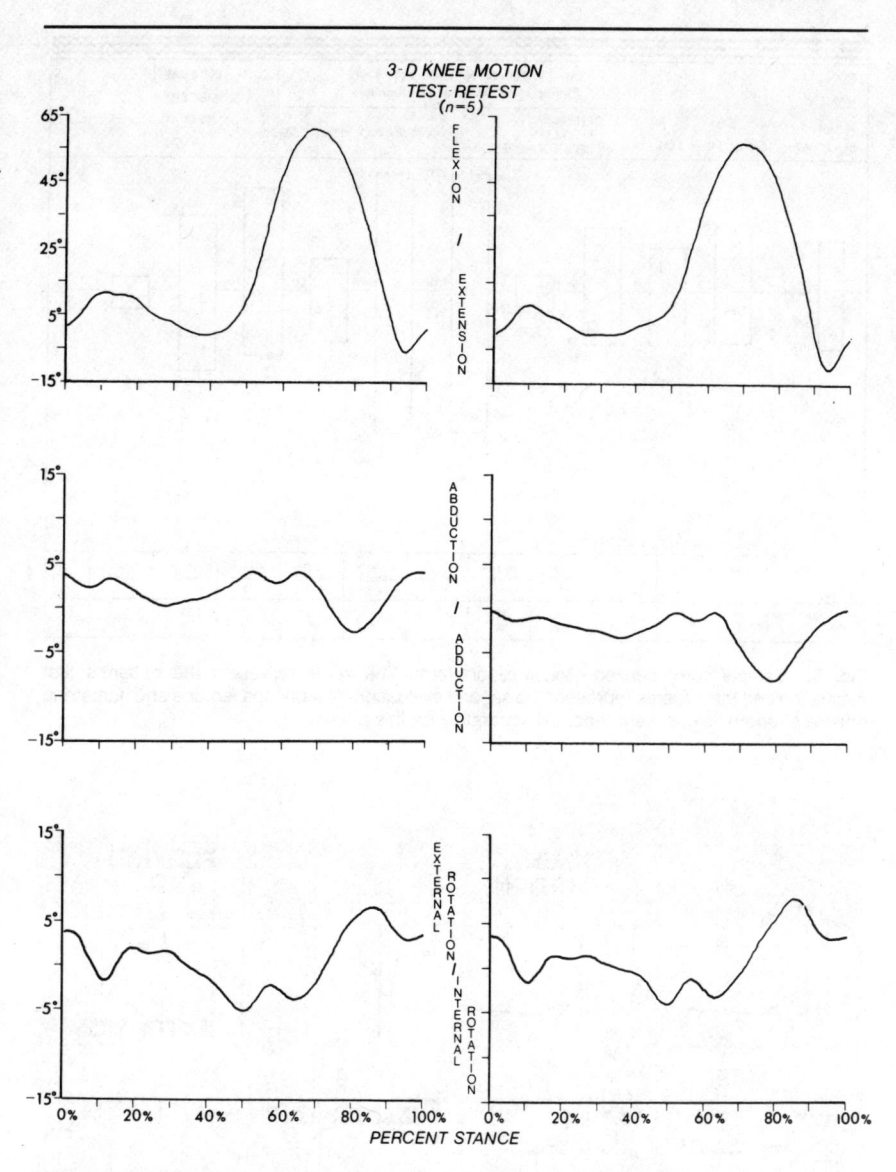

Fig. 4. Average three-dimensional knee motion data obtained from five subjects under test-retest conditions. First assessment on left, with the follow-up assessment on the right.

Figure 6 illustrates a vault that involves the replacement of normal stance phase knee flexion with stance phase knee extension. The footswitch pattern is also abnormal, demonstrating the prolonged metatarsal contact associated with this gait pattern.

Figure 7 illustrates the coronal plane hip motion associated with an abducted gait pattern. In this deviation, the normal stance phase hip adduction of the femur relative to the trunk is replaced by hip abduction. This gait problem is frequently seen in patients with weak hip musculature or a painful hip joint. It is also seen in amputees with pain at the stump-socket interface.

The instrumented mats embedded in the walkway measure both step length and width. Figure 8 illustrates an asymmetrical step length that is frequently observed in patients for a variety of reasons. We obtained these data early in the patient's rehabilitation and at the end of training. The marked improvement is easily documented. Figure 9 illustrates the broad-based gait (compared with normal) that was observed in an above-knee amputee.

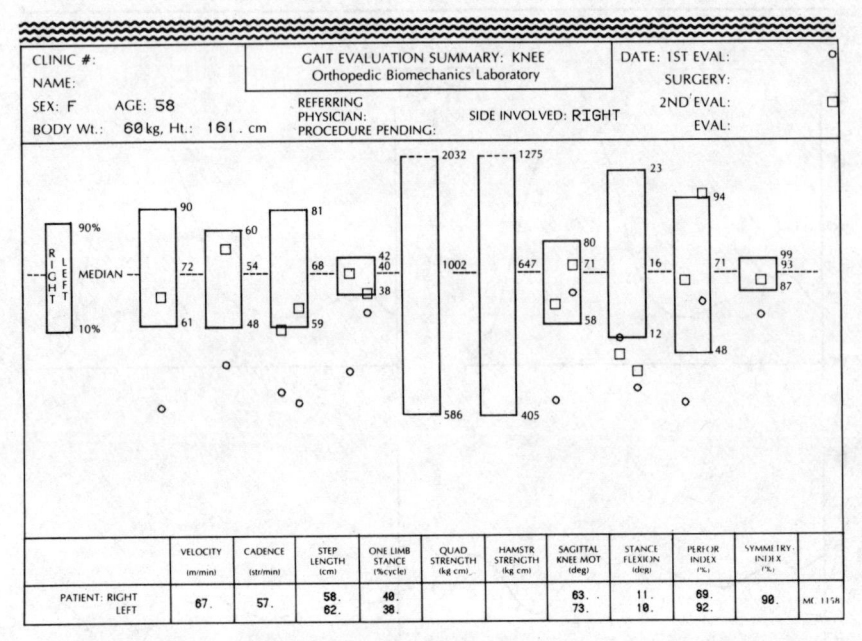

Fig. 5. Sample computerized clinical report form. The zeros represent the patient's first evaluation and the squares represent the second evaluation. Quadriceps femoris and hamstring muscle strength values were reported separately for this patient.

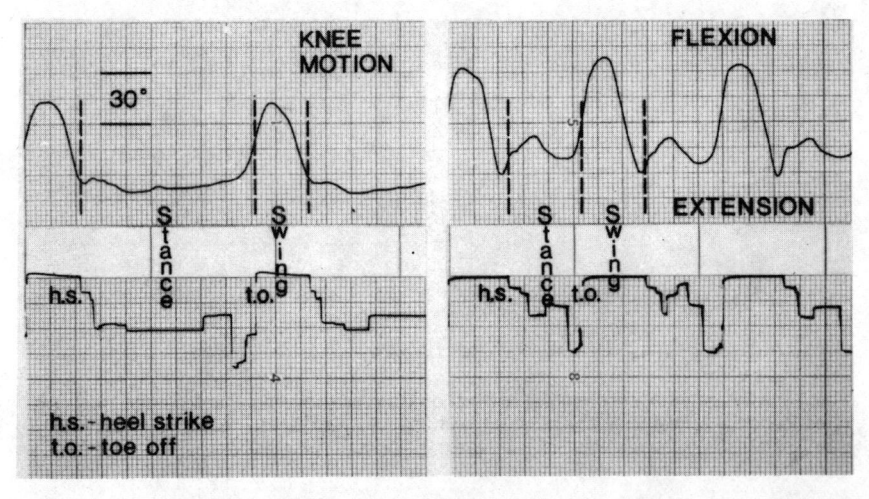

Fig. 6. Tracing on the left represents sagittal-plane knee motion associated with vaulting. The tracing on the right represents normal sagittal-plane knee motion.

The four footswitches under the plantar surface of the foot provide information on the patient's foot-floor contact sequence and the temporal and distance gait patterns. Figure 10 is an example of a patient who initially demonstrated a rather flatfooted gait pattern, making minimal contact with the anterior portion of the foot. After physical therapy intervention, the patient was rolling over the prosthetic foot more effectively with improved first metatarsal and great toe contact (indicated by the increased height of the footswitch pattern).

Instrumented gait analysis can aid in examining the effectiveness of a specific treatment and selecting an optimal mode of treatment. Figures 11 and 12 are sample tracings from a patient who had been using a prosthesis for many years when he came to us for treatment of his low back pain. We believed his back pain most likely was caused by his bilateral abducted gait pattern. The patient underwent symptomatic treatment for his back. He also received gait retraining. After seven treatments, the patient reported reduced back pain and, according to both the patient and therapist, his gait pattern had improved. At reevaluation, we could not identify any change in the patient's gait pattern. This points out a problem experienced by all clinicians: the difficulties associated with monitoring a patient's progress through subjective impressions. Objective data provide a way to measure actual changes in function and identify which patients may be benefiting from a particular treatment. This type of data could also provide justification for additional gait training during the early phases of prosthetic use in an attempt to prevent the development of habit patterns that may not be correctable at a later time.

Our largest group of patients are individuals who, secondary to arthritic changes (rheumatoid or degenerative joint disease), underwent total knee replacement. We assessed these patients before and one year after their surgical procedure. Based on the average performance index values, the patients benefited functionally from this surgical procedure. Postoperatively, the patients' performance indexes moved into a range slightly below the lower limits of normal knee function. The strong relationship that exists between the patients' preoperative to postoperative change versus their preoperative levels of function could prove helpful in determining the appropriate timing for total knee replacement in patients who have borderline results in a clinical exam (Fig. 13). Not all patients experience similar improvement in function as a consequence of total knee surgery, with progressively less improvement noted as the preoperative knee function approaches normal limits. The values noted by zeros in Figure 13 represent patients who, at the time of their postoperative evaluation, had increased symptoms resulting from an identifiable episode, such as a sprained joint capsule from a fall, flareup of rheumatoid arthritis, or malposition of prosthetic components.

The following is an example of a patient who had undergone an arthrotomy for medial meniscectomy of the right

Basic Research

knee. She had persistent symptoms after surgery and made little progress during rehabilitation. Seven months after the initial surgery, the patient underwent follow-up arthroscopy. At that time, no evidence existed of internal derangement, but we found a grade 3 to 4 chondromalacia involving the medial femoral condyle. The initial evaluation data (zeros) in Figure 5 illustrate her overall knee function was below normal. In addition, based on Cybex® strength testing, her quadriceps femoris muscle values on the right were approximately 47 percent of the left. With these assessment results in mind, we initiated a more aggressive rehabilitation protocol with increased emphasis on strengthening. Subsequent reevaluation two months later documented marked progress. Her right-to-left Cybex® torque ratios were in the 80 percent range. The second evaluation (squares) demonstrates considerable relative improvement in all gait characteristics except stance phase knee flexion (Fig. 5).

DISCUSSION

Although this paper has dealt with gait instrumentation and analysis, objective measurement techniques for the upper extremity are also available. Key characteristics for the upper extremities include muscle strength, joint orientation and range of motion, and ability to perform functional activities. We have developed methods that provide quantitative assessment of these characteristics within our laboratory and have used these methods to establish a normal data bank and to assess numerous patient groups.

The specific applications of instrumented patient assessment are as varied as the different patient populations and the characteristics of interest. Not all analysis systems are equally suited for a particular patient group; neither are all of the instrumentation of the large existing gait laboratories always necessary. Many simplified analysis methods are on the market or can be designed fairly easily in-house, which allows much useful information to be obtained with only a modest expenditure.

In this era of government reimbursement for medical services (Medicare-Diagnosis Related Groups), the ability to document the effectiveness of and the need for a particular treatment will assume an increasingly important role. Technology, in particular electronic instrumentation, is at a level where it is both feasible and affordable for virtually all physical therapy departments to provide, in some measure, an objective form of patient assessment. Once this capability is widespread, it should facilitate the identification of optimal treatment regimes, suitable patient populations, and provide a foundation for much needed clinical research.

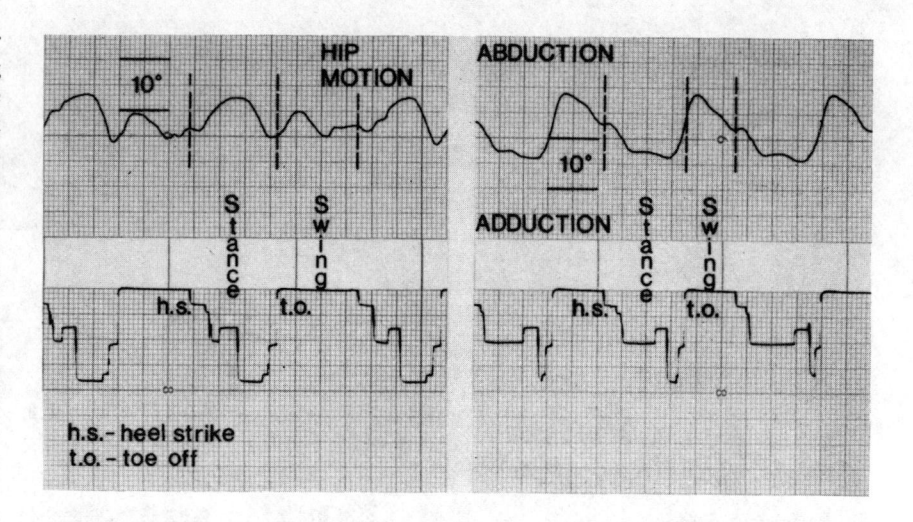

Fig. 7. Tracing on the left represents an abducted gait pattern. Tracing on the right illustrates normal coronal plane hip motion.

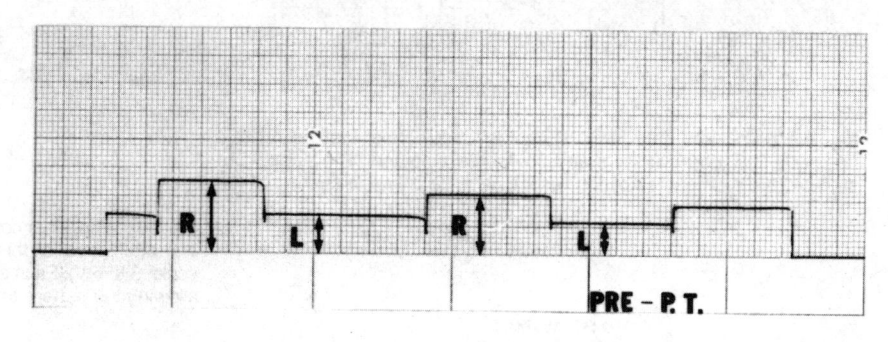

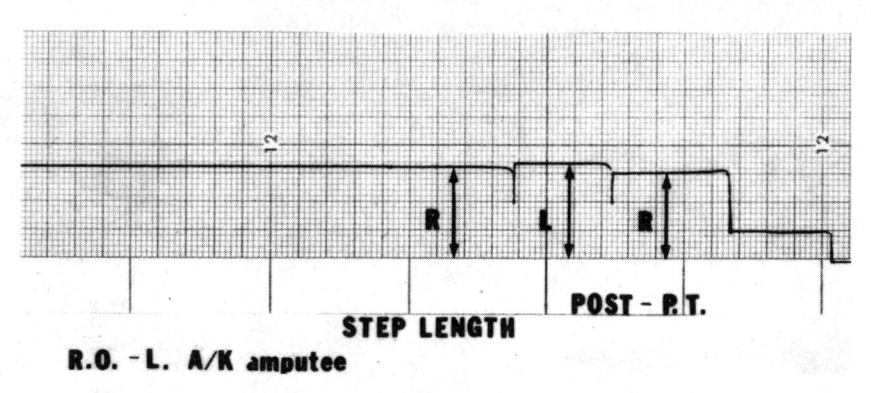

Fig. 8. Step-length data obtained from a patient before and after physical therapy. Vertical scale represents step length in centimeters (letters R and L referring to the patient's right and left step lengths, respectively—120-cm full scale).

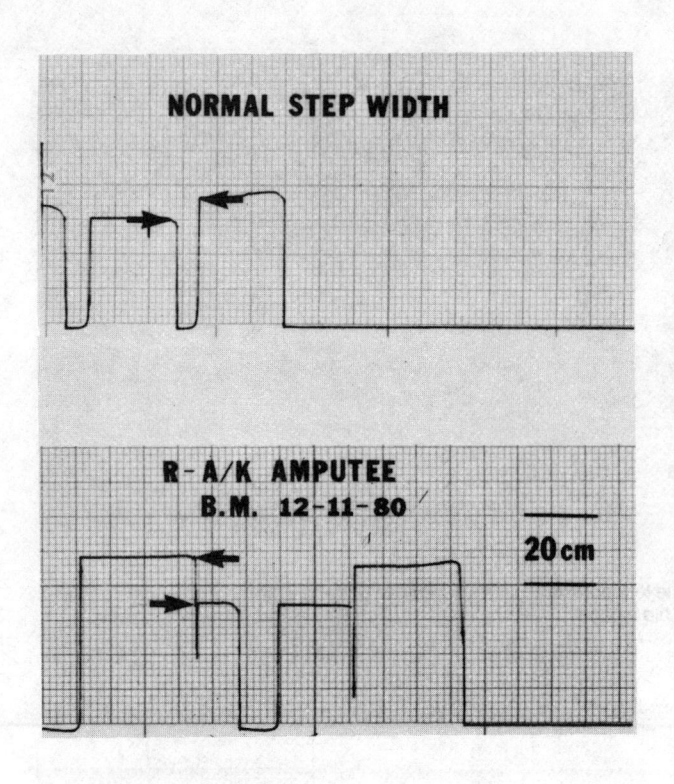

Fig. 9. Step-width data (arrows) comparing a normal step width (top) with that obtained from an above-knee amputee. The actual step width is represented by the vertical distance between arrows.

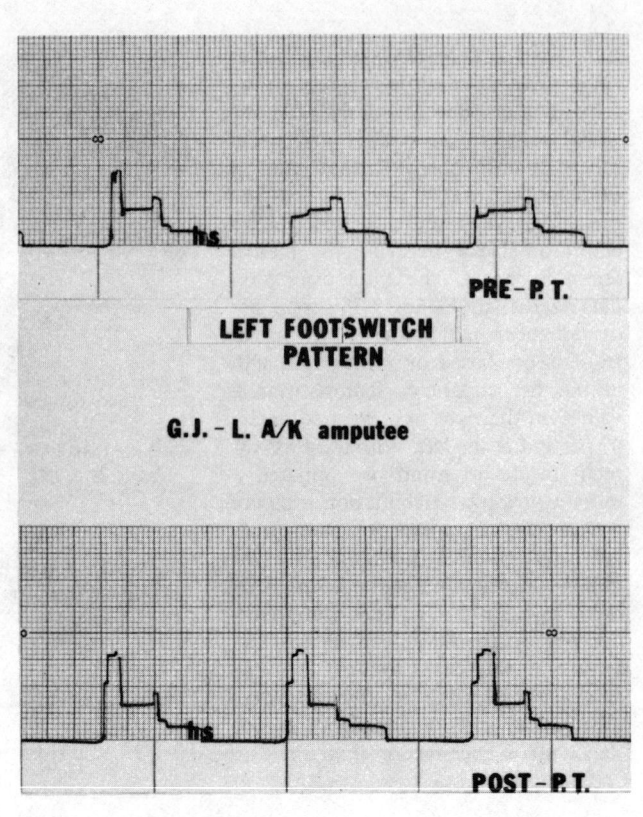

Fig. 10. Foot-floor contact pattern obtained from an amputee before and after physical therapy. Vertical scale representing voltage (full scale, 0 to 5 V). Each switch is represented by a specific voltage level allowing the pattern to be decoded (hs refers to heel strike).

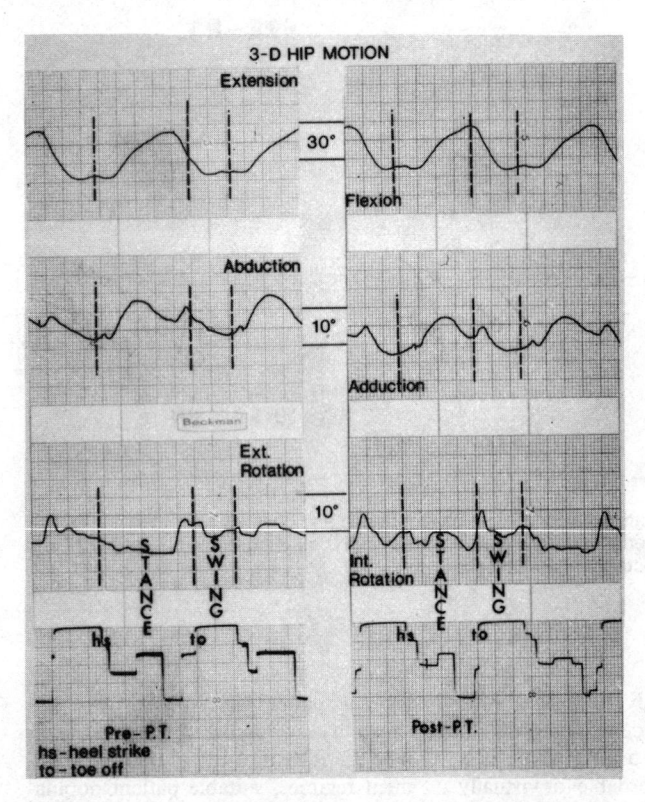

Fig. 11. Three-dimensional hip motion obtained from a patient before and after physical therapy.

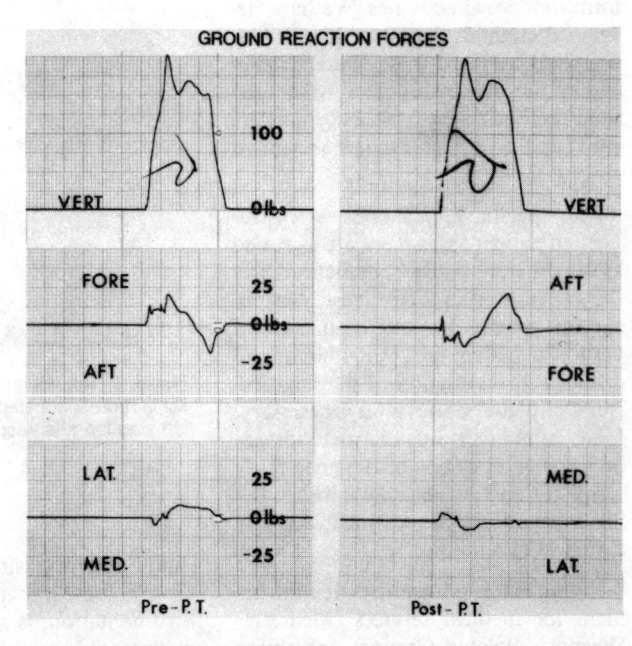

Fig. 12. Foot-floor reaction forces from a patient before and after physical therapy. These raw data are later normalized to percent body weight.

KNEE FUNCTION

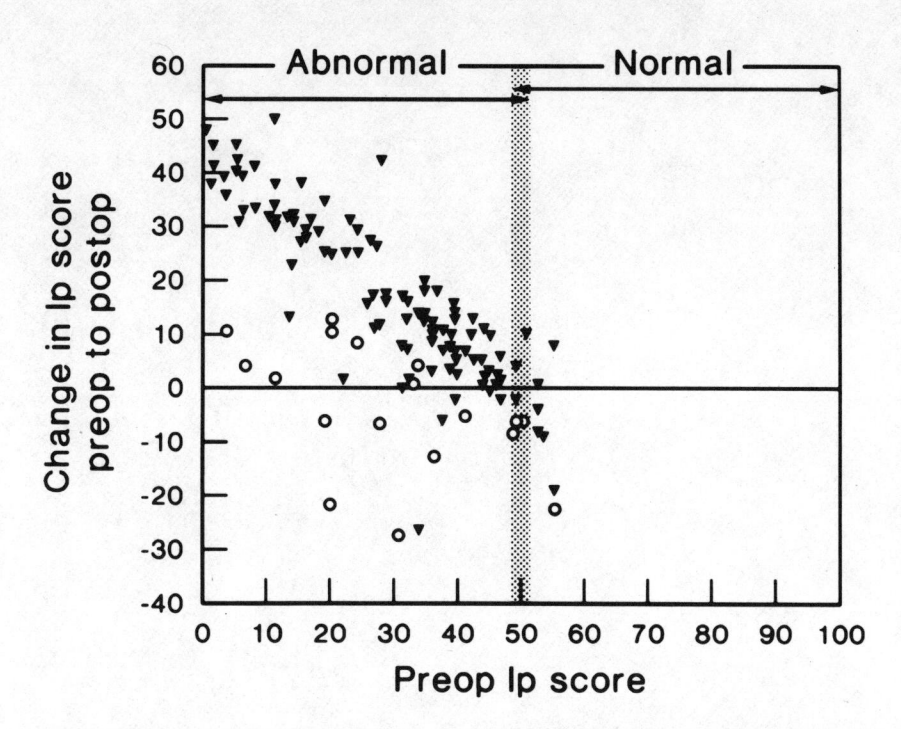

Fig. 13. Preoperative to postoperative change in knee function (Ip), plotted as a function of preoperative function. The 18 knees designated by zeros had reasons identified for their below average postoperative results.

REFERENCES

1. Andriacchi TP, Galante JO, Fermier RW: The influence of total knee-replacement design on walking and stair climbing. J Bone Joint Surg [Am] 64:1328–1335, 1982
2. Simon SR, Trieshmann HW, Burdett RG, et al: Quantitative gait analysis after total knee arthoplasty for monarticular degenerative arthritis. J Bone Joint Surg [Am] 65:605–613, 1983
3. Murray MP, Mollinger LA, Sepic SB, et al: Gait patterns in above-knee amputee patients: Hydraulic swing control vs constant-friction knee components. Arch Phys Med Rehabil 64:339–345, 1983
4. Taylor KD, Mottier FM, Simmons DW, et al: An automated motion measurement system for clinical gait analysis. J Biomech 15:505–516, 1982
5. Hershler C, Milner M: Angle-angle diagrams in above knee amputees and cerebral palsy gait. Am J Phys Med 59:165–183, 1980
6. Letts RM, Winter DA, Quanbury AO: Locomotion studies as an aid in clinical assessment of childhood gait. Can Med Assoc J 112:1091–1095, 1975
7. Waters RL, Frazier J, Garland DE, et al: Electromyographic gait analysis before and after operative treatment for hemiplegic equinus and equinovarus deformity. J Bone Joint Surg [Am] 64:284–288, 1982
8. Quanbury AO: Keynote Address. Read at the First International Congress of Biomechanics, Symposium on the Assessment of Pathological Gait, Waterloo, Ontario, Canada, August 1983
9. Brand RA, Crowninshield RD: Comment on criteria for patient evaluation. J Biomech 14:655, 1981
10. Chao EY: Justification of triaxial goniometer for the measurement of joint rotation. J Biomech 13:989–1006, 1980
11. Chao EY, Laughman RK, Stauffer RN: Biomechanical evaluation of pre- and postoperative total knee replacement patients. Arch Orthop Trauma Surg 97:309–317, 1980

Basic Research

Technical Report

Effects of Electromyographic Processing Methods on Computer-Averaged Surface Electromyographic Profiles for the Gluteus Medius Muscle

Rob FM Kleissen

The objective of this study was to demonstrate how two different linear envelope detectors, as used in the quantification of surface electromyographic (EMG) signals, can lead to differences in the properties of observed computer-averaged EMG profiles. Eight healthy male subjects, aged 24 to 32 years, participated in the study. Computer-averaged EMG profiles for the gluteus medius muscle were recorded at a free-walking cadence and at stepping frequencies of 78 and 120 steps/min, using detectors with 3- and 25-Hz cutoff frequencies simultaneously. The 3-Hz filtered EMG profiles proved to be smoother, to exhibit a significantly lower cycle-to-cycle variability, and to have a greater time lag with respect to the original unprocessed EMG signal than the 25-Hz filtered EMG profiles. The observed intersubject variability also was lower for the 3-Hz filtered EMG profiles than for the 25-Hz filtered EMG profiles. The results indicate that comparison of EMG profiles recorded with different detectors is difficult. A standard detector may solve this difficulty. [Kleissen RFM. Effects of electromyographic processing methods on computer-averaged surface electromyographic profiles for the gluteus medius muscle. Phys Ther. 1990;70:716–722.]

Key Words: Computers; Electromyography; Kinesiology/biomechanics, gait analysis; Muscle performance, general; Tests and measurements, electrodiagnosis.

During the past decade various reports[1-4] have described the use of computer-averaged electromyographic (EMG) profiles in the analysis of pathological gait. For instance, Knutsson[2] used computer-averaged EMG profiles to study muscle coordination in hemiparetic gait, Conrad et al[3] reported on the use of EMG profiles for investigating the gait patterns of paraspastic patients with a variety of diagnoses, and Winter and Sienko[4] used EMG profiles in an attempt to elucidate the mechanics of gait in patients with below-knee amputation. These studies show that computer-averaged EMG profiles provide potentially valuable information about the time course of myoelectric activity of muscles.

A muscle's EMG profile may be observed during a number of gait cycles, with the beginning of each new gait cycle defined by the moment of heelstrike. This "raw" EMG signal is usually full-wave rectified and low-pass filtered, an operation known as *linear envelope detection.*[5] The output signal of the detector represents the time course of the estimated intensity of the EMG signal and is stored in a computer to allow the second processing step. The intensity of the EMG signal, sometimes called the "amount of activity,"[5] indicates the muscle's myoelectric activity. The accuracy of the estimate in this first step is governed by the bandwidths of the raw EMG signal and the low-pass filter used[6] and is limited. In the second step, the time scale of the recorded intensity signal in each individual gait cycle is normalized to a standard cycle length. The resulting intensity signal frames, which correspond to the

---

R Kleissen, MSc, is Research Engineer, Research and Development Group, Rehabilitation Centre Het Roessingh, Roessinghsbleekweg 33, 7522 AH Enschede, the Netherlands.

This work was supported by a grant from St Jorisstichting, Bussum, the Netherlands.

This article was submitted January 24, 1989, and was accepted July 23, 1990.

individual cycles, are ensemble-averaged to form the final EMG signal intensity profile. This ensemble-averaging process improves the accuracy of the intensity information, because random fluctuations in the contributing signal frames tend to cancel each other. The choice of the low-pass filter in the linear envelope detector is an important consideration for the application of the EMG profiles. A wide variety of low-pass filters have been reported in the literature. Knutsson[2] reported time constants of both 0.2 and 0.02 second. Winter[1] used a critically damped filter with a cutoff frequency of 3 Hz. Arsenault et al[7] worked with a 6-Hz filter. Kleissen et al[8] used a third-order Butterworth filter with a cutoff frequency of 25 Hz. This lack of uniformity indicates that choosing a low-pass filter is not a trivial matter. It also makes comparison of the published results of different investigators difficult.

The purpose of this study was to demonstrate how the choice of the low-pass filter used in the linear envelope detector can affect the recorded EMG profiles. The results of this study may contribute to clinicians' understanding of the effects of EMG processing methods on computer-averaged EMG profiles.

Method

Instrumentation

In this study, computer-averaged profiles were recorded for the surface EMG activity of the gluteus medius muscle. Figure 1 schematically shows the instrumentation used for obtaining the EMG profiles. A K-Lab SPA-10 skin-mounted bipolar preamplifier,* which snap-connects to two Medi-Trace self-adhesive disposable silver-silver chloride pellet electrodes† picked up and amplified the EMG signal at the electrode location. This preamplifier provides a gain factor of 100 (± 1%) and has an input impedance greater than 100 MΩ, a common mode rejection ratio greater than 100 dB, and a passband of −3 dB between 20 Hz and 10 kHz. The rim-to-rim electrode distance was 22 mm, and the pickup area of each electrode was 10×10 mm^2. The low mass of the amplifier (15 g) and its low output impedance reduce the problems of movement and cable artifacts. A reference electrode was attached to the wrist of the subject.

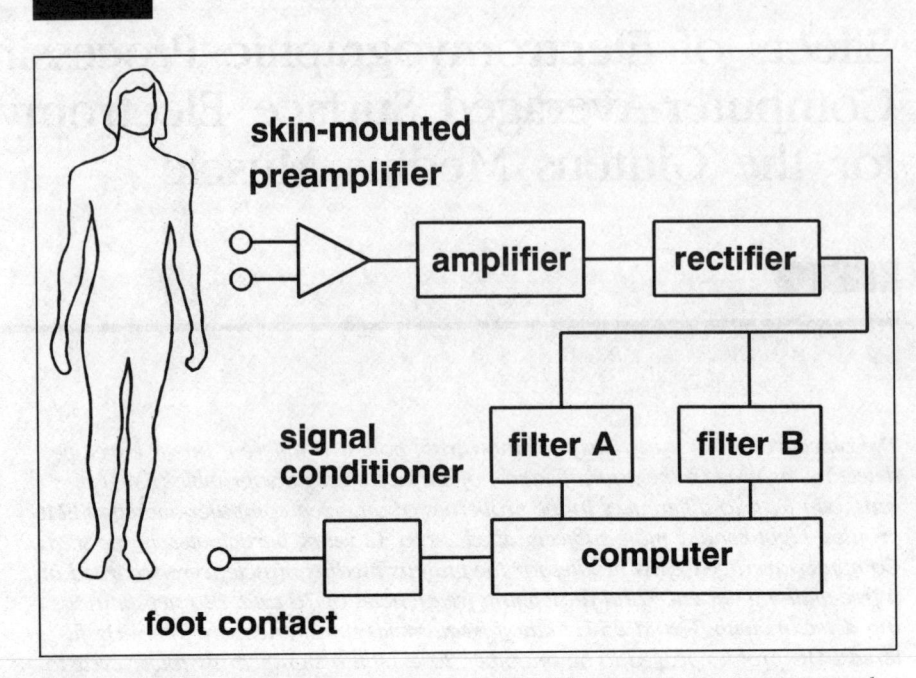

Figure 1. *Block diagram of instrumentation used for obtaining computer-averaged electromyographic profiles.*

The time of heel-strike was identified using footswitches. The footswitch system comprised four sheets of conductive rubber, covering the forefoot and the heel of the shoes, in conjunction with a conductive rubber mat. When one of these sheets touched the rubber mat, an electric signal indicated floor contact.

A cable transported the preamplified EMG and footswitch signals to the data-processing station where they were conditioned before computer processing. An electronic circuit removed occasional instability in the footswitch signals ("switch-bouncing") at heel-strike and toe-off. Processed signals representing the stance and swing phases of the gait cycle changed state only after the raw signals from the footswitches had been stable for 100 msec. The computer system took this 100-msec time delay into account.

The EMG signal first passed through a third-order high-pass Butterworth filter with a cutoff frequency of 20 Hz (± 5%) for suppression of remaining movements and cable artifacts. This cutoff frequency specifies the frequency at which the filter's attenuation has increased to 3 dB. Subsequently, after further amplification and full-wave rectification, the EMG signal passed through two different low-pass filters connected in parallel. Filter A was a third-order Butterworth filter with a cutoff frequency of 25 Hz (± 5%). Filter B was a critically damped second-order filter with two time constants of 47 msec (± 10%). This filter is identical to two first-order Butterworth sections connected in series, each section with a cutoff frequency of 3.4 Hz (± 10%). This filter will be referred to as the 3-Hz

---

*K-Lab, PO Box 70167, 1007 KD Amsterdam, the Netherlands.

†Graphics Controls Canada Ltd, Gananoque, Ontario, Canada K7G 2Y7.

filter. Footswitch signals and the EMG signal, thus processed with two linear envelope detectors having only different low-pass filters, were sampled at 200 Hz and converted automatically into computer-averaged EMG profiles using the procedure and equipment described by Kleissen et al.[8]

On the conductive rubber mat, two infrared beams demarcated a 10-m walkway. Crossing the beams automatically started and stopped data collection of footswitch and EMG signals. An HP 3310B signal generator[‡] and a small loudspeaker served as a metronome. Off-line data analysis was performed with a Tulip AT Compact 2 computer[§] using the LOTUS 1-2-3 spreadsheet package.[ǁ]

Protocol

Eight healthy male subjects, aged 24 to 32 years, participated in this study. Each subject provided informed consent prior to participation in the study. After thoroughly rubbing the skin over the gluteus medius muscle with alcohol, and after shaving the area when necessary, the electrodes were placed vertically on the muscle belly of the gluteus medius muscle, halfway between the iliac crest and the greater trochanter of the right leg. Before measurements were taken, the subjects were given a few minutes to practice the walking procedure so that they could become familiar with the experimental constraints. They practiced the procedure until they said they were steady in their walking pattern.

Each subject performed three walking trials. The first trial was at a comfortable, free-walking cadence. During the second and third trials, the metronome was set at 78 and 120 beats per minute (bpm), respectively, and the subjects were instructed to adjust their walking cadence accordingly.

The subjects were free to choose their own comfortable stride length. An assistant walking behind the subject carried the trailing signal cable to avoid obstruction by cable drag.

Data were collected during 20 to 25 gait cycles. Because the 10-m walkway did not allow this number of gait cycles to be observed in one pass, the subjects were instructed to walk back and forth across the walkway until the desired number was reached.

Data Analysis

The computer system automatically determined both the average EMG signal intensity profile of each trial and its standard deviation. Each stride period between two consecutive heel-strikes was normalized to 100% of gait-cycle duration, where 0% corresponded to heel-strike. The output signal of the linear envelope detector was interpolated at 0.125% gait-cycle intervals; thus, the normalized gait cycle contained 800 data points. Averaging the EMG data across the normalized strides for each of the 800 data points produced the ensemble-averaged profile. Similarly, the standard deviation for each of the data points of the final EMG signal intensity profile was calculated.

In off-line analysis, across-subject ensemble averages and their standard deviations were computed by averaging the average EMG profiles of the eight subjects. For a given ensemble average and its standard deviation, the coefficient of variation (CV) describes the variability. The CV is defined as the root-mean-square standard deviation over the stride period divided by the mean ensemble average over the stride.[1] Pearson's product-moment correlation coefficient (r) can be used to quantify the similarity of two EMG profiles.[9] The 800 data points defining one EMG profile are paired with the corresponding 800 data points from the other EMG profile. A high correlation coefficient indicates that the time courses of both EMG profiles are similar.

Because this coefficient does not necessarily express absolute differences between the EMG profiles, an additional measure for similarity (S), defined as the CV for the ensemble averages of the two EMG profiles to be compared, was computed. Statistical significance was set at the alpha level of .05.

Results

Figure 2 shows a typical result of one walking trial at a free-walking cadence. In this figure, A and B present the ensemble averages and standard deviations of EMG signal intensity for the gluteus medius muscle, as observed using a 25-Hz and a 3-Hz filter, respectively. It is evident that the 3-Hz filter produces a smoother EMG profile than the 25-Hz filter. The variability, as indicated by the standard deviation bars, also was lower for the 3-Hz filter than for the 25-Hz filter. Table 1 summarizes this effect of cutoff frequency on the variability in the EMG profiles of the individual trials. Averages and standard deviations for the CV were calculated over the group of eight subjects. Table 1 indicates that the CV for the 3-Hz-filtered EMG profiles was lower than the CV for the 25-Hz-filtered EMG profiles by a factor of two, which was statistically significant ($P < .05$, Student's paired t test).

In Table 2, the similarity of the EMG profiles recorded at 78 and 120 steps/min for each subject is indicated by the average value and standard deviation of correlation (r) and similarity (S) over the group of eight subjects. Comparison of the values for correlation (r) and similarity (S) of the 25-Hz-filtered EMG profiles with those of the 3-Hz-filtered EMG profiles for each subject revealed that values for correlation (r) were significantly higher and that the values for similarity (S) were significantly lower for the

---

[‡]Hewlett-Packard Co, Intercontinental Headquarters, 3495 Deer Creek Rd, Palo Alto, CA 94304.

[§]Tulip Computers, PO Box 3333, 5203 DH Den Bosch, the Netherlands.

[ǁ]Lotus Development Co, 55 Cambridge Pkwy, Cambridge, MA 02142.

Table 1. *Effect of Filter Cutoff Frequency on Gait-Cycle Variability[a] in Healthy Subjects (N = 8) at Three Stepping Frequencies*

	Cutoff Frequency			
	25 Hz		3 Hz	
Stepping Frequency	X̄	SD	X̄	SD
78 steps/min	.67	.12	.33	.05
Free-walking cadence	.70	.10	.31	.04
120 steps/min	.75	.11	.32	.05

[a]Variability described by coefficient of variation, defined as root-mean-square standard deviation over the stride period divided by the mean ensemble average over the stride.

3-Hz–filtered EMG profiles ($P < .05$, Student's paired t test).

The results of pooling the EMG profiles across all subjects for each walking cadence are presented in Figure 3. In this figure, A and B present these across-subject averaged EMG profiles for the cut-off frequencies of 25 and 3 Hz, respectively.

In an attempt to quantify the changes in across-subject averaged EMG profiles with changing stepping frequencies, the similarity of the EMG profiles for 78 and 120 steps/min was para-

Table 2. *Similarity of Electromyographic (EMG) Profiles Recorded at Stepping Frequencies of 78 and 120 Steps/Min for Healthy Subjects (N = 8) at Two Filter Cutoff Frequencies*

	Cutoff Frequency			
	25 Hz		3 Hz	
	X̄	SD	X̄	SD
Correlation (r)[a]	.75	.14	.87	.13
Similarity (S)[b]	.30	.08	.21	.10

[a]Pearson's product-moment correlation coefficient.

[b]Coefficient of variation for ensemble averages of two EMG profiles.

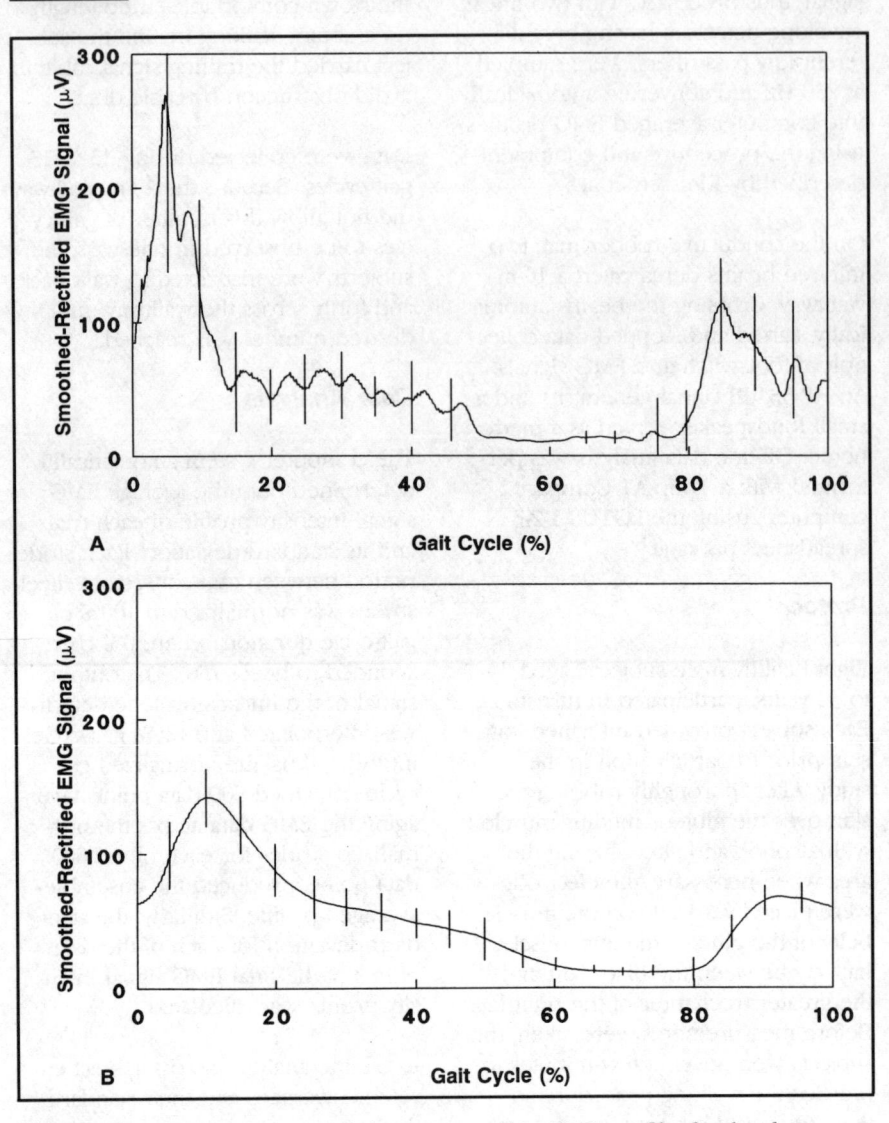

Figure 2. *Computer-averaged electromyographic (EMG) profiles for the gluteus medius muscle of a healthy subject at a free-walking cadence and a cutoff frequency of (A) 25 Hz and (B) 3 Hz.*

metrized with their correlation (r) values. The similarity (S) values for the averaged pair of these EMG profiles were also calculated. These values are presented in Table 3.

The CVs for the across-subject averaged EMG profiles for all walking cadences and filter frequencies are presented in Table 4. For the 3-Hz filter, CVs for these EMG profiles tended to decrease with increasing stepping frequency, whereas for the 25-Hz filter, the CVs tended to increase. Retrospectively, the average walking speeds were 0.83 (SD = 0.06), 1.41 (SD = 0.09), and 1.52 (SD = 0.08) m/sec for 78 steps/min, free-walking cadence, and 120 steps/min, respectively. The average stepping frequency over the group of eight subjects at a free-walking cadence was 109 steps/min (SD = 4).

Discussion

Figure 2 shows that for a given raw EMG signal, the type of low-pass filter used in the linear envelope detector considerably affects the resulting EMG profiles. It is not surprising that the 3-Hz filter produces a smoother EMG profile than the 25-Hz filter, because it damps rapid fluctuations in the

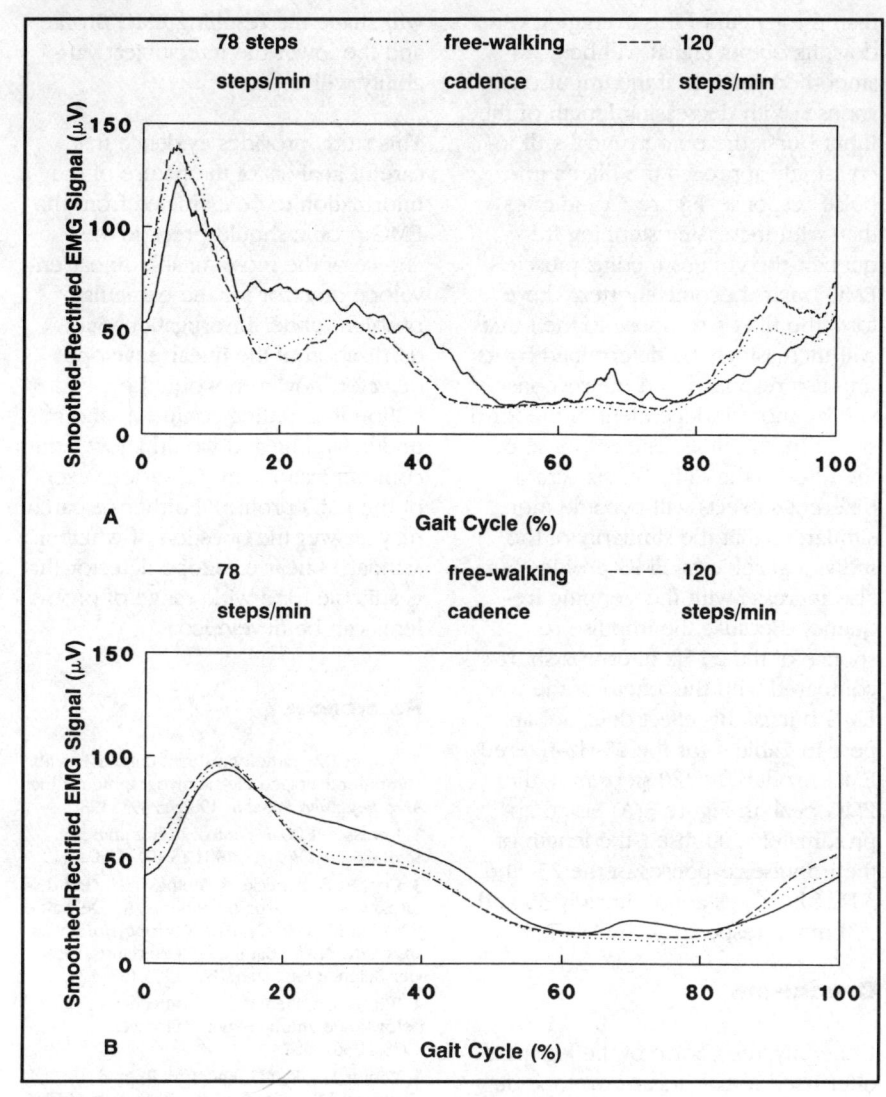

Figure 3. Across-subject computer-averaged electromyographic (EMG) profiles for the gluteus medius muscle for three walking cadences at a cutoff frequency of (A) 25 Hz and (B) 3 Hz.

EMG signal intensity estimate more strongly. Additionally, the 3-Hz–filtered EMG profiles lagged behind the 25-Hz–filtered EMG profiles. At a free-walking cadence, the peak in the EMG profile shown in Figure 2B appears approximately 50 msec later than the peak shown in Figure 2A. Filter theory states that the cutoff frequency and the time lag between filter input and output signal are closely related. Approximations show that this time lag would be inversely proportional to the cutoff frequency.[10]

The two different filters give two different representations of the same raw EMG signal. For the interpretation the EMG profiles, it is important to be aware of what is represented. Kleissen et al[8] viewed the linear envelope detector as an estimator of the intensity of the muscle activation that precedes the actual muscle contraction and force generation. The EMG profiles should provide detailed information about the effects of the motor control system on the muscle studied. The requirement that rapid fluctuations in the EMG signal intensity were to be observed led to the choice of the 25-Hz filter.

Winter[1] proposed that the linear envelope detector be designed so that its output will be closely related to the muscle force resulting from the observed EMG activity. He recognized that the time constants of a second-order critically damped low-pass filter could be tuned to achieve a dynamic input-output relationship that would parallel the dynamics of the EMG activity-force relationship of the muscle studied. For isometric contractions, there is both a theoretical ba-

Table 3. Similarity of Across-Subject Averaged Electromyographic (EMG) Profiles Recorded at Stepping Frequencies of 78 and 120 Steps/Min for Healthy Subjects (N = 8) at Two Filter Cutoff Frequencies

	Cutoff Frequency	
	25 Hz	3 Hz
Correlation (r)[a]	.85	.94
Similarity (S)[b]	.19	.11

[a]Pearson's product-moment correlation coefficient.
[b]Coefficient of variation for ensemble averages of two EMG profiles.

Table 4. Across-Subject Variability[a] Depending on Stepping Frequency and Filter Cutoff Frequency for Healthy Subjects (N = 8)

	Cutoff Frequency	
Stepping Frequency	25 Hz	3 Hz
78 steps/min	.45	.39
Free-walking cadence	.52	.38
120 steps/min	.50	.34

[a]Variability described by coefficient of variation, defined as root-mean-square standard deviation over the stride period divided by the mean ensemble average over the stride.

sis[11] and experimental evidence[12,13] to support this approach.

The effect of the low-pass filter on the individual EMG profiles propagates to the across-subject averaged EMG profiles shown in Figure 3. With increasing stepping frequency, the peak in the 25-Hz–filtered EMG profile shown in Figure 3(A) tends to grow and the EMG activity tends to concentrate progressively nearer to the time of heel-strike. In the 3-Hz–filtered EMG profile shown in Figure 3(B), this tendency is less distinct. The sensitivity of across-subject averaged EMG profiles for changes in the stepping frequency appears to be higher when filtered at 25 Hz. Data presented in Table 3 support this finding quantitatively. These findings confirm reported results from similar experiments.[14]

The across-subject averaged 3-Hz–filtered EMG profile for the gluteus medius muscle at a free-walking cadence agrees with those published by other investigators.[15,16] A review of the literature indicates that no EMG profiles for this muscle at other stepping frequencies have yet been published. The variability in the EMG profiles observed by Winter[15] (CV = 105%) is considerably higher than the variability observed in this study. Possible explanations may be a different electrode location used or a less homogeneous group of subjects.

Finally, it seems remarkable that the variability in the across-subject averaged 3-Hz–filtered EMG profiles decreases with increasing stepping frequency (Tab. 4), whereas this variability tends to increase for the 25-Hz–filtered EMG profiles. This finding means that the 3-Hz–filtered EMG profiles for the gluteus medius muscle of healthy individuals tend to become more similar with increasing stepping frequency.

The low-pass filter in real time forms the output signal as a moving average[5] over the rectified EMG signal where the shape of the averaging window is described by the filter's impulse response. However, if the filter's input signal is a burst in which the length is comparable to or shorter than the length of this averaging window, its output signal will be a smoothed version of the impulse response. With decreasing length of the input burst, the output signal will increasingly approach the filter's impulse response. Figure 3A indicates that, with increasing stepping frequency, the gluteus medius muscle's EMG bursts become shorter. Therefore, the filter's response to the bursts will increasingly be determined by its impulse response, and the response will be more independent of the form of the input bursts. The response of the filter to the EMG bursts across different subjects will become more similar, so that the similarity of the individual subjects' EMG profiles can also increase with the stepping frequency. Because the impulse response of the 25-Hz filter was short compared with the length of the raw EMG bursts, this effect does not appear in Table 4 for the 25-Hz–filtered EMG profiles. At 120 steps/min, the EMG peak in Figure 3(A) lasted approximately 200 msec; the length of the impulse responses of the 25- and 3-Hz filters was approximately 35 and 200 msec, respectively.

Conclusion

Generally, the choice of the low-pass filter used in the linear envelope detector considerably affects both the shape and the properties of computer-averaged EMG profiles. In this report, some effects on the EMG profiles for the gluteus medius muscle have been described as an example of this general principle. The results of this study show that EMG profiles recorded with a linear envelope detector with a low cutoff frequency generally are smoother and have a lower cycle-to-cycle variability and a greater time lag with respect to the raw EMG signal than a filter with a higher cutoff frequency. The interpretation of the EMG profiles also is closely related to the type of filter used. The observed intersubject variability in the EMG profiles can be influenced by the low-pass filter: The longer the impulse response of the filter and the shorter the EMG bursts to be observed, the more the filter will shape the resulting EMG profile and the lower the intersubject variability will appear.

This study provides evidence that careful analysis of the nature of the information to be extracted from the EMG profile should precede the choice of the most suitable linear envelope detector for the particular problem under investigation. Standardization of the linear envelope detector, however, would be a contribution to avoiding confusion and misunderstanding and would allow better communication among various users of the EMG profiles. Further research may answer the question of whether a standard linear envelope detector that is suitable for a wide range of problems can be developed.

References

1 Winter DA. Pathological gait diagnosis with computer-averaged electromyographic profiles. *Arch Phys Med Rehabil.* 1984;65:393–398.

2 Knutsson E. Gait control in hemiparesis. *Scand J Rehabil Med.* 1981;13:101–108.

3 Conrad B, Benecke R, Meinck HM. Gait disturbances in paraspastic patients. In: Delwaide J, Young RR, eds. *Clinical Neurophysiology in Spasticity.* Amsterdam, the Netherlands: Elsevier Science Publishers BV; 1985:155–174.

4 Winter DA, Sienko SE. Biomechanics of below-knee amputee gait. *J Biomech.* 1988;21:361–367.

5 Winter DA, Rau G, Kadefors R, et al. *Units, Terms and Standards in the Reporting of EMG Research.* Montreal, Quebec, Canada; International Society of Electrophysiological Kinesiology (report of ad hoc committee); August 1980.

6 Kadefors R. Myo-electric signal processing as an estimation problem. In: Desmedt JE, ed. *New Developments in Electromyography and Clinical Neurophysiology.* Basel, Switzerland: S Karger AG Medical and Scientific Publishers; 1973;1:519–532.

7 Arsenault AB, Winter DA, Martiniuk RG. Is there a "normal" profile of EMG activity in gait? *Med Biol Eng Comput.* 1986;24:337–343.

8 Kleissen RFM, Hermens HJ, den Exter T, et al. Simultaneous measurement of surface EMG and movements for clinical use. *Med Biol Eng Comput.* 1989;27:291–297.

9 Yang J, Winter DA. Surface EMG profiles during different walking cadences in humans. *Electroencephalogr Clin Neurophysiol.* 1985;60:485–491.

10 Humphreys DS. *The Analysis, Design and Synthesis of Electrical Filters.* Englewood Cliffs, NJ: Prentice-Hall Inc; 1970:38–45.

The Feet in Relation to the Mechanics of Human Locomotion

R. Plato Schwartz, M. D.*

Arthur L. Heath

John Hunter (1729-1796) believed that a knowledge of function was more important than the study of structure. In 1860, John Hilton urged the practical surgeon to "realize that his chief duty consists in ascertaining and removing those impediments which obstruct the reparative processes or thwart the effort of Nature, and thus enable her to restore the parts to their normal condition." Hilton was supported by a Liverpool contemporary, Hugh Owen Thomas (1834-1891). Both men were strong individualists. Their principles are now more commonly practiced than when they lived.

Although this is true, attention continues to be focused upon morphology and structure. We are nearing the end of that era. Osteology was completed when all bones became perfectly described separately and in relation to respective articulations, ligaments, muscle origins, and tendinous insertions. Gross anatomy defined joint motions from muscular relationships on the cadaver. Accurate drawings have long been available; at present, there are no frontiers related to gross circulation and innervation. This proper order has been followed. Today, anatomists and pathologists are exerting their efforts in the field of experimental physiology. Long before this change came, attention was directed to the fundamental laws of locomotion.

James B. Pettigrew[1] (1834-1908) of St. Andrews and Edinburg, was a visualist self-commissioned to analyze, assemble, depict, and correlate a major theme. "The most exquisite form loses much of its grace if bereft of motion, and the most ungainly animal conceals its want of symmetry in the co-adaption and exercise of its several parts . . . the great panorama of life is interesting because it moves. One change involves another, and everything which co-exists, co-depends. . . . Can it be that the animate and inanimate world in this, as in other things reciprocates, and that traveling organs of animals are made to impress the inanimate bodies in precisely the same manner as the inanimate bodies impress each other"?

"This much seems certain: The wind communicates to the water similar impulses to those communicated to it by the fish in swimming; and the wing, when it is made to vibrate, impinges on the air as ordinary sound does. The extremities of bipeds and quadrupeds, moreover, describe wave tracts on land when walking and running, so that one great law apparently determines the course of the bird in the air, the fish in the water, and the biped and quadruped on the land."

The breadth of this theme, expressed in 1867, is almost limitless. It appeals particularly to those of us who have a respect for the unity of all things. The greatest principles, although clearly expressed, have frequently had to wait long for application. Pettigrew's work applied to the living; his principles, therefore, were not necessary for understanding the gross anatomy of the dissecting room of the nineteenth century, nor was his theme an essential part of the information sought in the postmortem study of disease in later years.

Pettigrew had only one contemporary in America. Eadweard Muybridge[2] (1830-1904) was neither a physiologist nor an anatomist. He resigned his position with the United States Geodetic Survey to begin a photographic study of the locomotion of the horse at Sacramento, California, in May, 1872. With periods of interruption, he continued at Palo Alto until 1879. Thereafter, for more than twenty years, the work was continued at the University of Pennsylvania. One large volume of his work is devoted to photographic records of locomotion of male and female figures. They provided confirmation of the group action of muscles.

Marey[3] (1830-1904) was the French contemporary at the University of Paris. He applied pneumographic recording apparatus to the study of locomotion. In contrast with the respective fundamental contributions of Pettigrew and Muybridge, the value of Marey's work was chiefly in relation to history of the subject.

The symphony of muscle function correlated with osteo-articular form and structure was the essential theme presented by the work of these men. They revealed the relationship of all forms of locomotion, while Pettigrew associated the differences to the character of the respective media—land, water, and air. In this, therefore, we see emphasized the true principle of muscle function as related to the interaction of opposing groups, and applied to all three forms of locomotion.

Serious consideration has not been generously devoted to the study of human gait, except in such rare instances. The present treatment of deformities, etc., rests largely upon the training in morphology which dominated the past century. The general attitude of mind thus created was unattracted by the type of investigation required for penetrating the shield of unfamiliar difficulties associated with the study of the extremities in motion. Appreciating the propriety of disagreement, it seemed most evident that methods of precision were indicated for the improvement of that branch of surgery which is devoted to the treatment of impaired function of locomotion. Had not the advancement of medicine been defined by the mileposts of the clinical thermometer and the electrocardiograph? In every phase of human need, the requirements have been most satisfactorily met when impressions were replaced by facts. A rational premise relevant to human locomotion was dependent upon the interpretation of records resulting from the application of physical laws which have a constant and consistent behavior.

The record of personal failures in this effort has been previously acknowledged. These experiences were profitable in that each one more clearly defined the particular requirements. In the end, we succeeded in recording the duration of weight-bearing time on the heel, midfoot, and forefoot. This electrobasographic method of recording human gait has been previously described. Clinically, it is practical from the viewpoint of technical simplicity, economy of operation, and precision of inter-

---

From the Department of Surgery, Division of Orthopedics, Rochester University, School of Medicine and Dentistry.
*Associate Professor of Orthopedic Surgery.
Presented at Russell Sage College, February 14, 1936.

THE PHYSIOTHERAPY REVIEW

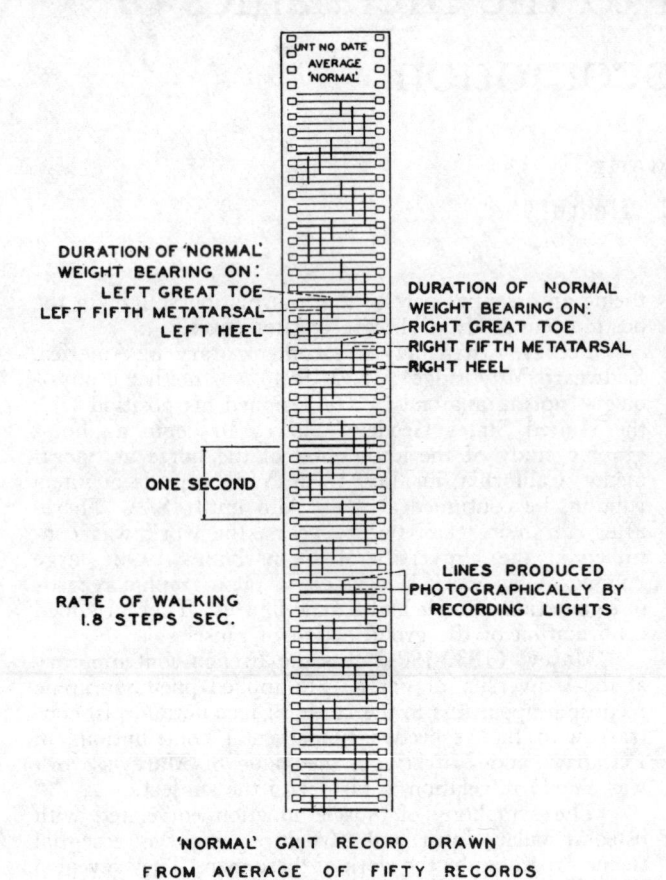

Fig. 1.

'NORMAL' GAIT RECORD DRAWN FROM AVERAGE OF FIFTY RECORDS

DURATION OF 'NORMAL' WEIGHT BEARING ON:
LEFT GREAT TOE
LEFT FIFTH METATARSAL
LEFT HEEL

DURATION OF NORMAL WEIGHT BEARING ON:
RIGHT GREAT TOE
RIGHT FIFTH METATARSAL
RIGHT HEEL

ONE SECOND

RATE OF WALKING 1.8 STEPS SEC.

LINES PRODUCED PHOTOGRAPHICALLY BY RECORDING LIGHTS

pretation, to the thousandth of a second. Without error, it reveals any disproportion in the duration of weight-bearing on the respective functional areas of each foot while the subject walks a distance of fifty feet. (Fig. 1)

For these reasons, the electrobasographic records of human gait reveal the presence of structural and functional abnormalities which alter normal walking. Such departures from normal are evident without reference to location or the cause of abnormally long or short durations of weight-bearing. This accomplishment was both gratifying and discomforting. Discomfiture arose from the fact that such precision records could not be explained by prevailing textbook statements and clinical practices related to foot function, in stance and gait. It became most evident that "Without principles deduced from analytical reasoning, experience is a useless and blind guide." (Cullen)

Attention was, therefore, directed to the structural characteristics as related to the observed and recorded function of the foot. For want of time, only those points related to pronation will be discussed.

Over a period of the past fifty years, the "arches" of the feet have been a pivot of discussion. This impression of longitudinal and transverse arches has dominated because of prevailing contours. Because of this associated impression, the osteo-articular structure of the foot has long been accepted with full assurance of understanding. If it has not become trite, the subject has been accepted without requiring further consideration for an interpretation which is more essential to a sound premise for rational treatment of symptoms resulting from weak, pronated feet.

Such is the fixed opinion. Unlearning is frequently more difficult than the acceptance of new facts pertaining to a recent discovery. Although previous efforts have been so well expressed, "longitudinal arches" are still suspended in adhesive slings, or propped up with steel plates and rigid shanks in shoes.

In 1895, Walsham,[4] surgeon in charge of the Orthopedic Department and Lecturer in Anatomy at St. Bartholomew's Hospital, London, stated that, "the mechanism of the instep is in no way comparable to that of an ordinary arch." There is no keystone. The weight from above is received wholly upon the posterior part of the foot, and that part of it which is transmitted forward acts in the anterior-posterior axis of the bones; whereas, in an arch, the weight is received from above along its whole span. Moreover, in the place of a rigid, passively resisting structure, we have an elastic, living mechanism, actively adapting itself to every change of position and function. Golding Bird also shows the weak points in the simile. He says (Guy's Hospital Reports, 1883): "There are no piers on which the extremities of the arch can rest, nor by which they can be prevented from separating from each other. The (so called) arch of the foot is kept in its position by the tying together of its two extremities by means of two distinct sets of bands, one of which is purely passive in its operation (the ligaments), and the other more or less active (the muscles)."

We must, for at least the present, agree with these men in their regard for the foot as a living structure with the astragalus resting directly upon the os calcis. This structural relationship is of primary importance when regarded as a living, osseous structure suspended in ligaments and muscles. Moreover, the bones of the foot retain their arch-like contour by virtue of normal neuromuscular and ligamentous function. We must, therefore, not assume full understanding of function from observation of dissection and the living foot in repose.

The functional significance of the os calcis is the fact that it is the structure first to receive weight in making a normal step. Normal function of the foot in completing the step is dependent upon a prevailing sensitivity of reflex balance in its multi-jointed, therefore, semi-mobile structure. This sensitive reflex balance, which normally controls the rhythmic motion and accurately times the position of the foot with respect to the leg in normal gait, is the function of a normal neuro-muscular mechanism in relation to

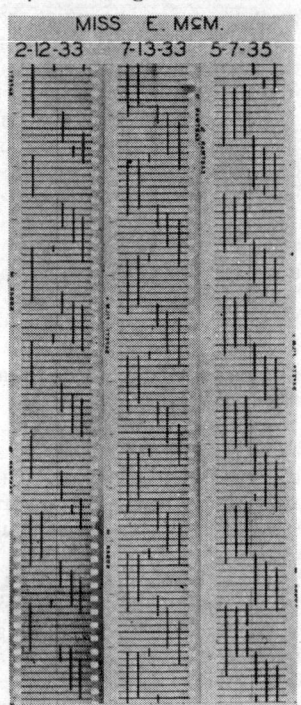

Fig. 2

MISS E. McM.
2-12-33   7-13-33   5-7-35

On 2/12/33, gait record revealed left calcaneol weight-bearing due to infantile paralysis involving only gastrocnemius and soleus.

On 7/13/33, heel and midfoot weight-bearing following sub-astragalar arthrodesis and perineal transplantation, posteriorly.

On 5/7/35, marked improvement in function on all three weight-bearing areas of the left foot fifteen months after operation. (Note corresponding improvement in function of right foot.)

Basic Research

normal osteo-articular and ligamentous structures. The prevailing instability of the foot, with or without deformities, following impairment of muscle function is glaring evidence in support of this statement. These abnormalities h a v e received much attention, chiefly, if not only, in relation to the operative procedures designed for their correction.

Only by means of electrobasographic records has it been possible to record any expression of functional limitation of the foot in walking. Moreover, this is the only method of recording the presence or absence of functional improvement resulting from any form of treatment including operation. (Fig. 2)

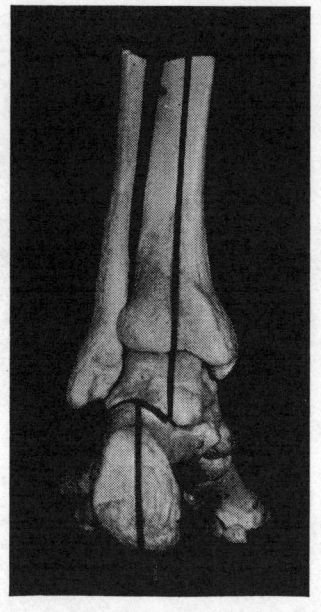

Most of our adult population have uncomfortable, if not abnormal f e e t. Without weight-bearing, such feet may

Fig. 3
The axis of weight-bearing passes from tibia through the middle of the body of the astragalus. The axis of weight-bearing of the os calcis is 1.2cm. lateral to the corresponding axis of the tibia. (8/16/35)

appear to be normal in contour. Under the influence of superincumbent weight, the medial aspect of the foot becomes prominent in company with a lowering of the medial side, while the outer aspect of the heel projects lateral to the external malleolus. These characteristic evidences of pronation may prevail in the first, second, or third degree. Beyond the middle of the third decade, they are invariably accompanied by the physical signs as well as the subjective symptoms of foot strain. Imbalance prevails, although there may be no demonstrable neuro-muscular pathology.

Attention is, therefore, directed to the osteo-articular relationship of the tibia, astragalus, and os calcis. The medial and lateral condyles of the ankle provide for the primary motions, dorsal and plantar flexion, of the astragalus. From the posterior view, it is evident that the axis of tibial weight-bearing passes through the middle of the astragalar body. (Fig. 3.) The ankle point is, therefore, doubly protected with relation to medio-lateral stability—(1) respective condyles which form a mortise for (2) the astragalus in most favorable alignment with the tibia. These are predominant characteristics.

The morphological differences between the articulations for the respective malleoli are less prominent, but of great importance with relation to function. They allow a small, but essential, degree of rotation of the astragalar body around the internal malleolus as a center of motion. The motion of the astragalus in walking is, therefore, compound instead of the simple dorsal and plantar flexion usually ascribed to the ankle joint.

Full consideration of the subastragalar (astragalo-calcaneal) joint would lead us into details foreign to our present purpose. Of major importance is the fact that medial and lateral motion (inversion and eversion) of the foot is initiated in this subastragalar articulation. Every osteo-articular characteristic essential for this function is expressed herein, just as the opposite findings prevail as described in relation to the function of the ankle.

These facts force upon us the realization that the os calcis is the foundation of the foot. Inquiry should, therefore, be directed to those characteristics which favor stability. Where its contours or its relationship indicate structural weakness under the influence of superincumbent weight, the compensatory agents must be defined. Answers to these questions must be given because it is functionally significant that the os calcis is the bony structure first to receive weight in making a normal step.

There are two major reasons why the os calcis is inherently an unstable weight-bearing structure. Comparison of a number of such bones reveals that the area which receives weight is variable in all of the characteristics which promote stability in relation to the axis along which the weight of the body is transmitted to it from the tibia and astragalus. (Fig. 4.) Second, this axis through the leg is 1.2 cm. medial to the respective axis which passes through the os calcis. (Fig. 3.) This malalignment of axial relationships is weakly compensated for by the medial projection of the sustentaculum tali. (Fig. 5.) This, together with the inferior calcaneo-scaphoid ligament, passively supports the head and neck of the astragalus. The active support which compensates for these fundamental structural weaknesses is expressed by the tibialis posticus and flexor longus hallucis (to a less extent by the tibialis anticus). These facts reveal that there are major intrinsic influences at work in favor of the os calcis rotating into the pronated or weak position of the foot. Congenital or acquired shortening of the calf muscles is a common extrinsic cause of pronation and prevails in both sexes, by far most frequently in women. With them, it is usually an acquired characteristic associated with the height of heels worn after adolescence.

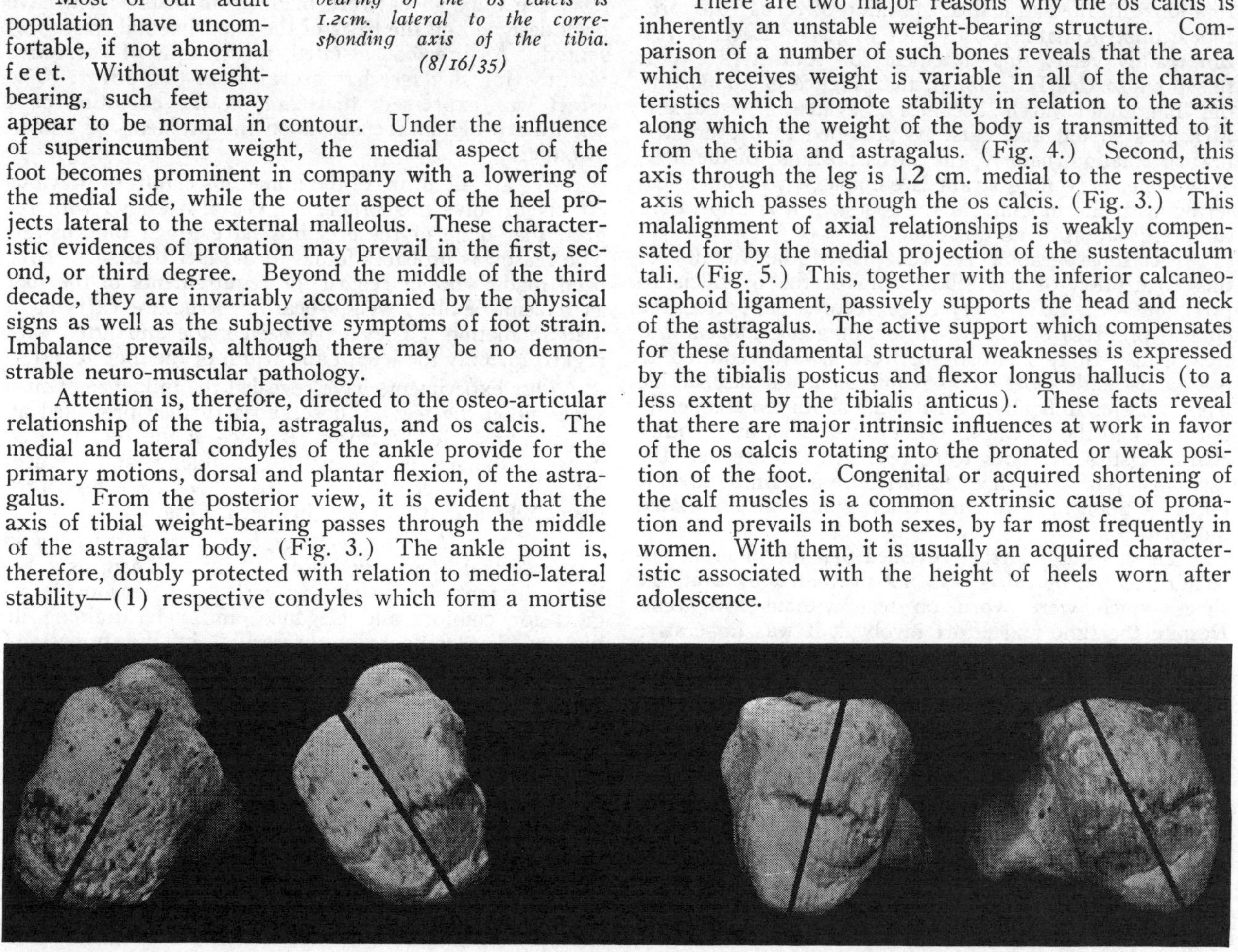

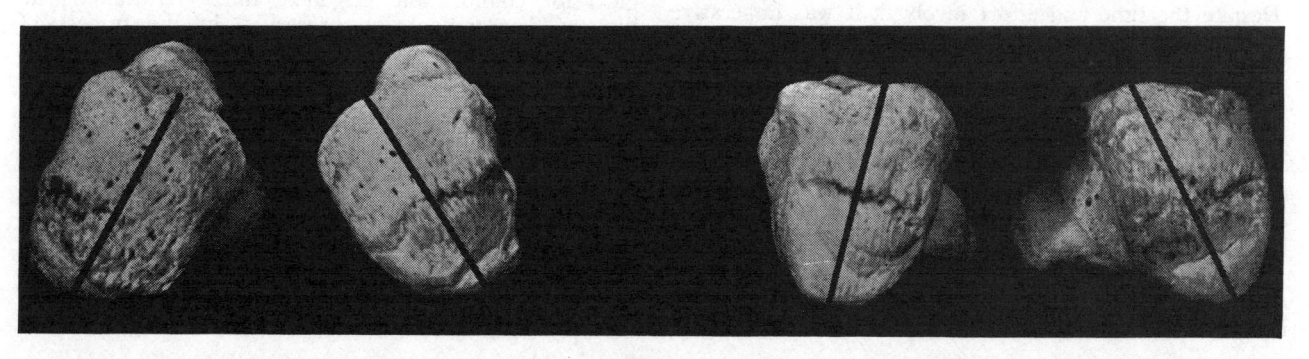

Fig. 4

Posterior view of two pairs of os calci. The rounder, plantar weight-bearing "surface" favor the weak, pronated position of the foot. (10-7-29)

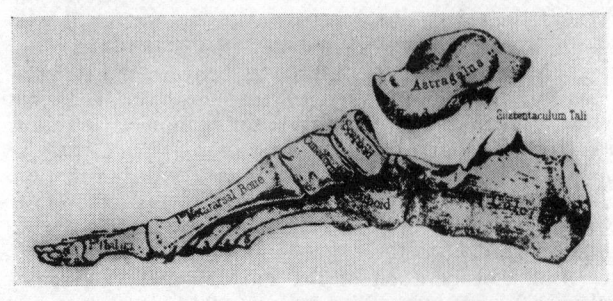

Fig. 5

(Ellis, 1889) Medial aspect of right foot showing relationship of astragalus to os calcis, sustentaculum tali and scaphoid.

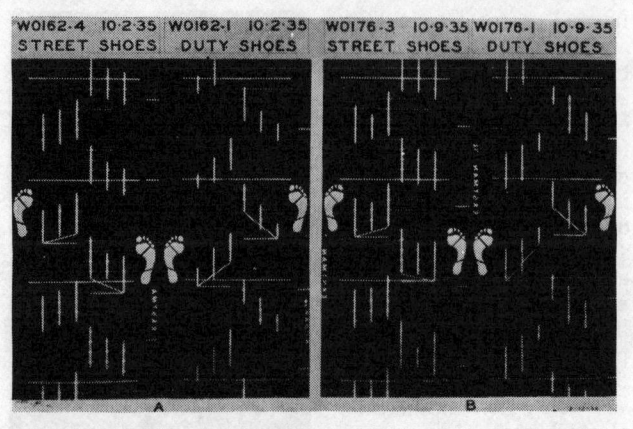

Fig. 6

The difference in the sequence and duration of weight-bearing on the three functional areas of each foot is revealed. With street shoes, pronation is present in every step. With duty shoes, in each step the feet are in normal balance during the period of weight-bearing in motion.

A weak foot caused by any one or a combination of these factors invariably has a normal medial contour when not under the influence of weight-bearing. With weight-bearing, the astragalo-scaphoid level of the foot rolls inward and downward. This is the weak, pronated position commonly, but erroneously, referred to as "flat feet." Such displacement cannot occur at this level without internal rotation of the os calcis under the astragalus (position of valgus).

This fact cannot be too strongly emphasized. The converse of the statement may more clearly reveal the reason for such emphasis. If the heel is held in external rotation with the astragalus, (position of varus) the forefoot cannot pronate, (position of valgus). Each of you can readily verify this statement for yourselves. With this premise clearly in mind, the practice of supporting the "longitudinal arch" becomes an evident gross fallacy. Measures should and must be provided to prevent the inward rotation, namely, the varus position, of the heel. This done, prevailing strain is eliminated without interfering with the mobility of the foot which is imperative for preservation of its function.

Electrobasographic records proved the validity of this conception of the relationship of the os calcis to pronated feet. Moreover, these records supported the clinical progress of patients in defining the efficiency of treatment. In this effort, we were always required to use the patient's shoes. This requirement was inseparable from a major difficulty, namely, the shoes too frequently did not hold the os calcis. Along with this, many other characteristics of shoes were recorded. Moreover, the relationship of such shoes to feet bore a definite adverse influence upon the local and remote symptoms associated therewith.

These studies further revealed that the inherent imbalance of the os calcis was not prevented by available shoes which were worn on feet without symptoms. Despite the time and effort involved, it was imperative to challenge this new, and in some respects, undesirable problem of mal-fitting shoes. All such shoes have been made without knowledge of the requirements of the foot in motion.

Through cooperative effort it was our purpose to design correct shoes; shoes that would meet the functional requirements of the foot in motion. Through the medium of electrobasographic records and slow-motion pictures, we directed the experimental work and verified the results of our efforts.

Twenty-four girls were admitted to the Training School for Nurses at the Strong Memorial Hospital in September of 1935. Electrobasographic records were made of each girl in the shoes which they wore, and in the duty shoes which were prescribed for them.

The duty shoes were made from lasts which have been designed on the basis of the experimental work presented. As previously stated, no thought of corrective features for sick feet has prevailed in this effort. The effort was expressed, therefore, in the evolution of a scientific program focused upon the problems of human locomotion.

We are all aware of the numerous claims of previous efforts to solve this problem. We also know, with regret, that none of these claims have given the answer. The reason is probably found in the fact that shoes have been made without regard for requirements of the foot as a semi-flexible, weight-bearing structure carrying a load in motion. They have been made and fitted with regard for only the length and width of the foot at rest.

Our experiments have revealed the influence of such shoes upon the feet. Those shoes further provoked an already unstable structure. Balance of the foot in each phase of the step is essential for confidence and a graceful gait. Our records reveal that shoes have been produced which actually meet this need. (Fig. 6)

. . . *"Let the form of boot-lasts correspond in shape to that of feet in action . . . Then, as I believe, the foot may be clothed with every reasonable regard for comfort and elegance, and yet, attaining its highest development, be preserved in the fulness of strength and of beauty."* (Ellis,[5] 1889)

REFERENCES

1. Pettigrew, James B.: "Design in Nature." Longmans, Green and Co., London. 1908, III: 1074.
2. Muybridge, Eadweard: "The Human Figure in Motion." Fourth Impression. Chapman and Hall, Ltd., 1913, London.
3. Marey, E. J.: "De la Locomotion Terrestre chez les Bipèdes et les quadrupsèdes." J. de l'Anat. et de la Physiol. IX, 42, 1873.
4. Walsham, W. J. and Hughes, W. K.: "The Deformities of the Human Foot with Their Treatment." Baillière, Tindall and Cox, London. 1895, p. 6.
5. Ellis, Thomas S.: "The Human Foot." J. and A. Churchill. London, 1889.

Basic Research

THE VERTICAL PATHWAYS OF THE FOOT DURING LEVEL WALKING

I. Range of Variability in Normal Men

M. P. MURRAY, Ph.D., and B. H. CLARKSON

OUR INTEREST in studying the vertical pathways of the foot of normal walking subjects was stimulated when, in a separate study, we noted curious pathways of the foot of patients with selected locomotor disabilities. A search of the literature disclosed that the several studies of the vertical trajectories of the heel, or of the toe, were done on a very limited number of subjects and provided no information on the ranges of normal variability for the vertical pathways of the foot.[1-6] The present study was designed to provide such standards which could then be used for comparison with the distorted pathways of pathological gaits.

From the Physical Therapy Section, Physical Medicine and Rehabilitation Service, Veterans Administration Center, Wood, Wisconsin, and the Curriculum in Physical Therapy, Marquette University School of Medicine, Milwaukee, Wisconsin.

This investigation was supported in part by Grant CD 00052-02 from the Division of Chronic Diseases, U.S. Public Health Service, Department of Health, Education, and Welfare.

MATERIAL AND METHODS

Subjects

Thirty normal males were chosen for study from the employees and professional staff of the Wood Veterans Administration Center, and from students and staff of the College of Engineering and the School of Medicine of Marquette University. Six subjects were selected in each of the five age groups: twenty to twenty-five, thirty to thirty-five, forty to forty-five, fifty to fifty-five, and sixty to sixty-five years. Each age group included two tall men, 71 to 73.5 inches in height; two men of medium height, 68.5 to 70 inches; and two short men, 61.5 to 67 inches. Each subject weighed within 12 pounds of the range listed as "desirable" by the height-weight table of the Metropolitan Life Insurance Company (1959).

Recording and Measuring Procedures

The displacement patterns of the walking subjects were recorded by means of interrupted-

VERTICAL PATHWAYS OF THE FOOT IN NORMAL LEVEL WALKING

light photography. The subjects, appropriately targeted with reflective fabric, walked before a camera in the illumination of a strobe-light. The resultant photographs from which the measurements were made showed the target positions as white images against a black background. The details of the photographic and walking procedure, as well as the measuring techniques for all excursions except the vertical pathways of the foot, have been reported previously.[7,8]

The pathway of the heel was measured from the most posterior point of a strip of reflective fabric fastened around the heel of the shoe; the pathway of the toe, from the most anterior point of a similar strip fastened around the outsole of the shoe. The vertical distance of these points from the floor at every 1/20 second interval was measured throughout the walking cycle. The *walking cycle* is defined as the interval between successive instants of initial foot-to-floor contact of the same foot. Within each cycle there is a period of *stance* when that foot is in contact with the floor, and a shorter period of *swing* when that foot is off the floor moving forward to create the next step. The mean cycle duration for the thirty subjects for the free speed walking trials was 1.06 ± 0.09 seconds *, as compared to 0.87 ± 0.06 seconds for the fast speed trials. The duration of both the stance and swing phases of the cycle were shorter with the more rapid cycles of the faster walking speed. The mean durations of stance and swing for free speed walking were 61 per cent and 39 per cent of the walking cycle, respectively, whereas the stance and swing durations for fast speed walking were 57 per cent and 43 per cent of the cycle, respectively.

RESULTS

Heel

The mean vertical trajectories of the heel plotted at 5 per cent intervals throughout the walking cycle for free and fast speed walking are shown in Figure 1. These patterns represent the means of sixty trials (two trials for each of the thirty men) at each of the two walking speeds. The zero reference on the ordinate represents the lowest vertical distance between the posterior tip of the heel target and the floor.

As seen in Figure 1, the vertical pathway of the heel showed one pronounced peak of ele-

---

* One standard deviation.

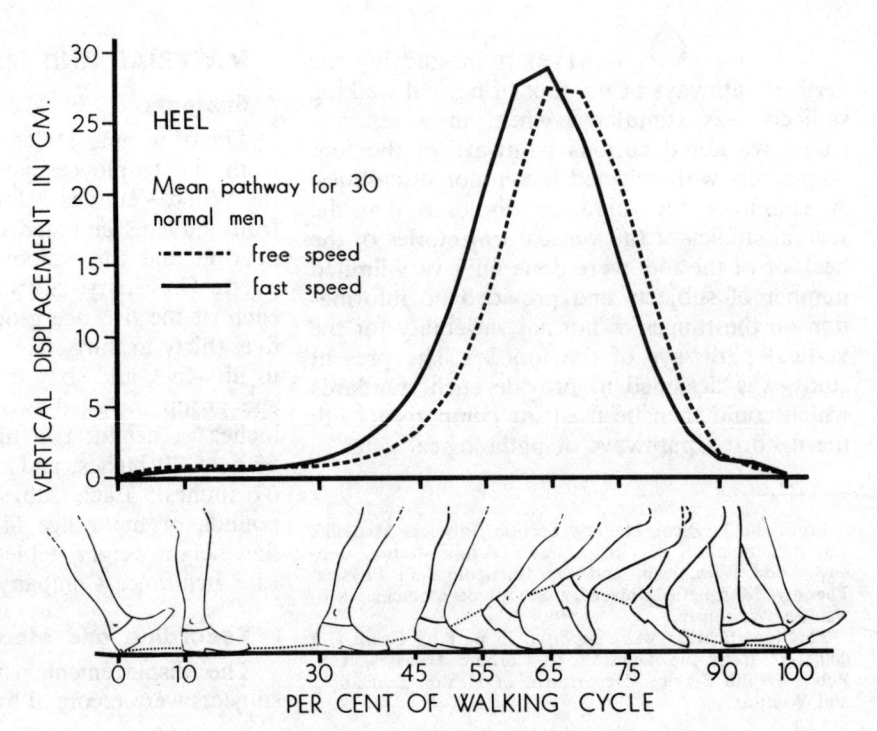

FIG. 1. Mean vertical pathway of the heel for free and fast speed walking for thirty normal men, two trials at each speed, plotted at 5 per cent intervals throughout the walking cycle. Zero reference on ordinate represents lowest vertical distance between heel target and floor.

Basic Research

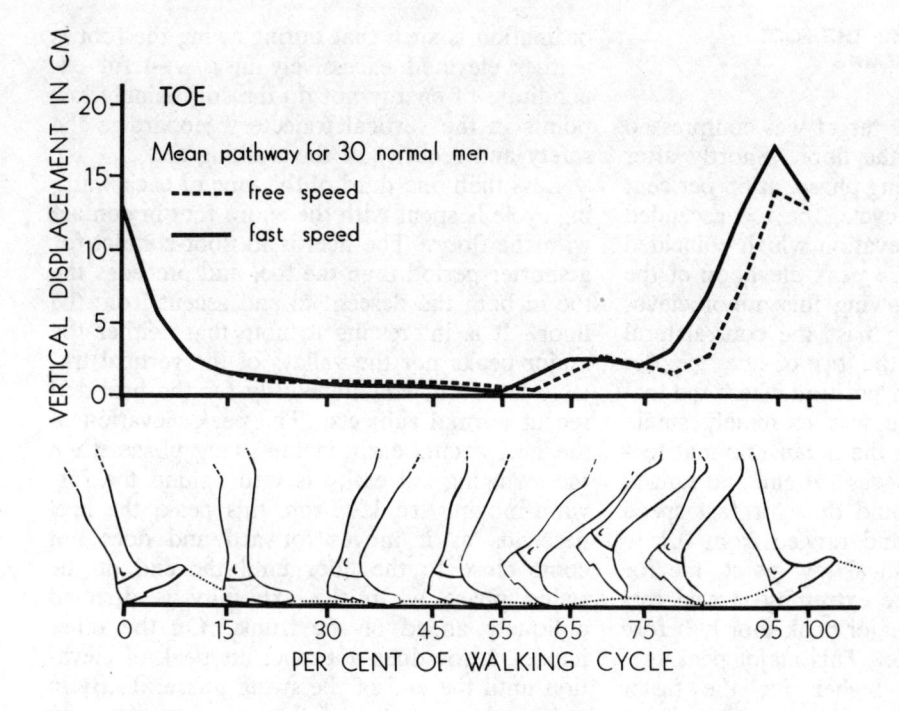

FIG. 2. Mean vertical pathway of the toe for free and fast speed walking for thirty normal men, two trials at each speed, plotted at 5 per cent intervals throughout the walking cycle. Zero reference on ordinate represents lowest vertical distance between toe target and floor.

vation within each walking cycle. As the walking cycle began at heel-strike, the posterior tip of the heel target was at its lowest vertical distance from the floor. The heel remained in contact with the floor for the first half of the stance phase as the body moved forward over this supportive foot. Later in stance, at about the 30 per cent period of the walking cycle, the heel began a gradual rise, shifting the floor-contact area from the entire foot to the forepart of the foot. The rate of ascent of the heel increased after the opposite foot had made floor contact at the 50 per cent point in the cycle. The peak of heel-rise was reached at about 65 per cent of the cycle, shortly after the end of ipsilateral stance when the foot had just begun to swing forward. Following this apex, the heel descended rapidly as the extremity continued its forward swing. Just before the end of the swing phase, the rate of descent of the heel showed a definite decrease for all subjects.

Although the patterns of the vertical pathway of the heel were similar for the two walking speeds, the ascent and descent of the heel occurred slightly earlier in the cycle for the faster walking speed. Moreover, the magnitude of the peak heel-rise was slightly but significantly greater for the faster walking speed ($P<0.01$). The peak heel-rise for free speed walking was 28.8 ± 1.7 cm., as compared to 29.9 ± 1.6 cm. for fast speed walking. There were also statistically significant differences in the magnitude of the peak heel-rise which were systematic with the three height groups for both walking speeds ($P<0.01$). For the combined free and fast walking speeds, the mean peak heel-rise for the tall subjects was 30.1 ± 1.7 cm., for the subjects of medium height, 29.4 ± 1.6 cm., and for the short subjects, 28.7 ± 1.6 cm. There were neither systematic nor significant differences in the magnitude of the peak heel-rise for the five age groups nor for repeated trials of the same subject. The magnitude of the peak heel-rise for the right and left extremity during the same walking trial was notably similar.

Toe

The mean vertical pathways of the toe for two trials of both free and fast speed walking for the thirty normal men are shown in Figure 2. Within each walking cycle, there were two waves of elevation for the pathway of the toe. A minor wave of elevation occurred between 65 and 70 per cent of the cycle, and a major elevation occurred near the end of the cycle.

At the heel-strike which initiated the walking cycle, the toe was elevated from the floor in a phase of rapid descent. With this descent completed, there was a short period when both heel and toe were in contact with the floor. The toe continued to be in contact with the floor for the remainder of the stance phase after the heel had begun to rise. At the end of the stance

VERTICAL PATHWAYS OF THE FOOT IN NORMAL LEVEL WALKING

phase, the tip of the toe target was compressed to its closest point to the floor. Shortly after the beginning of the swing phase, at 65 per cent or 70 per cent of the cycle, the toe ascended to its minor peak of elevation which coincided closely to the time of the peak elevation of the heel. Immediately following this minor elevation, as the foot swung past the contralateral supporting extremity, the tip of the toe descended to a critical low position where the toe-floor clearance distance was extremely small. For free speed walking the mean minimal toe-floor clearance distance was 1.4 cm. and ranged from 0.1 to 3.8 cm., and that for fast speed walking was 1.6 cm. and ranged from 0.1 to 3.4 cm. From this critical low point, the toe ascended sharply as the extremity moved forward, and reached its major peak shortly before the instant of heel-strike. This major peak toe-rise was significantly higher for the faster walking speed than for free speed walking ($P<0.01$). The mean major peak for fast walking was 18.3 ± 2.5 cm., as compared to 15.7 ± 1.9 cm. for free speed walking.

There were no significant differences in the magnitude of the major or minor toe elevations, nor for the critical toe-floor clearance distance, for the age or height groups at either walking speed. The pathways of the toe on repeated trials were strikingly similar.

DISCUSSION

An identification of the normal ranges of variability in walking patterns is essential for meaningful comparison and evaluation of the distorted gait patterns of patients with locomotor disturbances. The normal values reported in this study should provide useful standards, since these data represent a comparatively large number of observations of men in wide ranges of age and height under varying conditions of walking speed.

The vertical pathway of the foot is the resultant of the finely adjusted interaction of muscular contractions controlling many joints, particularly the hip, knee, and ankle. The similarity in the vertical pathway of the foot for repeated walking trials of the same subject and also among the subjects with various limb lengths suggests that these stereotyped excursions result from maximal efficiency in the coordinated patterns of movement. Normal coordination is such that during swing the foot is neither elevated excessively in a wasteful expenditure of energy nor do the surprisingly low points in the vertical trajectory jeopardize the safety and security of the walking act.

Less than one-third of the time of each walking cycle is spent with the entire foot in contact with the floor. The heel is in floor-contact for a shorter period than the toe, and precedes the toe in both the descent to and ascent from the floor. It is interesting to note that neither the major peaks nor the valleys of the vertical trajectories occur simultaneously for the heel and toe of normal subjects. The peak elevation of the heel occurs early in the swing phase when the swinging extremity is well behind the forward-moving trunk. From this peak, the heel descends as it moves forward and does not come close to the floor until the end of the swing phase when the extremity is directed obliquely ahead of the trunk. On the other hand, the toe does not reach its peak of elevation until the end of the swing phase. Early in swing when the toe of the swinging extremity is well behind the trunk, the toe-floor clearance distance is very small. Then, as the swinging extremity passes the opposite supportive extremity, the toe descends to an even more critical floor clearance distance. This explains why toe-stubbing, in contrast to heel-stubbing, results in a more hazardous loss of balance. The low points in the toe pathway occur when the swinging extremity is behind or adjacent to the opposite supportive extremity, whereas the low points in the heel pathway occur when the swinging extremity is outstretched ahead of the trunk in good position to receive the oncoming body weight.

Our results are in accord with other investigators, who found that the amplitude of peak heel-rise is greater with increased walking speed.[2,4,6] Moreover, we have found that the amplitude of peak heel-rise is greater for taller subjects. The influence of body height on the peak heel-rise is understandable, since the foot of any normal man is in a relatively vertical position early in the swing phase. Thus the longer feet of taller subjects require the heel to be in a higher position as the foot is lifted from the floor. The kinematic reasons for the increased heel-rise with the faster walking speed are that the flexion phases of hip, knee, and ankle rotation preparatory to the swing phase begin earlier in the walking cycle for the faster walking speed. Therefore, the earlier phasing of flexion for the faster walking speed results in a greater degree of flexion for all three joints at the instant of peak heel-rise, which contributes to the slightly higher lifting height of the heel. The kinetic determinant of the increased heel-rise is probably the more forceful muscular contraction at the end of the stance phase for the faster walking speed.

The peak elevation of the toe similarly is higher for the faster walking speed, although there are no significant differences in the peak toe elevation for the three height groups. The peak toe-rise which occurs at the end of swing as the extremity has kicked forward is a result of the slightly greater degree of hip and ankle flexion found for the faster walking speed.[8] It should be noted that although the normal hip and ankle show a greater amplitude of flexion during the swing phase for the faster walking speed, the peak amplitude of knee flexion is similar for normal free and fast speed walking.

In two previous studies of normal locomotion, we noted several differences in the walking patterns of the subjects in the sixty- to sixty-five-year age group which suggested a presenile pattern.[7,8] One of these differences was a tendency for these sixty-year-old men to show slightly less ankle extension at the end of the stance phase than the younger men. We, therefore, expected to find a slightly decreased lifting height of the heel for this oldest group. Spielberg, in his study of the gait of old people from sixty to one-hundred-and-four years of age, found that one of the first symptoms of old age in walking was a weakening of the "forward kick" at the end of the swing phase.[9] He also noted that the sixty- to seventy-two-year-old subjects had a tendency to initiate foot-floor contact with the entire foot flat on the ground, rather than in the normal heel-toe sequence. Therefore, in anticipation of finding presenile patterns, careful scrutiny was given to the vertical trajectories of the foot of the sixty- to sixty-five-year-old subjects in this study. We found no statistically significant differences in the vertical pathway of the heel or of the toe between our sixty-year-old subjects and the younger subjects at either walking speed, although the mean distance of the toe from the floor at the time of the initial heel-strike was less for the sixty- to sixty-five-year-old men than for the younger men at both walking speeds. However, all of the subjects in the age range included in this study made initial foot-floor contact in the normal heel-toe sequence, rather than with the entire foot flat on the ground.

SUMMARY

The vertical pathways of the heel and toe were measured for free and fast speed walking of thirty normal men aged twenty to sixty-five years, and heights from 61 to 73 inches.

The vertical pathway of the heel throughout each walking cycle consists of one peak of elevation which occurs early in the swing phase. The amplitude of this peak elevation is significantly higher for the faster walking speed and for the taller subjects.

The vertical pathway of the toe reaches one major peak of elevation late in the swing phase which is significantly higher for the faster walking speed. The toe-floor clearance distance throughout the early and midswing phase is extremely small.

Although the vertical pathways of the heel and toe do not relate systematically nor significantly with the five age groups studied, the vertical height of the toe at the time of heel-strike is lower for the sixty- to sixty-five-year age group than for the younger groups at both walking speeds.

The vertical pathways of the foot for repeated trials of the same subject are strikingly similar.

REFERENCES

1. Weber, W., and E. F. Weber, Mechanik der menschlichen Gehwerkzeuge. Dietrich, Göttingen, 1836.
2. Tiller, G., Die Hubhöhe des schwingenden Fusses wehrend der Gehbewegung und ihre Veränderung unter verschiedenen Umstanden. Arbeitsphysiologie, 9:332–340, 1936.
3. Wagner, E. M., and J. G. Catranis, New developments in lower-extremity prostheses. In Klopsteg, P. E., and P. D. Wilson, editors, Human Limbs and Their Substitutes. McGraw-Hill Book Company, Inc., New York, 1954, chap. 17.
4. Catranis, J. G., E. M. Wagner, and M. Reichen, Final report submitted by Catranis, Inc., Syracuse, New York, to the National Research Council, Committee on Artificial Limbs. November 1951.
5. Bresler, B., and J. P. Frankel, Forces and moments in the leg during level walking. Presented at Annual Meeting, American Society of Mechanical Engineers, New York, New York, November 28–December 3, 1948.
6. Report Number 115.15, New York University, College of Engineering Research Division. The Functional and Psychological Suitability of an Experimental Hydraulic Prosthesis for Above-the-Knee Amputees. Report to the National Research Council, Advisory Committee on Artificial Limbs, March 1953.
7. Murray, M. P., A. B. Drought, and R. C. Kory, Walking patterns of normal men. J. Bone Joint Surg., 46A: 335–360, 1964.
8. Murray, M. P., R. C. Kory, B. H. Clarkson, and S. B. Sepic, A comparison of free and fast speed walking patterns of normal men. Amer. J. Phys. Med., 45:8–24, February 1966.
9. Spielberg, P. I., Walking patterns of old people: Cyclographic analysis. In Bernstein, N. A., Investigations on the Biodynamics of Walking, Running, and Jumping. Central Scientific Institute of Physical Culture, Moscow, 1940, part II, chap. 15.

Basic Research

THE VERTICAL PATHWAYS OF THE FOOT DURING LEVEL WALKING

II. Clinical Examples of Distorted Pathways

M. P. MURRAY, Ph.D., and B. H. CLARKSON

In the preceding paper ranges of variability for the vertical pathways of the heel and the toe for free and fast speed walking of normal men were identified.[1] These values may serve as useful standards for comparison of the distorted pathways of patients with pathological gaits, since the normal subjects selected for this study represented wide ranges of age and height.

This present study identifies the distorted vertical pathways of the foot of disabled patients commonly referred for physical therapy or consultation. The gait patterns of an above-knee amputee and of a patient with flaccid paralysis of the ankle flexor muscles were selected to illustrate increased vertical motion of the foot during the swing phase. Those of a patient with hemiplegia illustrate decreased vertical motion.

The purpose of this paper is to demonstrate the relationship between the vertical pathway of the foot and the patterns of rotation of the hip, knee, and ankle for both normal and abnormal subjects.

RESULTS AND DISCUSSION

Above-Knee Amputee

Figure 1 contrasts normal displacement patterns with those of the prosthetic extremity of an above-knee amputee. The serial excursions throughout a fast speed walking cycle are shown: the sagittal rotation patterns for the hip, knee, and ankle, and the vertical pathways of the heel and toe. Rotation into extension for the hip, knee, and ankle is shown by downward

---

From the Physical Therapy Section, Physical Medicine and Rehabilitation Service, Veterans Administration Center, Wood, Wisconsin, and the Curriculum in Physical Therapy, Marquette University School of Medicine, Milwaukee, Wisconsin.

This investigation was supported in part by Grant CD 00052-02 from the Division of Chronic Diseases, U.S. Public Health Service, Department of Health, Education, and Welfare.

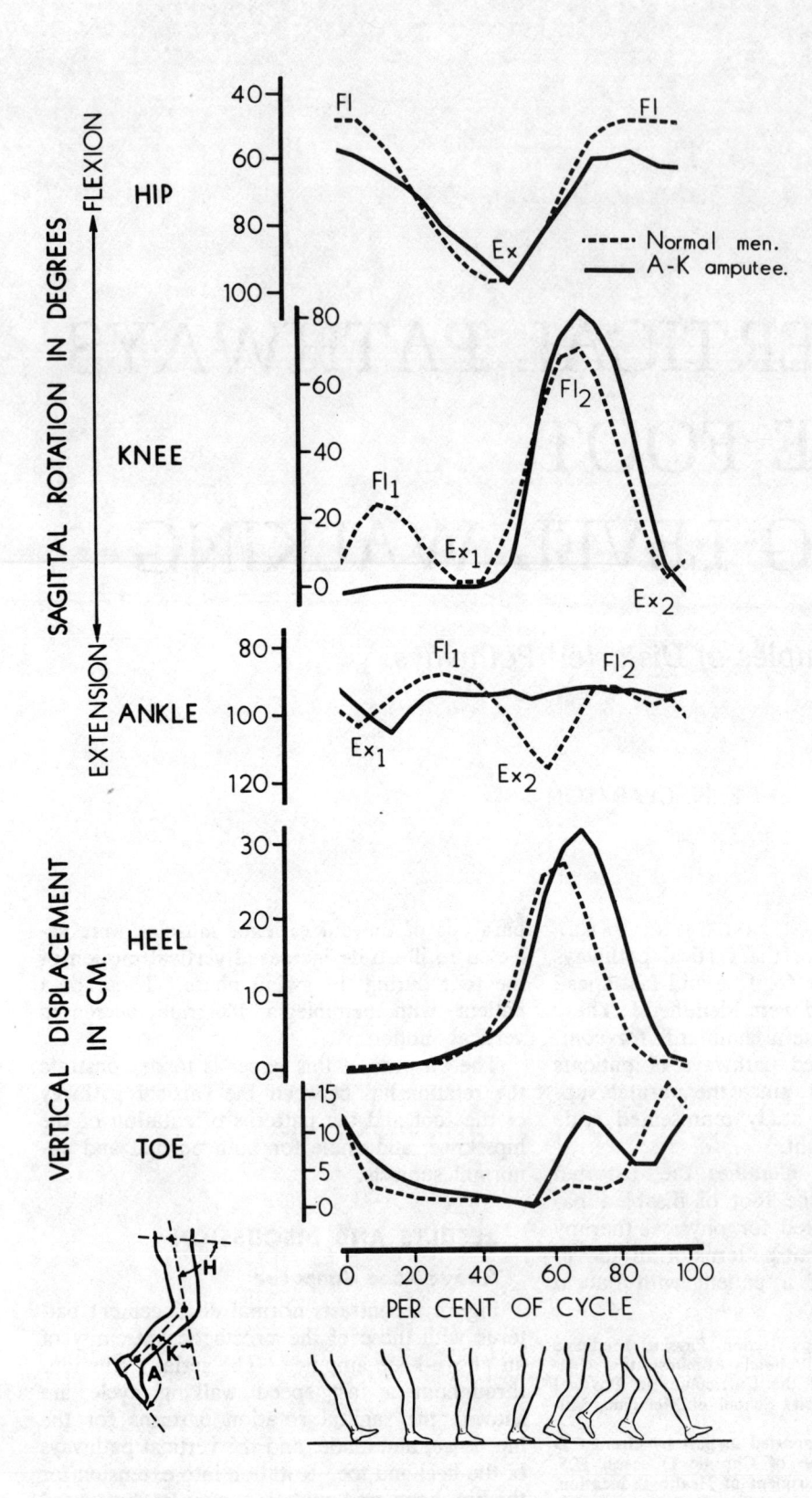

FIG. 1. Patterns of sagittal rotation of the hip, knee, and ankle, and vertical pathways of the heel and toe throughout the walking cycle for the prosthetic extremity of an above-knee amputee as compared to the mean patterns of thirty normal men. These patterns are for fast speed walking. Rotation of hip, knee, and ankle into flexion (Fl) is represented by upward deflections on the ordinate; extension (Ex), by downward deflections. Limb diagrams show the displacements of a normal lower extremity throughout the walking cycle.

DISTORTED VERTICAL PATHWAYS OF THE FOOT IN LEVEL WALKING

deflections on the ordinate; rotation into flexion, by upward deflections. The dashed lines represent the mean displacement patterns for thirty normal men [2]; the solid line, the displacement patterns of the prosthetic extremity. The mean walking speed for the normal subjects was 218±25 cm. per second, considerably faster than the 135 cm. per second speed for the amputee.

The amputee was thirty-six years of age and had been walking with his prosthesis for fifteen months. The prosthetic components were as follows: rigid pelvic band suspension, single-axis hip hinge, constant friction knee joint set for moderate friction, extension aid (kick-strap) under moderate tension, single-axis ankle, and a wooden foot with compressible heel and toe.

As can be seen in the vertical pathways of the heel and toe in Figure 1, the elevation of both the heel and the toe of the prosthesis early in the swing phase was decidedly greater than that for normal men. In addition, the descent of the toe to the floor, after the initial heel-strike, was delayed for the prosthetic extremity.

Figure 1—Hip shows that the patterns of hip flexion and extension throughout the walking cycle are similar for the amputee and normal subjects. The first half of the walking cycle is characterized by progressive extension as the body moves forward over the supportive foot (segment Fl-Ex, hip). After the contralateral foot makes floor-contact at 50 per cent of the cycle time, the hip reverses from extension to flexion for the ensuing swing phase. The apex of the extension wave which marks the reversal from hip extension to flexion was more abrupt for the amputated extremity than for normal subjects.

As can be seen in Figure 1—Knee, the normal pattern of knee rotation shows two flexion waves (Fl_1 and Fl_2) and two extension waves (Ex_1 and Ex_2) within each walking cycle. Initial floor-contact for normal subjects is made with the extended knee just beginning to flex.

The flexion proceeds as the body passes forward over this supportive extremity (Fl_1, normal knee), and then the knee extends obliquely behind the body (Ex_1, normal knee). Following this, the normal knee undergoes a maximum flexion wave (Fl_2) to provide the relative shortening of the extremity for the swing phase, and then extends to reach forward for the next step (Ex_2).

In contrast, the prosthetic knee pattern shown in Figure 1 was devoid of the normal initial flexion wave, Fl_1. Instead, the cycle was initiated with the prosthetic knee in complete extension, and the knee remained completely extended throughout most of the stance phase. Only after the contralateral foot had made floor-contact did the prosthetic knee begin to flex for the ensuing swing phase. The amplitude of the Fl_2 wave was greater for the prosthetic knee than for normal subjects. Following this peak flexion of swing, the prosthetic knee extended with greater angular velocity than that seen for normal men (segment Fl_2-Ex_2, prosthetic knee).

The normal pattern of ankle rotation similarly shows two waves of flexion and two of extension within each walking cycle. As can be seen in the patterns of ankle rotation of Figure 1, there is a striking difference between patterns for normal subjects and those for the prosthetic ankle joint. For normal men, the initial heel-floor contact, made with the ankle in flexion, is followed by a short sharp extension wave (Ex_1, normal ankle). This ankle extension allows rapid descent of the forefoot to the floor for normal subjects. Once full floor-contact is provided, the leg segment rotates forward over the fixed foot as the ankle progressively flexes (segment Ex_1-Fl_1, normal ankle). Following this, the ankle extends backward for the propulsive effort (Ex_2, normal ankle) and then reverses again into flexion to provide foot-floor clearance for the swing phase (Fl_2, normal ankle).

In contrast, the prosthetic ankle remained in relative flexion throughout most of the walking cycle. The first extension wave, Ex_1, caused by compression of the prosthetic heel and plantar bumper, was delayed which accounts for the delayed descent of the prosthetic toe to the floor. Moreover, the prosthetic ankle pattern was devoid of the normal Ex_2 wave, which is symbolic of the extension thrust at the end of the stance phase.

The excessive foot-floor clearance of the prosthetic extremity early in the swing phase can be explained on a relatively simple mechanical basis. As the hip of the amputated extremity reverses abruptly from extension to flexion, a force is applied to the thigh segment of the prosthesis by the stump. This force produces a forward (positive) acceleration of the thigh segment. Since the application of this accelerative force is eccentric to the mass center of the prosthetic leg-foot segment, a turning moment is produced. The effect of the moment is analogous to the initiation of the forward stroke of a double-pendulum in that forward acceleration of the proximal pendulum produces a clockwise rotation of the distal pendulum about the axis which connects the two segments.

In the same manner, the acceleration of the thigh segment produces backward rotation of the leg-foot segment about the prosthetic knee axis, thus excessive knee flexion and excessive heel-rise. The inertia of the leg-foot segment is ultimately counteracted by the forces of gravity, friction on the knee axis, and the extension bias of the kick-strap, which restore the desired direction of rotation of the prosthetic leg-foot segment. Although the excessive heel-rise per se may not be cosmetically objectionable, the double-pendulum action of the prosthesis does cause a prolongation of the swing phase which results in inequality in the timing of successive stance and swing phases and, more important, a deficit in walking speed.

In a separate study of the gait patterns of twelve above-knee amputees with constant friction knee components, we found the magnitude of prosthetic heel-rise was greatly influenced by walking speed.

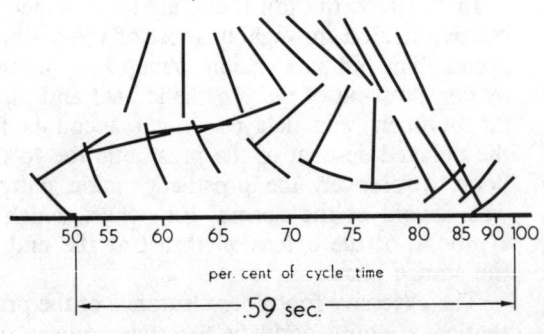

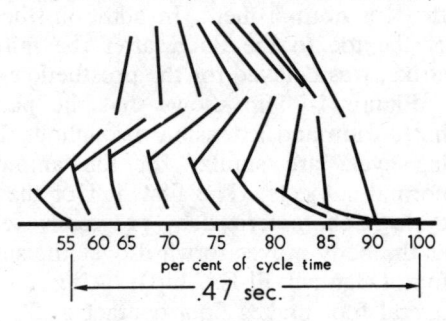

The increased heel-rise with increased speed was far more pronounced than that observed in normal men. These findings

DISPLACEMENT OF THE LOWER EXTREMITY DURING THE SWING PHASE OF WALKING

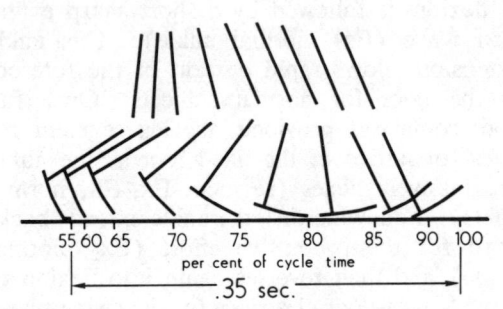

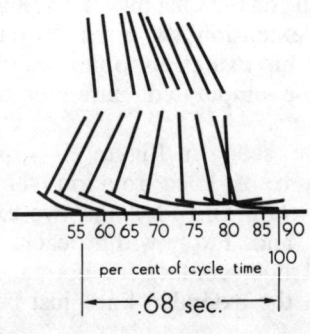

FIG. 2. Displacements of the lower extremity of four individual subjects for the same percentage intervals of the swing phase of the walking cycle. The stick diagrams, drawn to scale, show fast speed walking for a typical normal man and an amputee, and free speed walking for the two paralytic men.

The mean duration of swing for fast speed walking of thirty normal men is 0.38±0.03 seconds, 43 per cent of the walking cycle as compared to 0.59 seconds, 50 per cent for the amputee. The mean duration of swing for free speed walking of thirty normal men was 0.41±0.04 seconds, 39 per cent of the cycle as compared to 0.47 seconds, 45 per cent for the subject with the drop foot and 0.68 seconds, 45 per cent for the hemiplegic subject.

The forward displacement of the foot during swing, or stride length, for the fast speed walking of thirty normal men was 186 ±16 cm. as compared to 190 cm. for the amputee. Stride length for free speed walking of thirty normal men was 156±13 cm. as compared to 129 cm. for the subject with the drop foot and 61 cm. for the hemiplegic patient.

DISTORTED VERTICAL PATHWAYS OF THE FOOT IN LEVEL WALKING

suggest that the identification of the degree of "excessive heel-rise" listed for the analysis of prosthetic gait deviations [3,4] is of limited value except within circumscribed speeds of walking.

The contrast between the displacements of the lower extremity of a normal man and those for the prosthetic extremity are perhaps best visualized in the stick diagrams in Figure 2A and 2C. Although the duration of the swing phase for the prosthetic extremity is longer than normal, the positions of the extremities are shown for the same percentage intervals of the walking cycle to facilitate comparison. These figures illustrate the more abrupt hip flexion, the excessive knee flexion, and excessive heel- and toe-floor clearance distance, and consequently the prolonged swing phase for the prosthetic extremity as compared to normal subjects. These stick figures also depict the more rapid angular velocity for the prosthetic knee and the lack of the deceleration at the end of the swing phase which often results in an objectionable terminal swing impact for the constant friction prosthetic knee. For normal subjects, this decelerating force is provided by a restraining contraction of the "hamstring" muscles.[5]

Flaccid Paralysis of Ankle Flexor Muscles

Another example of excessive elevation of the foot during the swing phase can be observed in the free speed walking patterns of a patient with a drop foot shown in Figure 3. This figure shows the flexion and extension patterns of the hip, knee, and ankle, and the vertical pathways of the heel and toe for the paretic extremity of a patient with a drop foot in contrast to those of normal men. The patient was a forty-two-year-old male who had normal muscular strength except for unilateral flaccid paralysis of the ankle flexor muscles. The patterns of the paretic extremity without a brace are shown as a solid line; those of the paretic extremity with a spring-loaded ankle flexion-assist brace, in dash-dot lines; and the mean patterns of thirty normal men, in dashed lines. The walking speed for the patient was 121 cm. per second without the brace, and 123 cm. per second with the brace. The mean walking speed of the thirty normal men was 151 ± 20 cm. per second.

The excessive foot-floor clearance for the swing phase can be seen in the vertical pathways of both the heel and the toe for the unbraced paretic extremity in Figure 2. These distortions in the vertical pathway typify the classical high steppage gait. In addition, the height reached by the toe just before and at the time of initial foot-floor contact was markedly diminished for the paretic extremity. Instead of the normal initial floor-contact being made with the heel alone, the initial floor-contact for the paretic extremity was made almost simultaneously with the heel and toe, thus producing the characteristic and audible "foot slap."

Figure 3 shows that both the hip and knee of the unbraced paretic extremity flexed excessively during the swing phase (apex Fl_2 waves, paretic hip and knee). The excessive knee flexion for the paretic extremity provided the main compensatory mechanism for toe-floor clearance of the dangling foot early in the swing phase, whereas the excessive hip flexion provided compensation later in the swing phase.

In Figure 3, the ankle pattern for the unbraced paretic extremity shows distortion of both the Ex_1 and Fl_2 waves seen for normal subjects. The first ankle extension wave, Ex_1, which normally allows a rapid but controlled descent of the forefoot to the floor immediately after the initial heel-strike, was absent for the paretic extremity since the forefoot had already "slapped" down in uncontrolled descent. The diminished Fl_2 wave for the paretic ankle demonstrates the inability of the paralytic ankle flexor muscles to flex the foot against gravity to provide toe-floor clearance for the swing phase. However, close inspection of the last half of the Fl_2 wave for the unbraced paretic ankle shows that the ankle actually did flex slightly against gravity, particularly for the last 10 per cent of the cycle. This ankle rotation against gravity in the absence of forces applied by flexor muscles can be explained on a mechanical basis similar to that of the amputee.

For the patient with a drop foot, the deceleration or negative acceleration in forward motion at the end of the swing phase tends to produce a counter-clockwise rotation of the distal limb segments, similar to that produced at the termination of the forward stroke of the proximal segment of a double-pendulum. Thus, deceleration of the thigh at the end of the swing phase tends to rotate the knee into extension and deceleration of the leg tends to rotate the ankle into flexion. For the patient with the footdrop, the premature reversal of the hip of the paretic extremity from flexion to extension at 85 per cent of the cycle appears to provide the main

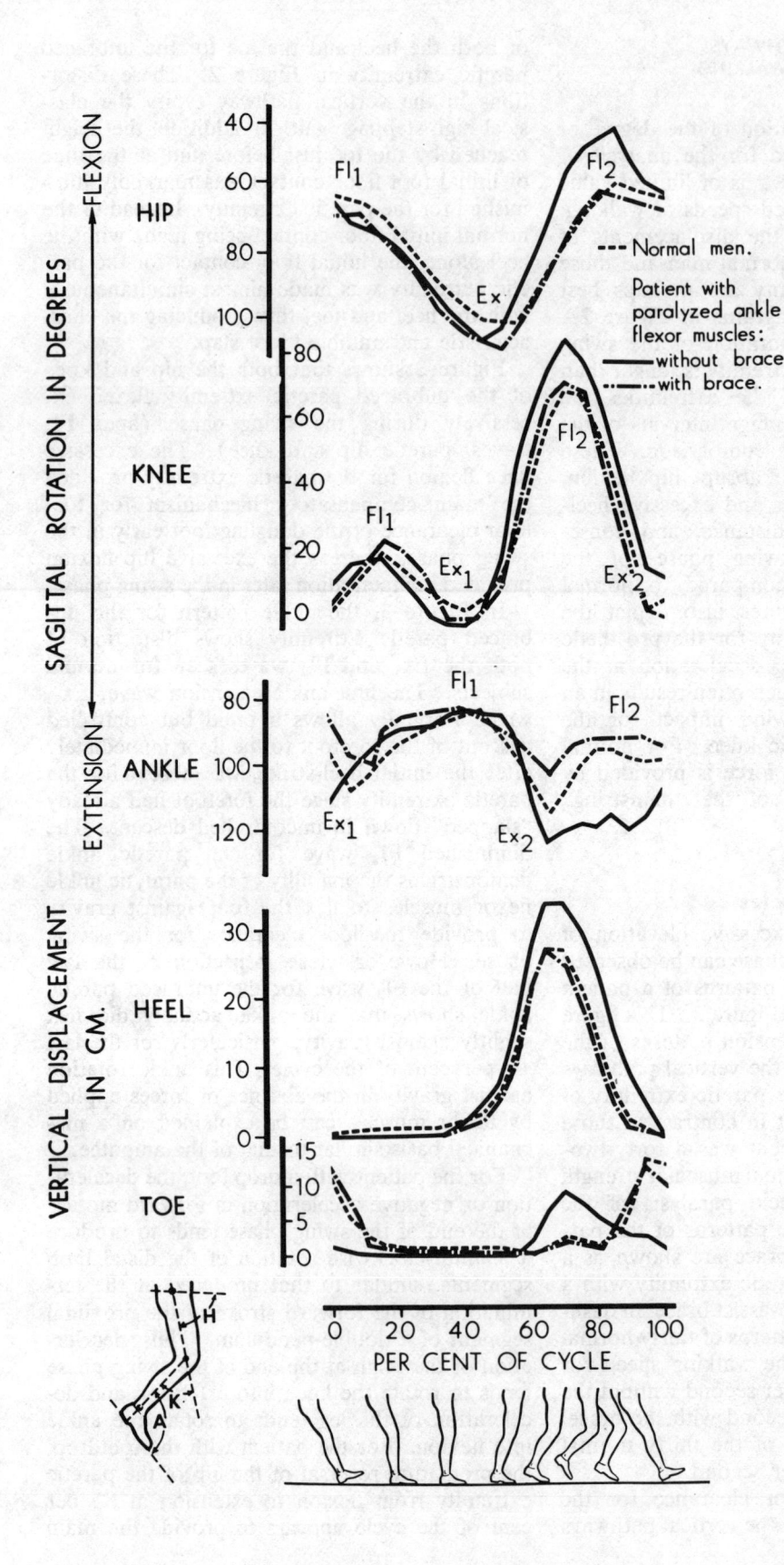

FIG. 3. Patterns of sagittal rotation of the hip, knee, and ankle, and vertical pathways of the heel and toe throughout a free speed walking cycle, for the paretic extremity of a patient with flaccid paralysis of ankle flexor muscles, as compared to the mean patterns of thirty normal men. Patterns for the paretic extremity are shown with and without a drop-foot brace. Rotation of hip, knee, and ankle into flexion (Fl) is represented by upward deflections on the ordinate; extension (Ex), by downward deflections. Limb diagrams show the displacements of a normal lower extremity throughout the walking cycle.

Basic Research

DISTORTED VERTICAL PATHWAYS OF THE FOOT IN LEVEL WALKING

impulse which decelerates the forward motion. This can be seen in the Fl_2 apex of the paretic hip pattern in Figure 3 and also in the stick diagrams of Figure 2B.

Another force moment which provides counter-clockwise rotation (flexion) of the paretic foot at the ankle is demonstrated in the distorted pattern of descent of the heel at the end of the walking cycle. The heel of the unbraced extremity does not show the deceleration in descent immediately before the terminal heel-floor contact which is typical of normal subjects. Deceleration in the descent would provide an impulse eccentric to the mass center of the paretic foot which would produce an undesirable clockwise turning moment rotating the foot into extension at the ankle. Conversely, acceleration in descent rotates the foot into flexion. Thus, the abnormal accelerated descent of the heel of the paretic extremity at the end of the walking cycle serves a useful purpose. The two maneuvers, lack of deceleration in the descending motion as well as the accentuated deceleration in the forward motion, were effectual in preventing the toe of the paretic extremity from making initial floor-contact before the heel.

It is apparent from the similarity of all five displacement patterns for the braced paretic extremity and those for normal subjects that the ankle flexion provided by the brace was effective in correcting the high steppage gait (Fig. 3).

Hemiplegia

In contrast to the two preceding examples of excessive vertical motion of the foot during the swing phase, the following displacement patterns of a hemiplegic patient illustrate diminished vertical amplitude. The patterns shown in solid lines of Figure 4 are those of a thirty-eight-year-old male with hemiplegia; those in dashed lines, the patterns of a normal man of similar age and height. The displacement patterns of a normal individual subject, rather than mean patterns, were selected for contrast with those of the hemiplegic subject in order to show the discrete sequence of reversal in direction of rotation for the joints of the normal lower extremity. The mean patterns of reversal show more rounded and less peaked apexes of the flexion and extension waves than the individual patterns because of the slight differences in walking speed among the subjects. The displacement patterns for both the normal man and the hemiplegic subject shown in Figure 4 represent free speed walking. The walking speed of the hemiplegic patient was 36 cm. per second, considerably slower than the 153 cm. per second walking speed of the normal subject.

The hemiplegia followed a cerebral vascular accident five weeks before the gait records were made. At the time of the recording, the strength of the patient's hip and knee muscles, without special facilitation methods, was Fair minus; the ankle muscle strength, except for eversion, was Poor minus; eversion was Trace. The hip and ankle extensor muscles were slightly spastic. Although the patient had been ambulatory with various supportive and assistive devices for two weeks, the displacement patterns shown represent his early attempts at ambulation without support or assistance.

The vertical pathways of the heel and toe of the hemiplegic subject in Figure 4 are both diminished in amplitude and also distorted in pattern of movement. Cursory observation of the patterns of hip, knee, and ankle rotation discloses that the hemiplegic patterns are more restricted in amplitude of rotation than those of the normal man. For ease of description, the walking cycles will be compared for the following phases: early, mid-, and late stance; and swing.

Early Stance. The normal walking cycle begins with the hip and ankle flexed and the knee extended, projecting the heel forward for initial floor-contact. Instead of the normal heel-toe sequence of initial floor-contact, the toe of the paretic extremity contacted the floor before the heel. As seen in Figure 4, this resulted from both insufficient ankle flexion and insufficient knee extension for the paretic extremity.

Midstance. Normally, both the supportive knee and ankle undergo a flexion wave during the stance phase as shown by the Fl_1 waves of the normal knee and ankle in Figure 4. In contrast, both the paretic knee and ankle extended, rather than flexed, in response to full-weight bearing. As a matter of fact, the paretic knee hyperextended during midstance (Ex_1, paretic knee).

Late Stance. During late stance, the normal lower extremity is directed obliquely backward for the propulsive effort. This backward direction is achieved normally by extension of the hip, knee, and ankle. In comparison, the lower extremity of the hemiplegic patient failed to achieve a backward direction late in stance be-

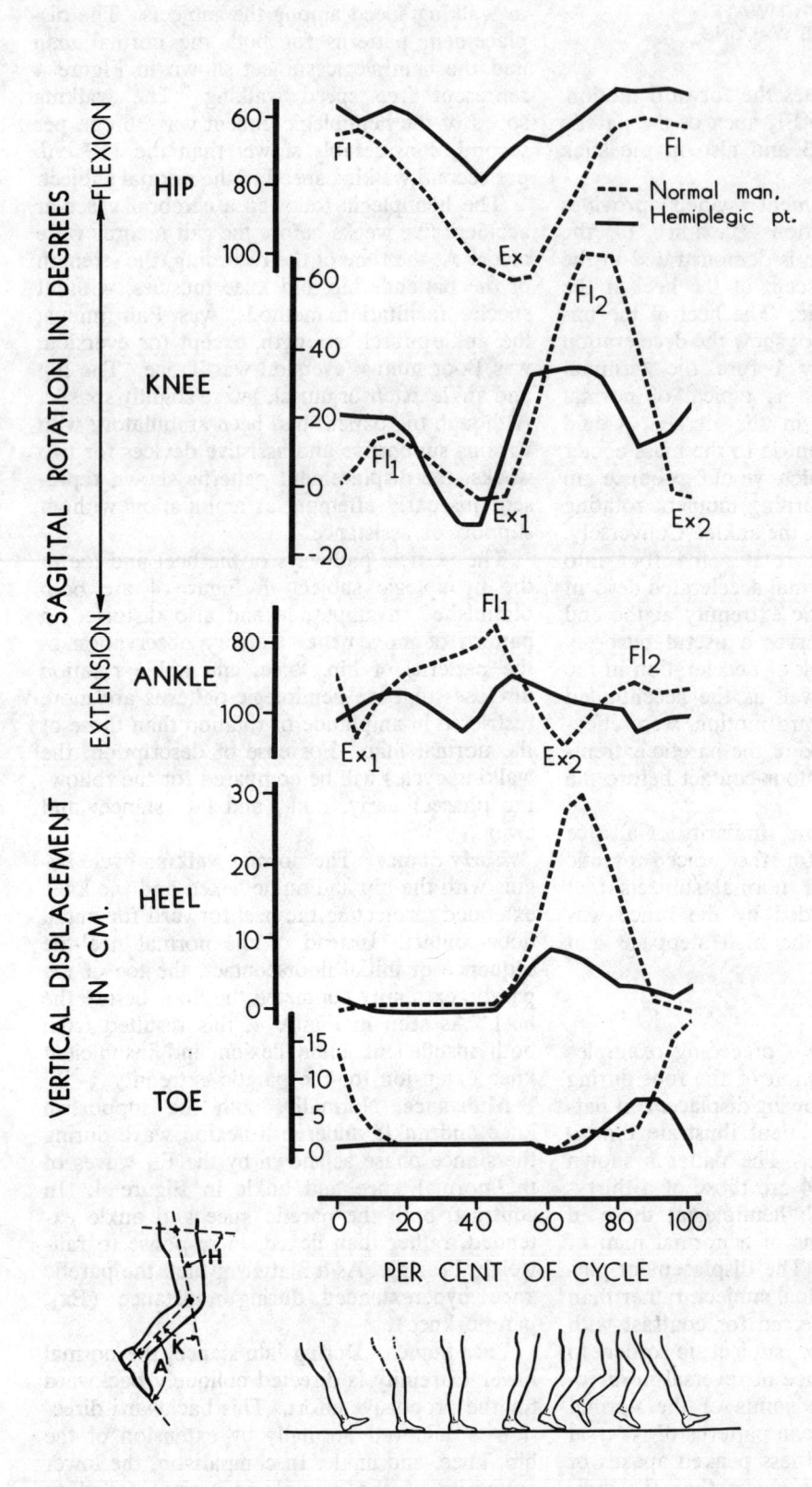

FIG. 4. Patterns of sagittal rotation of the hip, knee, and ankle, and vertical pathways of the heel and toe throughout a free speed walking cycle, for the paretic extremity of a hemiplegic male compared to those of a normal male of similar age and height. Rotation of hip, knee, and ankle into flexion (Fl) is represented by upward deflections on the ordinate; extension (Ex), by downward deflections. Limb diagrams show the displacements of a normal lower extremity throughout the walking cycle.

DISTORTED VERTICAL PATHWAYS OF THE FOOT IN LEVEL WALKING

cause of insufficient hip and ankle extension. Instead of being extended obliquely behind the trunk, the paretic extremity was in a more vertical position (Fig. 2D).

Also late in stance, the normal hip, knee, and ankle undergo an orderly sequence of reversal from extension to flexion preparatory to the swing phase. The orderly sequence of flexion for the normal subject in Figure 4 begins first for the knee at 40 per cent of the cycle time, then for the hip at 50 per cent, and finally, the ankle, at 65 per cent. Instead of the normal orderly sequence of flexion preparatory to the swing phase, the paretic extremity flexed almost simultaneously, at 40 per cent of the cycle for the hip and knee, and at 35 per cent for the ankle.

Swing. Normal foot-floor clearance during swing is provided by smooth interaction of the rotation excursions of the hip, knee, and ankle. The diminished amplitude of foot-floor clearance of the hemiplegic subject was the result of both the inability to produce an adequate range of rotation and to the inco-ordinated patterns of rotation, particularly for the knee and ankle. In order to provide relative shortening of the lower extremity early in swing, the normal knee rotates rapidly through a large range of flexion (Fl_2 knee, Fig. 4). Following this peak degree of flexion, the normal knee reverses into rapid extension to project the extremity forward as the hip and ankle sustain flexion for floor clearance.

In contrast, the peak degree of hemiplegic knee flexion was markedly diminished (Fl_2, paretic knee) and the ankle dropped progressively into extension as the swinging extremity moved forward (Fl_2, paretic ankle). Moreover, as the attempt was made to project the paretic extremity forward, the knee extension excursion (segment Fl_2-Ex_2, paretic knee) was halted at 85 per cent of the cycle time and actually reversed into flexion. As shown in the stick diagrams of Figure 2D, this reversal resulted in a backward displacement of the forward reaching extremity and consequently a decreased stride length. It is interesting to note that at 85 per cent of the cycle time when the paretic knee reversed abnormally from extension to flexion, the paretic ankle, which had been dropping into extension, echoed the flexion wave (Fl_2 paretic ankle, Fig. 4).

The patterns depicted for the example of the hemiplegic gait are quantitative demonstrations of components of disordered motor behavior which are frequently described in the literature as being typical of this type of cerebral lesion.

Most textbooks of clinical neurology describe the motor impairment of a hemiplegic subject as being more pronounced distally in the limb than proximally. The hip, knee, and ankle rotation patterns in Figure 4 show quantitative evidence of this type of motor behavior. The hemiplegic hip rotation pattern is not grossly distorted from normal, although it lacks the smoothness and amplitude of normal motion.

The pattern of hemiplegic knee rotation is considerably more distorted than the hip in both the time and amplitude of rotation. Although two flexion waves and two extension waves can be identified in the knee rotation pattern for the paretic extremity, the peak amplitudes of these waves during the stance phase are in excess of normal (Fl_1 and Ex_1, paretic knee) while those for the swing phase are markedly diminished (Fl_2 and Ex_2, paretic knee). The hemiplegic ankle rotation pattern shown in Figure 4 is more severely distorted than that for the hip or knee, and no semblance of the normal flexion and extension waves is discernible.

The hemiplegic knee rotation pattern for the swing phase illustrates the typically increased difficulties encountered when the hemiplegic subject attempts both rapid movements and movements of great amplitude. In normal walking, the excursion of knee rotation during the swing phase shows a larger amplitude and a greater angular velocity than any other joint of the lower extremity. It can be seen that the hemiplegic subject fails to provide sufficient amplitude of knee flexion for adequate foot-floor clearance during the swing phase (Apex Fl_2, paretic knee). Moreover, the usually rapid knee extension which projects the extremity forward at the end of swing is "blocked" prematurely (segment Fl_2-Ex_2, paretic knee).

Brunnstrom has clearly described the dominance of primitive patterns of movement and limb synergies in the motor performance of patients with hemiplegia.[6,7] We believe that the differences between refined normal co-ordination and the primitive motor patterns characteristic of hemiplegic patients are illustrated in quantitative form in several phases of the walking cycle in Figure 4. Despite the range of variability in walking patterns of normal subjects, we found the discrete sequence of reversals in direction of rotation for the various limb segments to be stereotyped.[2, 8] For example, in normal walking the flexion movements preparatory to the relative shortening of the limb for the swing phase occur in an orderly sequence, first for the knee, then the hip, and last, the ankle. In contrast, primitive patterns are seen to prevail for the paretic extremity, as the hip, knee, and ankle initiate flexion almost simultaneously for the swing phase at 40 per cent of the walking cycle.

Further dominance of limb synergies can be seen early in stance. This extension synergy appears to be a positive supporting reaction in response to full weight bearing. At the onset of full weight bearing in normal walking, the knee and ankle undergo flexion as the hip extends. In contrast, the hip, knee, and ankle of the paretic extremity extend simultaneously from 20 to 40 per cent of the walking cycle.

It is also frequently cited in the literature that hemiplegic motor behavior is characterized by the inability to perform isolated movements. Careful scrutiny of normal walking patterns reveals that, although the hip, knee, and ankle are constantly rotating throughout each walking cycle, these three joints rarely rotate simultaneously in the same direction. Although the examples of shifting directions of rotation can be identified throughout the entire walking cycle, the brief interval just before and after heel-strike is selected for illustrative purposes. As the normal extremity reaches forward for the new step (85–100 per cent of the cycle), the ankle and hip sustain flexion as the knee extends. Immediately following heel-contact (0 per cent of the cycle), the ankle extends and the knee flexes. Shortly after toe-contact (15 per cent of the cycle), the ankle flexes, and the knee and hip extend. Thus the movement patterns of normal walking entail the ability to flex one joint as adjacent joints are extending, and vice versa. Figure 4 demonstrates the inability of the hemiplegic subject to perform these fine adjustments in the direction of rotation.

SUMMARY

Examples of distorted vertical pathways of the heel and toe are shown for an above-knee amputee, a patient with a drop foot, and a hemiplegic patient. These patterns are shown to illustrate the relationship between hip, knee, and ankle rotation and the resultant vertical pathways of the foot for both normal and abnormal subjects.

Acknowledgment is given to Susan B. Sepic and Jerome Hensen for assistance with photographic, statistic, and illustrative procedures.

REFERENCES

1. Murray, M. P., and B. H. Clarkson, The vertical pathways of the foot during level walking: I. Range of variability in normal men. J. Amer. Phys. Ther. Ass., 46:585–589, June 1966.
2. Murray, M. P., R. C. Kory, B. H. Clarkson, and S. B. Sepic, A comparison of free and fast speed walking patterns of normal men. Amer. J. Phys. Med., 46:8–24, February 1966.
3. Prosthetics Education Program, Post-Graduate Medical School, New York University. Gait analysis. In Prosthetic Clinic Operations, New York University textbook, November 1956, sec. 4, pp. 9–11.
4. Prosthetic-Orthotic Education, Northwestern University Medical School. Gait analysis. In Lower Extremity Prosthetics for Physicians, Surgeons, and Therapists, special course textbook, p. 130C.
5. Eberhart, H. D., V. T. Inman, J. B. DeC. M. Saunders, A. S. Levens, B. Bresler, and T. D. McGowan, Fundamental studies of human locomotion, and other information relating to the design of artificial limbs. A report to the National Research Council, Committee on Artificial Limbs. Berkeley, University of California, 1947.
6. Brunnstrom, S., Recording gait patterns of adult hemiplegic patients. J. Amer. Phys. Ther. Ass., 44:11–18, January 1964.
7. Brunnstrom, S., Walking preparation for adult patients with hemiplegia. J. Amer. Phys. Ther. Ass., 45:17–29, January 1965.
8. Murray, M. P., A. B. Drought, and R. C. Kory, Walking patterns of normal men. J. Bone Joint Surg., 46A:335–360, March 1964.

Basic Research

The Base of Support in Stance*

Its Characteristics and the Influence of Shoes Upon the Location of the Center of Weight

Frances A. Hellebrandt, Doris Kubin, Winifred M. Longfield and L. E. A. Kelso

According to Morton[1] the ease and economy with which man meets the prime mechanical disadvantages of the biped stance represents the acme of his physical evolution. The unfavorable factors are the height of man's center of gravity and the smallness of his base of support. Morton postulates that as long as the line of gravity falls within a certain sector of the area of underpropping, rotatory stresses will be equilibrated by increase in tone without resorting to phasic contraction of the antigravity muscles. He locates the center of gravity of a subject standing in a comfortable posture as perpendicularly above a point between the feet in the region of the navicular bones. Steindler[2] believes there is an inherent tendency for each individual to maintain his center of weight at a constant point in relation to the structural supports of his body. However, we[3] observed that sway is inseparable from the upright stance and that the center of weight shifts incessantly during the maintenance of a natural comfortable posture.

The object of this study was (1) to observe the base of support characteristics in stance; (2) to project the average location of the shifting center of weight into the base of support; and (3) to determine the effect of shoes upon the location of the center of weight and upon postural stability.

METHODS

1. *Recording the Base of Support.* All observations were made in Morton's foot position of greatest stability in stance. In this posture the toes turn out so that the lateral diameter of structural support approximates the antero-posterior. The position was standardized by the insertion of a 30-degree wedge between the heels which were made to abut against a low vertical ledge. The subject stood upon absorbent paper after first moistening the soles of the feet with a concentrated solution of potassium permanganate. The contact area of the feet with the ground was recorded during five minutes of quiet standing. The resulting footprint was then analyzed in accord with the suggestions embodied in Morton's description of the functionally important diameters of support. The total area of underpropping was measured in square centimeters with a planimeter after introducing tangents to the heel, the lateral border of the feet and the toes. The diameter of antero-posterior stability was taken as the vertical distance between the heel tangent and a line through the heads of the first metatarsals, the latter being placed midway between the two in the presence of inequality in the length of the feet. The distance between the fifth metatarsals measured the lateral support. The accessory margin of anterior stability was taken as the distance between the first metatarsal line and the great toe tangent.

2. *Projecting the Center of Weight into the Base of Support.* Figure 1 illustrates the method and equipment. The subject stands upon the uppermost of two platforms placed at right angles to each other and supported on steel wedges to minimize friction. One knife-edge of each beam rests upon a hundred-pound platform scale modified to graphically record changes in weight upon a constant-moving kymograph. The subject faces the antero-posterior scale which records all forward and backward oscillations of the center of weight. A simultaneous record of sway in the transverse vertical plane is made by the auxiliary lever of the lateral scale. Knowing the weight of the subject, the distance between the knife-edges, the indications of the scales and the tares, the location of the center of gravity in the two vertical orientation planes may be determined at any instant by equating moments. Since the position of the feet is fixed in regard to the knife-edges in terms of which the center of gravity is located, the latter may be projected into the imprint of the base of support.

3. *Determining the Average Position of the Shifting Center of Weight.* The secondary of a slotted core transformer excited from a constant potential source was suspended from each auxiliary scale lever. Since the position of the scale lever is directly proportional to the former, the involuntary vibratory motion of the center of weight may be electrically integrated by placing the secondary of each transformer in series with the current coil of a rotating watt hour meter. From the reading of the meters the average load on the antero-posterior and lateral scales may be obtained and by simple calculation converted into the mean location of the center of gravity in the two cardinal vertical orientation planes.

4. *Modifying the Size of the Base of Support and the Height of the Center of Gravity by Wearing Shoes.* A series of consecutive 3 minute records was made standing in a comfortable posture in bare feet, wearing shoes with low (1.0 to 2.5 cm.), medium (2.8 to 4.5 cm.) and high (4.8 to 7.0 cm.) heels. Sufficient rest was allowed between each standing period to eliminate fatigue. Heel height was estimated by the difference in the tallness of the subject measured with and without shoes.

RESULTS AND THEIR DISCUSSION

1. *The Characteristics of the Base of Support in Stance.* Observations were made on 88 professional students in physical education accustomed to vigorous exercise and possessing, by that functional criterion, essen-

---

*From the Department of Physiology, University of Wisconsin, Madison, Wisconsin, supported in part by a grant from the Wisconsin Alumni Research Foundation.

ACKNOWLEDGMENTS—Our thanks are due Profs. Trilling and Bassett for the use of certain of the facilities of the Physical Education Department, Genevieve Braun for help in the conduct of the experiments and O. H. Brogdon for assistance in the preparation of the manuscript.

tially normal feet. The data are presented in Table I. These young adult women show a 63 per cent larger margin of static security laterally than antero-posteriorly in the foot posture of Morton, disregarding the accessory area of support which is functionless during stance in the average subject. The greatest lateral diameter more nearly approximates total foot length. Considerable variability was evident in the diameters studied, especially in the relative importance of the accessory area of support which is a margin of safety called into activity when the center of gravity shifts too far forward in the antero-posterior plane.

TABLE I
Characteristics of the Base of Support in Stance

	Mean	Standard Deviation	Coefficient of Variability
Number of subjects—88			
Antero-posterior foot support	16.43 ± 0.05 cm.	0.75 ± 0.04	4.6
Lateral foot support	26.78 ± 0.12 cm.	1.63 ± 0.08	6.08
Accessory foot support	6.10 ± 0.04 cm.	0.49 ± 0.02	8.03
Total area of underpropping	502.0 ± 3.37 sq. cm.	46.89 ± 2.38	9.3
Foot length	23.09 ± 0.07 cm.	1.00 ± 0.05	4.3

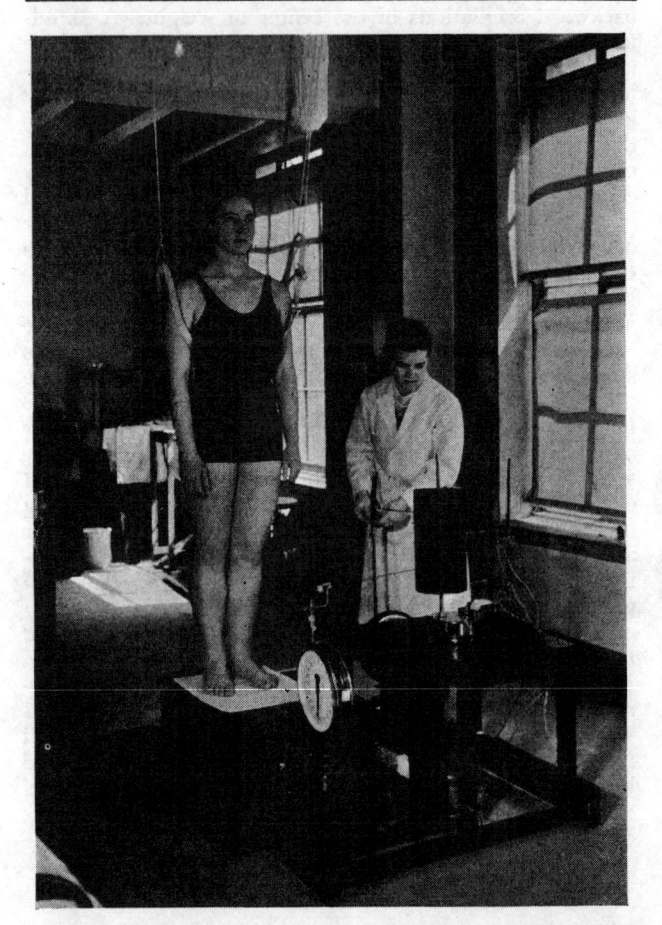

Fig. 1.—Photograph of the apparatus developed for the graphic registration of concurrent shifts of the center of gravity in the antero-posterior and transverse vertical planes during the maintenance of the upright stance. The subject assumes a natural comfortable posture with the gaze fixed upon a designated point. The arm straps are loosely adjusted and offer no restraint. Their function is protective in the event of syncope.

2. *The Location of the Center of Weight in the Base of Support.* The area of structural support in the foot position of Morton forms a trapezoid bounded by the intersection of the heel and lateral tangents and the extension of the horizontal line passing through the heads of the first metatarsals. The geometric center of the base of support may be approximately located by constructing an antero-posterior median from the midpoints of the heel tangent and the metatarsal line, and another, perpendicular to the first, bisecting it. The intersection of these medians marks the center of the active support. The projected average position of the center of weight within the base of support may then be mathematically related to the geometric center of the trapezoidal area approximating the superficial extent of the underprop space. The distance from the geometric center of the base to the center of gravity projection is called the eccentricity, $\frac{eAP}{\frac{1}{2} AP}$ and $\frac{eT}{\frac{1}{2} T}$ being the eccentricity ratios evaluating deviations as regards the antero-posterior and transverse areas respectively. (eAP = the perpendicular distance from the center of gravity to the transverse median. ½ AP = half the length of the antero-posterior median. eT = the perpendicular distance from the center of gravity to the antero-posterior median. ½ T = half the length of the lateral median.) Figure 2 illustrates the above points.

To support the center of gravity and uphold the body in the erect posture, the straight line joining the center of weight with the center of the earth must fall within the area enclosed by the feet. The average position of the center of weight falls relatively close to the geometric center of the base. That it is moderately eccentric was indicated by our[4] observation that the upright stance was asymmetrical in the majority of a group of 445 young adult women. The weight fell preponderantly to the left in a comfortable stance. However, the mean location of the center of weight is such that the margins of static security are roughly equal on all sides when the functionless accessory area of support is disregarded. We have also demonstrated that, as Steindler believes, there is an inherent tendency for each individual to maintain his center of gravity at a constant point in relation to his structural supports. The standard procedure for the graphic study of the trajectory of the center of gravity during three minutes of quiet standing in a natural and comfortable stance was followed on different days. The utmost care was exercised to assume the same posture. The involuntary sway of one of our subjects was unusually great. The maximum displacements of her center of gravity forward, back, right and left during a typical three-minute period of standing are projected into the base of support in Fig. 3. The maximal forward sway brings the center of weight dangerously near the fore extremity of the antero-posterior diameter of structural foot support. The subject was forced to grip with the toes to keep from falling forward, even though the center of gravity never moved beyond the heads of the first metatarsals. Backward motion was less great. The remarkable feat is that, in spite of involuntary postural sway of this extraordinary magnitude, the average position of the center of gravity on eight different days was confined to an area of 1.5 sq. cm. in a base of support measuring approximately 500 sq. cm. The data are illustrated in Fig. 3. Six consecutive trials on the same day yielded essentially identical results, the average centers of weight now being still more closely clustered. However capricious the behavior of the center of gravity may seem during so-called "quiet" standing, its oscillations are so accurately balanced that the average relation of the center of gravity to the base of support remains remarkably constant at least in vigorous, healthy young adults with a good equilibratory and kinesthetic sense.

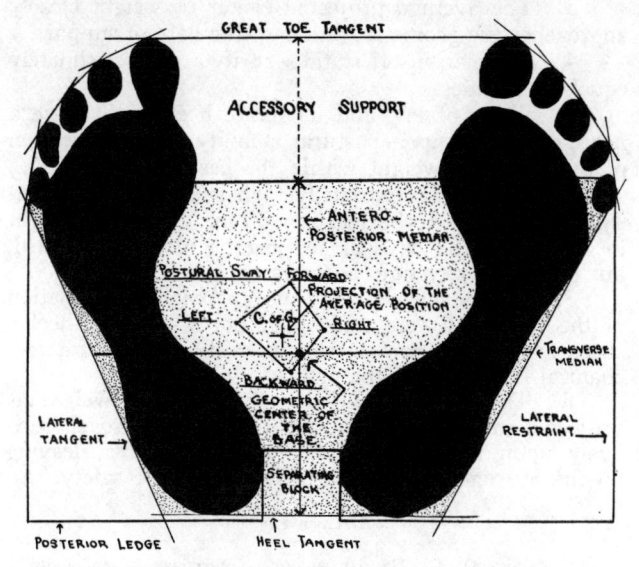

Fig. 2.—Planographic reproduction of the base of support of a normal young adult woman illustrating the slight eccentricity of the average position of the center of weight during three minutes of comfortable standing. The peripheral points FBRL indicate the maximal shift in the center of weight forward, back, right and left. The foot guide is removed before a record is made. If the projected center of weight falls anterior to the transverse median, the AP eccentric value is positive; if it falls posterior to it, it is negative. Similarly, if the projected center of weight falls to the right of the anteroposterior median, the lateral eccentric value is positive; if it falls to the left, the value is negative.

3. *The Effect of Shoes upon the Location of the Center of Weight and upon Postural Stability.* Fifty-one experiments were performed on a homogeneous group of 12 vigorous young adult women, habitually wearing low-heeled shoes. The area of underpropping varied with the heel height of the shoe being worn. The low-heeled shoes provided an average increase of 10.5 per cent in the total supporting area. The base of support furnished by shoes of medium heel height was slightly smaller than that of the bare feet. The highest heeled shoes decreased the contacting foot area on an average by 15.1 per cent.

The subjects responded similarly. In quiet standing without shoes the average position of the center of weight was close to the geometric center of the base of support. When a low-heeled shoe was worn, the center of weight remained in approximately the same position with respect to the new base. Heels of moderate height similarly failed to significantly shift the average location of the center of weight. When shoes with the highest heels to which the subject was accustomed were worn, the projection of the center of gravity into the base of support continued to fall near the geometric center of the area of underpropping in 7 of the 12 subjects. In the remaining 5, the center of weight was thrown forward. The data indicate that the center of gravity when projected vertically, typically occupies a position near the geometric center of the base even when the heels are elevated, raising the center of gravity, and further re-

ducing the already small base of support. To achieve this, the multijointed body must redistribute its parts, thus nullifying the derangement caused by uplifting the heels. This may be accomplished in a variety of ways, the most common being a compensatory flexion of the knees or hyperextension of the lumbo-dorsal spine, the latter conveying the upper portion of the body backward and thus balancing the forward pitch caused by elevating the heels.

Since the position of the projected center of gravity remained nearly constant when shoes were worn, involuntary postural sway was observed in order to determine the influence upon stability of augmenting the chief mechanical impediments to the orthograde stance, raising the center of gravity and narrowing the base of support. The margins of the area within the base allocated to shifts in the center of weight were found by projecting the maximum oscillations in the coronal and sagittal planes. Low-heeled shoes had little effect on postural stability, the area of sway being in some instances slightly decreased in magnitude in consonance with the enlargement of the base. An augmentation of the involuntary postural sway was evident with moderate elevation of the heels. In all but one subject sway was excessive when shoes of greatest heel height were worn. The degree of perturbation may be unduly accentuated by the fact that the subjects were unaccustomed to standing in high-heeled shoes, but the tendency to become less steady with each infringement of the security of a body already in unstable

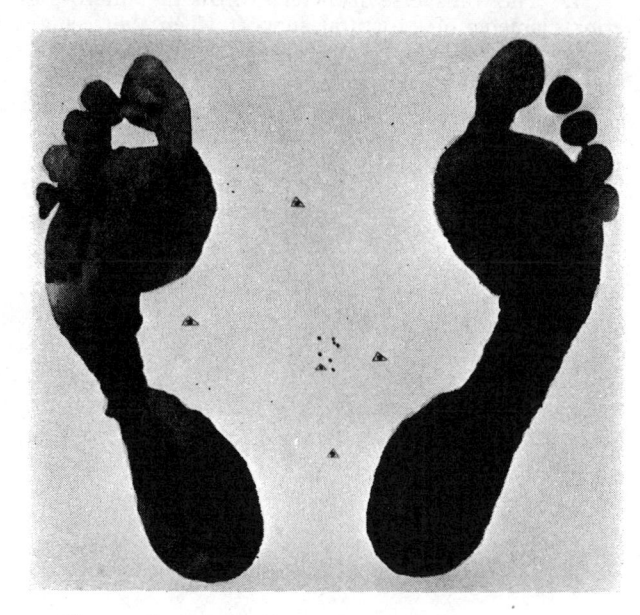

Fig. 3.—A photographic reproduction of the base of support in stance. The central cluster of eight dots represents the projection of the average location of the center of gravity during three minutes of standing in a natural comfortable stance on eight different days. The maximal shift of the center of weight to and fro in the two vertical orientation planes on the day differentiated by the triangular outline of the average spot is represented by the outlying similarly enclosed data.

equilibrium might be deduced on *a priori* grounds. Table II contains a typical set of data illustrating the trend of the results. From its examination it is evident that even though very high-heeled shoes heighten the center of gravity, lessen the base of support and magnify the unintentional postural sway, the greatest area within the pedal foundation to which shifts in the center of weight are apportioned is only a fraction of the total underprop space. A large margin of safety still remains, affording protective security in all planes.

Table II—Showing the Influence of Shoes With Heels of Various Heights upon the Base of Support, the Location of the Projected Center of Weight and Postural Stability.

Heel Height	Total Base of Support	Eccentricity in the Vertical Orientation Planes		Area of Maximal Sway
cm.	sq. cm.	Antero-Posterior	Transverse	sq. cm.
0.0	426.9	+ .129	— .063	4.4
2.5	429.9	+ .098	— .052	3.3
4.0	380.9	+ .027	— .014	5.5
5.2	369.3	.000	— .129	19.1

If the projected center of weight falls anterior to the transverse median or to the right of the antero-posterior median, the eccentric value is positive; if it falls posterior to or to the left of the respective medians, it is negative.

Summary and Conclusions

The characteristics of the base of support in stance were studied on young adult women possessing functionally normal feet. The average position of the incessantly shifting center of weight was located within the area of underpropping, and disturbances in static security induced by shoes were measured. From the evidence presented it may be concluded that:

1. The transverse diameter exceeds the antero-posterior diameter of structural support when the feet are separated by a 30-degree wedge, but closely approximates total foot lengths.

2. Irrespective of the magnitude of the involuntary shift in the center of weight which is inseparable from the upright stance, young adults of good equilibratory and kinesthetic sense maintain the *average* location of the center of gravity in a highly constant relation to the structural diameters of the base of support.

3. The average projected center of weight closely approaches the geometric center of the base of support.

4. The margins of static security are approximately equal on all sides.

5. Shoes of low and moderate heel height have a negligible effect upon postural stability and the location of the center of weight within the base of support.

6. Shoes with extremely high heels may either insignificantly effect the location of the center of weight in respect to the base of support or move it forward, but postural sway now becomes excessive.

7. It is suggested that the augmented vascillation of the center of weight is obligatory when high-heeled shoes are worn because they intensify the natural mechanical instability of the orthograde stance.

8. The maximal shifts in the center of weight as projected into the base of support rarely encroach seriously upon the limits of the underprop area, leaving on the average a relatively wide margin of safety.

Bibliography

1. Morton, D. J.: The Human Foot. Columbia University Press, N. Y., 1935.
2. Steindler, A.: Mechanics of Normal and Pathological Locomotion in Man. Charles C. Thomas, Springfield, Illinois, 1935.
3. Hellebrandt, F. A.: Standing as a Geotropic Reflex. In press.
4. Hellebrandt, F. A., R. H. Tepper, G. L. Braun and M. C. Elliott: The Location of the Cardinal Anatomical Orientation Planes Passing Through the Center of Weight in Young Adult Women. In press.

Walking Patterns of Healthy Subjects Wearing Rocker Shoes

MARGERY J. PETERSON,
JACQUELIN PERRY,
and JACQUELINE MONTGOMERY

We compared walking in rocker shoes with walking in athletic shoes at free and fast velocities in 15 healthy women who were between the ages of 21 and 30 years. Footswitches, electrogoniometers, surface electromyograms, and a force plate were used for data collection. No significant differences were found in velocity, cadence, gait-cycle duration, single-limb support, or swing-stance ratios in free and fast walking. Double-limb support was decreased by 9% in the rocker shoes in free walking. The ankle range of motion showed accommodation of the foot to the 8.5 degrees of plantar flexion built into the rocker shoes. Electromyographic responses appeared to be similar regardless of shoes or cadence.

Key Words: Gait, Physical therapy.

Clinical interest in the use of rocker shoes, more commonly known as clogs, has increased. These shoes have been commercially available for years and have three primary features: 1) open heel, 2) heel height of approximately 6 cm, and 3) a terminal toe rocker. Some patients have found these features to be very helpful in walking. Currently, at our pathokinesiology laboratory, an experimental program is in progress to identify the gait problems that are specifically improved by the rocker shoes. The actual effects of the specific features of the rocker shoes and the patients that may be candidates for this type of footwear need to be identified. The effects of rocker shoes on the normal gait pattern have not yet been determined. Once these effects have been identified, pathological gait patterns in patients wearing the rocker shoe can be compared with normal gait patterns. The purpose of this study is to determine the effects of the rocker shoe on the normal gait pattern.

METHOD

Subjects

The subjects for this study were 15 healthy adult women between the ages of 21 and 30 years with a mean age of 25.5 years. All subjects chosen for this study had no current or previous history of any neurological or orthopedic disorder affecting their lower extremities. Persons at the extremes of height and weight were excluded. The mean height of the subjects was 165.2 cm and the mean weight was 62 kg. All subjects were screened for muscle weaknesses, range-of-motion limitations, and leg-length discrepancies to eliminate obvious factors that would influence normal walking. Each subject signed an informed consent before participating in this research project.

Footwear

We used a rocker shoe with a rigid wooden sole, open heel, and a vamp made of a plastic material.* The heel of the shoe was undercut so that the posterior aspect of the heel was 0.9 cm off the ground. The total heel height from the ground was 5.7 cm. The terminal toe rocker began at about 63% of the total shoe length (about 1 cm posterior to the metatarsal heads). The arc formed by the resulting roll-off had a 30.5-cm radius. The anterior aspect of the shoe was 2.7 cm off the ground. The incline of the shoe heel to the roll-off point formed an angle of 8.5 degrees (Fig. 1).

The terminal toe rocker formed an angle of 7 degrees from the medial to the lateral edge of the shoe to conform with the alignment of the metatarsal heads (Fig. 2). The interior of the shoe was formed to the foot with a slight depression for the heel, a depression for the metatarsal heads, and a toe grip (Fig. 3).

The athletic shoes were made of canvas or leather and laced up. The soles were flat. No athletic shoes with negative heels were used. Heel height ranged from 0.05 to 1.3 cm with a mean of 0.7 cm.

Instrumentation

We used the following equipment to assess gait characteristics: 1) a pair of insole footswitches† to identify specific phases of gait, 2) knee and ankle electrogoniometers† to detect joint motion in the sagittal plane, 3) paired surface electrodes‡ to detect phasic electrical activity of muscles during walking,

---

Ms. Peterson was a master's degree candidate at the University of Southern California when this study was conducted. She is currently Supervisor, Marina Professional Services, Long Beach, CA 90814 (USA).

Dr. Perry is Director, Pathokinesiology Service, Rancho Los Amigos Medical Center, Downey, CA 90242, and Professor of Orthopaedic Surgery, University of Southern California, Los Angeles, CA.

Ms. Montgomery is Physical Therapy Supervisor II, Stroke/Head Trauma Service, Rancho Los Amigos Medical Center.

This article was submitted October 15, 1984; was with the authors for revision five weeks; and was accepted April 24, 1985.

* Impo International, Inc, 3510 Black Rd, Santa Monica, CA 93456.

† Pathokinesiology Laboratory, Rancho Los Amigos Medical Center, 7601 E Imperial Hwy, Downey, CA 90242.

‡ EMG Preamplifier, Motion Control Inc, 1005 Smith 300 West, Salt Lake City, UT 84101.

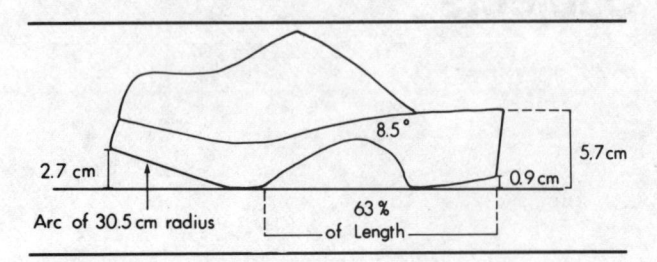

Fig. 1. Exterior dimensions of the rocker shoe.

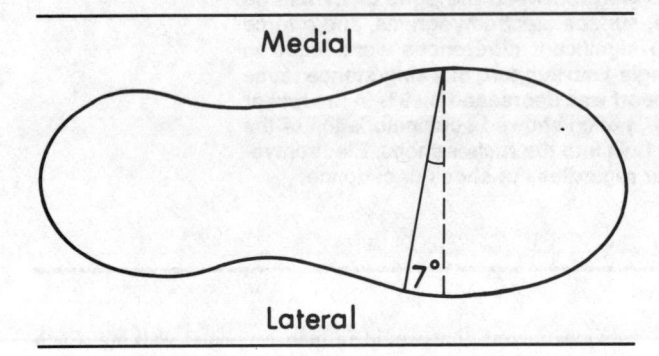

Fig. 2. Angle of terminal toe rocker.

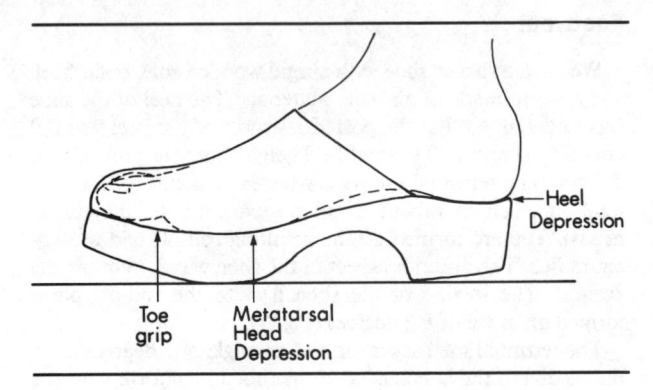

Fig. 3. Interior of rocker shoe.

4) a force plate** to transmit patterns and duration of force application during stance, and 5) a seven-channel Honeywell Visicorder oscillograph†† light beam recorder for paper printout and videotape display of footswitch, electrogoniometer, electromyograph, and force plate data.

Procedure

Subjects were informed that the purpose of this study was to obtain information about the effects of the rocker shoe on walking. Each subject was asked to bring a pair of athletic shoes for the test procedure.

The following subject information was obtained: name, age, height, weight, and any relevant physical history. Each subject was asked if she had worn rocker-style shoes previously and for how long. Bilateral leg-length measurements were taken of the subjects in the standing position in bare feet from the greater trochanter to the floor on the lateral aspect of the foot. We noted the type of athletic shoes, height of the heel, and condition of the shoes.

A gross test of range of motion and strength of the hip, knee, and ankle consisted of the subject performing a standing squat with her heels on the ground followed by 10 toe raises on each extremity. Any person who could not complete this test was not accepted for the study.

The following ranges of motion were measured with a goniometer: maximum knee extension in the standing position and active ankle plantar flexion and dorsiflexion in the sitting position.

We placed knee and ankle electrogoniometers on the right side of each subject. The thigh portion of the knee goniometer was placed above the bulge on the lower part of the vastus medialis muscle, and the leg cuff was placed below the tibial tubercle to avoid potential tilting by changes in patellar tension. The proximal cuff of the ankle goniometer was placed about 5 cm above the malleoli, and the distal end was taped on the dorsum of the foot just proximal to the tarsometatarsal joints (Fig. 4).

We placed bilateral insole footswitches inside the athletic shoe and the rocker shoe. The paired surface electromyogram electrodes were placed over the right gastrocnemius muscle in the center of the thickest aspect, over the bulge of the vastus lateralis muscle on the lateral aspect of the distal thigh, and over the soleus muscle on the lateral bulge. We carried out EMG, electrogoniometer, and footswitch applications according to the procedures established by the Pathokinesiology Laboratory at Rancho Los Amigos Medical Center.[1]

Each subject was tested to obtain bilateral footswitch, simultaneous right knee and ankle electrogoniometer, and surface EMG recordings during free and fast walking. The subject walked only in one direction to facilitate interpretation of force plate data and to allow videotaping of the right leg. To collect all data necessary for the data analysis, we asked the subjects to perform eight test runs.

For each test involving the force plate, the run was repeated until a complete stance-phase record was obtained for the right foot. The purpose of the force plate was not explained, and the trials were discarded when the foot did not hit fully on the force plate.

We made resting and strength recordings before conducting the functional EMG tests. Using a Cybex® isokinetic dynamometer‡‡ for the knee and a tensiometer for the ankle, we recorded maximum and 50% maximum efforts. The strength of the vastus lateralis muscle was measured with the knee at 60 degrees of flexion, and the gastrocnemius and soleus muscles were measured with the knee flexed at 70 degrees.

We instructed the subject to walk at her normal pace for free walking runs and to walk at a comfortable fast pace for fast walking runs. After the equipment was applied, we allowed the subject to walk casually around to familiarize herself with the equipment and the walkway. The subject also completed a formal preliminary run identical to the run to be tested. No data were collected on this run. To avoid artifacts of starting and stopping, we recorded data on the test runs

---

** Measuring Platform, Kristal Instrument Corp, 2475 Grand Island Blvd, Grand Island, NY 14072.
†† 2106 Visicorder, Honeywell, Inc, 7037 Havenhurst Ave, Van Nuys, CA 91407.

‡‡ Cybex, Div of Lumex, Inc, 2100 Smithtown Ave, Ronkonkoma, NY 11779.

only when the subject traversed the center 6 m of the 15-m walkway.

Data Analysis

Mean and standard deviations were calculated for velocity; cadence; stride length; gait-cycle duration; swing and stance duration; single-limb and double-limb support; percentage of stance duration occupied by heel contact, foot flat, and forefoot only; peak vertical force plate values; rate of loading and unloading of the foot; range of motion of the knee and ankle at heel contact, beginning of mid-stance and ending of preswing; maximum knee flexion and ankle dorsiflexion during stance; and maximum knee flexion and ankle plantar flexion during swing. These values were determined for all walking tests on all subjects.

Rate of loading was calculated from the ascending arm of the first vertical force peak, and rate of unloading was calculated from the descending arm of the second vertical force peak (Fig. 5).

A paired t test was used to determine any significant level of difference between the gait patterns of subjects in athletic shoes and in rocker shoes.

To account for laboratory error, we considered only data results with a difference greater than 1% on stride characteristics (footswitch data) and greater than 3% on change in velocity (footswitch and force plate data).[2,3]

The EMG recordings on the gastrocnemius, soleus, and vastus lateralis muscles were reviewed for the timing of muscle activity during the gait cycle. To avoid erroneous conclusions from surface electrodes, no attempt was made to measure the EMG responses or to differentiate their responses.[4]

RESULTS

Footswitch Data

No significant difference was found between walking in athletic shoes or rocker shoes in velocity, cadence, gait-cycle duration, single-limb support, and swing-stance ratio during free walking or fast walking (Tab. 1). The temporal components of free rate of walking were found to be compatible with those reported in the literature (Tab. 2).[4-10]

Double-limb support was 9% greater in athletic shoes than in rocker shoes during free walking ($p < .01$). Stride length was 2% greater during fast walking in athletic shoes than in rocker shoes ($p < .025$).

No significant difference was seen in the percentage of stance occupied by heel contact between athletic shoes and rocker shoes during free or fast walking (Tab. 3).

During fast walking in rocker shoes, the subjects' stance time in the foot-flat position was 22% ($p < .05$) greater than in athletic shoes. No significant difference was seen in foot-flat time during free walking.

The time on the forefoot was only 19% greater ($p < .005$) during free walking in athletic shoes than in the rocker shoes. We found a similar result during fast walking ($p < .005$); stance time on the forefoot in athletic shoes increased 15% in comparison with stance time in the rocker shoes.

Electrogoniometer Data

A significant difference ($p < .005$) in ankle plantar flexion was shown during free and fast walking at heel contact,

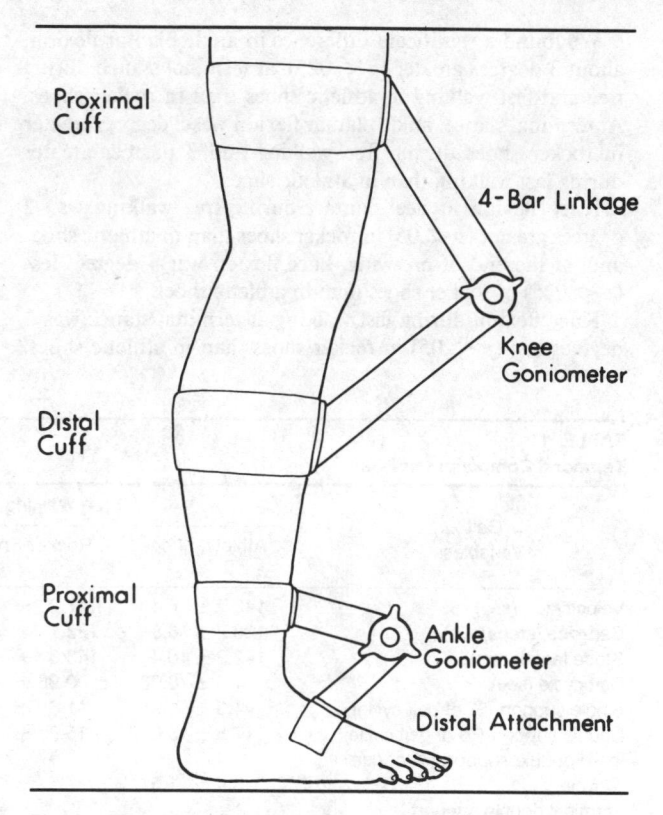

Fig. 4. Placement of knee and ankle electrogoniometers.

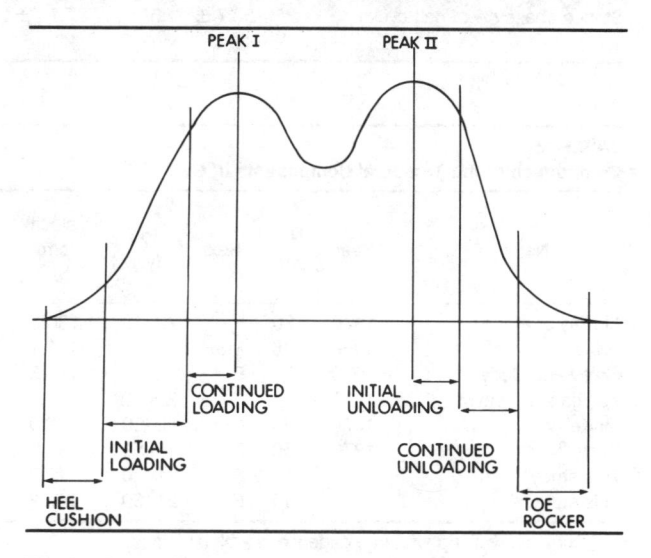

Fig. 5. Rate of loading and unloading on the vertical force plate.

beginning mid-stance and at the end of preswing (Tab. 4). Ankle plantar flexion in the rocker shoes was about 7 degrees greater at heel contact and beginning mid-stance and about 12 degrees less at the end of preswing in comparison with athletic shoes. At beginning mid-stance, ankle plantar flexion was 6.8 degrees greater in rocker shoes during free walking and 7.1 degrees greater during fast walking than in athletic shoes. At the end of preswing, ankle plantar flexion was 9.6 degrees less in the rocker shoe during free walking and 13.6 degrees less during fast walking than in athletic shoes.

We found a significant difference in ankle plantar flexion, about 3 degrees greater ($p < .025$), at terminal stance during free and fast walking in athletic shoes than in rocker shoes. At terminal stance, ankle plantar flexion was 2 degrees greater in rocker shoes during free walking and 3 degrees greater during fast walking than in athletic shoes.

Knee flexion at heel contact during free walking was 2 degrees greater ($p < .05$) in rocker shoes than in athletic shoes and, at the end of preswing, knee flexion was 4 degrees less ($p < .025$) in rocker shoes than in athletic shoes.

Knee flexion during fast walking at terminal stance was 2 degrees less ($p < .05$) in rocker shoes than in athletic shoes.

At the end of preswing, knee flexion was 4 degrees less ($p < .05$) in rocker shoes than in athletic shoes.

Maximum dorsiflexion occurred during terminal stance (Tab. 5). In free walking, dorsiflexion was 3 degrees greater ($p < .01$) in athletic shoes than in rocker shoes. During fast walking, dorsiflexion was 5 degrees greater ($p < .05$) in athletic shoes than in rocker shoes.

Maximum plantar flexion occurred at initial swing. During fast walking, ankle plantar flexion was 4 degrees greater ($p < .01$) in athletic shoes than in rocker shoes.

Maximum knee flexion was achieved during initial swing. During free and fast walking, a significant difference was

TABLE 1
Temporal Components of Gait

Gait Variables	Free Walking				Fast Walking			
	Athletic Shoes $\bar{X} \pm s$	Rocker Shoes $\bar{X} \pm s$		p	Athletic Shoes $\bar{X} \pm s$	Rocker Shoes $\bar{X} \pm s$		p
Velocity (cm/sec)	142.7 ± 16.4	147.2 ± 15.2		...	201.0 ± 19.8	198.0 ± 17.2		...
Cadence (steps/min)	120.2 ± 6.8	123.1 ± 6.8		...	145.2 ± 12.3	146.1 ± 10.7		...
Stride length (cm)	142.2 ± 10.4	143.3 ± 9.3		...	166.2 ± 9.3	162.7 ± 8.1		< .025
Gait cycle (sec)	1.0 ± 0.06	0.98 ± 0.05		...	0.83 ± 0.07	0.82 ± 0.06		...
Single support (% of gait cycle)	41.3 ± 1.9	41.8 ± 1.2		...	43.3 ± 1.7	43.8 ± 1.5		...
Double support (% of gait cycle)	17.5 ± 3.0	15.9 ± 2.5		< .01	14.1 ± 4.0	12.5 ± 2.7		...
Initial double support (% of gait cycle)	8.4 ± 1.5	7.5 ± 1.1		< .25	6.5 ± 1.7	5.9 ± 1.7		...
Terminal double support (% of gait cycle)	9.1 ± 1.8	8.3 ± 1.7		< .05	7.6 ± 3.5	6.6 ± 1.2		...
Swing phase (% of gait cycle)	41.3 ± 1.6	42.3 ± 1.4		< .005	42.6 ± 3.6	43.7 ± 1.8		...
Stance phase (% of gait cycle)	58.7 ± 1.6	57.7 ± 1.4		< .005	57.4 ± 3.6	56.3 ± 1.8		...

TABLE 2
Comparison of the Temporal Components of Gait

Name	Year	No. subj	Sex	Age (yr)	Velocity (cm/sec)	Gait Cycle (sec)	Cadence (steps/min)	Double Support (% of gait cycle)	Stride (cm)	Swing-Stance Ratio (% of gait stance)
Murray et al[5]	1970	30	F	20–70	130.0	1.03	117.0	22.0	133.0	39.0/61.0
Drillis[6]	1958	936	M&F	...	145.1	1.05	112.5	...	152.6	...
Finley and Cody[7]	1970	572	F	...	123.3	1.03	116.5	...	126.8	...
Zuniga and Leavitt[8]	1973	20	F	$\bar{X} = 27$	...	1.12	106.0	17.2	...	39.3/60.7
McKelvy[9]	1974	40	F	40–69	122.0	1.03	117.0	15.0	125.0	42.0/58.0
Curry[10]	1976	50	F	19–59	134.8	1.04	116.0	20.4	139.4	39.8/60.1
This study[a]		15	F	21–30	142.7	1.00	120.2	17.5	142.2	41.3/58.7
This study[b]		15	F	21–30	147.2	0.98	123.1	15.9	143.3	42.3/57.8

[a] Subjects walking with free cadence in athletic shoes.
[b] Subjects walking with free cadence in rocker shoes.

TABLE 3
Percentage of Stance Time on Segments of Foot

Foot Segments	Free Walking				Fast Walking			
	Athletic Shoes $\bar{X} \pm s$	Rocker Shoes $\bar{X} \pm s$		p	Athletic Shoes $\bar{X} \pm s$	Rocker Shoes $\bar{X} \pm s$		p
Heel-only (% of stance)	21.2 ± 7.6	25.8 ± 14.3		...	25.6 ± 11.0	25.4 ± 12.1		...
Flat foot (% of stance)	30.5 ± 10.6	33.1 ± 16.8		...	26.8 ± 12.8	32.7 ± 15.1		< .05
Forefoot only (% of stance)	48.3 ± 7.9	40.7 ± 10.5		< .005	48.2 ± 6.6	41.9 ± 8.9		< .005

found ($p < .005$). In free walking, knee flexion was 4 degrees greater in athletic shoes than in rocker shoes. During fast walking, knee flexion was 3 degrees greater in athletic shoes than in rocker shoes.

Force Plate Data

Initial and terminal peak vertical force showed no significant difference (Tab. 6). The terminal vertical force peak occurred 2% of the gait cycle later ($p < .005$) during free walking in rocker shoes than in athletic shoes. A similar difference ($p < .05$) was seen during fast walking.

The rate of loading the foot at initial contact (heel cushion) showed a significant difference ($p < .005$) in free and fast walking. The rate of loading at initial contact during free walking was 4.5 times greater in rocker shoes than in athletic shoes. During fast walking, the rate of loading was 5 times greater in rocker shoes than in athletic shoes (Tab. 7).

Initial rate of loading revealed a significant difference ($p < .025$) in free walking with rocker shoes and was 1.7 times greater than in athletic shoes. Fast walking showed a similar difference ($p < .005$). The rate of continued loading was not compared because of the high variation in the ascending arm of the first vertical peak. Initial unloading as well as continued unloading of the foot showed no significant difference.

Unloading the forefoot (toe rocker) was significantly different ($p < .005$) in free and fast walking. The rate of unloading the forefoot in athletic shoes was 1.2 times slower during free walking than during fast walking. In rocker shoes, however, only two subjects during free walking and one subject during fast walking demonstrated a toe rocker. All remaining subjects

TABLE 4
Degrees of Ankle Plantar Flexion and Knee Flexion During Stance

Degrees	Free Walking			Fast Walking		
	Athletic Shoes $\bar{X}$ s	Rocker Shoes $\bar{X}$ s	p	Athletic Shoes $\bar{X}$ s	Rocker Shoes $\bar{X}$ s	p
Heel Contact						
Plantar flexion	8.5 ± 4.7	14.9 ± 4.8	< .005	5.8 ± 4.9	12.5 ± 6.2	< .005
Knee flexion	1.4 ± 2.6	2.9 ± 3.5	< .05	3.6 ± 3.1	5.0 ± 3.9	...
Beginning mid-stance						
Plantar flexion	13.9 ± 3.9	20.7 ± 4.2	< .005	12.1 ± 4.9	19.2 ± 5.3	< .005
Knee flexion	6.9 ± 4.1	8.0 ± 4.9	...	7.8 ± 4.3	9.0 ± 5.0	...
Terminal stance						
Plantar flexion	1.9 ± 4.2	4.3 ± 4.1	< .025	2.4 ± 3.9	5.0 ± 3.1	< .025
Knee flexion	2.2 ± 2.6	2.0 ± 2.5	...	2.6 ± 3.0	1.1 ± 3.2	< .05
Preswing						
Plantar flexion	23.4 ± 7.7	13.8 ± 6.0	< .005	30.2 ± 7.6	16.6 ± 6.1	< .005
Knee flexion	28.2 ± 6.3	23.7 ± 5.2	< .025	23.7 ± 8.1	19.8 ± 5.4	< .05

TABLE 5
Degrees of Maximum Dorsiflexion, Plantar Flexion, and Knee Flexion in the Gait Cycle

Degrees	Free Walking			Fast Walking		
	Athletic Shoes $\bar{X}$ s	Rocker Shoes $\bar{X}$ s	p	Athletic Shoes $\bar{X}$ s	Rocker Shoes $\bar{X}$ s	p
Dorsiflexion (terminal stance)	3.1 ± 5.4	.4 ± 5.5	< .01	.3 ± 4.9	−2.1 ± 5.5	< .05
Plantar flexion (initial swing)	35.0 ± 4.9	34.5 ± 7.9	...	40.7 ± 5.7	36.5 ± 8.5	< .01
Knee flexion (initial swing)	50.1 ± 5.5	46.0 ± 5.4	< .005	51.0 ± 4.7	47.5 ± 5.0	< .005

TABLE 6
Peak Vertical Forces

Vertical Force	Free Walking			Fast Walking		
	Athletic Shoes $\bar{X}$ s	Rocker Shoes $\bar{X}$ s	p	Athletic Shoes $\bar{X}$ s	Rocker Shoes $\bar{X}$ s	p
Peak I (% of BW[a])	122.0 ± 9.9	119.2 ± 11.2	...	144.3 ± 14.6	142.7 ± 11.2	...
Peak II (% of BW)	121.3 ± 6.4	120.7 ± 6.4	...	126.2 ± 8.9	122.8 ± 7.5	...
Peak I (% of stance)	19.7 ± 4.0	19.7 ± 3.8	...	16.3 ± 4.7	15.5 ± 6.5	...
Peak II (% of stance)	75.1 ± 1.1	76.5 ± 1.6	< .005	75.8 ± 1.1	77.2 ± 2.6	< .05

[a] Body weight.

TABLE 7
Rate of Loading and Unloading the Foot

Gait Variable	Free Walking			Fast Walking		
	Athletic Shoes $\bar{X}$ s	Rocker Shoes $\bar{X}$ s	p	Athletic Shoes $\bar{X}$ s	Rocker Shoes $\bar{X}$ s	p
Loading						
Heel cushion (% of BW[a]/sec)	788.3 ± 790.8	3,484.4 ± 2,341.6	< .005	931.2 ± 851.1	4,750.0 ± 2,046.3	< .005
Initial (% of BW/sec)	2,483.1 ± 899.1	4,208.4 ± 2,464.6	< .025	4,947.2 ± 1,262.7	8,439.1 ± 4,036.5	< .005
Unloading						
Begin (% of BW/sec)	400.4 ± 86.6	402.5 ± 108.1	...	512.7 ± 86.6	486.3 ± 120.2	...
Continued (% of BW/sec)	1,143.1 ± 267.7	1,053.5 ± 245.3	< .025	1,512.2 ± 497.4	1,451.5 ± 482.2	...
Toe rocker (% of BW/sec)	454.8 ± 105.4	82.7 ± 219.3	< .005	548.0 ± 261.2	40.0 ± 155.0	< .005

[a] Body weight.

had already unloaded the foot abruptly and did not demonstrate the toe rocker.

Electromyograph Data

The vastus lateralis muscle was active primarily from terminal swing through loading response. The gastrocnemius and soleus muscles were primarily active during single-limb support with increased activity as the heel rose. Electromyographic responses increased with fast walking.

DISCUSSION

The significant difference in ankle range of motion during the swing and stance phases was directly related to the foot accommodating to the 8.5 degrees of plantar flexion built into the rocker shoe. The greatest effect of the rocker shoe on free walking occurred during the stance phase. The impact of loading at heel contact was four times greater in the rocker shoe than in the athletic shoe because of no cushioning from the wooden heel. To compensate for this impact at heel contact, knee flexion was also increased approximately 2 degrees to act as a buffer during free walking. None of the subjects demonstrated a foot slap in spite of this high impact because the undercut heel effectively decreased the heel lever.

The rate of loading the limb at loading response with the rocker shoes was still twice as great as with athletic shoes. The decreased double support time in the rocker shoes correlated with the rapid loading in the rocker shoes and the rapid roll-off that was occurring on the contralateral side.

The toe rocker in terminal stance caused a fast roll-off; however, the very controlled roll-off resulted from the subsequent double support period showing on the force plate record decreased rate of unloading of the limb during free walking. This decrease indicated a need for a high degree of extensor muscle activity to maintain a controlled roll-off.

At preswing, the terminal toe rocker failed to provide a smooth transition from stance to swing phase as evidenced by the force plate data. The lack of this smooth transition was a result of the rigid wooden sole. Knee flexion was also significantly less by about 5 degrees. Because preswing is a passive phase of gait, the contour of the sole did not allow the foot to fall into plantar flexion when the knee passively flexed. The rigid sole acted as a lever to restrain the passive advancement of the tibia during normal passive knee flexion. The decreased, rather than increased, knee flexion during the initial swing to achieve toe and foot clearance reflected the decreased need for normal preswing knee flexion with the rocker shoe.

Clinical use of the rocker shoe should be limited to those patients who have enough hip and knee extensor muscle control to buffer the impact of initial loading and to control the rate of roll-off. For the rocker shoe to be used clinically, a softer heel to help absorb the impact of heel contact would be valuable.

The rocker shoe appears to be a valuable aid in the accommodation of plantar flexion contractures in those patients who have good extensor muscle control. Normal stride characteristics were maintained with an average of 9 degrees plantar flexion throughout the gait cycle.

To analyze further the effects of the rigid rocker shoe, future studies should be directed toward identifying the effects of the rocker shoe on men and on specific patient groups. For patient groups, a custom rocker shoe that attaches a rocker sole to a lace-up oxford style shoe may be appropriate. Common shoe modifications can be made in the oxford shoe, such as addition of an ankle-foot orthosis, metatarsal bars, and shoe inserts.

CONCLUSION

Walking in rocker shoes was compared with walking in athletic shoes at free and fast velocities in 15 healthy women between the ages of 21 and 30 years. The effect of the heel height and the terminal toe rocker of the rocker shoe was related to the various phases of gait by using data derived from electrogoniometry, surface EMG, foot switches, and force plate studies.

No significant differences were seen in velocity, cadence, gait-cycle duration, single-limb support, or swing-stance ratio in free and fast walking. Double-limb support was significantly decreased by 9% in the rocker shoes compared with athletic shoes in free walking. This decreased time was due primarily to the rapid loading of the limb. The rate of loading of the foot was twice as fast in the rocker shoes.

Vertical force at mid-stance and terminal stance was not significantly different. Significant difference in ankle range of motion showed accommodation of the foot to the 8.5 degrees of plantar flexion built into the rocker shoes. Electromyographic responses appeared to be similar in subjects walking in athletic shoes or in rocker shoes at either free or fast cadence.

Basic Research

Technical Report
Reliability of Kinematic Measurements of Rear-Foot Motion

Michael J Mueller
Barbara J Norton

The purpose of this study was to examine the ability of a video-based, computer-interfaced motion analysis system to provide reliable data. Ten subjects with no significant orthopedic or neurological dysfunction and ranging in age from 22 to 45 years ($\bar{X}=29.6$, $SD=7.8$) were tested. Retroreflective markers were placed on the posterior shank and foot of each subject. Footswitches were attached to the plantar forefoot and rear foot. A video camera was placed behind the subject, and video data were collected while the subject walked on a treadmill. One representative gait cycle for each subject was selected and processed 10 times with a video processor and analysis software. Three intraclass correlation coefficients (ICCs) were calculated for variables generated by the analysis software, one for two individual measures and one each for the mean of three and five repeated measures. Except for temporal variables, processing data introduced additional variability into the measurement process, particularly for angular velocity data. Measurement of all variables was highly reliable (ICC values $\geq .95$) when based on the mean of at least three repeated measures. Although a single measure of temporal and angular position variables may be considered reliable, we recommend using a mean of three trials for angular velocity variables. Additional research is needed to determine tester and subject variability and validity of the measures. [Mueller MJ, Norton BJ. Reliability of kinematic measurements of rear-foot motion. Phys Ther. 1992;72:731–737.]

Key Words: *Equipment, general; Kinesiology/biomechanics, gait analysis; Movement.*

Recently, there has been considerable interest in quantifying motion at the subtalar joint because dysfunction at this joint is believed to contribute to foot and ankle overuse injuries. Several investigators[1–4] have analyzed the kinematics of the rear foot during walking and running with the use of one or more cameras positioned to the rear of the subject. Such studies primarily have been conducted in laboratory settings, but as equipment that provides automated analysis becomes more readily available and easier to use, greater application in clinical settings is feasible. Prior to the use of such equipment in any setting, however, the reliability and validity of the obtained measurements must be demonstrated.

In any measurement, error potentially can be attributed to a variety of sources including the subject, the tester, and the test equipment. Ideally, the equipment would introduce no additional error, but the reliability of the obtained measurements should be tested, not assumed. Kinematics of the rear foot have been measured using high-speed cinematography with digitization of the individual frames.[1–4]

MJ Mueller, PT, is Instructor, Program in Physical Therapy, and Doctoral Candidate, Interdisciplinary Program in Movement Science, Washington University School of Medicine, St Louis, MO 63110. Address all correspondence to Mr Mueller at Program in Physical Therapy, Washington University School of Medicine, Box 8083, 660 S Euclid Ave, St Louis, MO 63110 (USA).

BJ Norton, PT, is Coordinator, Applied Kinesiology Laboratory, and Instructor, Program in Physical Therapy, Washington University School of Medicine.

This study was approved by the Washington University School of Medicine Institutional Review Board.

This article was submitted January 13, 1992, and was accepted May 18, 1992.

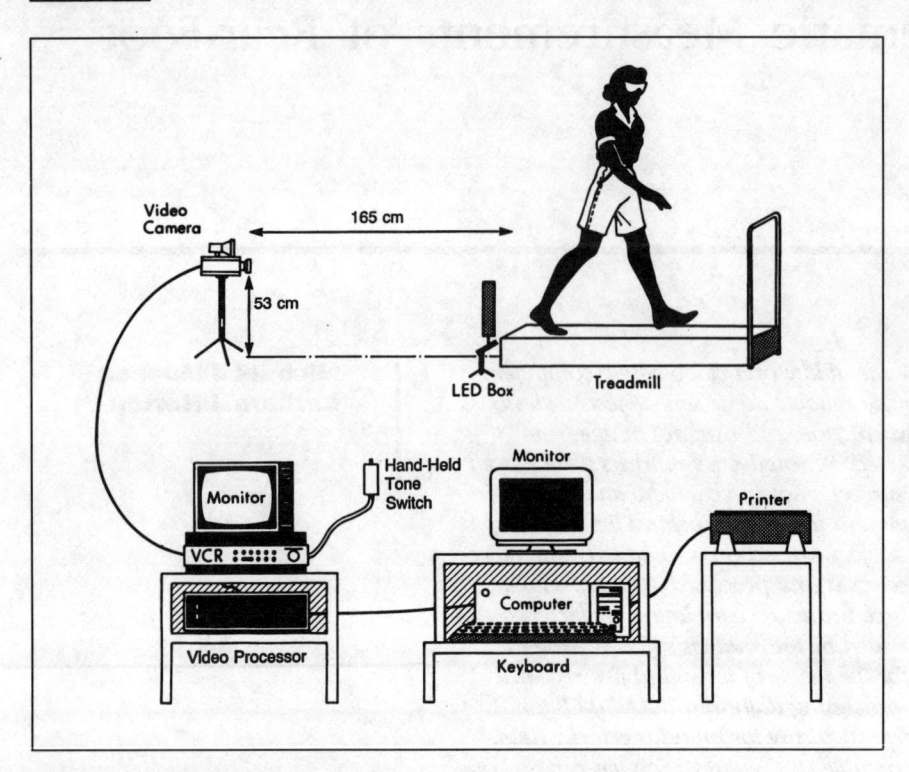

Figure 1. *Motion analysis equipment and procedural setup for collecting video data on subjects. (LED=light-emitting diode, VCR=videocassette recorder.)*

Engsberg and Andrews[4] reported a Pearson Product-Moment Correlation Coefficient of .99 when performing a test-retest experiment on a series of six consecutive cinematographic frames digitized on five separate occasions. Frame-by-frame digitization, however, is very tedious and time-consuming. Recently, microcomputer-based systems (hardware and software) that provide various kinematic measures of rear-foot motion have become available commercially.

The purpose of this study was to assess the ability of a specific video-based, computer-interfaced motion analysis system to provide reliable data. We hypothesized that equipment-related measurement error would be minimal, as indicated by acceptably high reliability coefficients.

Method

Sample

Ten volunteers (8 women, 2 men) with no significant orthopedic or neurological dysfunction were tested. The absence of significant orthopedic or neurological dysfunction was determined by the tester (MJM) and operationally defined as no pain, weakness, or instability resulting in an inability to perform normal activities of daily living. The subjects ranged in age from 22 to 45 years ($\bar{X}$=29.6, SD=7.8). No effort was made to determine each subject's static foot and ankle alignment because we did not believe that either these or other subject-related factors would influence the ability of the system to process data reliably. All subjects gave informed consent to participate in the study.

Measurement System

The measurement system studied has both hardware and software components. The main hardware component, the Motion Analysis™ ExpertVision™ system,* is a general purpose, video-based, microcomputer-interfaced, two-dimensional motion analysis system. The ExpertVision™ system includes the following components: a video camera and lens,[†] a spotlight, a video monitor, a video processor (VP-110),* a videocassette recorder (VCR),[‡] and a microcomputer (Fig. 1). The video camera uses a CCD (charge-coupled device) solid-state imaging device to capture visual images and convert them to electrical signals. The camera's electronic shutter operates at 30 frames per second, and the scanning mode is set at 2:1 noninterlaced to allow a sampling frequency of 60 fields per second. The camera was equipped with a video lens that has a 12.5-mm focal length.

The spotlight, a 30-W reflector bulb with a reflector hood, was attached to the tripod next to the camera (to maximize the amount of light reflected back to the camera from the retroreflective markers* secured to the subject). The video monitor was a conventional black and white monitor with a 50.5-cm (12-in) diagonal screen. The video processor has a maximum sampling rate of 60 Hz. The VCR is an industrial grade, 1.27-cm (0.5-in) VHS-type recorder/player. The VCR has several features that aid in locating appropriate data for processing, including pause, frame-by-frame advancer, search modes, and audio event channels. The microcomputer is an IBM-PC[§]–compatible system that incorporates an Intel 80836 processor chip[∥]

---

*Motion Analysis Corp, 3650 N Laughlin Rd, Santa Rosa, CA 95403.

[†]TI-23A, NEC Corp, NEC Building, 33-1, Shiba 5-chrome, Minato-ku, Tokyo 108, Japan.

[‡]AG-6300, Panasonic, Audio-Video Systems Division, One Panasonic Way, Secaucus, NJ 04094.

[§]International Business Machines Corp, PO Box 1328-W, Boca Raton, FL 33429.

[∥]Intel Corp, Robert Noyce Bldg, 2200 Mission College Blvd, Santa Clara, CA 95052.

running at 16 MHz. The system contains 640K of random access memory (RAM), two floppy disk drives, a 40M hard disk drive with a 28-millisecond access time, and a VGA color graphics monitor[#] and adaptor card. The operating system for the microcomputer is DOS, version 3.3.

The software component of the measurement system is FootTrak,* a set of computer programs designed to process the video data and produce various measures of rear-foot kinematics during standing and walking. In addition to FootTrak and the standard hardware components of the ExpertVision™ system,* the motion analysis system included a black metal box containing light-emitting diodes (LEDs), which were connected by wires to footswitches* on the plantar surface of the subject's foot. The footswitches are on-off devices that close when the foot makes contact with the floor. A specific LED is activated when the connected footswitch closes.

Basic System Operation

Retroreflective markers are placed on the subject in locations that permit measurement of specific variables of interest. The LED box and the subject are placed in the camera's field of view. The video camera, which is focused on both the subject and the LED box, captures a series of images that represent both movement of the markers and timing of foot-floor contact. The series of images is sent either directly to the video processor for processing or first to the VCR for recording and then to the video processor for processing. The video processor detects the edges of the markers and LEDs in the analog video signal and digitizes them. On command from the FootTrak software, the video processor sends the digital data representing the x and y coordinates of the edges of the markers and LEDs to the microcomputer. After the digitized data have been collected by the microcomputer, the FootTrak software locates the centroid (center) of each marker (and LED) in each image. The FootTrak software then creates a time-based representation of movement, referred to as a "path," by joining successive locations of each centroid (taken from successive images). Finally, the path data are processed further using conventional mathematical techniques to produce measures of temporal, angular position, and angular velocity variables.

Procedure

One tester (MJM) performed all procedures. Pairs of retroreflective markers were placed at the midline of the calcaneus and the distal one third of the leg bilaterally using the following procedure. Subjects were positioned prone on a firm plinth with their leg in neutral (neither medially nor laterally rotated). The calcaneus was held with the thumb on one side and the index finger on the other. The calcaneus was bisected visually, and two retroreflective markers were placed over the midline, one just above the plantar surface of the heel and one at the superior-posterior aspect of the heel just below the axis of the subtalar joint (determined visually by pronating and supinating the subtalar joint). The lower leg was then bisected visually, and two markers were placed over the midline, one each at 5 and 20 cm proximal to the malleoli (Fig. 2).

Thin, relatively pliable on-off footswitches were secured to the plantar surface of the forefoot (at the first metatarsal head) and rear foot (at the posterior lateral heel) bilaterally with double-sided adhesive tape and wrapped with paper tape (Fig. 2). The footswitches were connected via a thin, flexible cable to the LED box. The LED box was placed to the right of the posterior aspect of a treadmill in the camera's field of view (Fig. 1).

The camera and attached spotlight were placed 165 cm (65 in) behind the subject and 53 cm (21 in) above the ground. Although the FootTrak manual recommends placing the camera 140 cm (55 in) behind the subject and 38 cm (15 in) above the ground, preliminary testing on 10 subjects (by the first author) prior to the study indicated this camera placement often failed to record data during the late stance phase. Increasing the height and distance of the camera from the subject increased the field of view and allowed improved measuring during the late stance phase.

Video data required for calibration were obtained prior to testing each subject. A bar with a pair of retroreflective markers was placed across the rear of the treadmill, adjacent to the LED box. The camera was adjusted so that its field of view included the horizontal reference bar and the LED box. Room lighting, camera aperture, and threshold control of the video processor were adjusted to optimize the detection of the targets (ie, the retroreflective markers and the LEDs). Three seconds or more of calibration video data were then recorded on the VCR for subsequent processing.

Subjects practiced walking on the level treadmill with the footswitches secured to their feet. Subjects were allowed to take as much time as necessary to increase their walking speed slowly to 3.2 mph. During the acclimation period, the tester checked to ensure that all markers remained attached to the posterior leg and calcaneus, footswitches were attached to the plantar surface of the foot, and appropriate images of the reflective markers were displayed on the video monitor. When the subject was acclimated to walking on the treadmill, video data were collected on the VCR. The videotape was allowed to run for at least 8 seconds prior to the sampling period. A hand-held tone switch was used to place an audio event tone on the tape to mark the beginning of each sample. Data were collected for at least 2 minutes, with audio event marks placed approximately every 30 seconds.

[#] Quadrant Components Inc, 4378 Enterprise St, Fremont, CA 94538.

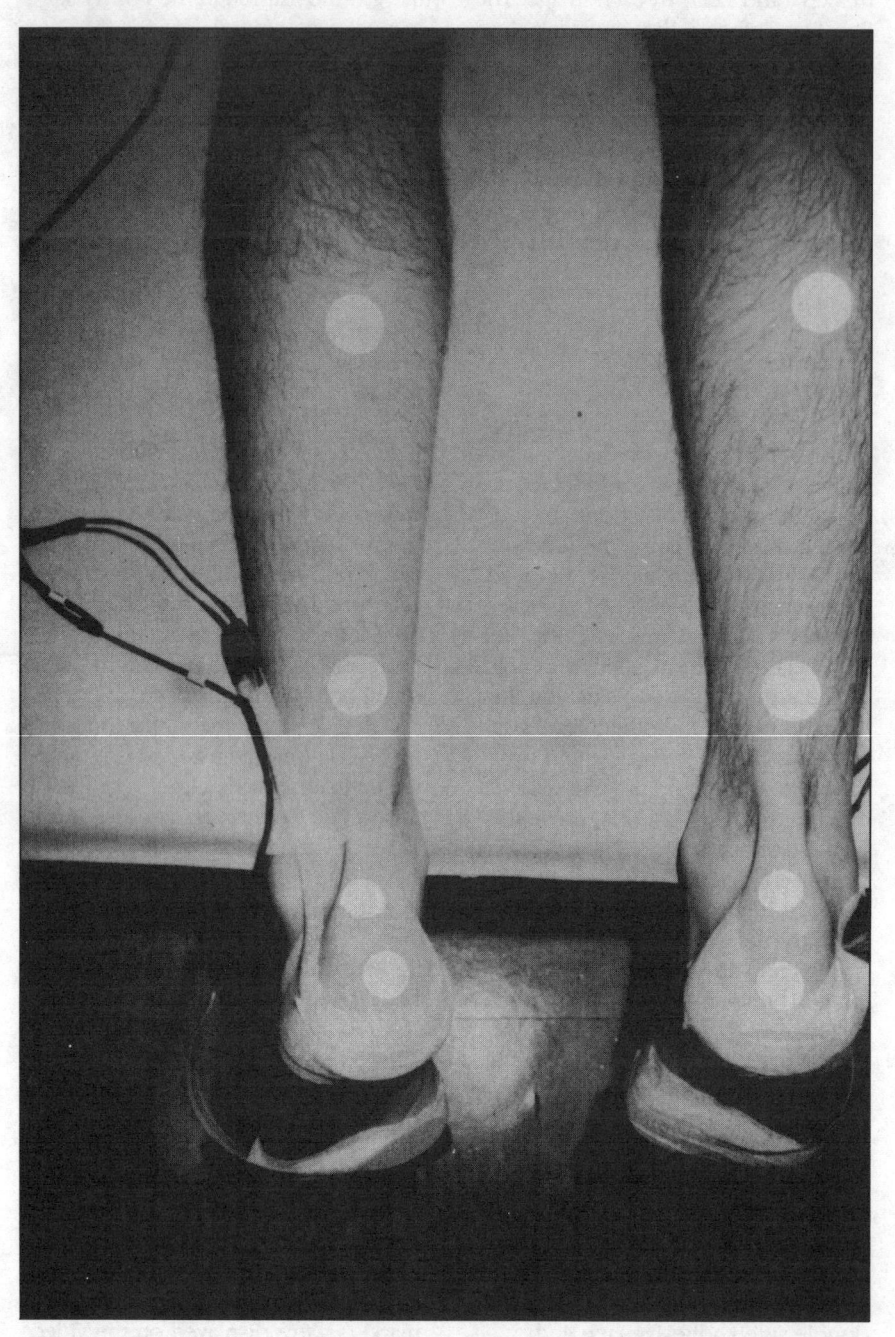

Figure 2. *Subject positioned prone on plinth showing placement of retroreflective markers and footswitches.*

Data Processing

Recorded video data were processed using the video processor and the FootTrak software. First, the videotape was rewound to a point at least 8 seconds prior to a specified audio tone. Next, as the videotape was played, the video processor (1) detected the audio event tone, (2) began extracting data representing the edges of the markers and the LEDs, and (3) sent the data to the microcomputer. Then, using the dynamic treadmill analysis mode of the FootTrak software, one representative gait cycle from the data collected on a subject was selected to be used for repeated processing. A "representative gait cycle" was defined as a cycle that (1) contained all necessary footswitch data and (2) visually showed similar kinematics to the mean of the multiple gait cycles displayed on the monitor. The selected gait cycle was processed 10 times; each time, processing began with rewinding the videotape to the same starting point (initiated by the audio tone) and playing the same segment of videotape to be processed.

Data used for analysis were taken from ASCII files generated by the FootTrak software instead of from the standard FootTrak printout because data on the FootTrak printout (1) were rounded to the nearest whole degree (which seemed to introduce additional error in preliminary testing) and (2) did not include standard deviations. Raw data (to 0.01° precision) from the ASCII files were collected and summarized in a separate file. The variables examined in this study and their operational definitions are presented in the Appendix. Names of variables are those provided by the FootTrak software.

Data Analysis

Intraclass correlation coefficients (ICC[3,1]) and standard deviations were used as indicators of the ability of the equipment to provide reliable data. As described earlier, the representative gait cycle for each subject was processed 10 times to yield 10 measures for each variable. Three ICCs were calculated for each variable, one based on 2 individual measures (first and second) and one each based on the mean of 3 (first 3 and second 3) and of 5 (first 5 and second 5) repeated measures of the variable. The standard deviation of all 10 measures of each variable was also determined. The criterion used for judging the acceptability of the reliability coefficients was ICC ≥ .95. This relatively high criterion was chosen because we believed computer-assisted video analysis should introduce little error into the measurement.

Table. Means, Standard Deviations, Coefficients of Variation, and Intraclass Correlation Coefficient Values

Variable[a]	$\bar{X}$[b]	SD[c]	CV	ICC[d] SM	3M	5M
Calcaneus-to-tibia angle (°)						
Angle at touch-down	8.10	0.86	10.62	.93	.99	1.00
Maximum pronation	12.06	0.42	3.48	1.00	1.00	1.00
Total pronation range of motion	5.36	0.86	16.04	.86	.99	.99
Time to maximum pronation (% of gait cycle)	23.31	3.64	15.61	.95	.98	.99
Toe-off angle (n=8)	5.55	0.80	14.41	.96	.99	1.00
Calcaneus-to-tibia angular velocity (°/s)						
Initial velocity	36.57	27.63	75.55	.50	.95	.97
Maximum pronation velocity	125.57	22.10	17.60	.88	.96	.96
Maximum supination velocity	203.64	34.18	16.78	.93	.98	.99
Temporal variable						
Swing						
Percentage	44.79	0.00	0.00	1.00		
Time (ms)	416.33	0.00	0.00	1.00		
Stance						
Percentage	55.21	0.00	0.00	1.00		
Time (ms)	513.19	0.00	0.00	1.00		

[a] Left leg only.
[b] Grand mean of 10 measures on 10 subjects.
[c] Mean standard deviation for 10 measures on 10 subjects.
[d] Intraclass correlation coefficients (ICC[3,1]) on repeated single measures (SM), means of three repeated measures (3M), and means of five repeated measures (5M).

Results

The Table presents the means, standard deviations, coefficients of variation (CVs), and ICC values for the variables analyzed. Standard deviations were 0.00 and ICC values were 1.00 for all temporal variables. The ICC values for calcaneus-to-tibia angle measures were .86 to 1.00 for repeated single measures, and .98 to 1.00 for means of three and five repeated measures. The ICC values for calcaneus-to-tibia angular velocity measures were .50 to .93, .95 to .98, and .96 to .99 for repeated individual measures, the mean of three repeated measures, and the mean of five repeated measures, respectively. The standard deviations were <1.00 degree for angular position variables and 22.1° to 34.2°/s for angular velocity variables.

Discussion

The values of the ICC (1.00) and standard deviation (0.00) for all of the temporal variables indicate that the motion analysis system did not introduce any error into the measurements of those variables; that is, the system processed temporal data from the same gait cycle identically all 10 times. The high level of reproducibility for the temporal measures probably is related to the fact that the system has to identify the time at which the LEDs (the source in the video images of the temporal data produced by the footswitches) turn on and off, not their exact location in terms of Cartesian coordinates.

There was more variability in the measures of angular position than in the measures of temporal variables. Although standard deviations were low (≤0.86) for repeated single measures, the ICC values of some repeated measures were unacceptable according to our a priori criterion (ICC≤.95). Using a mean based on three repeated measures increased all ICC values (ICCs≥.98).

In comparison with the other variables, there was notable variability in the measurements of the angular velocity variables. The ICC values were as low as .50, and standard deviations were as high as 34.2 (CV= 17%-76%) for calcaneus-to-tibia angular velocity. Errors in angular velocity measures would be expected to be greater than errors in angular position measures because angular velocity measures are derived. They are derived from the angular position measures through the use of difference scores and the formation of ratios. Velocity is equal to position

2 minus position 1, divided by the given time interval. Because each position measure may contain error, this process apparently compounded the original error and rendered the derived measure less reliable than the data from which it was derived. Indexes of reliability improved markedly when the mean of either three (ICC≥.95) or five (ICC≥.99) repeated measures was used. Based on these data, we recommend using a mean of three repeated measures for all angular velocity variables.

A benefit of FootTrak's printed report is an option that allows the user to choose kinematic data from either individual cycles or the mean of up to six trials. Unfortunately, the report values of each cycle are rounded to the nearest whole degree, and preliminary testing (by the first author) seemed to indicate that this rounding also introduced error. In addition, the printed report does not contain standard deviations for the trials. We believe the report would be much more useful to researchers and clinicians if rounding error were reduced and standard deviations were included.

Besides the dynamic treadmill report, FootTrak provides options for color graphic displays of angular position and velocity data and for animated stick figures on the monitor of the computer. One to six cycles (or a mean of all cycles) of left or right tibia-to-calcaneus angle, tibia-to-calcaneus velocity, calcaneus-to-vertical angle, and tibia-to-vertical angle may be viewed with or without event markers (ie, heel-off or toe-on). Figures on the screen can be printed at any time. These features may be useful in a clinical setting to visualize kinematic patterns and to assist in patient education or training situations. Further research is needed to determine meaningful parameters of these kinematics in pathological and nonpathological populations.

This study assessed the reliability of data generated by the dynamic treadmill analysis option of FootTrak. The FootTrak software also contains options for measuring resting calcaneus stance position and neutral calcaneus stance position. Although we did not analyze these measures, we would expect reliability for repeated processing of these static angular position measures to be similar to the reliability of the angular position measures examined in this study.

Given certain constraints regarding lighting and marker placement, the Motion Analysis™ system allows the kinematic analysis of essentially any type of movement. Although flexible, use of the system requires some basic programming skills. A benefit of FootTrak software (especially to clinicians) is that no computer programming is required of the user. A limitation of this preprogrammed software is that it cannot be modified easily.

FootTrak software is designed to be used with the subject walking or running on a treadmill. The benefits of treadmill analysis are that it requires minimal space in a clinic or research laboratory and the field of view remains essentially stationary. Some patients, however, have difficulty adjusting to treadmill ambulation. In addition, ambulating on a treadmill may or may not be considered a legitimate simulation of normal walking. Kinematics of rear-foot motion may be different than kinematics of normal walking. Finally, integration of the FootTrak system with other movement analysis systems (ie, force platform or kinematics in the sagittal plane) would be dependent on the ability of the other systems to use a treadmill.

This study examined equipment performance. The same data (ie, the same gait cycle) were processed by the equipment multiple times. Placement of markers, footswitches, and foot and ankle alignment of the subjects were all constant and should have no influence on processing the same gait cycle repeatedly. Therefore, any variability in the measurements obtained in this study must be due to error introduced at intermediate stages of processing. Factors that are important in relation to accuracy, either at the level of acquisition of the raw video signal (eg, camera frame rate and resolution, speed of target movement, variations in lighting conditions) or at the later stages of processing (eg, type of filtering, methods for deriving values), cannot be addressed by our data. Our data also cannot address the errors that would be associated with the use of the equipment in applied settings.

The only relevant potential sources of error in this study would include the following: (1) reproduction by the VCR of the original analog video signal from the videocassette tape, (2) data acquisition and digitization by the video processor of the analog signal received from the VCR, and (3) acquisition and initial processing by the computer of the digitized data from the video processor (ie, creation of the video file). Walton[5] reported a loss of system precision (defined as agreement among repeated observations made under identical conditions) when comparing data collected directly from the system camera with data collected from prerecorded videotape. He concluded that although there was some loss of precision, "these losses are insignificant when compared to other sources of error" and "storing raw video images on video cassettes has little or no impact on overall system performance." We are unaware of any data that address the precision of either data acquisition and digitization by the video processor or data acquisition and initial processing by the computer system. Therefore, the degree to which each of the factors noted contributes to random error cannot be determined at this time.

In general, the primary purpose of these measures is to infer motion occurring at the subtalar joint. Variables generated by the FootTrak software and similar variables reported in the literature[1,2] purport that the equipment is measuring pronation and supination. With this type of two-dimensional, video motion analysis system, however, data are taken only from the plane of motion that is parallel to the lens of the camera. Error (as related to actual joint motion) will be introduced if the axis of the joint is not perpendicular to the midline of the camera lens. Because the axis of the subtalar joint is oblique and there is considerable variation among subjects,[6] only a portion of the actual triplanar pronation or supination motion is being analyzed. The portion of motion viewed by a single camera would more closely approximate the frontal-plane motion of calcaneal inversion and eversion at the subtalar joint. Even this component of actual subtalar joint motion will be distorted as the leg rotates in the transverse plane or moves medially or laterally in the frontal plane.[7,8] Additional research is needed to determine the reliability, validity, and clinical usefulness of these measurements.

Conclusion

Except for measures of temporal variables, the Motion Analysis™ system and the FootTrak software introduced additional variability into the measurement process, particularly for angular velocity data. The reliability of all angular position and angular velocity measurements increased with the number of measures used. Measurement of all variables was highly reliable (ICC values ≥ .95) when taking a mean of at least three repeated measures. Although this study documents the reliability of measurements obtained under controlled conditions (ie, the equipment was the only source of error), additional research is needed to determine (1) the error attributable to tester and subject factors and (2) the validity of the measures.

Acknowledgment

The primary software used in this study, FootTrak, was furnished by Motion Analysis Corporation.

References

1 Kernozek TW, Ricard MD. Foot placement angle and arch type: effect on rearfoot motion. *Arch Phys Med Rehabil.* 1990;71:988–991.

2 Stacoff A, Denoth J, Kaelin X, et al. Running injuries and shoe construction: some possible relationships. *International Journal of Sports Biomechanics.* 1988;4:342–357.

3 Nigg BM, Herzog W, Read LJ. Effect of viscoelastic shoe insoles on vertical impact forces in heel-toe running. *Am J Sports Med.* 1988;16:70–76.

4 Engsberg JR, Andrews JG. Kinematic analysis of the talocalcaneal/talocrural joint during running support. *Med Sci Sports Exerc.* 1987;19:275–284.

5 Walton JS. The accuracy and precision of a video-based motion analysis system. In: *Proceedings of the 30th International Technical Symposium on Optical and Optoelectric Applied Sciences and Engineering.* 1986;693:17–22.

6 Manter JT. Movements of the subtalar and transverse tarsal joints. *Anat Rec.* 1941;80:397–410.

7 Areblad M, Nigg BM, Ekstrand J, et al. Three-dimensional measurement of rearfoot motion during running. *J Biomech.* 1990;23:933–940.

8 Soutas-Little RW, Beavis GC, Verstraete MC, et al. Analysis of foot motion during running using a joint co-ordinate system. *Med Sci Sports Exerc.* 1987;19:285–293.

Appendix. *Variables Assessed in This Study and Their Operational Definitions*

Calcaneus-to-tibia angle (CTA): Angle formed between the bisect of the posterior shank and the bisect of the posterior heel, as determined by placement of the retroreflective markers. The angle is measured in degrees.

Angle at touch-down: The CTA measured at the instant of initial contact of the foot.

Maximum pronation: The greatest value of the CTA corresponding to the position of pronation.

Total pronation range of motion: The CTA excursion between initial contact and maximum pronation.

Time to maximum pronation: The instant (measured as percentage of gait cycle) at which maximum pronation occurs.

Toe-off angle: The CTA measured at the instant of toe-off.

Initial velocity: The mean of the CTA velocities measured from the instant of initial contact to 10% of stance phase.

Maximum pronation velocity: The greatest value corresponding to CTA velocity in the direction of pronation.

Maximum supination velocity: The greatest value corresponding to CTA velocity in the direction of supination.

Temporal variables: The time (measured in milliseconds) and percentage of total gait cycle of the swing and stance phases of the selected gait cycle.

Hip Motion and Related Factors in Walking

GARY L. SMIDT, Ph.D.

Thirty-one subjects were used for this study of hip motion and related factors during walking. An electrogoniometric method was used for measuring hip motion and a newly devised method was used for obtaining stride-length measures.

Conclusions based on the findings are as follows: (1) The electrogoniometric method for measuring hip motion employed in this study was sufficiently reliable for experimental use. (2) Hip motion in the sagittal plane is positively related to walking velocity. (3) The stride-length/lower-extremity-length ratio may be a useful index in the study of hip motion during walking.

Hip-motion measurements are useful in the study of human gait. One use in the study of gait is to obtain normative data for hip-motion measurements and subsequently to compare these data with similar measurements obtained from patients who walk in some abnormal manner. Another function is to evaluate objectively a treatment regimen in which hip-motion measurements are obtained for patients exhibiting abnormal gait. These measurements may be obtained before, during, and after the treatment regimen. Still another use for hip-motion measurements obtained during walking is to compare two or more patients who walk in some abnormal fashion.

With respect to the aforementioned uses for hip-motion measurements in the study of gait and the practical utilization of such data in the clinical situation, are there factors which should receive special consideration?

Several writers have seen the need to control cadence for walking,[1-3] some have controlled velocity,[4,5] while others have allowed the subjects to walk at a "natural pace."[6-10] Stride length and forward velocity for walking are among the basic components of gait outlined by Drillis.[11] Stride length is defined by Drillis as the distance between two consecutive ipsilateral heel contacts on the walking surface. Walking velocity is the time rate of the linear forward motion for the body. The writer defines

---

Dr. Smidt is Director and Assistant Professor, Master's Degree Program in Physical Therapy, University of Iowa, Iowa City, Iowa 52240.

Adapted from a paper presented at the Forty-Seventh Annual Conference of the American Physical Therapy Association, Washington, D.C., July 1970.

the functional lower extremity length as the distance between the center of the femoral head and a point on the floor located slightly anterior and medial to the medial malleolus on the ipsilateral side.

Murray and others found that the composite group mean for the stride length was greater for subjects who walked at a fast forward linear velocity, but the within-group variation was large.[12] Furthermore, they reported a positive relationship ($r=.46$) between step-length measurements obtained for walking and standing-height measurements.

The question posed earlier is now expanded. When hip-motion measurements are obtained in the study of gait, should forward velocity be controlled and should consideration be given to functional lower extremity length and stride length? If forward walking velocity, functional lower extremity length, and stride length are variables which significantly relate to the requirements of hip motion for walking, then certainly these variables should be given primary consideration in gait analysis in either the clinical or the research setting. Conversely, the consideration of these variables is obviated if they do not significantly relate to the requirements of hip motion for walking.

The purposes of this study were 1) to determine the relationships among the stride-length/lower-extremity-length ratio and the hip-motion measurements in the sagittal, coronal, and transverse planes for slow, moderate, and fast forward linear walking velocities, and 2) to determine the differences among the respective hip-motion measurements in the sagittal, coronal, and transverse planes for slow, moderate, and fast forward linear walking velocities.

For this study, the stride-length/lower-extremity-length ratio was obtained by dividing the mean stride-length measurement for walking by the lower-extremity-length measurement.

Slow, moderate, and fast forward linear walking velocities were approximately 90, 135, and 188 centimeters per second respectively. These criteria were based on information from previous studies.[12, 13]

A walking sequence refers to the event in which a subject walks from one end of the walkway to the other. A set of measurements consisted of data from four walking sequences.

REVIEW OF RELATED LITERATURE

Eberhart and others obtained the hip-motion measurements for walking in the sagittal and coronal planes by using a photographic technique.[1] For the three normal subjects, the hip-motion measurements in the sagittal plane were 43, 39, and 44 degrees (0.75, 0.68, and 0.77 rad) respectively. For the same three subjects, the hip-motion measurements in the coronal plane were 10, 9, and 16 degrees (0.17, 0.16, and 0.28 rad).

To determine the magnitude of the femoral rotation in the transverse plane relative to the pelvis during walking, Levens and his associates used three synchronized thirty-five-millimeter motion picture cameras, one camera for each of the primary axial planes of the body.[14] Hip-motion measurements in the transverse plane were obtained from photographs of the targets which extended laterally from the distal femur and the pelvis. The mean hip motion in the transverse plane during walking was 8.7 degrees (0.15 rad) with a range of 4.1 degrees to 13.3 degrees (0.07 to 0.23 rad). The measurements were taken from the right side of the body.

Murray, Drought, and Kory employed the interrupted-light photographic technique to study the gait of sixty normal men.[9] Before the subjects were photographed during free-cadence walking, they walked several trials at a cadence of 112 steps per minute in time with a metronome. This preliminary activity was done in an attempt to familiarize the subjects with the walking area and to reduce variation of the walking velocity of the subject. The following results are pertinent to this study: 1) the mean for the left hip-motion measurements in the sagittal plane was approximately 45 degrees (0.79 rad), 2) the mean stride length was 156.5 centimeters with a standard deviation of 11.4 centimeters, and 3) the coefficient of correlation between stride length and maximum hip-motion measurement in the sagittal plane was .38.

Another study was conducted by Murray and others to compare subjects' walking patterns for two different forward linear velocities.[12] A total of thirty subjects was involved: two tall (180–187 cm), two short (156–170 cm), and two subjects of medium height (173–179 cm) for each of the following age groups: twenty to twenty-five, thirty to thirty-five, forty to forty-five, fifty to fifty-five, and sixty to sixty-five years. For free-cadence walking, the mean forward linear walking velocity was 151 centimeters per second with a standard deviation of 20 centimeters per second, the mean maximum excursion for hip motion in the sagittal plane was 48 degrees (0.84 rad) with a standard deviation of 5 degrees (0.09 rad), and the mean stride length was 156 centimeters with a standard deviation of 13 centimeters. For the fast forward linear walking velocity of the subjects the mean was 218 centimeters per second with a standard deviation of 16 centimeters per second. For this fast velocity the mean maximum excursion for hip motion in the sagittal plane was 52 degrees (0.91 rad) with a standard deviation of 6 degrees (0.10 rad), and the mean stride length was 185 centimeters with a standard deviation of 16 centimeters.

Murray, Kory, and Clarkson also studied the walking patterns of sixty-four normal men who ranged in age from twenty to eighty-seven years.[15] The average free-walking velocity was 139 centimeters per second and the average fast-walking velocity was 195 centimeters per second. For free walking the average total hip motion used in the sagittal plane was 46 degrees (0.80 rad) and 50 degrees (0.87 rad) of motion was the average motion utilized for fast walking.

Twenty-three elderly women ranging in age from sixty-four to eighty-six years and twelve young women ranging in age from nineteen to thirty-eight years were studied by Finley, Cody, and Finizie.[16] The average walking velocity for the young and elderly groups was 80 and 75 centimeters per second respectively. Total hip motion in the sagittal plane averaged 24 degrees (0.42 rad) for the elderly group and 20 degrees (0.35 rad) for the young group, while motion in the coronal plane averaged 16 degrees (0.28 rad) for the elderly group and 9 degrees (0.16 rad) for the young group.

PROCEDURE

Thirty-one subjects from the state penal institution at Fort Madison, Iowa, served as subjects for this study. The mean age was thirty-six years with a range of twenty-three to fifty-six, the mean height was 172 centimeters with a range of 161 to 192, and the mean weight was 79.8 kilograms with a range of 54 to 104. Not included among the thirty-one subjects were eleven subjects used in a pilot study (Appendix).

An electrogoniometric method was used to obtain hip-motion measurements exclusively for the right side. The electrogoniometric apparatus consisted of three structures (Fig. 1).

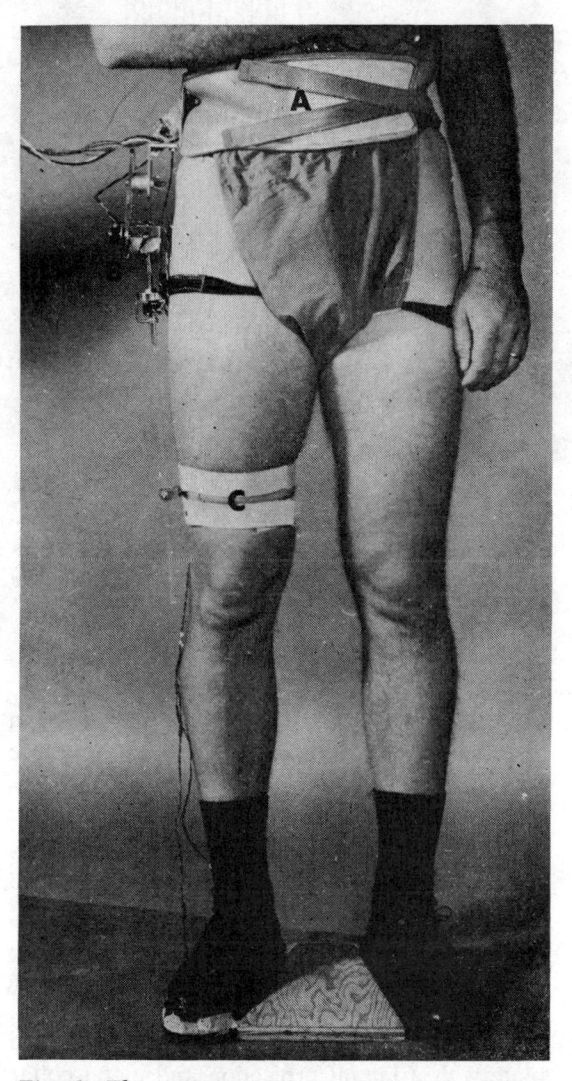

Fig. 1. Electrogoniometric apparatus attached to subject. A. Leather belt which was the primary supportive structure. B. Electrogoniometric assembly which includes the potentiometers for each plane of hip motion. C. Suprapatellar cuff which served as the distal attachment to receive the plexiglass rod.

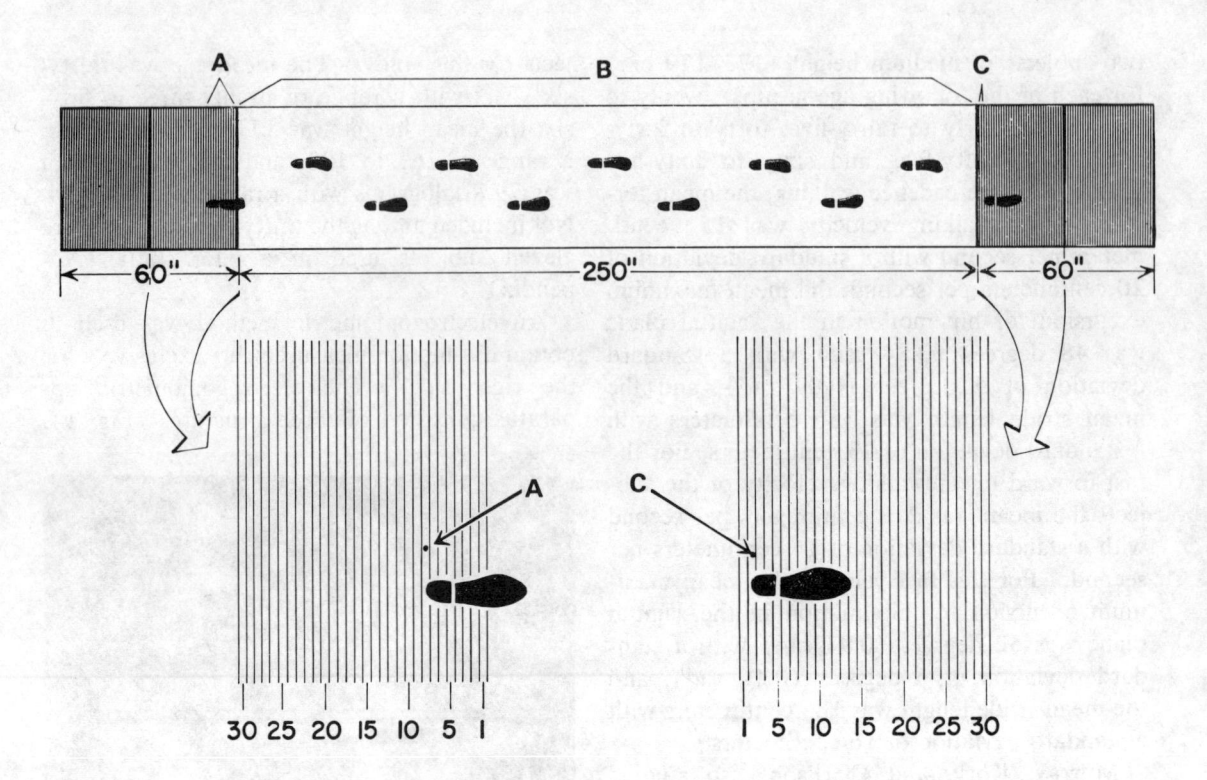

Example: Calculation of stride length

$$\frac{A + B + C}{\text{number of strides}} = \frac{8'' + 250'' + 2''}{5} = 52''$$

Fig. 2. Stride-length measurement method. A. Distance from heel contact to edge of first grid encounter by the subject. B. Distance between the inside edges of the two grids. C. Distance from heel contact to the edge of the second grid encountered by the subject.

One structure was a wide leather belt with an attached lateral flange. A second structure was an electrogoniometric assembly which included a stainless steel linkage system connecting the axes of three potentiometers, one situated for each plane of hip movement. A third structure, included with the electrogoniometric apparatus, was a suprapatellar cuff which consisted of a piece of heavy leather attached to a pair of elastic Velcro® straps. Also affixed to the piece of heavy leather was a small metal plate with a lateral projection and metal collar which supported the plexiglass rod extending from the electrogoniometric assembly.

An eight channel ink-writing recorder * was utilized to obtain recordings for the hip-motion data. An electronic "klockounter" † and two photo cell switches were utilized to monitor the subjects' walking velocity. Cadence was obtained from information provided by footswitches attached to the undersurface of the right shoe. Johnston and Smidt have reported the standardized procedure for applying the electrogoniometric apparatus, plus the details and the validity of the electrogoniometric method.[13]

Method Used to Obtain Stride-Length Measurements

To obtain stride-length measurements for walking, a grid pattern was placed at each end

---

* Beckman Type R Direct Writing Recorder. Beckman Instruments, Inc., Schiller Park, Illinois 60176.

† Hunter "Klockounter." Hunter Manufacturing Company, Coralville, Iowa 52240.

Basic Research

of the walkway in the laboratory. The grid patterns consisted of strips of masking tape which were placed parallel to each other at 2.54 centimeter intervals and perpendicular to the direction of walking. The strips of masking tape were numbered consecutively from one through sixty. These integers made identification possible of the distance between the inside border of each grid pattern and the point of right heel strike upon the grid pattern. The distance between the grid patterns was 635 centimeters. For one walking sequence the mean right stride-length measurements were obtained by summing the distance between the points of right heel strike and the inside borders of each grid pattern and the predetermined distance between the two grid patterns, and then dividing this sum by the total number of strides (Fig. 2). For two sets of measurements taken on the same day, test-retest reliabilities were obtained for the mean stride-length measurements at slow, moderate, and fast forward walking velocities.

Sequence of Events for the Experiment

The right lower-extremity-length measurements were obtained and the electrogoniometric apparatus and foot switches were applied. With the aid of an electronic "klockounter" and two photo cell switches (one toward each end of the walkway), each subject was trained to walk at three different forward velocities. First, the subject was trained to walk at a slow forward velocity in which he traversed the length of the walkway within the range of 10.2 to 10.8 seconds. The slow forward walking velocity was approximately 90 centimeters per second. Subsequently, each subject was trained to walk at a moderate forward velocity in which he traversed the length of the walkway within a range of 6.7 to 7.3 seconds. The moderate forward walking velocity was approximately 135 centimeters per second. Finally, each subject was trained to walk at a fast forward velocity in which he traversed the length of the walkway within a range of 4.7 to 5.3 seconds. The fast forward walking velocity was approximately 188 centimeters per second. Hip-motion recordings and stride-length measurements were obtained for the slow, moderate, and fast forward walking velocities. If the subject failed to satisfy the criterion for each of the predetermined forward walking velocities, the walking sequence was disregarded. Measures were obtained for a total of four walking sequences for each forward walking velocity: slow, moderate, and fast.

Explanation and Justification for Stride-Length/Lower-Extremity-Length Ratio

For purposes of analysis, a stride-length/lower-extremity-length ratio was calculated by dividing the right stride length by the functional right lower extremity length. Simultaneous consideration of the right stride length and the right lower extremity length in the form of a ratio includes individual differences among subjects which might be expected to relate to hip-motion measurements for walking. The magnitudes of the ratios are discrete but not cumbersome and the ratios are meaningful since a value of 1.00 would indicate equality between stride length and lower extremity length. For example, two individuals may have equal lower extremity length and different stride length (Fig. 3). Conversely, two other individuals may

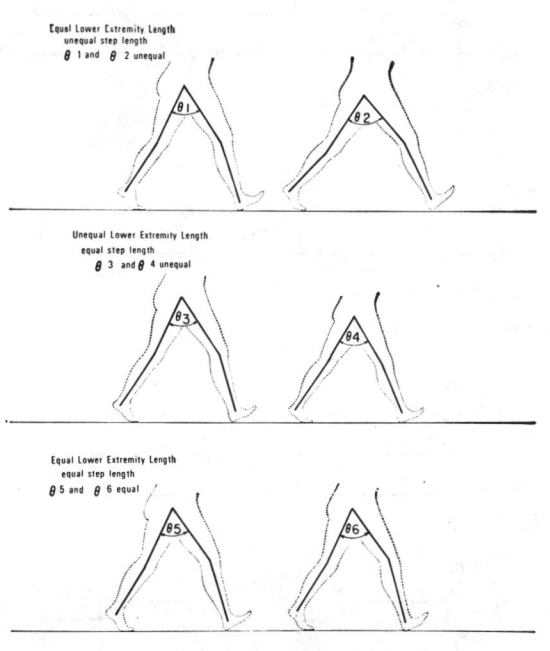

Fig. 3. Rationale for stride-length/lower-extremity-length ratio. (For simplification of illustration step length is used instead of stride length.)

possess unequal lower extremity length and equal stride length.

RESULTS

Patterns of Hip Motion for Walking

Maximum excursions for hip motion commonly occurred in the following order during the gait cycle: external rotation and abduction during early swing phase, and flexion during late swing or early stance phase. These motions were followed by maximum internal rotation, adduction, and extension during early, mid, and late stance phase respectively (Figs. 4–6). One exception was for slow walking where maximum abduction and external rotation tended to occur simultaneously just after toe off.

Reliability of Measurements

Reliability is the degree to which consistent results are obtained from the use of the same method by the same person. When test and retest measurements are consistent, the results should be trustworthy.

For reliability, the rs were obtained between two sets of hip-motion measurements for each forward linear walking velocity taken from eight subjects on the same day and ranged from .92 to .99 for the sagittal plane, .70 to .87 for the coronal plane, and .74 to .94 for the transverse plane. The reliability of the stride-length measurements is indicated in Table 1. An r of .99 was obtained between two sets of lower-extremity-length measurements.

From fifteen subjects rs of .93, .94, and .83 were obtained between two sets of mean stride-length measurements taken on the same day for slow, moderate, and fast forward linear walking velocities respectively (Tab. 1).

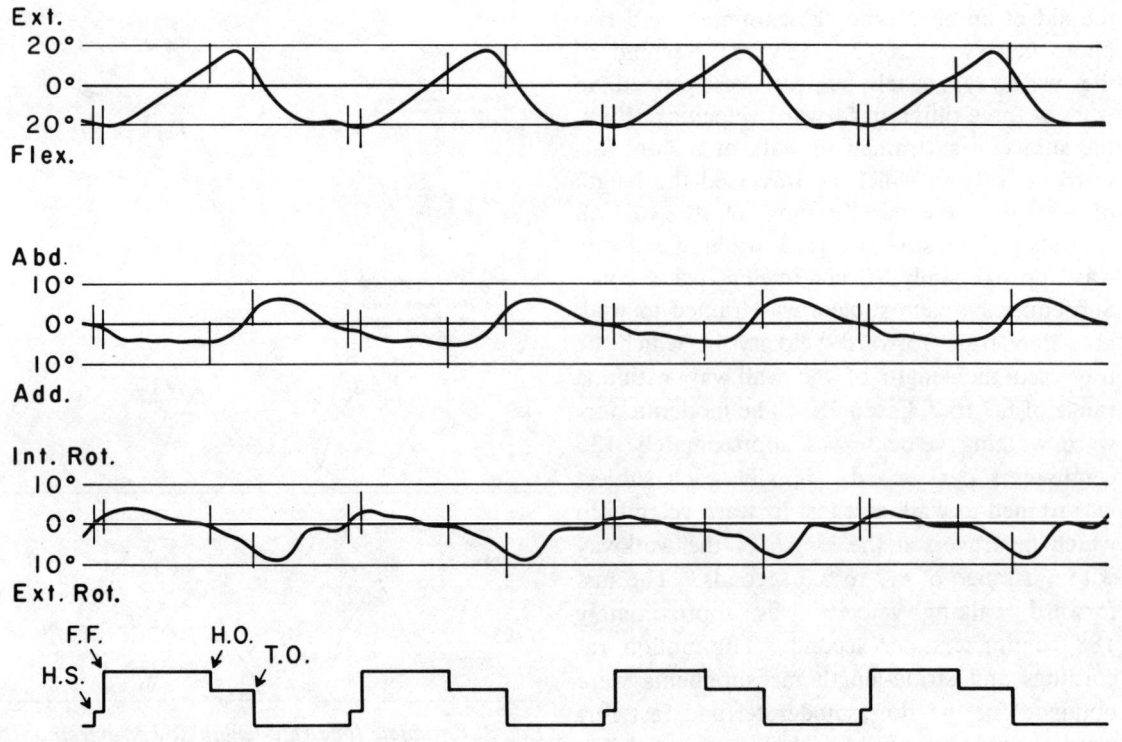

Fig. 4. Patterns of hip motion for slow walking velocity. H.S.=Heel Strike. F.F.=Foot Flat. H.O.= Heel Off. T.O.=Toe Off.

Basic Research

TABLE 1

MEANS, STANDARD DEVIATIONS, STANDARD ERRORS AND *r*S OF RELIABILITY FOR
STRIDE-LENGTH MEASUREMENTS (CM)

Stride-Length Measurement (N=15)	Mean	S.D.	S.E. of the Mean	r
Slow Velocity				
Test	127	7.39	2.01	
Retest	127	7.36	1.96	.93
Moderate Velocity				
Test	152	5.69	1.52	
Retest	152	6.83	1.83	.94
Fast Velocity				
Test	180	9.04	2.41	
Retest	180	7.14	1.91	.83

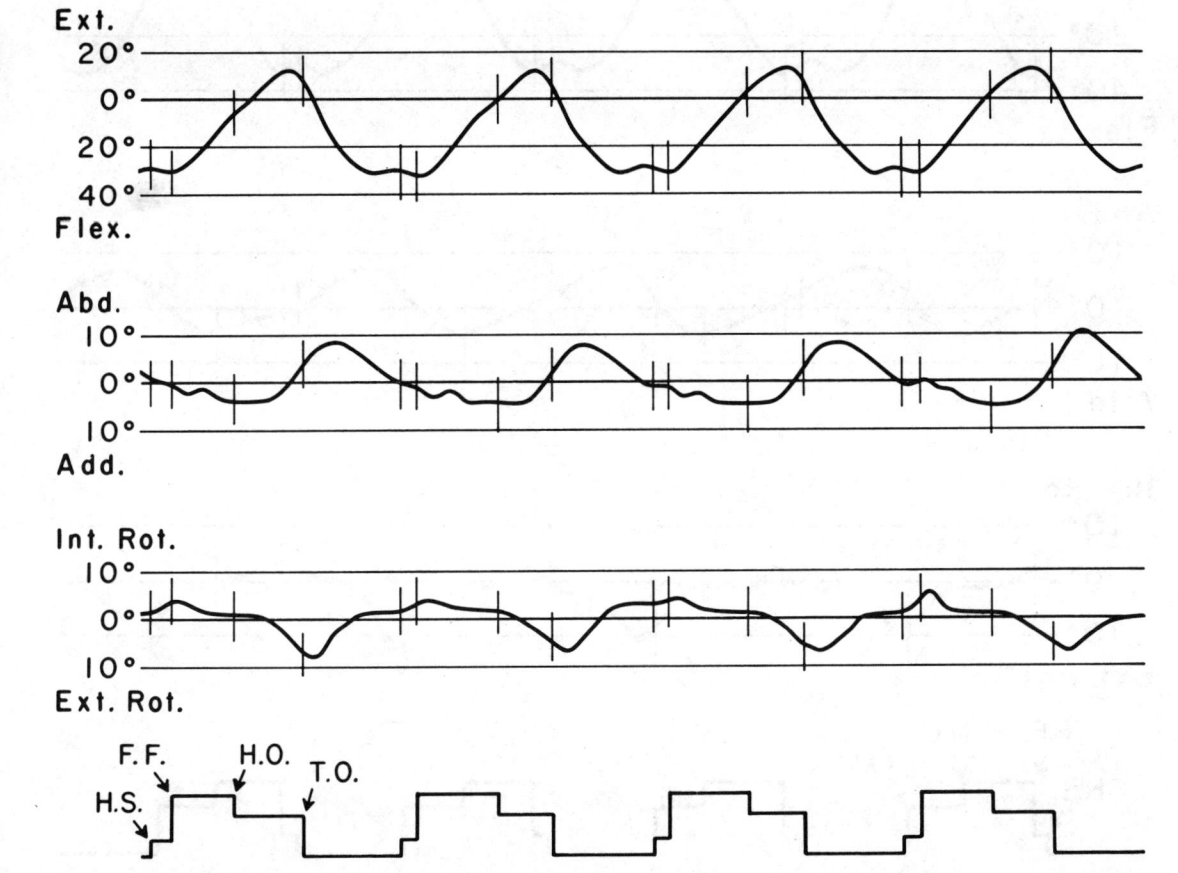

Fig. 5. Patterns of hip motion for moderate walking velocity. H.S.=Heel Strike. F.F.=Foot Flat. H.O.=Heel Off. T.O.=Toe Off.

TABLE 2

MEANS, STANDARD DEVIATIONS, AND STANDARD ERRORS
FOR VELOCITY (CM/SEC) AND CADENCE (STRIDES/MIN)
FOR SLOW, MODERATE, AND FAST WALKING

Measurement	Mean	S.D.	S.E. of Mean
Slow Velocity	91	1.26	0.23
Moderate Velocity	134	4.38	0.88
Fast Velocity	190	6.72	1.23
Slow Cadence	43	2.22	0.40
Moderate Cadence	53	2.37	0.43
Fast Cadence	63	2.89	0.53

Velocity and Cadence for Walking

The data for velocity and cadence appear in Table 2. Although the group means for cadence and velocity increased for slow, moderate, and fast forward linear walking velocities, the two are not highly related to one another when their values are compared for individuals. The coefficients of correlation between cadence and velocity were .30 for slow walking, .12 for moderate walking, and .28 for fast walking. These relatively low rs demonstrate that some walking subjects used longer strides and decreased cadence while others used shorter strides and increased cadence to accomplish the same velocity.

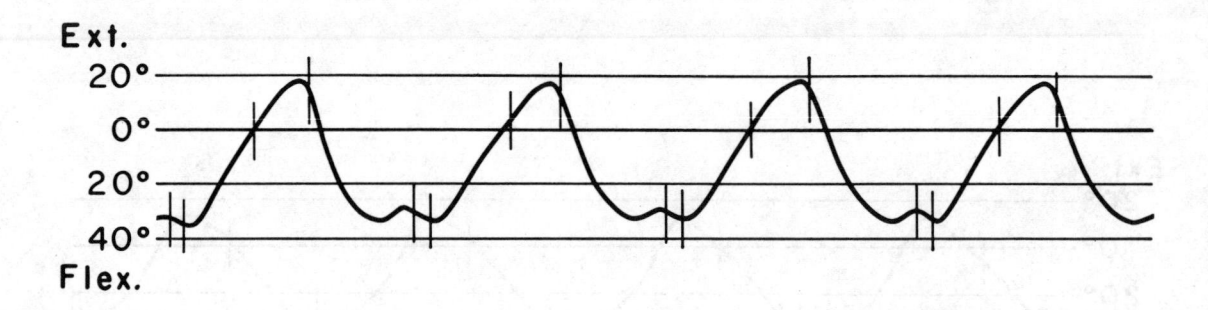

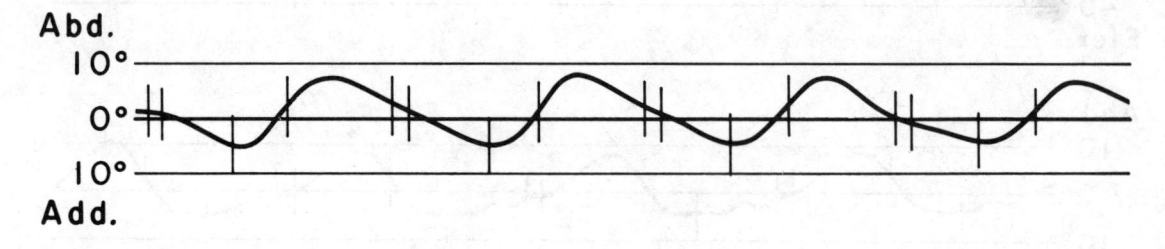

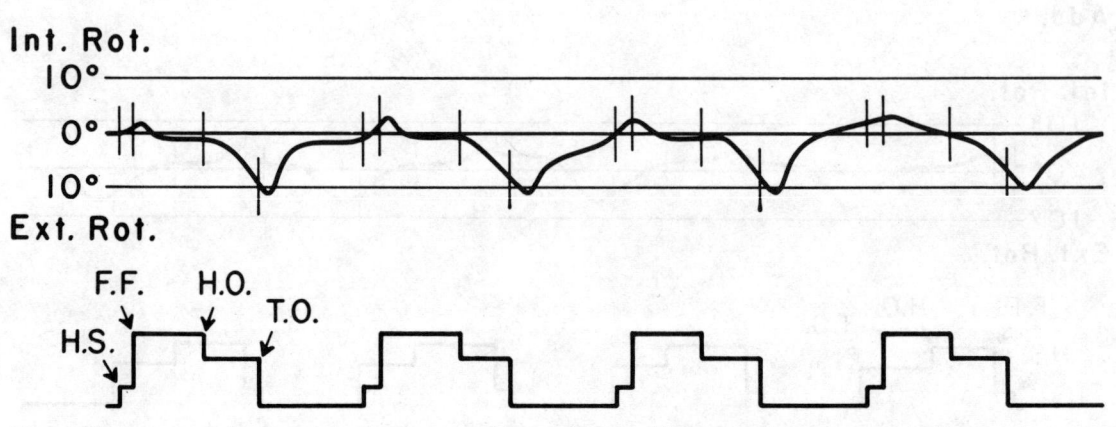

Fig. 6. Patterns of hip motion for fast walking velocity. H.S.=Heel Strike. F.F.=Foot Flat. H.O.=Heel Off. T.O.=Toe Off.

Basic Research

Relationship Between Measurements for Walking

For each walking velocity, the relationships between a set of stride-length/lower-extremity-length ratios and a set of hip-motion measurements for each axial plane were significant for the sagittal plane only, while the relationships for the coronal and transverse planes were low and positive. These data appear in Tables 3 to 5. The range of rs between hip motion and

TABLE 3

rs BETWEEN STRIDE-LENGTH/LOWER-EXTREMITY-LENGTH RATIOS AND HIP-MOTION MEASUREMENTS (DEG) FOR SLOW WALKING VELOCITY

Measurement ($N=31$)	Mean	S.D.	S.E. of Mean	r^a
[b] SL/LEL Ratio	1.36	.084	.015	.53
Hip Motion: Sagittal Plane	40.00	4.180	.760	
[b] SL/LEL Ratio	1.36	.084	.015	.26
Hip Motion: Coronal Plane	11.00	2.810	.510	
[b] SL/LEL Ratio	1.36	.084	.015	.28
Hip Motion: Transverse Plane	11.00	2.990	.550	

[a] r for p .05 (29 df) = .43.
[a] r for p .01 (29 df) = .52.
[b] SL/LEL = Stride-Length/Lower-Extremity-Length.

TABLE 4

rs BETWEEN STRIDE-LENGTH/LOWER-EXTREMITY-LENGTH RATIOS AND HIP-MOTION MEASUREMENTS (DEG) FOR MODERATE WALKING VELOCITY

Measurement ($N=31$)	Mean	S.D.	S.E. of Mean	r^a
[b] SL/LEL Ratio	1.64	.095	.017	.66
Hip Motion: Sagittal Plane	47.00	5.240	.960	
[b] SL/LEL Ratio	1.64	.095	.017	.19
Hip Motion: Coronal Plane	12.00	2.930	.540	
[b] SL/LEL Ratio	1.64	.095	.017	.29
Hip Motion: Transverse Plane	13.00	3.300	.600	

[a] r for p .05 (29 df) = .43.
[a] r for p .01 (29 df) = .52.
[b] SL/LEL = Stride-Length/Lower-Extremity-Length.

TABLE 5

rs BETWEEN STRIDE-LENGTH/LOWER-EXTREMITY-LENGTH RATIOS AND HIP-MOTION MEASUREMENTS (DEG) FOR FAST WALKING VELOCITY

Measurement ($N=31$)	Mean	S.D.	S.E. of Mean	r^a
[b] SL/LEL Ratio	1.91	.109	.020	.75
Hip Motion: Sagittal Plane	54.00	4.800	.880	
[b] SL/LEL Ratio	1.91	.109	.020	.20
Hip Motion: Coronal Plane	13.00	2.970	.540	
[b] SL/LEL Ratio	1.91	.109	.020	.24
Hip Motion: Transverse Plane	14.00	2.800	.510	

[a] r for p .05 (29 df) = .43.
[a] r for p .01 (29 df) = .52.
[b] SL/LEL = Stride-Length/Lower-Extremity-Length.

stride length only was from .02 to .09 for eight relationships; the remainder was .25 for the slow velocity and hip motion in the transverse plane.

Difference Among Hip-Motion Measurements

The differences among the means for hip-motion measurements, in each plane, for each forward linear walking velocity were subjected to an analysis of variance test with a subjects by treatments design. The selected level of significance was .05 which, for these data, requires an F value of >3.34. The differences were significant in each case (Tab. 6–8).

TABLE 6
Summary of F: Differences Among Hip-Motion Measurements (Sagittal Plane) for Slow, Moderate, and Fast Walking Velocities

Sources	df	SS	MS	F[a]
Subjects	30	1829.50	60.98	.....
Treatments	2	3138.25	1569.13	336.92
Error	60	259.44	4.66	.....
Total	92	5247.19	....	.....

[a] F for p .05 = 3.34.
[a] F for p .01 = 5.45.

TABLE 7
Summary of F: Differences Among Hip-Motion Measurements (Coronal Plane) for Slow, Moderate, and Fast Walking Velocities

Sources	df	SS	MS	F[a]
Subjects	30	691.95	23.04	.....
Treatments	2	111.37	55.69	35.55
Error	60	93.99	1.57	.....
Total	92	897.31	.....	.....

[a] F for p .05 = 3.34.
[a] F for p .01 = 5.45.

TABLE 8
Summary of F: Differences Among Hip-Motion Measurements (Transverse Plane) for Slow, Moderate, and Fast Walking Velocities

Sources	df	SS	MS	F[a]
Subjects	30	749.96	25.00	.....
Treatment	2	124.79	62.40	34.49
Error	60	108.56	1.81	.....
Total	92	983.31	.....	.....

[a] F for p .05 = 3.34.
[a] F for p .01 = 5.45.

Post-Hoc Comparisons

Since the F tests showed overall statistical significance, post-hoc comparisons were required to identify the locations of differences among group mean hip-motion measurements for each axial plane at slow, moderate, and fast forward linear walking velocities. The Tukey Test was chosen for this purpose.

For each of the three planes, the differences between group means for the hip-motion measurements for the slow, moderate, and fast walking were statistically significant at the .05 level.

DISCUSSION

The electrogoniometric method is subject to errors because the linkage system is not located in the hip joint. Therefore, differences exist between actual hip motion and recorded hip motion. Since the external linkage system is placed near the thigh, the errors are less than 3 degrees (0.05 rad) for the sagittal plane and less than 1 degree (0.02 rad) for the coronal and transverse planes. These differences, as determined by a 4 x 4 matrix method, were considered negligible.

A new method for obtaining stride-length measurements is presented in this study. This method is simple and reliable, and measurements can be obtained immediately following a walking sequence by a subject. Only right stride-length measurements are necessary for this method because, if a subject walks on a straight line of progression, the stride-length measurements for the right and left sides are equal. However, stride length and step length are not synonymous. Stride length was defined as the distance between two consecutive ipsilateral heel contacts on the walking surface. This method would not be appropriate for obtaining right and left step lengths as they are unequal for many pathological conditions which affect the lower extremities. However, when combined with utilization of a stop watch, the method for obtaining stride length used in this study can be used to determine walking velocity, cadence, and stride length. Therefore this method can be used in the clinical situation to accrue meaningful and objective information which should be helpful in evaluating gait.

The findings regarding significant differences among means for each plane of hip motion during slow, moderate, and fast forward linear ambulatory velocities support the thesis that the magnitude of hip motion varies directly with forward walking velocity. Despite statistical significance, the differences were small among the means for hip-motion measurements for the coronal plane (1 degree–0.02 rad) and among the means for hip-motion measurements for the transverse plane (1 to 2 degrees–0.02–0.03 rad). However, the differences among the sagittal means were considerably larger (7 degrees–0.12 rad). Since the individual's stride length increased progressively for slow, moderate, and fast forward linear walking velocities, hip motion in the sagittal plane might be the factor which contributes most to stride length increase. Some possibilities for increased stride length without increased motion at the hip joint are 1) greater forward thrust by plantar flexing the foot at toe off, 2) greater amount of knee extension at heel strike and toe off, and 3) greater amount of dorsiflexion of the foot at heel strike and plantar flexion at toe off.

The correlations between the stride-length/lower-extremity-length ratio and hip-motion measurements in the sagittal plane for slow, moderate, and fast forward walking velocities were statistically significant ($p<.05$) exclusively for the sagittal plane. In this case the dimension of lower extremity length complements the findings outlined in the previous paragraph. Therefore, increases in stride length and decreases in lower extremity length are directly related to increases in hip-motion measurements for the sagittal plane. The converse relationship is also true.

The findings regarding significant differences among means for hip motion for each plane during slow, moderate, and fast forward linear walking velocities also demonstrate the need for controlling velocity in the study of gait. If, for example, velocity is not controlled, the differences between hip-motion measurements taken before and after a treatment procedure may be attributed to changes in velocity irrespective of treatment.

Furthermore, the findings regarding differences among means for hip motion for each plane during slow, moderate, and fast forward linear walking velocities might be practically applied for gait training. Empirically, the therapist should recognize that if he encourages a patient to walk at a faster rate, demands for increased hip motion are probable. The writer postulates that persons use a slower forward walking velocity for moving about the home than they do for walking along a corridor or street. If this postulate is true, hip-motion requirements for walking will vary and each patient should be trained accordingly.

The rs among stride-length/lower-extremity-length ratio and hip motion during slow, moderate, and fast forward linear walking velocities were .53, .66, and .75 for the sagittal plane; .26, .19, and .20 for the coronal plane; and .28, .29, and .24 for the transverse plane. The rs between stride-length measurements and hip motion were near zero. The distinct differences among the rs relative to 1) stride-length/lower-extremity-length ratio and hip motion, and 2) stride length and hip motion clearly show that stride length in conjunction with lower extremity length is worthy of consideration in the study of hip motion during gait. Perhaps the stride-length/lower-extremity-length ratio should be deemed one of the components of gait.

During pilot work for this study an attempt was made to control cadence with a metronome. This method was deemed unsatisfactory because the subjects had a tendency to shorten their stride to accommodate the fast cadence dictated by the metronome, in which case walking velocity did not increase at a rate commensurate to cadence. This problem was eliminated when the predetermined velocity requirements were met by the subjects.

The results of this study, and other investigations concerned with the magnitude of hip-motion requirements for walking, are included in Table 9. The mean values for walking velocity of female subjects reported by Finley, Cody, and Finizie were less than the slow velocity for this study.[16] Stride length and hip motion in the sagittal plane were also reduced. However, for the elderly subjects, hip motion in the coronal plane was greater than the values for this study, while young females used less motion. Levens, Inman, and Blosser found less hip mo-

TABLE 9
SUMMARY OF GAIT STUDIES INVOLVING HIP MOTION

Investigator(s)	Velocity Category	Number of Subjects	Cadence (steps/min)	Velocity (cm/sec)	Stride Length (cm)	Hip Motion (deg.) Sag. Plane	Cor. Plane	Trans. Plane	Side Inspected	Sex
Finley, Cody, Finizie	Free	12 young	105	80	94	20	9	...		F
	Free	23 elderly	109	75	76	24	16	...		F
Murray, et al.	Free	30	113	151	156	48	...	...	Left	M
	Fast	30	138	218	186	52	...	...	Left	M
Murray, Drought, Kory	Free	...	117	...	157	45	...	...	Left	M
Murray, Kory, Clarkson	Free	64	...	139	146	46	...	...		M
	Fast	64	...	195	172	50	...	...		M
Levens, Inman, Blosser	Free	12	...	...	...	...	...	9	Right	M
Eberhart, et al.	Free	3	90+	122	...	42	12	...	Right	M
Johnston, Smidt	Free	33	120	...	...	52	12	13	Right	M
Smidt	Slow	31	86	91	127	40	11	11	Right	M
	Moderate	31	104	134	152	47	12	13	Right	M
	Fast	31	126	190	178	54	13	14	Right	M

tion for the transverse plane.[14] For fast walking, Murray et al. reported a mean velocity of 218 centimeters per second and a mean stride length of 186 centimeters.[12] Each of these values exceeded the values for the same parameters in this study. Other measurements compared favorably with previous studies.

SUMMARY AND CONCLUSIONS

The purposes of this study were 1) to determine the relationships among the stride-length/lower-extremity-length ratio and the hip-motion measurements in the sagittal, coronal, and transverse planes for different walking velocities, and 2) to determine the differences among the respective hip-motion measurements in the sagittal, coronal, and transverse planes for different walking velocities.

Thirty-one subjects were used for this study. The electrogoniometer used to obtain the hip-motion measurements was designed and constructed at the University of Iowa and the method for obtaining stride-length measurements was designed by the investigator. By means of an electronic timer, the subjects were trained to walk at three predetermined forward linear velocities: slow, moderate, and fast.

The rs of reliability and objectivity ranged from .70 to .99 for the electrogoniometric method. The rs of reliability for test-retest measures of stride-length are .93, .94, and .83 for the slow, moderate, and fast walking velocities respectively.

The rs (.53, .66, .75) between the stride-length/lower-extremity-length ratio and hip-motion velocities are statistically significant ($p < .05$). The rs between the stride-length/lower-extremity-length ratio and hip-motion measurements for the coronal and transverse planes during the three walking velocities are not statistically significant ($p > .05$).

There were statistically significant ($p < .05$) differences between the means of any two of the three walking velocities for hip-motion measurements obtained in the sagittal, coronal, and transverse planes.

The findings of this study seem to warrant the conclusions that 1) the electrogoniometric method for measuring hip motion employed in this study is appropriate for experimental use,

2) hip motion in the sagittal plane is positively related to forward linear walking velocity, and 3) the stride-length/lower-extremity-length ratio is a useful index in the study of hip motion during walking.

APPENDIX

PILOT STUDY

Differences for Right Stride-Length Measurements

By the use of the *t* test for correlated samples, the difference between the mean of the means for the right stride-length measurements taken for eleven subjects who walked at a slow forward velocity and the mean of the means for the right stride-length measurements taken for the same eleven subjects who walked at a moderate forward velocity is statistically significant ($p < .0001$). The difference is in favor of the stride length for the moderate forward walking velocity.

The difference between the mean of the means for the right stride-length measurements taken for eleven subjects who walked at a moderate forward velocity and the mean of the means for the right stride-length measurements taken for the same eleven subjects who walked at a fast forward velocity is statistically significant ($p < .002$). The difference is in favor of the stride length for the fast forward walking velocity.

REFERENCES

1. Eberhart HD, Inman VT, Saunders JB, et al: Fundamental Studies of Human Locomotion and Other Information Relating to the Design of Artificial Limbs. A report to the National Research Council, Committee of Artificial Limbs, University of California, Berkeley, 1947
2. Gersten J, Orr W, Sexton AW, et al: External work in level walking. J Appl Physiol 26:286–289, 1969
3. Ralston HJ: Comparison of energy expenditure during treadmill walking and floor walking. J Appl Physiol 15:1156, 1960
4. Corcoran PJ, Brengelmann GL: Oxygen uptake in normal and handicapped subjects, in relation to speed of walking beside velocity-controlled cart. Arch Phys Med 51:78–87, 1970
5. Corcoran PJ, Jebsen RH, Brengelmann GL, et al: Effects of plastic and metal leg braces on speed and energy cost of hemiparetic ambulation. Arch Phys Med 51:69–77, 1970
6. Finley FR, Cody KJ, Finizie RV: Locomotion patterns in elderly women. Arch Phys Med 50:140–146, 1969
7. Joseph J: The activity of some muscles in locomotion. Physiotherapy 50:180–183, 1964
8. Liberson WT: Biomechanics of gait: A method of study. Arch Phys Med 46:37–48, 1965
9. Murray MP, Drought AB, Kory RC: Walking patterns of normal men. J Bone Joint Surg 46A:335–360, 1964
10. Tipton CM, Karpovich PV: Electrogoniometric records of knee and ankle movements in pathologic gaits. Arch Phys Med 46:267–272, 1965
11. Drillis RJ: Objective recording and biomechanics of pathologic gait. Ann NY Acad Sci 74:86–109, 1958
12. Murray MP, Kory RC, Clarkson BH, et al: A comparison of free and fast speed walking patterns in normal men. Amer J Phys Med 45:8–24, 1966
13. Johnston RC, Smidt GL: Measurement of hip joint motion during walking. J Bone Joint Surg 51A:1083–1094, 1969
14. Levens AS, Inman VT, Blosser JA: Transverse rotation of the segments of the lower extremity in locomotion. J Bone Joint Surg 30A:859–872, 1948
15. Murray MP, Kory RC, Clarkson BH: Walking patterns in healthy old men. J Geront 24:169–178, 1969
16. Finley FR, Cody KA, Finizie RV: Locomotion patterns in elderly women. Arch Phys Med 50:140–146, 1969
17. Chao E, Rim K, Smidt GL, et al: The application of 4 x 4 matrix method to the correction of the measurements of hip joint rotations. J Biomechanics 3:459–471, 1970

---

the author

Gary L. Smidt, Ph.D., received his B.A. degree from Kearney State College in 1959. He received his certificate in physical therapy, M.A., and Ph.D. from the University of Iowa in 1960, 1967, and 1969 respectively. From 1960 to 1963 he was staff physical therapist with the U.S. Public Health Service in New York and was in private practice and a consultant physical therapist in Illinois from 1963 to 1966. Recently he was nominated Outstanding Young Man of America for 1970. He is currently director and assistant professor of the master's degree program in physical therapy at the University of Iowa.

Comments by Discussant
George F. Hamilton

Far too frequently physical therapists fall into the trap of accepting the value of a measurement tool without first testing what it truly measures and how significant its output may be to answering the question under study. As a profession, we need more evaluation of research instrumentation if we hope to gather the data essential to better understanding of the procedures utilized in the practice of physical therapy.

In formulating his study, Dr. Smidt asked his questions specifically and set about answering them in an acceptable and objective fashion. There should be little question in your mind, after reading this article, that to obtain valid measures of hip motion one may feel confident in the use of the electrogoniometer, that forward velocity is a variable of significant importance if you are measuring hip motion in the sagittal plane, and that if you do not take into consideration leg length and length of stride you may misinterpret the findings of a study.

It is most commendable that Dr. Smidt preceded his study by a preliminary, or pilot, study. By this aproach, he was able to confirm the apparent fact that stride length varied significantly when tested at each of the three speeds of walking, and, undoubtedly, he obtained an insight into the design features of this study.

I would like to raise a few areas of question which occur to me.

In the analysis of variance, a test was made of differences among the group means for hip motion measurement for each forward velocity in each of the three planes. Do "treatments" represent velocity? Would it not be interesting and informative to look at within-subject variance to determine the amount of *individual* variance occurring with replications of the test pattern. Although the F values obtained would suggest that individual variance would be significant, it is often valuable to test that assumption.

When the Tukey test was performed to identify locations of differences found in the analysis of variance, statistically significant differences were obtained in the hip motions in all planes for all speeds. Without benefit of raw data for review, I would question the practical significance of one degree differences expressed in the transverse and coronal planes. This is especially so in light of the indicated one degree error inherent within the measurement device. This factor was identified by Dr. Smidt and elaborated upon in his discussion. Perhaps a look at pair-wise differences would have assisted in the effort to pick out which means are different.

I was particularly pleased when Dr. Smidt extended his findings to areas of possible application to the practice of physical therapy. His suggestion of the importance of considering the velocity with which one walks when programing for a given individual's need for hip motion and exercise was excellent.

There are many aspects relative to the ambulatory process which one could imagine may be investigated. Although Dr. Smidt has not presented material which provides inference concerning gait deviation, he has most certainly contributed a significant link in the chain of factors essential to the validation of our tools of investigation. I hope that he will continue his inquiry, and that this information will stimulate others toward greater use of the electrogoniometer in quest of answers concerning the ambulatory process which will carry us further into the understanding of basic principles and their application to our treatment programs.

Basic Research

Knee Flexion During Stance as a Determinant of Inefficient Walking

DAVID A. WINTER

A biomechanical analysis of normal walking assessed the mechanical work cost in joules per unit mass and distance walked. For 21 walking trials—seven subjects at slow, natural, and fast cadences—these work costs (min = .73 J/kg.m, max = 1.65 J/kg.m) were correlated with maximum knee flexion during stance (min=6°, max= 33°). The results were contrary to the predictions of previous researchers who claimed that the energy cost would increase as the knee became more rigid during stance. This study showed a significant positive correlation between work cost and maximum knee flexion. The implications of these findings and the predicted increase of bone-on-bone forces as knee flexion increases are discussed relative to the gait training of certain patient populations.

Key Words: Energy expenditure, Gait, Knee.

In training patients with pathological gait, some efforts are being made to ensure that the patient does not walk with a stiff leg during weight bearing.[1] This approach is based on the known relationship between stiff-legged stance and the resultant increased rise in the body's center of gravity and on the related assumption that walking with a stiff leg consumes more energy than walking with normal knee flexion. The research most often quoted is that of Saunders et al who found that "the knee decreases vertical motion of the body by flexion ... the overall energy requirement is less than would be necessary in walking over a rigid knee."[2] Similar statements were made by Liberson.[3] Unfortunately, however, their inferences did not take into account the alterations of the trunk's kinetic energy that occurred simultaneously with these increases and decreases of potential energy, and they did not consider the work required to move the lower limbs.

Other researchers have discussed or analysed the conservation of energy in the total body or the trunk segment.[4–6] Cavagna and Margaria, using a point mass model of the total body, showed considerable exchange of energy between the potential and the kinetic energy components.[4] Winter et al demonstrated that in normal walking the trunk is a highly conservative system.[6] They found evidence of trade-off between the potential and the kinetic energy components, such that they cancel each other out. Consequently, the total energy requirement is drastically reduced, and the muscular activity necessary to raise the center of gravity is low. They also showed that the trunk's center of gravity rises during the first half of stance while the forward velocity decreases and that the center of gravity falls during the second half of stance while the forward velocity increases. In other words, the rise of the trunk's center of gravity is largely due to passive exchanges and not to muscle activity. Winter et al also found that the lower limbs accounted for up to 80 percent of the work done by the total body.

The prediction made by Saunders et al[2] in effect says that the energy cost of walking should decrease as midstance knee flexion increases over the range from stiff-legged gait to normal knee flexion. The purpose of this article is to show not only that this prediction is not valid but also that there is in fact a significant positive correlation between the mechanical work cost and maximum knee flexion during stance.

METHOD

A total of 21 walking trials of seven subjects without gait problems (five men, two women) were analysed. The subjects' mass varied from 55.8 kg to 86.5 kg and their age ranged from 20 to 26 years. Each subject walked at three voluntarily controlled cadences: slow, natural, and fast, varying from 79 to 129 steps/min. No effort was made to control knee flexion

---

Dr. Winter is Professor of Kinesiology, Department of Kinesiology, University of Waterloo, Waterloo, Ontario, Canada N2L 3G1.

This study was supported by the Medical Research Council of Canada, Grant No. MT4343.

This article was submitted September 1, 1981, and accepted April 26, 1982.

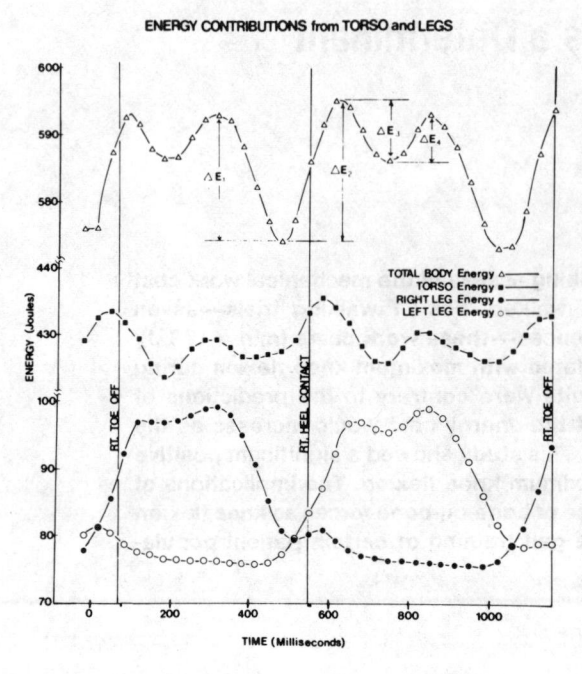

Fig. 1. Mechanical energy analysis of one stride of a walking subject without gait problems. Total energy of the right and left legs and the torso (head, arms, and trunk) are summed at each point in the stride to yield the total body energy.

or any other variable. Each subject had reflective markers attached to the following anatomical landmarks: toe, 5th metatarsal, heel, lateral malleolus, epicondyle of femur, greater trochanter, and midtrunk. They wore their own footwear and walked on a walkway while a tracking cart, carrying a TV and cinematographic camera (50 frames/sec) was guided along a track at a distance of 4 m from the centerline of the walkway. Background markers on the wall beside the walkway gave a "yardstick" reference so that the body coordinates could be properly scaled and used as absolute coordinates.[7]

Coordinates of the body and background markers were extracted from the film using a Numonics Digitizer.* Raw coordinate data were scaled and then corrected for parallax error between the plane of progression and the plane of the background. The absolute coordinates were transferred to the University of Waterloo IBM 370 Computer for analysis in a versatile biomechanics package developed over the past eight years.

The "noise" in these coordinate data, mainly caused by the digitizing process, was calculated to have a root mean square error of 2 mm. This random error in the x (horizontal) and y (vertical) coordinates

---

* Numonics Digitizer, Model 224–236, Numonics Corp, Hancock Stand Route 202, North Wales, PA 19454.

were removed by digital filtering,[8] resulting in smoothed data that better approximated the "true" position. Without filtering, the presence of random error in the displacement pattern would have been accentuated when the velocity and acceleration were calculated from the slope of the displacement and velocity curves, respectively. Subsequent validation of this filtering and the finite difference calculation of velocities and accelerations is supported by the study of Pezzack and colleagues.[9] Anthropometric data were obtained using tables provided by Dempster, derived from each subject's height and weight.[10]

Using the displacement and the linear and angular velocities of each body segment, the potential and kinetic energies were calculated and summed to give the total energy of each segment. These, in turn, were summed to yield the total body energy.[6] This summation gives a time history of the total energy of the body that accounts for the following: 1) all kinetic (translational and rotational) and potential energy components, 2) all energy exchanges within each segment, and 3) all energy exchanges between adjacent segments. This total body energy curve showed increases when net positive work was done by muscles of the body and showed decreases when there was negative work done as muscles absorbed energy. Over a given stride at constant speed, the body returns to the same energy level once per stride; thus, positive work per stride equals negative work per stride. The sum of this positive and negative work was calculated[11] and is called internal work.[11,12]

The work cost of walking, defined as the work per body mass per distance walked, was then calculated and correlated with the maximum knee flexion during stance. A check was made to see if there was any correlation between velocity and maximum knee flexion to eliminate velocity or cadence as variables that may have influenced the results.

RESULTS

A typical energy curve for one walking stride is shown in Figure 1. The energy of the left and right legs and of the torso, when summed at each point in time, gives the total body energy curve. Increases in body energy, as seen by ΔE_2 and ΔE_4, represent the net positive work being done by all active muscles during that time. Similarly, the decreases, ΔE_1 and ΔE_3, indicate net negative work as muscles absorb mechanical energy. A detailed description of this analysis technique and interpretation has been reported by Winter.[11]

Summing the measures of these positive and negative work components over the stride gives the work per stride. The work cost (defined as work per stride divided by stride length and body mass) was calculated. The work cost for each of the subject trials is plotted in Figure 2 against the maximum knee flexion measured during stance. A significant ($p < .01$) positive correlation ($r = .76$) between energy cost and maximum knee flexion is evident. There was neither a significant correlation between velocity and knee flexion nor between cadence and knee flexion.

DISCUSSION

The results show that the prediction of Saunders et al[2] that stiff-legged weight bearing is more energy consuming than normal flexed-knee stance does not appear to be true. In fact the reverse appears to be the case: the more we flex our knees during weight bearing, the greater the energy cost. Consequently, there seems to be no energy-related justification to train a patient to walk with a flexed knee.

Other kinetic factors should be considered in the gait training of patients. As knee flexion increases during stance, the extensor muscles at the knee, hip, and ankle must contract to limit knee flexion and to cause the knee to extend again during midstance. Such muscle activity further increases the metabolic cost of walking and also results in a drastic increase in the bone-on-bone forces at the knee and hip. These bone-on-bone forces increase from approximately body weight when the subject bears weight on a stiff knee to about five times body weight when the knee is flexed normally.[13-15] Thus, the potential for wear damage at the knee and hip joint also increases.

CONCLUSIONS

The generally held opinion that stiff-legged stance is more energy consuming than normal stance does not appear to be valid. Conversely, it appears that the energy cost increases as the angle of knee flexion during stance increases. Therefore, the reasons for training patients with pathological gait to walk with a flexed knee during stance must be reconsidered; the rationale for this approach cannot be based on energy conservation. The findings of this study suggest that in patients with certain pathological conditions, such as arthritis, knee arthroplasty, and patellectomy, knee flexion during stance will increase the bone-on-bone forces drastically and increase the potential for damage at the joint that is being rehabilitated.

Acknowledgment. The author is indebted to Paul Guy for professional assistance.

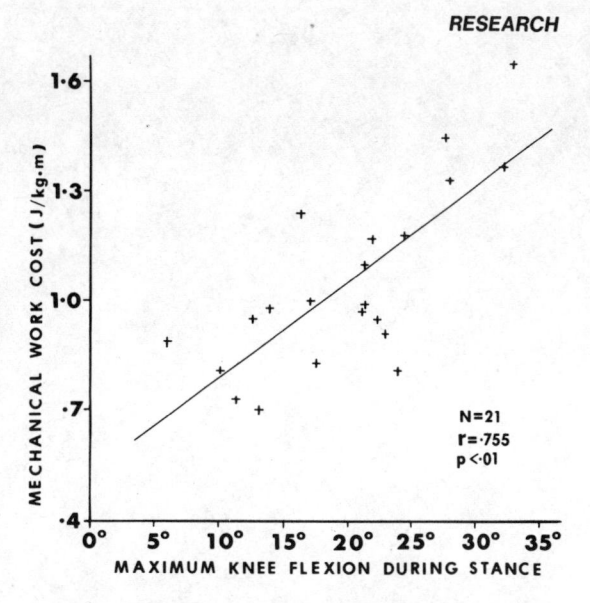

Fig. 2. Mechanical work cost of walking versus the maximum knee flexion recorded during stance during 21 trials for seven subjects without gait problems.

REFERENCES

1. Brunnstrom S: Movement Therapy in Hemiplegia. New York, NY, Harper & Row, Publishers Inc, 1970, p 123
2. Saunders JB, Inman VT, Eberhart HD: Major determinants in normal and pathological gait. J Bone Joint Surg [Am] 35:543-558, 1953
3. Liberson WT: Biomechanics of gait: A method of study. Arch Phys Med Rehabil 46:37-48, 1965
4. Cavagna GA, Margaria R: Mechanics of walking. J Appl Physiol 21:271-278, 1966
5. Inman VT: Human locomotion. Can Med Assoc J 94:1047-1054, 1966
6. Winter DA, Quanbury AC, Reimer GD: Analysis of instantaneous energy of normal gait. J Biomech 9:253-257, 1976
7. Winter DA, Greenlaw RK, Hobson DA: Television-computer analysis of kinematics of human gait. Comput Biomed Res 5:498-504, 1972
8. Winter DA, Sidwall HG, Hobson DA: Measurement and reduction of noise in kinematics of locomotion. J Biomech 7:157-159, 1974
9. Pezzack JC, Norman RW, Winter DA: An assessment of derivative determining techniques used for motion analysis. J Biomech 10:377-382, 1977
10. Dempster WJ: Space Requirements of the Seated Operator. Wright Patterson Air Force Base WADC-TR-55-159, US Government Printing Office, 1955
11. Winter DA: A new definition of mechanical work done in human movement. J Appl Physiol 46:79-83, 1979
12. Ralston HJ: Energetics of human walking. In Herman R, Grillner S, Stein RB, et al (eds): Neural Control of Locomotion. New York, NY, Plenum Press, 1976, pp 77-98
13. Morrison JB: Bioengineering analysis of force actions transmitted by the knee joint. Biomedical Engineering 3:164-170, 1960
14. Perry J, Antonelli D, Ford W: Analysis of knee joint forces during flexed knee stance. J Bone Joint Surg [Am] 57:961-967, 1975
15. Paul JP: The biomechanics of the hip joint and its clinical relevance. Proceedings of the Royal Society of Medicine 59:943-948, 1966

Basic Research

Bilateral Analysis of the Knee and Ankle During Gait: An Examination of the Relationship Between Lateral Dominance and Symmetry

This study examined the relationship between lower extremity dominance and kinematic symmetry during gait. Fourteen healthy volunteers without any observable gait deviations participated in the study. The subjects (8 male, 6 female) ranged in age from 19 to 56 years. Lower extremity lateral dominance was determined using an assessment method developed by Carol Coogler. Retroreflective spherical markers were placed bilaterally at points over the greater trochanter, the lateral joint line of the knee, the lateral malleolus, and the metatarsal break. A video-based data-acquisition instrument interfaced with a PDP 11/73 computer measured 12 kinematic variables while the subjects walked at self-selected speeds along a 10-m walkway. A multivariate analysis of variance with one repeated measure revealed significant differences between limbs, across subjects, for stance time and maximum knee extension. A within-subject analysis demonstrated significant differences for 10 variables; however, lateral dominance could not be related predictably to these variations. Our results indicate that symmetry cannot be generalized in view of intrasubject variability for these variables. [Valle DR, Gundersen LA, Barr AE, et al: Bilateral analysis of the knee and ankle during gait: An examination of the relationship between lateral dominance and symmetry. Phys Ther 69:640–650, 1989]

Key Words: Gait; Kinesiology/biomechanics, gait analysis, Kinetics; Lower extremity, general.

Lori A Gundersen
Dianne R Valle
Ann E Barr
Jerome V Danoff
Steven J Stanhope
Lynn Snyder-Mackler

L Gundersen, MSPT, is Staff Physical Therapist, Washoe Medical Center, 77 Pringle Way, Reno, NV 89520.

D Valle, MSPT, is Staff Physical Therapist, Medical College of Virginia Hospitals, Virginia Commonwealth University, 401 N 12th St, Richmond, VA 23298.

A Barr, MSPT, is Clinical Evaluator and Staff Physical Therapist, Hospital for Special Surgery, 535 E 70th St, New York, NY 10021. She was Staff Therapist, Department of Rehabilitation Medicine, Warren G Magnuson Clinical Center, National Institutes of Health, 9000 Rockville Pike, Bethesda, MD 20892, when this study was conducted.

J Danoff, PhD, is Research Consultant, Department of Rehabilitation Medicine, Warren G Magnuson Clinical Center, National Institutes of Health, and Associate Professor of Physical Therapy, Howard University, Washington, DC.

S Stanhope, PhD, is Director, Biomechanics Laboratory, Department of Rehabilitation Medicine, Warren G Magnuson Clinical Center, National Institutes of Health.

L Snyder-Mackler, MS, PT, is Assistant Professor, Department of Physical Therapy, Sargent College of Allied Health Professions, Boston University, 1 University Rd, Boston, MA 02215 (USA).

Ms Gundersen and Ms Valle were students in the Master of Science in Physical Therapy Program, Sargent College of Allied Health Professions, Boston University, when this study was conducted in partial fulfillment of the requirements of their Master of Science in Physical Therapy degree.

Address all correspondence to Ms Snyder-Mackler.

The opinions presented in this article reflect the views of the authors and not necessarily those of the National Institutes of Health or the US Public Health Service.

This article was submitted October 13, 1988; was with the authors for revision for six weeks; and was accepted March 3, 1989.

Historically, symmetry between the right and left lower extremities during gait has been assumed, usually for the sake of simplicity in data collection and analysis.[1] Numerous gait studies have relied on unilateral data collection.[2–10] Some gait experiments have been performed where both lower extremities were observed simultaneously and comments were made regarding symmetry of kinematic and kinetic variables.[11–16] Recent advancements in the methodology of motion analysis instrumentation have afforded investigators the ability to examine gait more closely than previously possible.[17] The precision of these tools has also facilitated research into such issues as the relationship between lower extremity lateral dominance and gait symmetry.[11,17] Although some authors have used these new developments to challenge the assumption of symmetry during gait, many investigators nevertheless attribute observed differences to experimental error or use data analysis methods that conceal individual variations. This study's intent was to reexamine this assumption of symmetry and to determine what role, if any, lower extremity lateral dominance played in any asymmetries observed for the variables studied.

Research on gait symmetry has focused both on kinematic and electromyographic variables. In a study by Chodera suggesting that each lower limb has a predominantly different function of control and propulsion to perform, asymmetries of stride width and foot placement angle were reported.[13] Barr et al reported asymmetries in step length, maximum knee flexion during stance, and maximum knee flexion during swing in five healthy men.[11] Hannah et al detected asymmetries in the transverse and coronal plane motions at the knee, but inexplicably attributed the reduced symmetry to the lower signal-to-noise ratio of the data for these planes (lower signal-to-noise ratio could falsely suggest symmetry, not asymmetry).[1] Leavitt et al reported slight variations in measurements of the same foot in gait as well as varying degrees of difference between both extremities. The detection of these asymmetries, however, was attributed to the increased sensitivity of gait analysis methods and the authors' recognition that gait characteristics in healthy subjects have ranges rather than fixed values.[15]

In EMG studies of muscles during gait, specific variations in amplitude and recruitment profiles, both between and within limbs, are often minimized in data analysis. Shiavi et al used ensemble averages of the linear envelopes, focusing on group trends rather than individual variability, to document normative childhood gait patterns in relation to age and walking speed.[14] Yang and Winter also analyzed ensemble averages of EMG profiles for varied cadences in adults.[5] These authors stated, "Whereas the EMG pattern changes averaged across subjects revealed a seemingly systematic trend, the individual subject responses varied greatly."[5(p490)] In describing overall EMG activity during self-selected walking speeds, Milner et al used an averaging technique to eliminate variability between successive steps.[6]

Arsenault et al reported asymmetries manifested as large differences in EMG amplitude measurements between left and right leg muscles.[12] This study also revealed occasional precise differences in the muscle activity profiles. Arsenault and colleagues chose to pool data across subjects, allowing them to report nearly perfect similarity between the profiles for a given muscle for different subjects while noting that EMG outcome (both profile and amplitude) is highly repeatable within subjects.[8] Although they did not cite so in their previous work,[12] these investigators acknowledged that pooling of data smoothes out individual peculiarities and that these ensemble averages cannot be perceived as the real profile of EMG activity.[8] Ounpuu's conclusions corroborate this finding.[18]

A few investigators have suggested that an association exists between lateral dominance and observed asymmetries in gait. Leavitt et al attempted to correlate handedness with complete footswitch closure after observing that one of the extremities took on a major portion of the body weight during double limb support.[15] Ounpuu reported a relationship between the plantar-flexion EMG profile and dominance.[18] As noted previously, Chodera suggested a difference in function between the lower limbs in gait that might result in asymmetrical recruitment profiles between the right and left rectus femoris muscles during gait.[13] Chodera reported that the dominance observed in some learned motor tasks does not appear to apply to gait (a more automatized motor task). The lack of a comprehensive definition of lateral preference is a limitation of these studies. Many authors have relied on the Harris Tests of Lateral Dominance, which involve only kicking and stamping activities (nonweight-bearing mobility functions).[19] Coogler notes in her work that a definition of lateral preference in the lower extremities must incorporate stability as well as mobility functions.[20]

In summary, several problems are apparent in the existing literature on gait analysis. First, the presence of symmetry is often assumed, as demonstrated by the preponderance of unilateral lower-limb studies. Second, authors of bilateral limb studies frequently support the notion of symmetry using data analysis techniques that minimize differences both within and across subjects. This notion is particularly true in studies that use ensemble averages. Pooling data across subjects may not be accurate or appropriate when attempting to establish the extent of gait symmetry between limbs because individual differences are concealed. Although we did not examine EMG activity in this study, we believe that these problems apply to kinematic as well as kinetic variables. Third, although several authors suggest that lateral preference or dominance may be associated with observed asymmetries, an examination of this issue based on a thorough operational definition of these concepts has yet to be completed. Studying the influence of lateral dominance, as defined by Coogler,[20] may be helpful in answering the question of gait symmetry. If lateral dominance can be correlated with documented asymmetries, then it may serve to identify differences between limbs in individuals. This may be an important consideration for the patient requiring rehabilitation of either or both lower extremities.

The purposes of this study were 1) to determine whether asymmetries exist between limbs of healthy individuals during gait and 2) to examine the relationship between lower extremity lateral dominance and any observed differences. Twelve kinematic variables of gait were measured bilaterally: step time, stance time, total gait cycle time, step length, maximum knee flexion and knee extension angles during the gait cycle, maximum ankle plantar-flexion and dorsiflexion angles during the gait cycle, and the percentage of total gait cycle time at which these latter four angles occurred. *Asymmetry* was defined as a statistically significant difference between limbs for each variable measured. Each variable was considered individually; that is, no preestablished number or combination of variables was required to determine the presence of an asymmetry. These measurements were used to test the following null hypotheses: 1) There would be no statistically significant difference between the dominant and nondominant lower limbs across subjects for each of the variables, and 2) there would be no within-subject difference between limbs.

Method

Subjects

Data were collected at the Biomechanics Laboratory, Department of Rehabilitation Medicine, National Institutes of Health in Bethesda, Md. Fourteen healthy volunteers (8 male, 6 female), aged 19 to 56 years ($\overline{X} = 30.4$ years), participated in the study. All subjects signed an informed consent form and completed a medical information questionnaire. Subjects were selected based on the absence of obvious gait deviations (as observed by the investigators), clearance from a full orthopedic and neurological screening examination, and demonstration of strong lower extremity lateral dominance.

Dominance was assessed for the mobility task of "kick-the-ball," the stability task of "unilateral-balance-with-eyes-closed," and the combined mobility-stability task of "hopping."[20] Performance in the kicking activity was rated based on the subject's ability to kick a 20-cm diameter rubber ball at a target bounded by four vertical lines taped to the wall at a distance of 5 m (Fig. 1). Unilateral balance was timed based on the subject's ability to place and maintain the dorsum of his or her nonweight-bearing foot on the popliteal space of the stance limb while keeping both hands on the hips and eyes closed (Fig. 2). Hopping ability was assessed by timing the subject as he or she hopped through a 10-m staggered course (Fig. 3). Each subject performed all tasks first with his or her preferred limb (that lower extremity chosen without instruction by the investigators); the tasks were then repeated with the opposite limb. Lateral dominance was determined based on an agreement between the subject's preferred lower extremity and his or her performance rating for each task. "Strong dominance" was indicated by two or more agreement scores. Agreement scores are summarized in Table 1. No subjects had complete agreement.

Instrumentation

The 12 variables were observed at a sampling rate of 50 Hz using a three-dimensional, video-based data-acquisition instrument* interfaced with a PDP 11/73 computer.† The *field of view* (including a calibration volume of $0.606 \times 1.813 \times 1.2$ m) was defined as one complete stride of each limb (heel-strike to ipsilateral

Fig. 1. *Dominance screen—kick-the-ball task.*

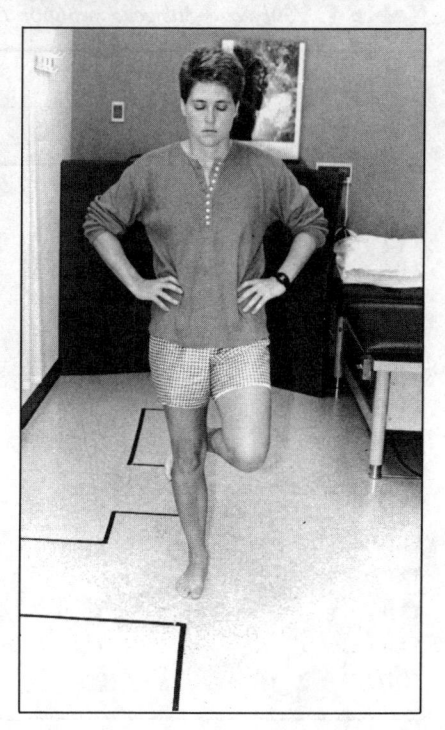

Fig. 2. *Subject demonstrating unilateral balance task.*

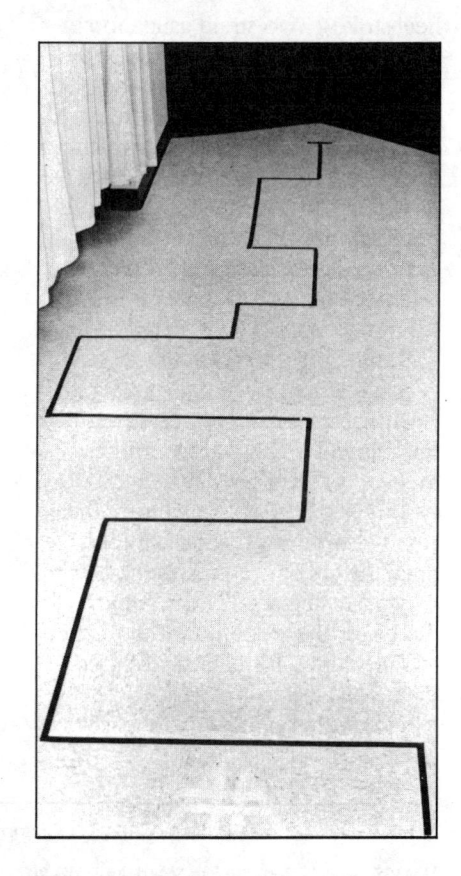

Fig. 3. *Dominance screen—hopping course task.*

---

*Oxford Metrics Ltd, Unit 8, 7 West Way, Botley, Oxford, England OX20JB.

†Digital Equipment Corp, 146 Main St, Maynard, MA 01754.

Table 1. Subject-×-Subject Scoring on Dominance Screen[a]

	Task		
Subject Number[b]	Kicking	Balance	Hopping
1 (M)	Y	Y	N
2 (F)	Y	Y	N
3 (M)	N	Y	Y
4 (M)	N	Y	Y
5 (F)	Y	N	Y
6 (M)	N	Y	Y
7 (M)	Y	N	Y
8 (M)	N	Y	Y
9 (F)	Y	N	Y
10 (F)	Y	N	Y
11 (M)	Y	Y	N
12 (M)	N	Y	Y
13 (F)	Y	N	Y
14 (F)	Y	N	Y
TOTAL	9	8	11

[a] Y = yes, N = no.
[b] Subject's sex in parentheses.

Fig. 4. Path of progression of gait analysis with force plates.

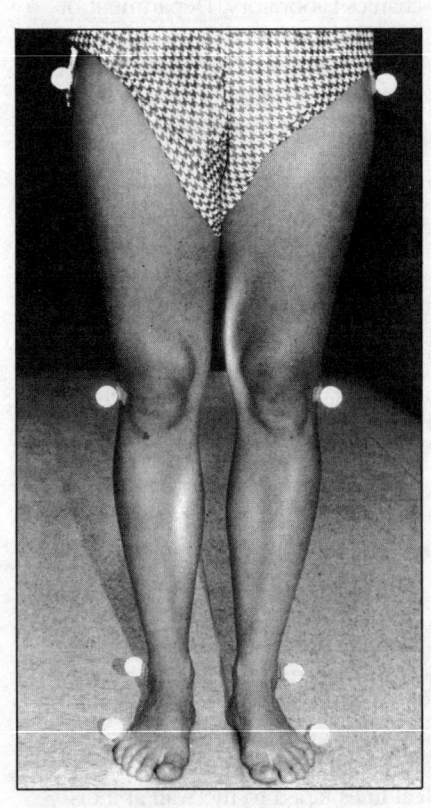

Fig. 5. Marker placement shown by infrared camera.

heel-strike). Two strain gauge force plates[‡] located 0.1 m apart along the path of progression within the calibration volume were used to define the time (within 0.02 second) at which gait cycle events occurred for each limb (Fig. 4). Five cameras were calibrated to maximal errors (the camera's average accuracy in locating a reference marker in space) less than 5.0 mm ($\bar{X}$ = 3.21 mm, range = 1.39–4.99 mm) prior to data collection. Each camera also went through a linearization process. The resolution of this particular configuration relative to the field of view is estimated to be 1 part in 3,000. In absolute terms, this resolution means that for a 3-mm field, the system can resolve to 1 mm. Obviously, the smaller the volume, the better the resolution. The theoretical resolution for angular displacement will obviously vary with the length of the segment and the calibration error. Conservatively, for an average thigh length of 500 mm and a maximum error of 5.0 mm, the resolution of angular displacement would be approximately 0.5 degree. Once collected, data sets were transferred to a VAX 11/750 computer system[†] for processing. Adtech reduction software[§] was used to convert raw data recorded from the video system into three-dimensional trajectories and to process analog signals, which allowed automatic determination of all measurements.

Procedure

Following subject selection, 25.4-mm diameter spherical retroreflective markers[‖] were affixed bilaterally to points over the greater trochanter, the lateral tibiofemoral joint line just proximal to the fibular head, the lateral malleolus, and the lateral aspect of the head of the fifth metatarsal (Fig. 5). Reference angles in standing were recorded by the system as 180 degrees at the knee and 0 degrees at the ankle. A universal goniometer,

---

[‡] Advanced Mechanical Technology, Inc, 141 California St, Newton, MA 02158.

[§] AMASS, Adtech, 2002 Ruatan St, Adelphi, MD 20783.

[‖] 3M, Medical-Surgical Div, Bldg 225-5S, 3M Center, St Paul, MN 55144-1000.

positioned with the arms aligned with the markers, was used to measure the difference between marker alignment and the 0-degree goniometric (anatomical) position at the knee and ankle. This procedure also served as a check on marker placement. The mean offset was 2.92 ± 1.49 degrees for the right knee and 3.00 ± 1.52 degrees for the left knee. The mean difference between the right and left offsets was 0.71 ± 0.61 degree. The mean offset was 10.78 ± 4.73 degrees for the right ankle and 10.00 ± 4.49 degrees for the left ankle. The mean differences between right and left sides were not significant at either the knee or the ankle. Marker placement, therefore, appears to be standard bilaterally. Each subject was then instructed to walk at his or her preferred cadence for 10 m along a walkway. Arms were allowed to swing freely. After a two- or three-minute warm-up period, 8 to 10 trials per subject were recorded and measured for all variables. A trial consisted of one complete stride of each lower limb within the field of view with a full stride hitting the force plates. Subjects were requested to walk at a self-selected speed. They were not instructed regarding the force plates. The knee and ankle were treated as a single axis joint in this study. All data were compressed into the sagittal plane, and a bilateral sagittal-plane analysis was performed.

A total of five trials per subject were selected from the recorded trials to be used in the data analysis. Selection was predicated on identification of a complete data set.

Data Analysis

Corresponding data points between limbs of each subject were compared using a two-way multivariate analysis of variance (MANOVA) with one repeated measure (trials). We used a MANOVA because the variables tested were clearly not independent of one another. Means for subject and limb (main effects) and for subject × limb (interaction effect) were tested by the MANOVA. A Newman-Keuls *post hoc* examination was performed on subject means for variables showing a significant within-subject difference between limbs. We used the Newman-Keuls *post hoc* test because it is a very conservative test. Pearson product-moment correlations (r) were calculated to examine the relationships among the 12 measured variables. A .05 level of confidence was used in all cases to determine significance.

Table 2. *Sample Mean Values by Variable*

Variable	Limb	
	Dominant	Nondominant
Gait cycle time (sec)	1.08	1.08
Maximum knee flexion (°)[a]	120.51	120.05
% gait cycle—maximum knee flexion	73.86	74.16
Maximum knee extension (°)[a]	184.59[b]	182.81[b]
% gait cycle—maximum knee extension	5.39	8.97
Maximum ankle dorsiflexion (°)[c]	90.33	89.98
% gait cycle—maximum ankle dorsiflexion	42.67	41.29
Maximum ankle plantar flexion (°)[c]	122.34	121.27
% gait cycle—maximum ankle plantar flexion	60.51	63.26
Step time (sec)	0.54	0.54
Stance time (sec)	0.70	0.70
Step length (mm)	696.52	696.66

[a] Measured as the posterior knee angle.
[b] Significant at $p < .05$.
[c] Measured as the anterior ankle angle.

Results

The MANOVA revealed significant F ratios in the main effect of limb for maximum knee extension. The first null hypothesis was rejected for this variable, but it was not rejected for the 11 variables showing no significant difference. Pooling the data, therefore, suggests no difference between limbs for a majority of the variables studied. Mean sample values for the 12 variables are listed in Table 2. A representative MANOVA summary is shown in Table 3.

Table 3. *Representative Multivariate Analysis with One Repeated Measure for Maximum Knee Extension During Gait Cycle*

Source	df	SS	MS	F	p
Subject	13	1962.22	150.94	32.34	.0001
Error	52	242.72	4.67		
Limb	1	122.02	122.02	21.35	.0099
Error	4	22.86	5.72		
Subject × limb	13	1184.55	91.12	31.44	.0001
Error	52	150.69	2.90		

Table 4. *Distribution for Intrasubject Differences by Variable as Determined by Newman-Keuls Post Hoc Analysis*[a]

Variable	Subject Number
Gait cycle time	1, 4, 5, 8, 11
Maximum knee flexion	4, 5
% gait cycle—maximum knee flexion	
Maximum knee extension	1, 4, 5, 8, 9, 10, 11, 12, 14
% gait cycle—maximum knee flexion	
Maximum ankle dorsiflexion	1, 2, 4, 5, 7, 10, 11
% gait cycle—maximum ankle dorsiflexion	3, 7, 11
Maximum ankle plantar flexion	3, 5, 12, 14
% gait cycle—maximum ankle plantar flexion	1, 6
Step time	5, 7, 8, 9, 12
Stance time	1, 6, 8, 12
Step length	1, 3, 4, 6, 8, 12, 14

[a] Significant at $p < .05$.

Significant differences in the subject-×-limb interaction were demonstrated for total gait cycle time, maximum knee flexion, maximum knee extension, maximum ankle dorsiflexion, percentage of the total gait cycle time at which maximum ankle dorsiflexion occurred, maximum ankle plantar flexion, percentage of the total gait cycle time at which maximum ankle plantar flexion occurred, step time, stance time, and step length. These results lead to rejection of the second null hypothesis stating there would be no within-subject difference for these variables. This hypothesis was not rejected for percentage of total gait cycle time at which maximum knee flexion and maximum knee extension occurred.

Newman-Keuls *post hoc* analysis of the 10 variables demonstrating a difference in the subject-×-limb interaction showed that only 3 of these variables' differences occurred in 7 or more of the subjects (Tab. 4). Furthermore, these differences within subjects were never unidirectional. That is, the direction of these differences varied between the dominant and nondominant limbs of the subjects showing a difference for a particular variable (Tab. 5). These results accept the second null hypothesis.

Cross-correlations were calculated among the 12 variables. The correlations and probabilities are listed in Table 6.

Discussion

Although our results did not support the existence of systematic asymmetry during gait, interesting observations regarding asymmetry can be made based on examination of intrasubject data. Symmetry may be disputed based on the frequency of within-subject differences occurring between limbs for each variable. In an attempt to draw meaningful conclusions from these results, two points should be noted: 1) Differentiation between clinical and statistical significance was required, and 2) normative baselines derived from pooled subject data were incomplete without recognition of the variability revealed by analysis of gait within subjects.

Examination of the variables depicting a difference in the subject-×-limb interaction indicates that symmetries occurred in an unpredictable fashion. Some subjects demonstrated a difference in stance time but not in other temporal events; other subjects demonstrated a difference in maximum knee angle measurements but not in

Table 5. *Newman-Keuls Post Hoc Analysis Mean Values for Subject-×-Limb Interaction for Maximum Knee Extension During Gait Cycle*[a]

Subject Number	Limb[b]	$\bar{X}$
1	d	184.65[c]
	n	180.78
2	d	176.84
	n	178.97
3	d	181.15
	n	179.89
4	d	189.60[d]
	n	192.46
5	d	187.06[c]
	n	171.56
6	d	185.46
	n	185.99
7	d	181.37
	n	183.49
8	d	186.49[c]
	n	182.35
9	d	180.92[c]
	n	184.42
10	d	185.94[c]
	n	181.12
11	d	191.53[c]
	n	188.10
12	d	194.61[c]
	n	182.95
13	d	182.73
	n	182.84
14	d	176.82[c]
	n	183.78

[a] Based on total of five trials per subject.
[b] d = dominant limb, n = nondominant limb.
[c] $p < .01$.
[d] $p < .05$.

maximum ankle angle measurements. Twelve of the 14 subjects, however, did demonstrate a significant difference for at least two of the variables. These variations possibly represent subtle adjustments that are part of the continual fine-tuning of gait. This accommodation may occur without dramatically disrupting the kinematic chain in a predictable fashion. The

Basic Research

weak correlation between revealed asymmetries for range of motion at the knee and ankle supports this possibility. This finding in itself was surprising. Cross-correlations resulted in significant correlation coefficients of less than .32 for combined angle-angle comparisons. One consideration is that the actual ROM through which the knee and ankle pass during gait is a small percentage of the total ROM available to both joints. More dramatic changes in one joint's ROM, as in the case of contracture or paralysis, may be required for another joint to be affected. Sutherland et al concurred with this hypothesis when they noted an increase in both ankle dorsiflexion and knee flexion during gait in limbs receiving a tibial nerve block to temporarily inhibit ankle plantar-flexion

Table 6. *Temporal and Angular Variable Correlations*

Variable	Combined		Dominant Limb		Nondominant Limb	
	r	p	r	p	r	p
Gait cycle time × step time	.88	.001	.90	.001	.87	.001
Gait cycle time × stance time	.93	.001	.94	.001	.93	.001
Step time × stance time	.79	.001	.80	.001	.79	.001
Maximum knee flexion × maximum ankle plantar flexion	−.23	.01			−.55	.001
Maximum knee flexion × maximum ankle dorsiflexion						
Maximum knee extension × maximum ankle plantar flexion						
Maximum knee extension × maximum ankle dorsiflexion	.32	.0001	.32	.007	.32	.005

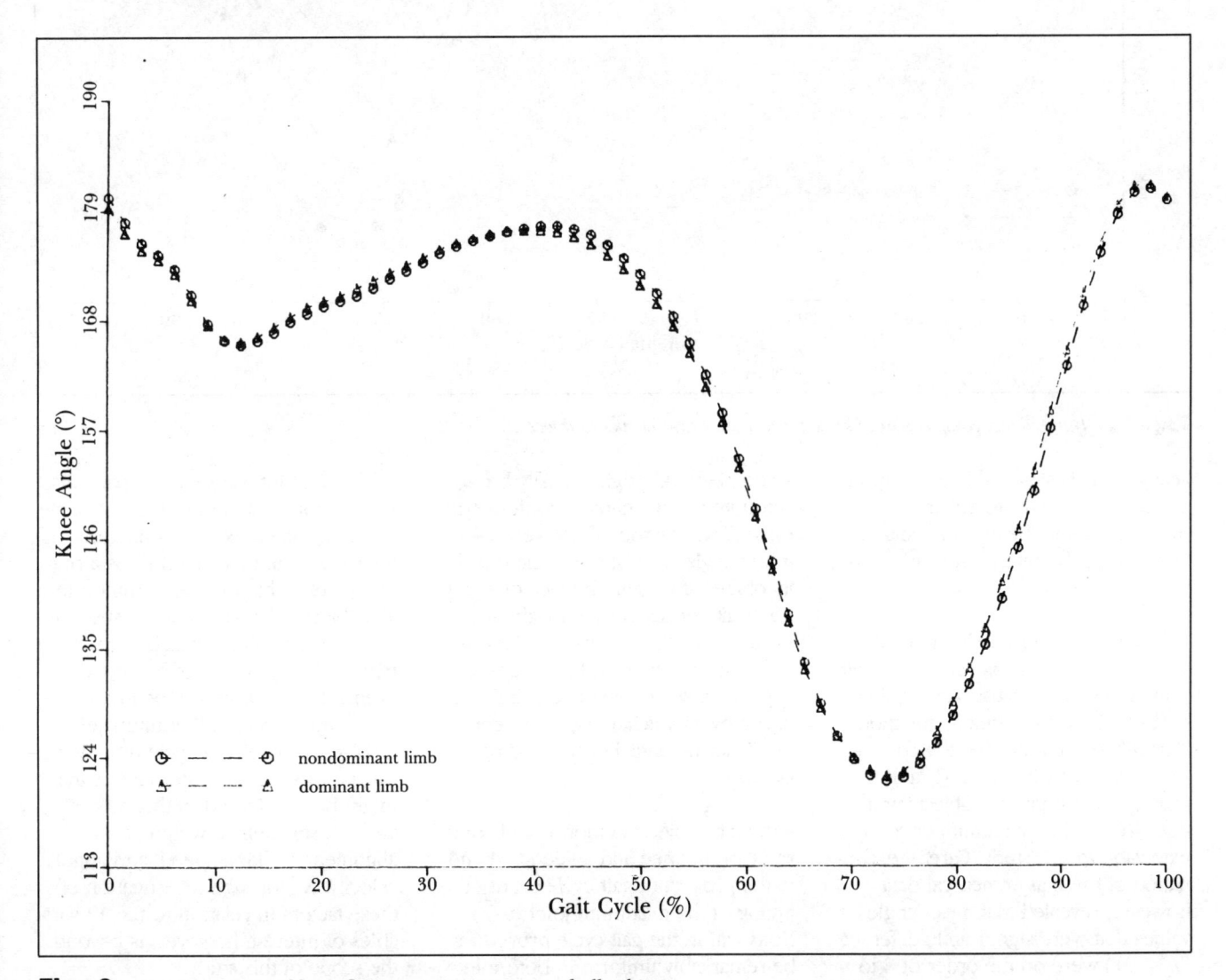

Fig. 6. *Sample means of knee flexion-extension curves of all subjects.*

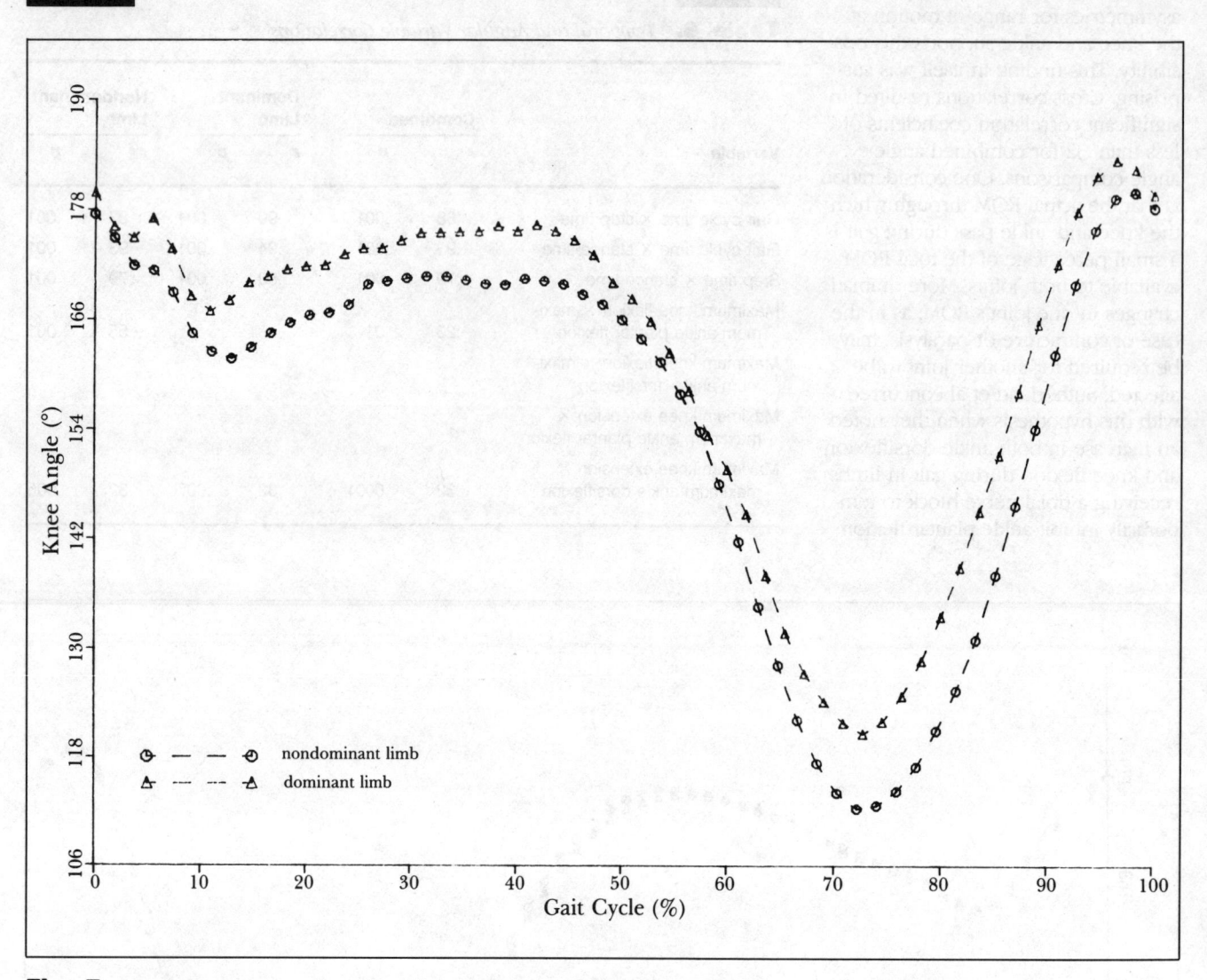

Fig. 7. *Intrasubject comparison of knee flexion-extension curves (Subject 10).*

activity.[21] This simulated pathological condition appeared to enhance the interrelationship between knee and ankle angles by interfering with control of the kinematic chain.

The physical therapist should distinguish between what is clinically significant and what is statistically significant when studying human function. This difference may alter the conclusions we derive from our data. For example, greatest intrasubject variation occurred in maximum knee extension in this study. Careful examination of these asymmetrical data, however, revealed that mean angle values that were significantly different ($p < .01$) were on the order of 8 to 10 degrees and might actually be accounted for by clinical goniometric error (goniometric offsets were used in the angle calculations). The equality of the offsets and the lack of significant differences between right and left adjustments, however, make this possibility unlikely. Interpretation of any asymmetry's relevance is influenced by a clinician's ability to control other potentially confounding factors.

Although subjects demonstrated some variation in knee and ankle ROM and in the phasing of gait cycle events, the timing of the maximum joint excursions within the gait cycle proved to be remarkably uniform in both inter-subject and intrasubject comparisons. Regardless of variations in ROM at the ankle and the knee, the timing of their maximum excursions has a regular pattern that may be common to all subjects. This observation suggests a model for walking based, in part, on temporal sequencing of gait cycle events that allow variations in the spatial components but that ultimately resolve to a total gait cycle time typical across subjects. Other factors that might be considered in this model are age, sex, height, weight, bony alignment, walking speed, and angular velocity. A thorough investigation of these factors in relation to the 12 variables of interest, however, is beyond the scope of this study.

Basic Research

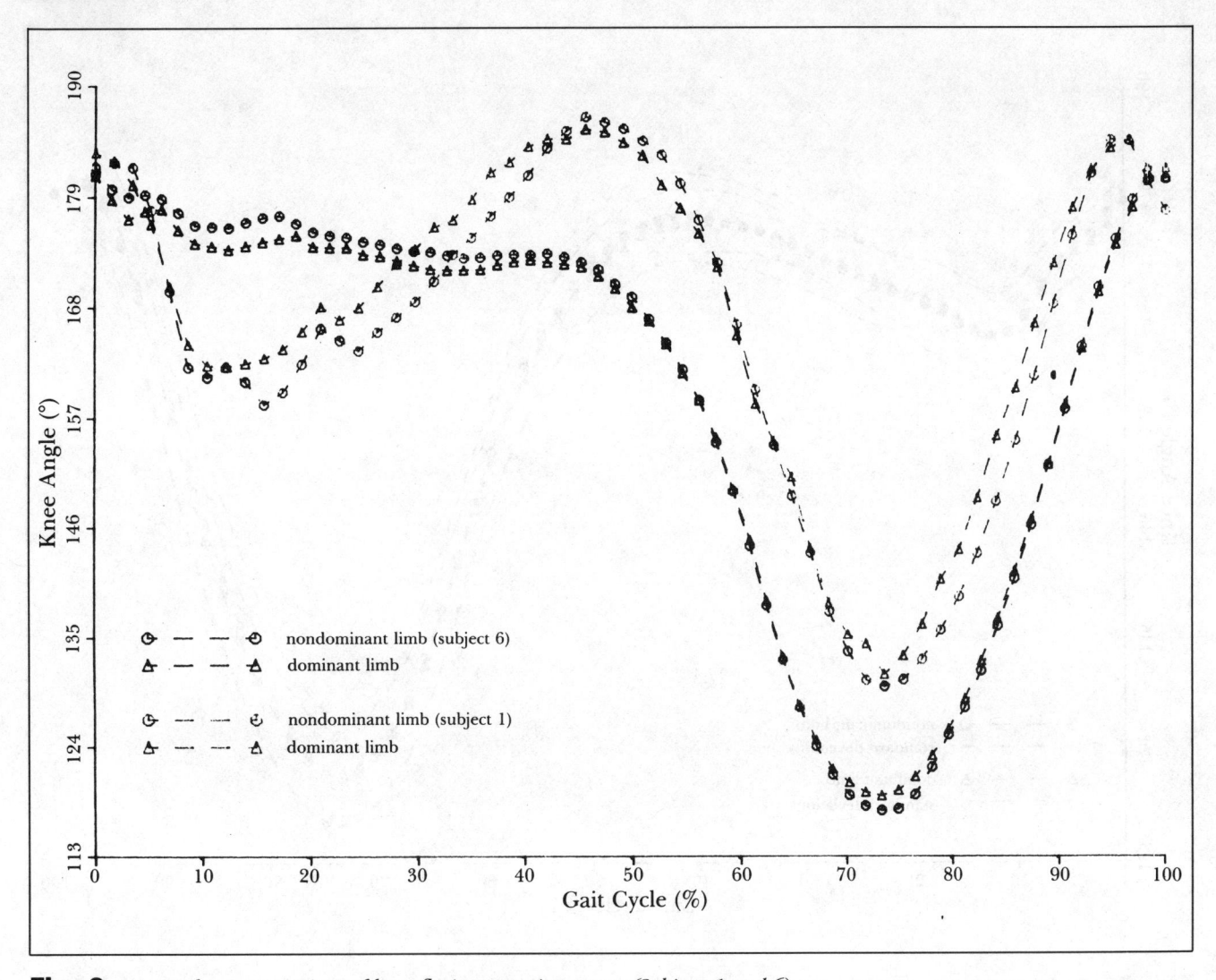

Fig. 8. *Intersubject comparison of knee flexion-extension curves (Subjects 1 and 6).*

Lateral dominance is one factor that was thought to play an important role in understanding asymmetries in gait. Several investigators have suggested that the lower extremities serve varied functional purposes.[11,18] Arsenault et al hypothesized that different functional uses might lead to different recruitment profiles for the dominant leg versus the nondominant leg.[12] Our study was also based on the assumption that a subject pool demonstrating strong lateral dominance in the lower extremities would most likely show differences related to this descriptor. Although some asymmetries during gait were demonstrated, however, they could not be correlated with lateral dominance.

As Arsenault et al later suggested, dominance may not play a role in tasks that are as automatized as gait.[12] Although our test for dominance was more thorough in including both stability and mobility tasks, these are not actual events occurring in the gait cycle. Although they may serve to indicate dominance in some individuals, dominance may not influence noncortical automative activities for which both legs are required to carry out the same sequence of tasks. An alternative experience, however, may be that a healthy person's typical gait pattern may not be stressful enough to emphasize lateral dominance in the lower extremity. No deprivation of visual, proprioceptive, or tactile feedback occurs in healthy subjects while they walk, yet each of these systems was taxed when the subjects kicked, hopped, and balanced during the lateral dominance screening. Lateral dominance in the lower extremity, therefore, may be regarded as a reflection of a person's ability to better accommodate with one leg than the other to such systemic disruption. As Kerstein et al noted, lower extremity amputees tend to rehabilitate faster if the nondominant limb is the affected limb.[22] Apparently, the dominant limb is better able to compensate for the interrupted proprioception and tactile input. Reproducing similar results in healthy subjects during gait may require a simulated

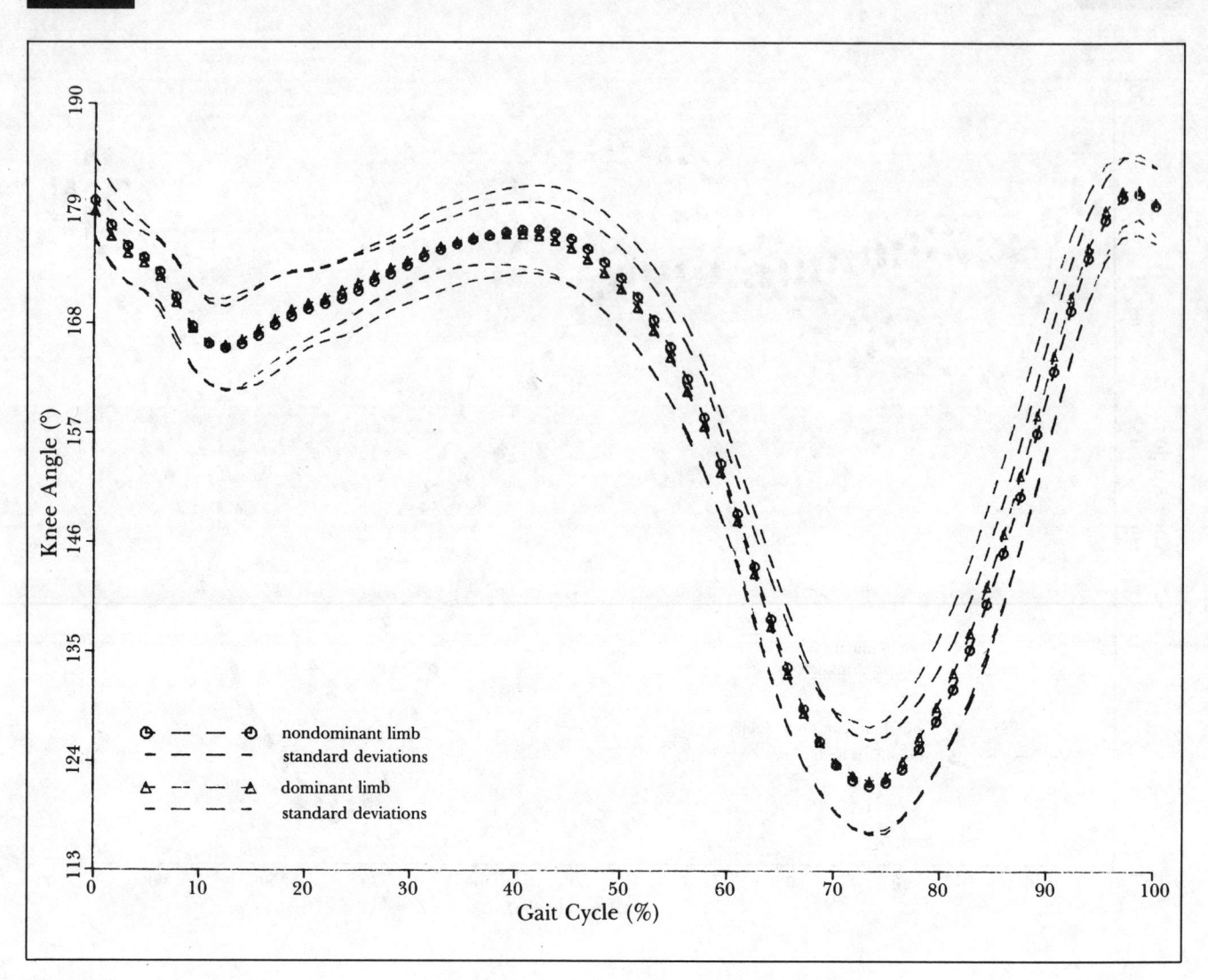

Fig. 9. *Sample means with standard deviations of knee flexion-extension curves of all subjects.*

pathological condition such as sensory deprivation.

Overall, our results suggest that neither symmetry nor asymmetry in these variables can be generalized across subjects. Symmetry should not be used in absolute terms when applied to human function. Rather, a model for gait may be described that incorporates a continuum of symmetry in which asymmetries result from accommodation to changing environmental factors. This model challenges the assumption made by investigators who have conducted unilateral limb studies or pooled data across subjects that right and left legs present symmetrical outcomes during gait.[2–10] As

Yang and Winter pointed out, averaging data across subjects may reveal an apparent systematic trend toward symmetry despite considerable individual variability.[5] Arsenault et al concluded that such differences could carry meaningful biological information, but they usually are easily averaged out when the data are pooled across subjects.[12] Averaging tends to produce symmetry across subjects. Figure 6, which illustrates sample means for knee ROM throughout the gait cycle, shows that when data are pooled, all subjects' combined dominant limbs appear to be similar to the nondominant limbs for this variable. Hannah et al[1] and Murray et al[16] have used these data analysis methods to describe nor-

mative baselines to serve as reference data for pathological gait study. In our estimation, however, such descriptions are incomplete without individual subject analysis. Figures 7 and 8, for example, demonstrate that both intrasubject and intersubject variability are prevalent. Furthermore, the degree of standard deviation in the sample mean curves is great among healthy subjects (Fig. 9); these ranges may be greater for subjects with pathological gait. Such information serves to further refine a working model of gait.

Clinically, our study implies that normative baselines derived from data pooling are not completely representative of all "normal" individuals and, therefore, should be used cautiously as reference data for pathologic gait study. Additional research is necessary to create a method for comparing normative data with data from subjects with pathological conditions. Other gait variables such as angular velocity and walking speed should be considered. Additionally, studies comparing samples of individuals demonstrating lateral dominance with those who do not are needed to clearly delineate the role of lower extremity lateral dominance in bilateral activities such as gait.

Conclusion

Observations from individual subject analysis should be used to increase our understanding of gait in healthy subjects. Symmetry can neither be assumed nor generalized for the variables studied but must be examined in view of idiosyncratic subject variability.

References

1 Hannah RE, Morrison JB, Chapman AE: Kinematic symmetry of the lower limbs. Arch Phys Med Rehabil 65:155–158, 1984

2 Elliot BC, Blansky BA: Reliability of averaged integrated electromyograms during running. Journal of Human Movement Studies 2:28–35, 1976

3 Sutherland DH, Olshen R, Cooper L, et al: The development of mature gait. J Bone Joint Surg [Am] 62:336–353, 1980

4 Crowinshield RD, Brand RA, Johnston RC: The Effects of walking velocity on hip kinematics and kinetics. Clin Orthop 132:140–144, 1978

5 Yang JF, Winter DA: Surface EMG profiles during different walking cadences in humans. Electroencephalogr Clin Neurophysiol 60:485–491, 1985

6 Milner M, Basmajian JV, Quandry AO: Multifactorial analysis of walking by electromyography and computer. Am J Phys Med 50:235–258, 1971

7 Dubo HIC, Peat M, Winter DA, et al: Electromyographic temporal analysis of gait: Normal human locomotion. Arch Phys Med Rehabil 57:415–420, 1976

8 Arsenault AB, Winter DA, Marteniuk RG: Is there a "normal" profile of EMG activity in gait? Med Biol Eng Comput 24:337–343, 1986

9 Arsenault AB, Winter DA, Marteniuk RG, et al: How many strides are required for the analysis of electromyographic data in gait? Scand J Rehabil Med 18:133–135, 1986

10 Cavanagh PR, Gregor RJ: Knee joint torque during the swing phase of normal treadmill walking. J Biomech 8:337–344, 1975

11 Barr A, Andersen JC, Danoff JV, et al: Symmetry of Temporal, Spatial and Kinematic Events During Gait. Read at the Third Annual East Coast Gait Laboratory Conference, Bethesda, MD, 1987

12 Arsenault AB, Winter DA, Eng P, et al: Bilateralism of EMG profiles in human locomotion. Am J Phys Med 65:1–16, 1986

13 Chodera JD: Analysis of gait from footprints. Physiotherapy 60:179–181, 1974

14 Shiavi R, Green N, McFadyen B, et al: Normative childhood EMG gait patterns. J Orthop Res 5:283–295, 1987

15 Leavitt LA, Zuniga EN, Clavert JC, et al: The development of mature gait. South Med J 64:1131–1138, 1971

16 Murray MP, Drought AB, Kory RC: Walking patterns of normal men. J Bone Joint Surg [Am] 46:335–360, 1964

17 Woltring HJ: Data Acquisition and Processing Systems in Functional Movement Analysis. Read at the Third National Congress of the Italian Society of Biomechanics in Orthopaedics and Traumatology, Ancona, Italy, 1986

18 Ounpuu S: Bilateral Analysis of the Lower Limbs During Walking in Normal Individuals. Master's Thesis. Waterloo, Ontario, Canada, University of Waterloo, 1986

19 Harris AJ: Harris Tests of Lateral Dominance. New York, NY, The Psychological Corp, 1955

20 Coogler CE: Lateral Differences in Lower Extremity Preference and Performance of Healthy Right-Handed Women. Doctoral Dissertation. Boston, MA, Boston University, 1983

21 Sutherland DH, Cooper L, Daniel D: The role of the ankle plantarflexors in normal walking. J Bone Joint Surg [Am] 62:354–363, 1980

22 Kerstein MD, Zimmer H, Dugdale FE, et al: Successful rehabilitation following amputation of dominant versus nondominant extremities. Am J Occup Ther 31:313–315, 1977

Basic Research

Study of Normal Men During Free and Fast Speed Walking

Patterns of Sagittal Rotation of the Upper Limbs in Walking

M. P. MURRAY, Ph.D., S. B. SEPIC, B.S. and E. J. BARNARD, M.S.*

FOR THE PAST several years we have been involved in extensive studies of walking patterns of normal men under various conditions of walking speed. The patterns studied include the excursions of the head, neck, trunk, and lower limbs.[1-5] Our purpose has been to determine the ranges of normal values for these excursions which could serve as standards for comparison of the disordered gait patterns of patients with locomotor disabilities. Recently we have begun to study the gait patterns of patients with selected bone and joint, or neurological disorders. In the course of appraising the gait of patients with Parkinson's Disease, and those with antalgic patterns due to hip pain, we noted that their arm swings appeared

From the Kinesiology Research Laboratory, Physical Therapy Section, Physical Medicine and Rehabilitation Service, Veterans Administration Center, Wood, Wis.; the Curriculum in Physical Therapy, Marquette University School of Medicine, Milwaukee, Wis.; and the School of Physical Therapy, University of Connecticut, Storrs, Connecticut.

This investigation was supported in part by Grant CD 00052-03 from the Division of Chronic Diseases, United States Public Health Service, Department of Health, Education and Welfare.

* Author supported by a Graduate Education Traineeship awarded by the Vocational Rehabilitation Administration, U. S. Department of Health, Education and Welfare.

grossly distorted. In an attempt to find normal standards for reference, the literature was reviewed. This disclosed very limited information on the nature of normal arm swing during locomotion.

In 1895 Braune and Fischer reported on the rotation patterns of the shoulder and elbow of one normal man during three walking trials.[6,7] In 1939 Elftman, using the data from Braune and Fischer's subject, calculated the torques exerted by muscles acting on the upper limbs during walking in an attempt to determine whether arm swing was brought about mainly by muscle activity or by pendular action.[8] In 1940 several Russian investigators under the direction of Bernstein reported on various aspects of upper limb movement during walking of children, unilateral amputees, and of men from sixty to one-hundred-and-four years of age.[9-11] In 1961 Drillis showed the shoulder rotation pattern of one normal subject and one subject in the early stages of aging.[12]

Most recently, the excellent study of Fernandez-Ballestreros, et al. in Copenhagen reported the electromyograms of shoulder muscle activity of twenty-three subjects during natural walking.[13] In addition they described the angular excursion of a line from the shoulder to the wrist of three subjects walking on a treadmill at a cadence of 100 steps per minute.

Since adequate standards for normal arm swing patterns during walking were not found in the literature, our present study was undertaken. Its purpose is to identify the ranges of normal variability in the upper limb displacement patterns. It is hoped that these data will provide useful baselines for comparison of the disordered gait patterns of patients with various disabilities.

MATERIAL AND METHODS

Photographic and Targeting Procedure

The patterns of sagittal rotation of the upper limbs were recorded by means of interrupted-light photography. The subjects, with reflective targets secured to specific anatomic landmarks, walked in the illumination of a strobe-light flashing at 0.05 second intervals. The details of the photographic and targeting procedures have been described in previous papers.[1-3] For this study the upper limb rotation was measured from images of targets marking the following three points: (1) the lateral aspect of the upper part of the humerus in line with the gleno-humeral joint center; (2) the lateral epicondyle of the humerus; and (3) the base of the third metacarpal bone. Shoulder flexion and extension were measured from the angles formed between a vertical reference line and a line joining the two targeted points on the humerus representing the mechanical axis of the arm. Elbow flexion and extension were measured from the angles formed between an inferior projection of the mechanical axis of the arm and a line joining the targeted points on the lateral epicondyle and the base of the third metacarpal bone. The reference diagrams of Figure 1A and 1B depict the angles measured.

ROTATION OF UPPER LIMBS IN WALKING

Subjects

The subjects were thirty normal men ranging in age from twenty-one to sixty-six years, and in height from 61.5 to 73 inches. There were six men in each of five age groups: twenty-one to twenty-five; thirty-two to thirty-five; forty-two to forty-eight; fifty-three to fifty-eight; and sixty-three to sixty-six years of age. Each age sub-group included two subjects in each of the following three height categories: short, from 61.5 to 67 inches; medium, from 68.5 to 70 inches; and tall, from 71 to 73 inches. Twenty-six of the men were right-handed; four were left-handed.

Walking Conditions

Upper limb rotation patterns were recorded for both free and fast speed walking trials. The mean velocity for free speed walking of the thirty men was 154 ± 23 cm. per second and their velocity for fast walking was 214 ± 26 cm. per second. All measurements were taken throughout one complete walking cycle, which is defined as the time interval between successive instants of heel-strike of the same lower limb (lt.-to-lt., or rt.-to-rt.). The mean duration of the walking cycles for the thirty subjects was 1.05 ± 0.08 seconds for free speed walking, and 0.88 ± 0.59 seconds for fast speed walking. This is a mean cadence of 114 steps per minute for free speed walking, and 136 steps per minute for fast speed walking.

RESULTS

Mean Patterns of Upper Limb Rotation for Free and Fast Speed Walking

The mean patterns of shoulder and elbow rotation throughout a walking cycle for the thirty normal men are shown in Figure 1. Upward deflections on the ordinate scale of Figure 1 represent flexion, and downward deflections, extension. The upper limbs swing forward and backward throughout each walking cycle. The direction of the upper limb motion coincides closely to that of the contralateral lower limb and is opposite to the direction of the motion of the ipsilateral lower limb.

1. Shoulder: Figure 1A shows that at the beginning of the walking cycle when the ipsilateral lower limb is projected forward for initial heel-strike, the shoulder is extended maximally in the backward direction. The first half of the walking cycle is characterized by forward flexion of the shoulder (segment Ex-Fl) as the ipsilateral lower limb extends progressively backward with respect to the trunk. The peak of shoulder flexion is reached at approximately the midpoint of the walking cycle, the time that the contralateral heel strikes the floor. The latter half of the cycle is characterized by shoulder extension (segment Fl-Ex) as the ipsilateral lower limb flexes forward for the swing phase. By the time the ipsilateral limb reaches its next heel-strike position, completing the walking cycle, the shoulder has returned to maximum extension.

The shoulder pattern in Figure 1A shows that the arm is extended behind the vertical reference line throughout most of the time in the walking cycle and is flexed in front of the reference line for relatively brief periods. Moreover, the amplitude of peak shoulder extension (0 and 100% of the cycle) is greater than the amplitude of peak shoulder flexion (50% of the cycle).

The amplitude of the total excursion of shoulder rotation was highly variable from one subject to another at both walking speeds. Despite this high degree of variability among the subjects, the total excursion of shoulder rotation for each subject was greater for fast speed walking than for free speed walking. The mean total excursion of shoulder rotation was 39 ± 12 degrees for fast speed walking, significantly greater than 32 ± 10 degrees for free speed walking (P<0.01). However, the increased amplitude of shoulder rotation seen for faster walking was provided almost exclusively by increased shoulder extension on the backward arc of the swing rather than by increased shoulder flexion on the forward arc of the swing (Fig. 1A). The mean amplitude of peak shoulder extension was 31 ± 8 degrees for fast walking in contrast to 24 ± 6 degrees for free speed walking; whereas the mean amplitude of peak shoulder flexion was 8 ± 11 and 8 ± 10 degrees for fast and free speed walking, respectively. The small differences between these mean values and those shown in Figure 1 are the result of slight individual variations in the times of occurrence of peak flexion and extension.

2. Elbow: Figure 1B shows that the elbow pattern is similar to the shoulder pattern in

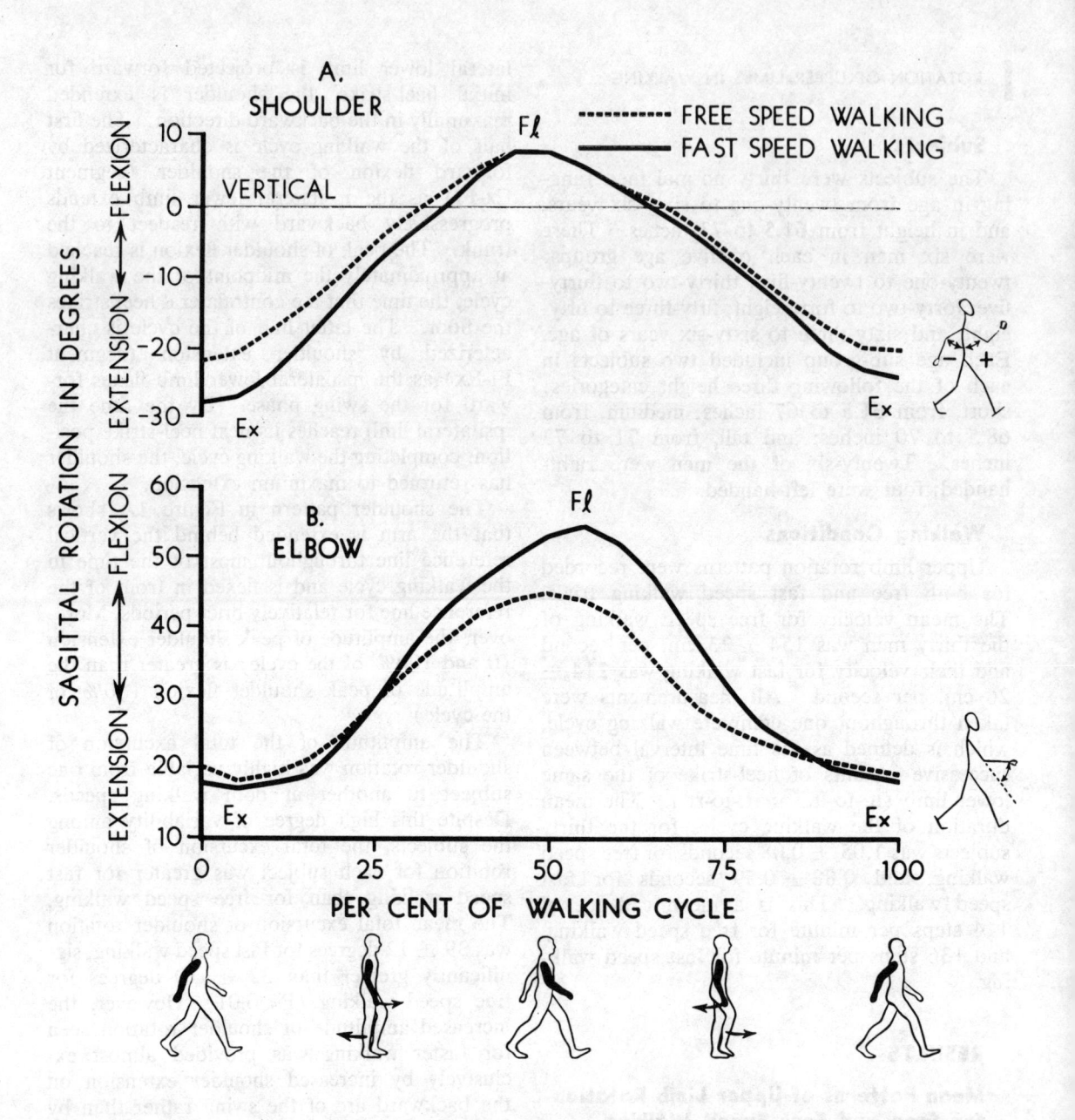

FIG. 1. Mean patterns of sagittal rotation of the shoulder and elbow throughout the walking cycles for free and fast speed walking of thirty normal men. The values shown represent the means of two patterns of left upper limb rotation, and one of the right upper limb from three successive walking trials of each subject at each of the two walking speeds.

Upward deflections on the ordinate scale represent flexion (Fl); downward deflections, extension (Ex). The reference diagram of Fig. 1A shows that when the arm was directed obliquely behind the vertical (0 reference line) the angles of relative shoulder extension were recorded as negative values on the ordinate: when the arm was directed forward from the vertical, the relative flexion angles were recorded as positive values. The reference diagram of Figure 1B depicts the angle measured for elbow rotation.

Basic Research

ROTATION OF UPPER LIMBS IN WALKING

that there is one flexion excursion (segment Ex-Fl) and one extension excursion (segment Fl-Ex) within each walking cycle. For the first half of the cycle the elbow flexes forward with the shoulder, and for the last half of the cycle, the elbow extends with the shoulder.

The amplitude of the total excursion of elbow rotation was also highly variable among the walking subjects and was greater for fast speed walking than for free speed walking ($P<0.01$). The mean total excursion of elbow rotation for fast speed walking was 40 ± 13 degrees, as compared to 30 ± 11 degrees for free speed walking. However unlike the shoulder, the significant increase in the total excursion of elbow rotation for the faster walking speed was provided mainly by increased flexion on the forward end of the arc of arm swing rather than by increased extension on the backward end of the arc. The mean amplitude of peak elbow flexion was 55 ± 12 degrees for fast speed, in contrast to 47 ± 11 degrees for free speed walking. The amplitudes of peak elbow extension were 15 ± 8 degrees and 17 ± 8 degrees for fast and free speed walking, respectively.

Examples of Upper Limb Rotation Patterns of Individual Subjects

In view of the high degree of variation seen in the amplitude of the shoulder and elbow rotation patterns among the thirty men tested, it seems appropriate to show some of the different patterns seen among the individual subjects. The patterns of shoulder and elbow rotation for three successive walking trials of four individual subjects are shown in Figure 2. Differences in these excursions can be seen in both the configuration of the patterns of rotation and in the amplitudes of flexion and extension.

Some subjects showed rapid reversals from flexion to extension as can be seen in the sharply peaked flexion waves in the central part of the cycle (Subject #4, Fig. 2). Others sustained their flexion position, delaying the reversal into extension, as shown by the plateaus in the flexion wave (shoulder patterns, Subjects #2 and #3). There were similar differences in the shape of the extension waves among the subjects although these are more difficult to demonstrate since the peak of extension occurred at the heel-strike times which marked the beginning or end of the cycles selected for study.

An unexpected finding was a small accessory wave of flexion superimposed on the extension excursion (segment Fl-Ex) of the elbow patterns of some of the individual subjects. Examples of this accessory flexion wave can be seen in the elbow patterns of the first and third walking trials of Subject #2. The accessory flexion wave occurred more frequently for free speed walking than for fast speed walking. It appeared in thirty-eight of the ninety observations for free speed walking and in only nine for fast walking. The accessory flexion wave was not highly reproducible. It occurred in all three free speed walking trials of five subjects; in two of three trials for eight subjects; and in only one of three trials for seven subjects.

Pronounced differences in the total amplitude of shoulder and elbow rotation can also be seen among the individual patterns in Figure 2. Some men showed relatively large amplitudes of both shoulder and elbow rotation as exemplified in the patterns of Subject #4. Others, as can be seen for Subject #2, showed small amplitudes of both shoulder and elbow rotation. Moreover, some men showed large amplitudes of shoulder rotation and small amplitudes of elbow rotation (Subject #1); and others showed small amplitudes of shoulder rotation and large amplitudes of elbow rotation (Subject #3).

Pearson product moment correlation coefficients were calculated to identify relationships between the amplitudes of shoulder and elbow rotation within the same walking trials. These values are based on sixty walking trials (two for each of the thirty men) at each of the two walking speeds. The coefficient values between the amplitudes of the total excursion of shoulder rotation and total excursion of elbow rotation within a given walking trial were close to zero, confirming the lack of relationship between these two components for the individual subjects.

An unexpected positive relationship was seen for the peak degrees of rotation between the two joints at the opposite ends of the arc of the arm swing. The subjects who showed a greater degree of shoulder flexion on the forward sweep tended to show a greater degree of elbow extension on the backward sweep ($r = .60$ and $.54$ for free and fast speed walk-

FIG. 2. Patterns of sagittal rotation of the shoulder and elbow from three successive free speed walking trials of four normal men. Upward deflections on the ordinate scale represent flexion; downward deflections, extension.

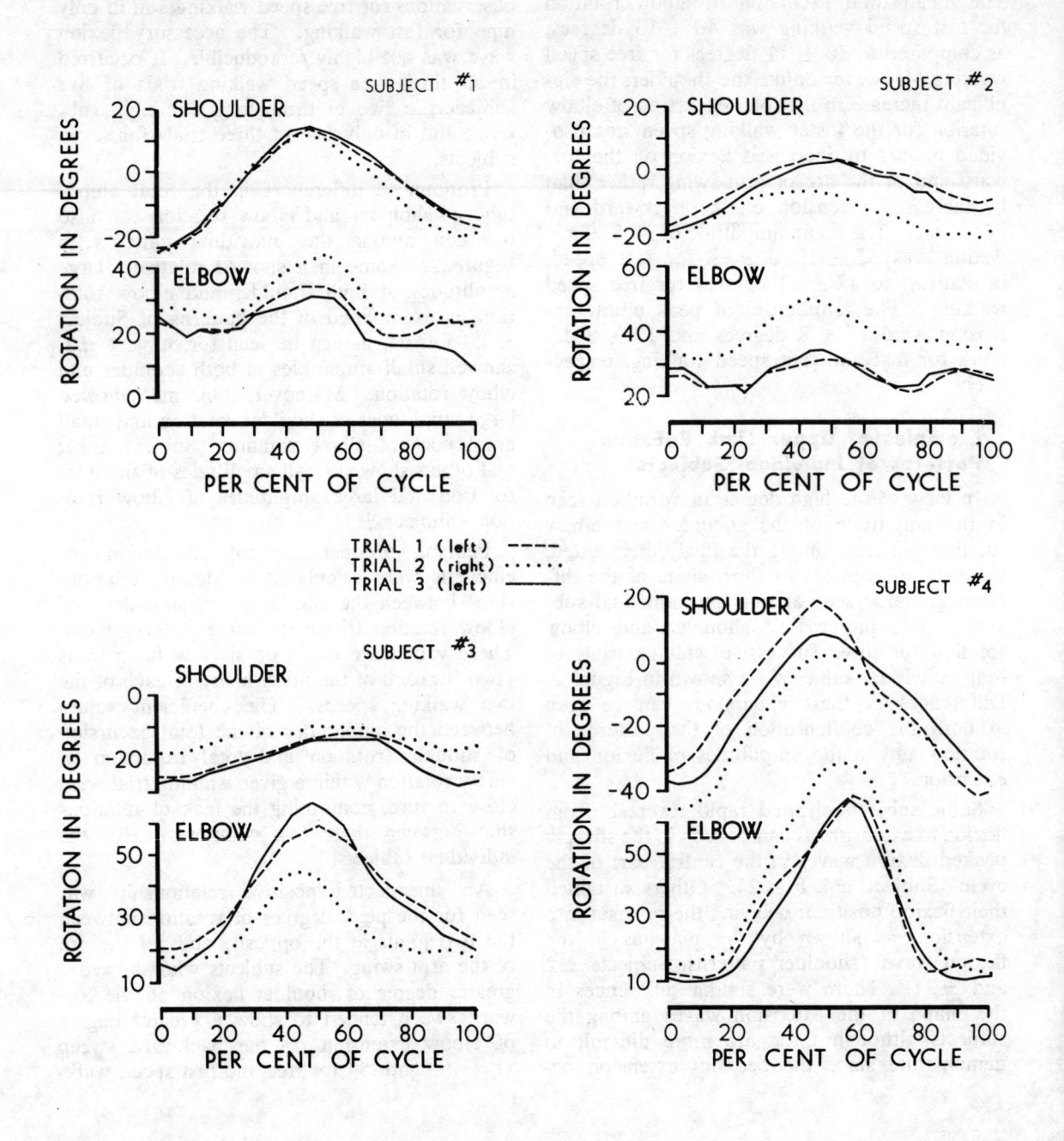

TRIAL 1 (left) -----
TRIAL 2 (right) .....
TRIAL 3 (left) ———

ROTATION OF UPPER LIMBS IN WALKING

ing, respectively). Similarly the subjects who showed a greater degree of shoulder extension on the backward sweep, tended to show a greater degree of elbow flexion on the forward sweep (r = .67 and .46 for free and fast speed walking, respectively).

Reproducibility of Shoulder and Elbow Rotation for Successive Walking Trials of Individual Subjects

The individual arm swing patterns of three successive walking trials (first, left; second, right; and third, left) were evaluated in order to assess the reproducibility of shoulder and elbow rotation. Successive trials for the same (left) limb were more highly reproducible than successive trials of the right and left limbs. Elbow rotation on repeated walking trials tended to be slightly more similar than shoulder rotation.

For free speed walking the mean difference between the amplitudes of the total excursion of left and right shoulder rotation was 3.1 ± 12.9 degrees, and the difference between successive measures of left shoulder rotation was 0.5 ± 5.1 degrees. For fast speed walking, the mean differences in total excursion of rotation were 0.0 ± 14.8 degrees between the right and left shoulders, and 1.6 ± 5.4 degrees for successive trials of the left shoulder.

For free speed walking, the mean difference in the amplitudes of the total excursion of rotation between the left and right elbows was 1.7 ± 9.7 degrees and the difference between successive trials of left elbow rotation was 0.4 ± 5.4 degrees. For fast speed walking, the mean difference in total excursion of rotation between the right and left elbows was 0.6 ± 12.2 degrees; and the difference between successive measures of left elbow rotation was 0.9 ± 7.2 degrees.

The mean total excursions of *shoulder* rotation were slightly less for the dominant than non-dominant limbs for free speed walking, averaging 30 and 33 degrees, respectively. Conversely the mean total excursions of *elbow* rotation were slightly greater for the dominant than non-dominant limbs, averaging 31 and 29 degrees, respectively. For fast speed walking the total excursions of rotation for the dominant and non-dominant upper limbs were notably similar.

Temporal Relationships of Upper Limb Rotation

Three aspects of arm swing patterns were studied to identify the nature of the temporal relationships of upper limb movements during walking. These were: (1) the sequence of reversal in direction of rotation for the shoulder and elbow of the *same* upper limb; (2) the relationship between the time of maximum flexion of the upper limb and the instant of heel-strike of the contralateral lower limb; and (3) the relationship between the time of maximum extension of the upper limb and the instant of heel-strike of the ipsilateral lower limb. For each study, the patterns from three successive walking trials (two for the left upper limb and one for the right) were analyzed for each of the thirty subjects at each of the two walking speeds.

1. The Sequence of Reversal in Direction of Rotation for the Shoulder and Elbow of the Same Extremity: In order to identify the sequence of reversal in direction of rotation, it was necessary to evaluate both the termination of the flexion excursion and the initiation of the extension excursion because of the plateaus seen in many of the flexion waves. The shoulder preceded the elbow in reversal in direction of rotation most frequently for both the free and fast speed walking trials. For the combined walking speeds the flexion excursion ended first for the shoulder in 111 observations; first for the elbow in thirty-two observations; and for the shoulder and elbow simultaneously in thirty-seven observations. Similarly the initiation of the extension excursion occurred first for the shoulder in 114 observations; first for the elbow in thirty-eight; and simultaneously for the shoulder and elbow in twenty-eight observations.

2. The Relationship Between the Time of Maximum Flexion of the Upper Limb and the Instant of Contralateral Heel-Strike: The histograms in Figure 3A show the relationship between the times of maximum shoulder and elbow flexion and the instant of contralateral heel-strike as measured at 0.05 second intervals. It can be seen that both the shoulder and elbow were most frequently in maximum flexion on the forward arc of the arm swing at the instant of contralateral heel-strike. However, this temporal relationship was more precise for the shoulder than for the elbow. For the combined

FIG. 3. Histograms showing the temporal relationships of the times of maximum upper limb flexion with respect to contralateral heel-strike; and the times of maximum upper limb extension with respect to ipsilateral heel-strike. The values are based on 180 observations for both the shoulder and elbow from three walking trials for each of the thirty normal men at each of the two walking speeds.

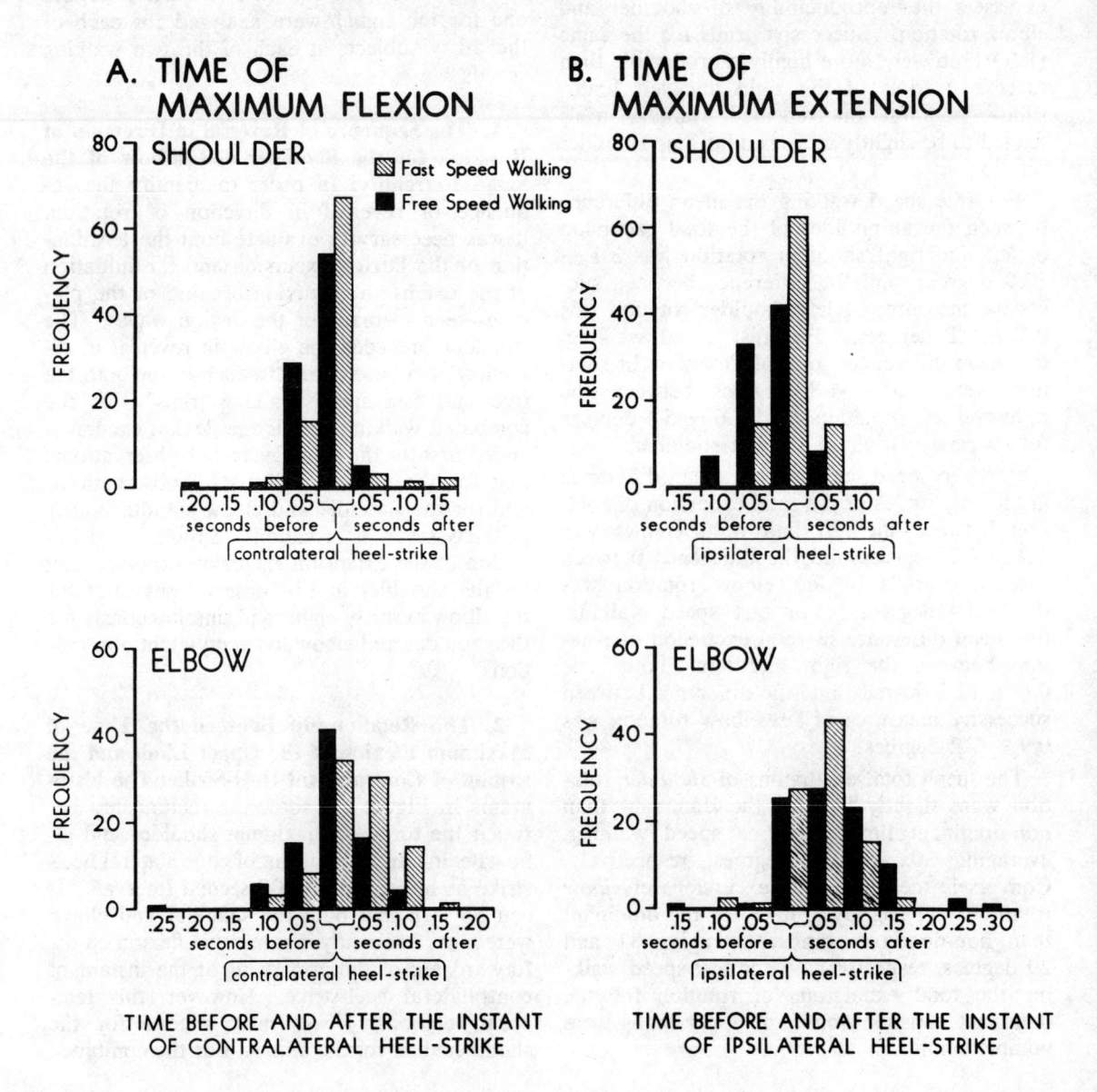

ROTATION OF UPPER LIMBS IN WALKING

free and fast speed walking trials, in 67 per cent of the observations the shoulder was in maximum flexion at the same instant as contralateral heel-strike; whereas in 42 per cent the elbow was in maximum flexion at the instant of contralateral heel-strike. Figure 3A also shows that the frequency distribution of the time of maximum flexion with respect to the instant of contralateral heel-strike is skewed for both the shoulder and elbow. However, the skewness is in opposite directions for the two joints. As expected, maximum shoulder flexion occurred more frequently before the time of heel-strike than after heel-strike; and maximum elbow flexion occurred more frequently after the time of heel-strike than before heel-strike. For the combined free and fast walking speeds in 27 per cent of the observations the maximum shoulder flexion occurred before the time of contralateral heel-strike as compared to 6 per cent after heel-strike. In contrast, only 18 per cent of the observations of maximum elbow flexion occurred before contralateral heel-strike as compared to 40 per cent after. In 173 of 180 observations the shoulder was in maximum flexion during the interval from 0.05 second before contralateral heel-strike to 0.05 second after it. In 146 of 180 observations, the elbow was in maximum flexion during the interval from the instant of contralateral heel-strike to 0.10 second after it.

3. The Relationship Between the Time of Maximum Extension of the Upper Limb and the Instant of Ipsilateral Heel-Strike: The histograms in Figure 3B show that there is also a close relationship between the times of maximum shoulder and elbow extension on the backward end of the arc of arm swing and the instant of ipsilateral heel-strike. Again, the temporal relationship was more precise for the shoulder than the elbow. In addition it can be seen that the skewness of the frequency distributions of the shoulder and elbow is even more pronounced for the extension end of the arc than for the flexion end (Fig. 3). For combined free and fast speed walking, the shoulder was most frequently in maximum extension at the instant of ipsilateral heel-strike (58% of the observations) whereas the elbow was most frequently in maximum extension 0.05 second after ipsilateral heel-strike (39% of the observations). In 173 of the 180 observations the shoulder was in maximum extension during the interval from 0.05 second before the instant of ipsilateral heel-strike to 0.05 second after it. In 160 of the 180 observations the elbow was in maximum extension during the interval from the instant of ipsilateral heel-strike to 0.10 second after it.

Amplitudes of the Total Excursions of Upper Limb Rotation for the Age and Height Groups

The mean amplitudes of the total excursions of shoulder and elbow rotation for the three height groups and five age groups are shown in Table 1. These values are based on two walking trials for each of the thirty men, at each of the two walking speeds.

1. Age: There were no statistically significant differences in the amplitudes of the total excursions of shoulder and elbow rotation for the five age groups. Although not statistically significant, the total excursions of both shoulder and elbow rotation was less for the sixty-year-old men than for the younger men for fast speed walking (Table 1).

2. Height: There were no statistically significant differences in the amplitudes of the total excursions of shoulder rotation for the three height groups at either walking speed. However, the total excursions of elbow rotation related systematically and significantly with the three height groups but only for the free speed walking trials ($P<0.05$). For free speed walking the tall subjects showed the greatest total excursion of elbow rotation, and the short subjects showed the least. We found that these differences in the total excursions of elbow rotation were due more to differences in the amplitude of elbow flexion on the forward swing of the arm than to elbow extension on the backward swing.

DISCUSSION

Since man is able to walk effectively with the upper limbs encumbered or clasped uselessly in front of or behind the trunk, arm swing has the appearance of a facultative component of locomotion. However when the upper limbs are free, their orderly participation in the swinging movement is an integral part of the total pattern of gait. Even though arm swing is not a prerequisite, some degree of flexion and extension of the shoulder and elbow oc-

TABLE 1

MEAN TOTAL EXCURSIONS OF UPPER LIMB ROTATION
FOR FREE AND FAST SPEED WALKING OF 30 NORMAL MEN

Subject Groups	Number of Observations	Amplitudes of the Total Excursion of Shoulder Rotation		Amplitudes of the Total Excursion of Elbow Rotation	
		Free Speed	Fast Speed	Free Speed	Fast Speed
Total	60*	32±11**	40±13	30±11	41±13
Age Groups					
Twenty-year-old men	12	28±11	40±16	35±10	42±14
Thirty-year-old men	12	30±9	39±9	26±9	36±9
Forty-year-old men	12	39±8	44±13	34±13	44±9
Fifty-year-old men	12	30±9	40±8	30±8	46±18
Sixty-year-old men	12	32±12	37±14	27±9	34±7
Height Groups					
Tall	20	31±10	35±12	34±8	41±10
Medium	20	33±9	42±12	32±10	42±13
Short	20	31±12	42±14	25±11	38±14

* Two observations at each of the two walking speeds.
** One standard deviation.

curred during every walking trial of every subject selected for study. Other evidence of the built-in nature of these associated movements is presented by Fernandez-Ballestreros who found step-related activity in the shoulder muscles when the arms of walking subjects were bound at their sides. Thus, she suggests that the innervation of the muscles during arm swing is part of a centrally determined pattern.[13]

The nature of arm swing is such that the two upper limbs rotate forward and backward reciprocally and concurrently. The direction of the forward and backward swing of the upper limbs is similar to the direction of movement of the contralateral lower limb and opposite to the direction of the ipsilateral lower limb. The forward swing is characterized by concurrent flexion of both the shoulder and elbow, and the backward swing by concurrent extension of the two joints. There is a tendency for the shoulder to precede the elbow in reversal of direction of rotation. The time of change in direction of rotation of both joints of the upper limb is closely related to the instants of contralateral and ipsilateral heel-strike.

The patterns of upper limb rotation are the most variable of the twenty gait components we have studied for normal men,[1-5] and the variability involves mainly the amplitude of rotation. Some of our normal subjects characteristically showed large excursions of arm swing whereas others showed small excursions. Moreover, some subjects showed large excursions of shoulder rotation and small excursions of elbow rotation; whereas for others the reverse was true. These differences are not surprising since arm swing is not a prerequisite component of normal gait. In view of the high degree of variability in amplitude among the subjects tested, the arm swing patterns of the individual subjects were surprisingly reproducible. However, the reproducibility was more pronounced for successive patterns of the same extremity than between the patterns of the right and left extremities. Thus normal subjects tend to show limb preference in upper extremity patterns. This limb preference has no apparent relationship to limb dominance. In previous studies we found no evidence of limb preference for the lower extremity patterns of walking.

In contrast to the high degree of variation in the amplitude of arm swing, we found the temporal relationships between the upper and lower limb movements to be stereotyped. The nature of the association is such that both the shoulder and elbow are in maximum flexion on the forward end of the arc of arm swing at the time the *contralateral* heel strikes the floor and in maximum extension on the backward end of the arc at the time the *ipsilateral* heel strikes the floor. In 355 of the 360 measurements the shoulder and elbow were in maximum flexion within 1/10 of a second of the instant of contralateral heel-strike; and in 334 of

ROTATION OF UPPER LIMBS IN WALKING

the 360 measurements the shoulder and elbow were in maximum extension within 1/10 of a second of the instant of ipsilateral heel-strike. This relatively precise temporal relationship is one of the few distinctive characteristics of normal arm swing.

In view of the consistency in occurrence of arm swing during walking, one might question the functions that this motion subserves. In certain sports activities such as broad jumping, the forward upward thrust of the arms augments the momentum of the body. Although the momentum-producing mechanism may be an important function for such bilateral activities as jumping, it is doubtful that arm swing produces this function in walking. If the forward-swinging upper limb augmented the forward momentum of a walking subject, the simultaneous backward swing of the opposite upper limb would negate this influence. It is interesting to note that Fernandez-Ballestreros reported absence of electrical activity in the shoulder flexor muscles whereas the shoulder extensors showed consistent activity before the end of the forward swing and during the backward swing.

We agree with other investigators who suggest that one of the functions of arm swing is to counteract excessive trunk rotation in the horizontal plane.[8, 13, 14] That the arms may serve as compensatory balancing beams in the sagittal and frontal planes is apparent from watching a child's earliest attempts at unassisted walking, or from watching a subject in the process of losing his balance. Admittedly such precarious imbalance is rarely present in normal walking. However, during normal walking, the pelvis and thorax do rotate simultaneously in opposite directions in arcs parallel to the horizontal plane.[1, 2] The pelvis rotates forward on the side of the forward-swinging *lower* limb as the thorax simultaneously rotates forward on the side of the contralateral forward-swinging *upper* limb. Thus arm swing during normal walking could provide a useful counterbalancing force against excessive horizontal rotation of the entire trunk in the same direction as the rotation of the pelvis. This supposition is further substantiated by the relationships between the trunk and extremities with increased walking speed. The longer step length of faster walking is accomplished by increased obliquity of the out-stretched lower limbs as well as increased horizontal rotation of the pelvis.[2] With the faster walking speed the upper limbs and thorax also rotate through greater arcs but in directions opposite to that of the pelvis and lower limbs. Thus the greater excursions of the lower limbs and pelvis may be counterbalanced by the more vigorous arm swing of faster walking. It appears that the greater the stress, the greater is the contribution of arm swing to the biomechanical needs of locomotion.

We would like to suggest that normal arm swing may also function to modulate the amplitude of the vertical excursion of the center of gravity of the entire body during walking. Previous studies have indicated that the mass center of the body oscillates vertically through two peaks and two valleys during each walking cycle.[7] The peaks occur during the time of single-limb support when the upper limbs are in relatively vertical positions passing the trunk: the valleys occur during double-limb support when the upper limbs are outstretched obliquely forward and backward. Using the mean values for body weight and for the rotational patterns of our thirty test subjects, we calculated the mean vertical trajectory of the mass center of the entire upper limb throughout the walking cycle. We found that the mass center of each upper limb also oscillates vertically through two peaks and two valleys within each walking cycle. However, there was an inverse relationship between the directions of the vertical trajectory of the mass center of the entire body and that of the upper limbs. The summits of the upper limb trajectories occurred when the trajectory of the body was at a valley; and similarly valleys in the upper limb trajectories occurred when the trajectory of the body was at a summit. Therefore, arm swing may decrease the total vertical excursion of the center of gravity of the entire body by decreasing its height during the peaks and elevating its height during the valleys of the trajectory. The decreased vertical excursion of the total body may decrease the energy expenditure of gait.

The appearance of the small accessory elbow flexion wave occurring during the elbow extension excursion was totally unexpected. When we first observed its occurrence, we were concerned that it might be an artifact caused by either parallactic distortion or by wrist motion because of the target placement on the third metacarpal bone. However, we plotted the elbow rotation patterns from the tables of Braune and Fischer and found that this subject also showed an accessory flexion wave.⁶ Their data were obtained from elaborate photographic records taken simultaneously from three cameras set at various angles in order to correct for photographic distortion. Moreover, his distal target was placed proximal to the wrist, and therefore ruled out wrist motion. Whether this accessory elbow flexion wave is a rebound due to passive forces or to a reflex contraction of the stretched elbow flexor muscles awaits further study.

Several investigators have reported changes in arm swing patterns with advanced age. Spielberg reported that for his subjects in ages sixty to one-hundred-and-four years, the upper limb activity was progressively reduced and the synergy between arm and leg movement was disturbed.¹¹ Specifically, his subjects in the presenile age group, from sixty to seventy-two years, showed a retarding of movement of the upper parts of the body with respect to the lower parts. On the other hand, Drillis illustrated the shoulder pattern of one subject in an early stage of aging which showed no disturbance in the temporal relationship between upper and lower limb movement, but this subject did show a markedly diminished amplitude of shoulder flexion.¹²

In three previous gait studies we found that certain patterns of the trunk and lower extremities of our sixty-year-old men differed from those of younger men, suggesting a presenile pattern.¹⁻³ In view of our previous studies as well as the reports of Spielberg and Drillis, careful scrutiny was given to the arm swing patterns of our oldest subjects. Neither the timing nor the amplitude of the arm swing patterns of our sixty-year-old men were found to differ significantly from those of the younger men. However, the total amplitudes of both shoulder and elbow rotation for our sixty-year-old men were less than those of the younger men, particularly for fast speed walking. Although these differences were not statistically significant, the decreased amplitude of arm swing may represent another component of a presenile pattern. We are presently evaluating the gait patterns of males in their eighth and ninth decades of life to determine if the presenile patterns previously suggested are consistent and progressive with advanced age.

SUMMARY

The patterns of sagittal rotation of the shoulder and elbow during free and fast speed walking were recorded by means of interrupted-light photography. The subjects selected for study were thirty normal men in age groups from twenty-one to sixty-six years and height groups from 61 to 73 inches.

Some excursion into flexion and into extension of both the shoulder and the elbow occurred for every walking trial of every subject tested. Although the amplitudes of the arm swing patterns were highly variable from one subject to another, the patterns for successive walking trials of the individual subjects tended to be similar. The reproducibility of the individual patterns was more pronounced for repeated measures of the same limb than between the right and left limbs.

The amplitude of rotation was significantly greater for the faster walking speed. This greater amplitude was achieved mainly by increased shoulder extension on the backward end of the arc of arm swing and increased elbow flexion on the forward end of the arc.

Another distinctive characteristic of normal arm swing was a close temporal relationship between upper and lower limb movements. Normal arm swing is characterized by the shoulder and elbow being in maximum flexion at the time of contralateral heel-strike and in maximum extension at the time of ipsilateral heel-strike.

The sixty-year-old subjects showed decreased amplitudes of shoulder and elbow rotation, particularly for fast speed walking, which may suggest a presenile pattern.

REFERENCES

1. Murray, M. P., A. B. Drought, and R. C. Kory, Walking patterns of normal men. J. Bone Joint Surg., 46A:335–360, March 1964.
2. Murray, M. P., R. C. Kory, B. H. Clarkson, and S. B. Sepic, A comparison of free and fast speed walking patterns of normal men. Amer. J. Phys. Med., 45:8–24, February 1966.
3. Murray, M. P., and B. H. Clarkson, The vertical pathways of the foot during level walking: I. Range of variability in normal men. J. Amer. Phys. Ther. Ass., 46:585–589, June 1966.
4. Murray, M. P., and B. H. Clarkson, The vertical pathways of the foot during level walking: II. Clinical examples of distorted pathways. J. Amer. Phys. Ther. Ass., 46:590–599, June 1966.
5. Murray, M. P., Gait as a total pattern of movement. Amer. J. Phys. Med., (in press).
6. Braune, C. W., and O. Fischer, Der Gang des Menschen: I. Teil. Abhandl. d. Math.-Phys. Cl. d. k. Sächs Gesellsch. Wissensch., 21:153–322, 1895.
7. Fischer, O., Der Gang des Menschen: II. Teil. Die Bewegung des Gesammtschwerpunktes und die Äussern Kräfte. Abhandl. d. Math.-Phys. Cl. d. k. Sächs Gesellsch. Wissensch., 25:1–163, 1899.
8. Elftman, H., The functions of the arms in walking. Hum. Biol., 11:529–536, 1939.

ROTATION OF UPPER LIMBS IN WALKING

9. Popova, T., The biodynamics of the child's independent walking. In N. A. Bernstein, ed.: Investigations on the Biodynamics of Walking, Running, and Jumping. Central Scientific Institute of Physical Culture, Moscow, 1940, part II, chap. 3, pp. 29–34.
10. Salzgeber, O., Co-ordination of movements of the amputated walking with prosthesis. In N. A. Bernstein, ed.: Investigations on the Biodynamics of Walking, Running, and Jumping. Central Scientific Institute of Physical Culture, Moscow, 1940, part II, chap. 13, pp. 68–70.
11. Spielberg, P. I., Walking patterns of old people: Cyclographic analysis. In N. A. Bernstein, ed.: Investigations on the Biodynamics of Walking, Running, and Jumping. Central Scientific Institute of Physical Culture, Moscow, 1940, part II, chap. 15, pp. 72–76.
12. Drillis, R. J., The influence of aging on the kinematics of gait. In The Geriatric Amputee. A report on a conference sponsored by the Committee on Prosthetics Research and Development, National Research Council. NAS–NRC Publication 919, 1961.
13. Fernandez-Ballestreros, M. L., F. Buchthal, and P. Rosenfalck, The pattern of muscular activity during the arm swing of natural walking. Acta Physiol. Scand., 63:296–310, March 1965.
14. Weber, W., and E. F. Weber, Mechanik der menschlichen Gehwerkzeuge. Dietrich, Gottingen, 1836.

Acknowledgement

We wish to thank Betty Clarkson, B.S. and Jerome Hensen for assistance in the photographic and analytical procedures, and F. A. Hellebrandt, M.D. for suggestions in the preparation of this manuscript. The illustrations were prepared by Carole Russell. The Medical Illustration Service of the V. A. Center, Wood, Wisconsin, prepared the enlarged photographs from which the measurements were made. The Marquette University Computer Center provided facilities for data processing.

Basic Research

Upper-extremity Muscular Activity at Different Cadences and Inclines during Normal Gait

RAYMOND E. HOGUE, Ph.D.

▶ *Electromyography was used to determine the activity of fifteen upper-extremity muscles at the cadences of 70, 95, and 120 steps per minute on inclines of zero and 15 degrees. An intramuscular copper wire served as the active electrode. The activity was transmitted via telemetry. Specific action potentials were integrated and recorded on graph paper. Major activity occurred in the posterior deltoid, middle deltoid, and teres major muscles. The fifteen subjects increased their activity as they walked faster, but not as they ascended steeper inclines. Although the upper extremities act as pendulums, the pendular action is caused both by muscular activity and by gravity.* ◀

THIS INVESTIGATION attempted to determine by electromyography the amount of upper-extremity muscular activity during normal gait.

Professional journals in the fields of physical therapy, physical education, orthopedics, and physical medicine have published articles on the function of the lower extremities during gait. However, little material is available on the role of the upper extremities. Kinesiology texts offer little help; some authors consider the upper extremities as pendulums during gait, and others hold the opposite view.

Dr. Hogue is Chief of Physical Therapy and Assistant Professor, University of Missouri Medical Center, Columbia, Missouri 65201.

Elftman calculated the angular momentum of the arms during gait.[1] He concluded that arm swing is not an example of pendular action. Murray and her associates identified the range of normal variability in upper-limb displacement patterns during normal walking.[2]

The first major study of upper-extremity muscular activity was conducted by Fernandez-Ballesteros and her associates.[3] Buchthal and Fernandez-Ballesteros recorded the muscular activity of arm swing in the patient with parkinsonism.[4]

However, no completed electromyographic study of the function of the upper extremities at different cadences and inclines has been reported, and no quantified data are available regarding the action of the arms during gait.

METHOD

Subjects

The subjects of this study were fifteen normal college students (twelve women and three men), ranging in age from twenty-one to twenty-five. The muscles tested were the upper, middle, and lower trapezius; the teres major; the infraspinatus; the anterior, middle, and posterior deltoid; the short head of the biceps brachii; the lateral head of the triceps brachii; the flexor carpi ulnaris; and the extensor carpi radialis brevis.

Equipment

Electromyography was used to obtain data about the muscular activity. A multichannel instrument was utilized to obtain electromyographic tracings.

An imbedded intramuscular copper-wire electrode served to pick up the electrical activity of each muscle. Because the electrode was comfortable and did not move during movement of the extremities, interference was eliminated and accuracy was assured. The wires were autoclaved at 250 degrees Fahrenheit for one hour before using.

The wire used was nylon karma, 10 inches long and .0011 inches in diameter. The lead length was 18 inches, and the barb length was one and one-half to two millimeters. The recording tip was one millimeter.

The electrical activity was led by way of telemetry to a preamplifier, where it was magnified. From the preamplifier the electrical activity was led by shielded wire to an amplifier. A pen motor received the electrical signals and drove the desired recording pen. Specific electromyographic recordings were obtained on one channel, and integrated potentials were shown on a separate channel. The integrator recorded the total activity in the electromyogram rather than specific amplitude, wave shape, and frequency (see Fig. 1).

Preparation of Subjects

Each subject was seated in a chair, and the muscle to be tested was palpated by the examiner. The active wire electrode was then inserted by a 22-gauge hypodermic needle. The needle was withdrawn slowly. A hook at the end of the wire enabled it to remain in place.

The transmitter for the telemetry system was applied to sponge rubber, which had been affixed to the skin. A small jack (Fig. 2), con-

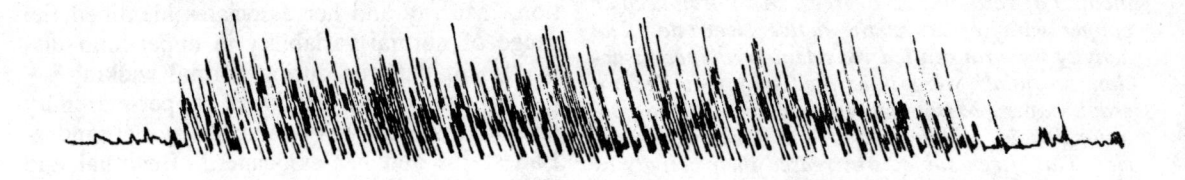

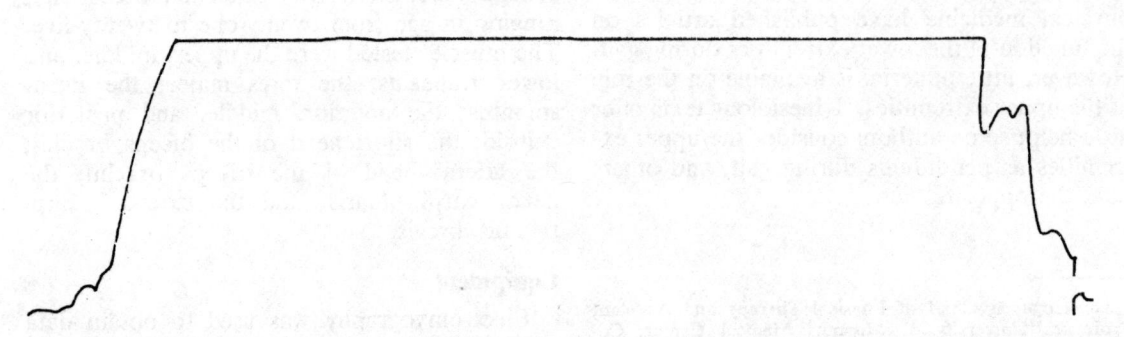

Fig. 1. (Top) Specific electromyographic action potentials. (Bottom) An integrated electromyogram.

taining one red receptacle for the active wire electrode and one black receptacle for the ground electrode, was plugged into the transmitter. The lead from the active electrode was inserted into the active receptable. A one-half-inch electroencephalographic needle served as a ground electrode, which was implanted into the dermis. This lead was introduced into the ground receptacle of the jack.

After each subject was prepared, the telemetry receiver was tuned into the transmitter. The receiver lead was infixed into the preamplifier. The recording instrument had been readied beforehand.

Normal Concentric Contraction

Before testing each muscle under the experimental conditions, a normal concentric contraction was obtained and recorded on the graph paper. This contraction served to determine whether the active electrode had been implanted properly. In addition, this contraction was considered a normal one and was to be compared later with the muscular activity obtained during each test condition.

Instructions and demonstrations for the movements were given by the examiner. The oral instructions were as follows:

Upper trapezius muscle. "Bring your shoulders to your ears and hold for a count of three."

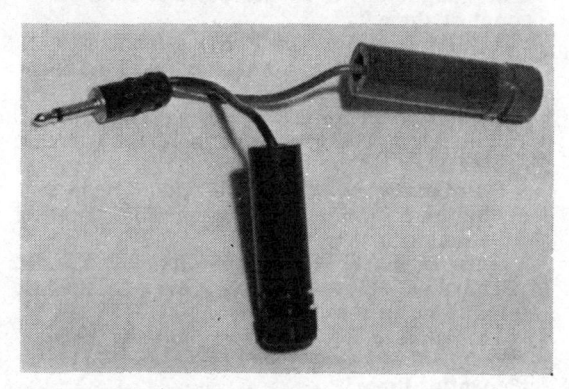

Fig. 2. Receptacles for active and ground electrodes.

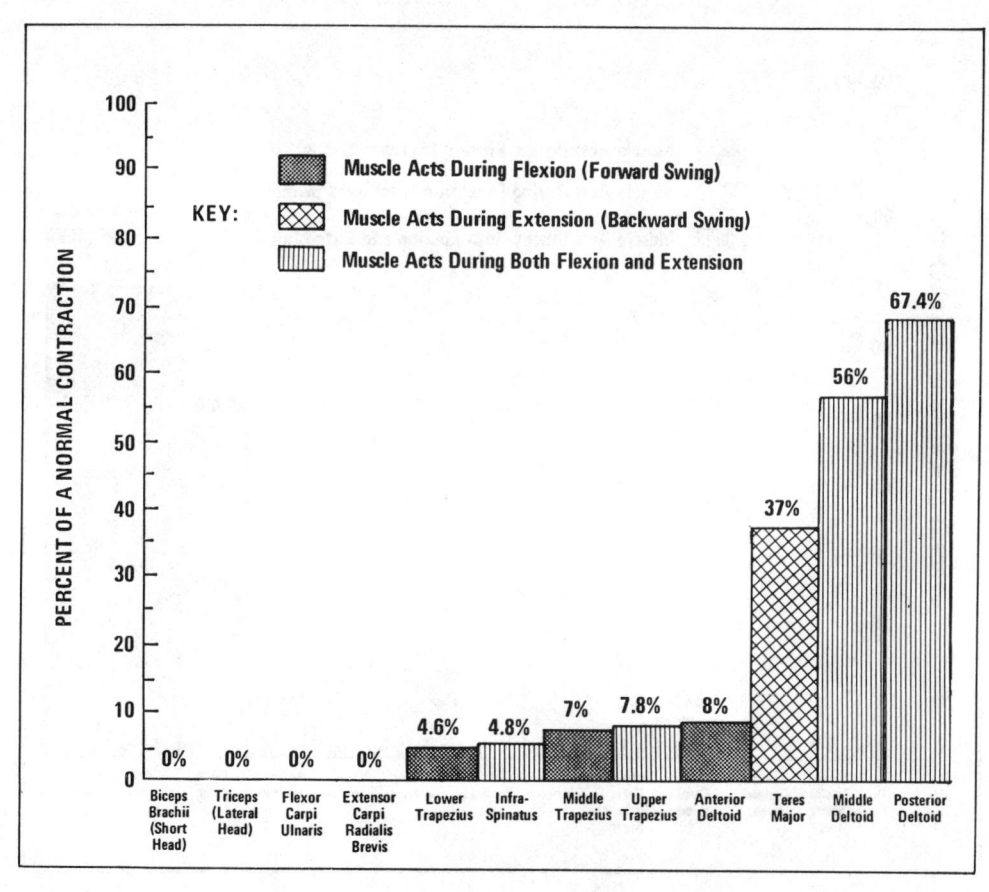

Fig. 3. Function of upper-extremity muscles at 70 steps per minute on a level surface.

Middle trapezius muscle. "Pull your shoulder blades together and hold for a count of three."
Lower trapezius muscle. "Put your hand on this chair. Push down and hold for a count of three."
Infraspinatus muscle. "Roll your arm out and at the same time pull it to your side. Hold for a count of three."
Anterior deltoid muscle. "Bring [flex] your arm straight up to 90 degrees and hold for a count of three."
Middle deltoid muscle. "Bring [abduct] your arm out to the side to 90 degrees and hold for a count of three."
Posterior deltoid muscle. "Bring [extend] your arm back to 45 degrees and hold for a count of three."
Teres major muscle. "Roll your arm in and at the same time pull it to your side. Hold it for a count of three."
Short head of the biceps brachii muscle. "Bend your elbow to 90 degrees and hold it for a count of three."
Lateral head of the triceps brachii muscle. "Straighten out your elbow [180°] and hold for a count of three."
Flexor carpi ulnaris muscle. "Bend your wrist forward [flexion] and at the same time bring it in [ulnar deviation]. Hold for a count of three."
Extensor carpi radialis brevis muscle. "Bend your wrist backward [extension] and at the same time bring it out [radial deviation]. Hold it for a count of three."

Testing Conditions

After the normal contraction was obtained, the subject was instructed to walk 20 steps on a level floor surface at 70, 95, and 120 steps per minute. The process was repeated on a treadmill at an incline of 15 degrees. A metronome served to guide the cadences for each subject.

INTERPRETATION OF DATA

After completion of the study, the recording paper was assembled and the ink writing analyzed. Major consideration was given to the integrated potentials.

Each muscle was considered separately at each incline and cadence. The maximum height

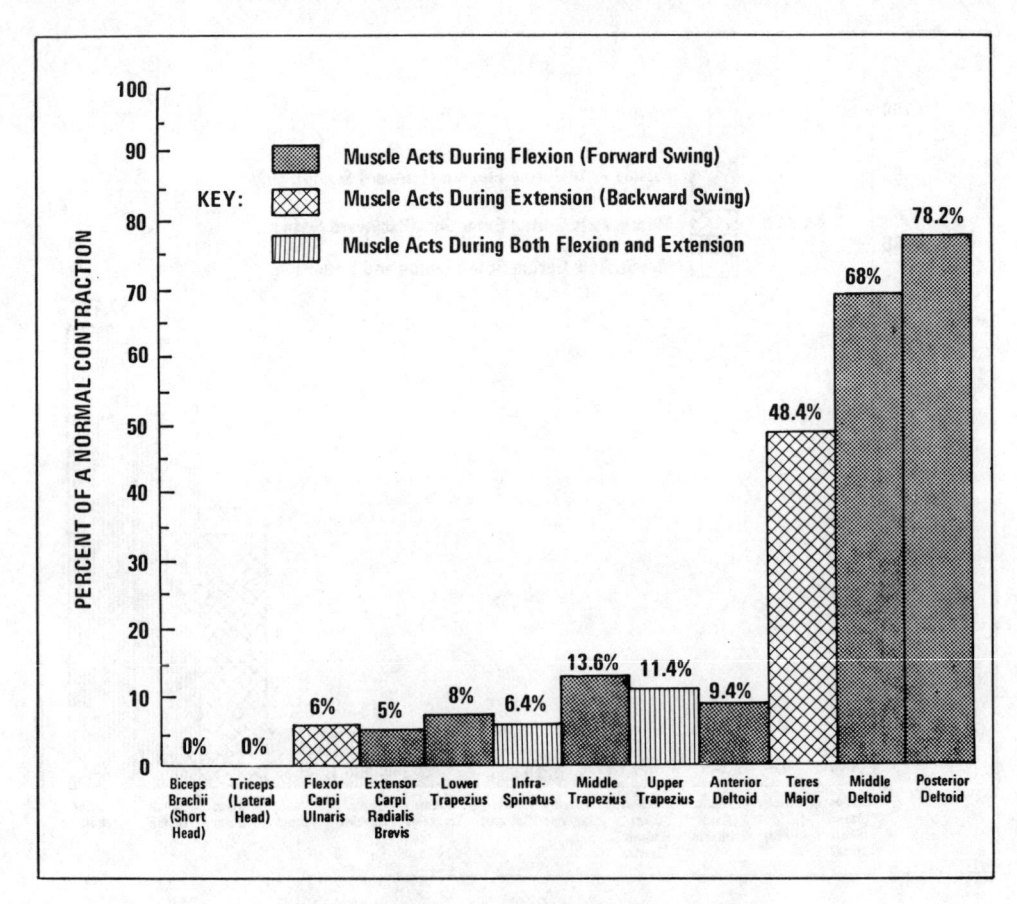

Fig. 4. Function of upper-extremity muscles at 95 steps per minute on a level surface.

of the integrated normal contraction was expressed as 100 percent. Then the integrated activity which occurred during each test condition was measured and couched as a certain percentage of normal. Assigning a percentage was more significant than merely stating that the muscle was or was not active at certain phases of the gait cycle. The histograms accompanying this paper are based on the percentage method of calculation.

Relationship of Activity to Position

Figures 3 through 8 indicate the function of the twelve muscles at each incline and cadence. They also show whether each muscle acted during flexion or extension, or both. This fact was determined after the author had studied films of gait, observed the subjects in this project, and listened to the sound of the muscular activity during the experiment. When forward swing began, the examiner activated an external-event marker pushbutton switch. This caused a momentary upward deflection of the recording pen. The same process was used when backward swing began. Because the first deflection on the recording paper always represented forward swing, it was easy to determine whether the activity between any two deflections represented forward or backward swing.

Levels of Muscular Activity

The percentages in the tables represent means of the muscular activity. A muscle which showed an activity of 40 percent or more was considered to have contributed significantly toward actively swinging the arm during gait. Any activity below 40 percent was not significant. The classification system was:

Nil	0	to 5	percent
Negligible	6	to 15	percent
Slight	16	to 39	percent
Moderate	40	to 69	percent
Marked	70	to 89	percent
Maximum	90	to 100	percent

This classification was an attempt to interpret the muscular activity in quantitative terms.

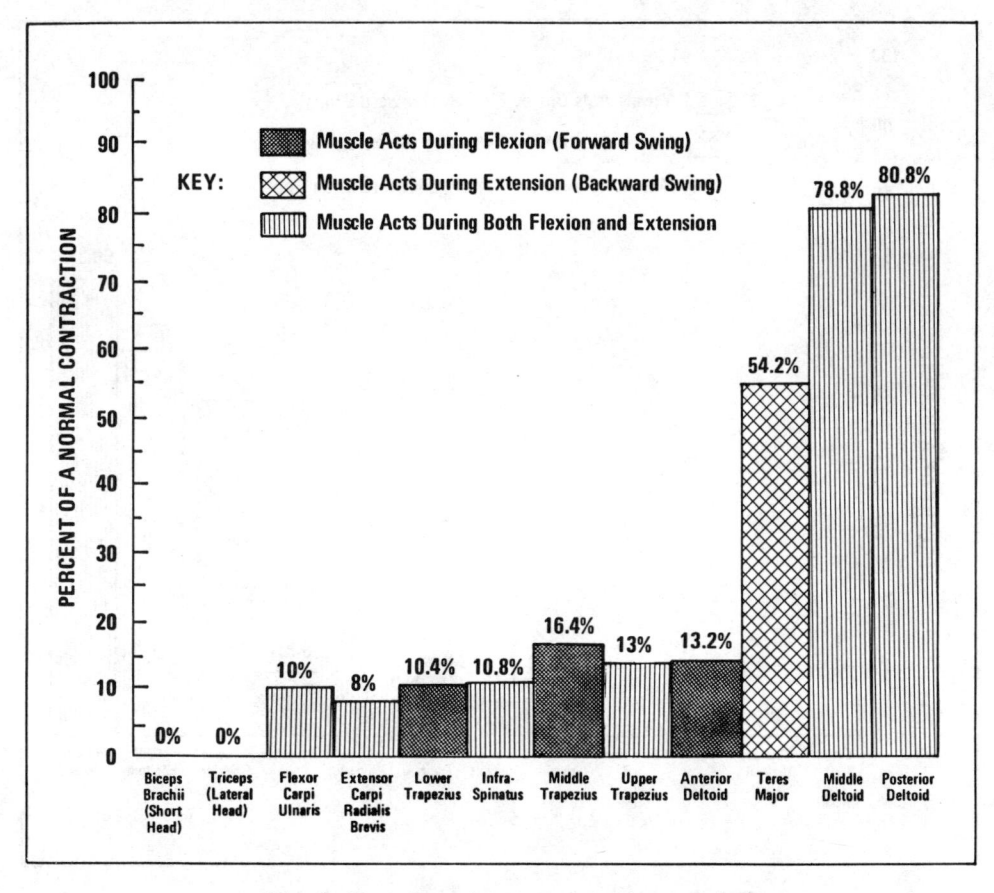

Fig. 5. Function of upper-extremity muscles at 120 steps per minute on a level surface.

RESULTS

Short head of the biceps brachii and the lateral head of the triceps brachii muscles. No activity was found in any subject at any cadence or incline.

Flexor carpi ulnaris and flexor carpi radialis brevis muscles. These muscles did not act at a cadence of seventy steps per minute on a level incline. The activity was nil or neglible during the remaining test conditions. For all practical purposes, these muscles did not function during normal gait.

Lower trapezius muscle. This muscle acted during both forward and backward swing, but its activity was nil or negligible throughout the investigation.

Infraspinatus muscle. The activity of this muscle was nil or negligible throughout the investigation.

Middle trapezius muscle. Slight activity was noted during arm flexion at the shoulder joint at 120 steps per minute both on a level surface and on a 15-degree incline.

Upper trapezius muscle. Negligible activity was recorded during all test conditions.

Anterior deltoid muscle. No activity was reported in two subjects during all the test condi-

TABLE 1

Mean Percentage Increase of Muscular Activity from a Level Surface to a 15-degree Incline at Varying Steps per Minute

Muscle	Steps per Minute		
	70	95	120
Biceps brachii	0.0	0.0	0.0
Triceps brachii	0.0	0.0	0.0
Flexor carpi ulnaris	1.8	4.0	5.0
Extensor carpi radialis brevis	1.0	1.0	2.0
Lower trapezius	0.8	2.4	1.4
Infraspinatus	0.0	0.0	4.0
Middle trapezius	1.6	−0.2	1.6
Upper trapezius	3.6	0.2	1.0
Anterior deltoid	0.8	2.7	0.2
Teres major	12.0	13.4	11.6
Middle deltoid	0.0	6.0	1.8
Posterior deltoid	−1.4	2.0	4.0

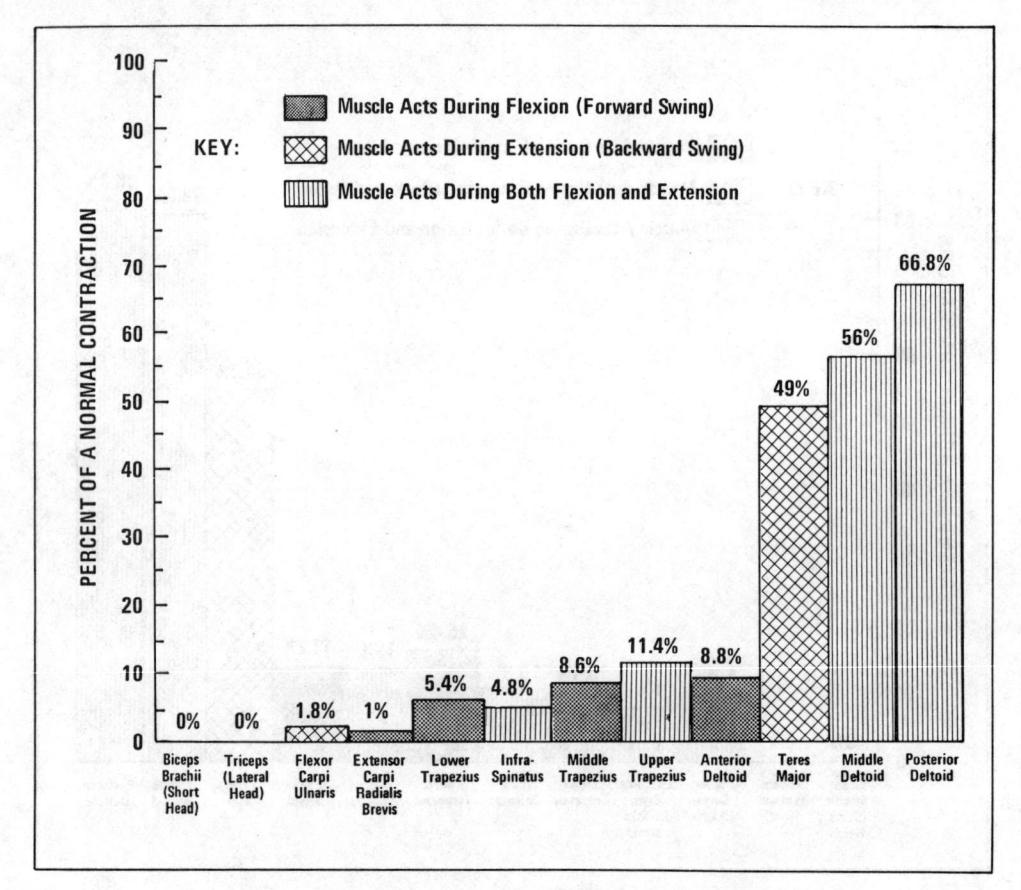

Fig. 6. Function of upper-extremity muscles at 70 steps per minute on incline of 15 degrees.

tions. Negligible activity, or none, was recorded during all conditions in the remaining subjects.

Teres major muscle. Slight activity was demonstrated on a level surface at a cadence of seventy steps per minute. Moderate activity was reported for the rest of the conditions during backward swing of the arm.

Middle deltoid muscle. Consistent activity was observed during both flexion and extension of the arm at the shoulder joint. Moderate activity was shown on a level surface at cadences of seventy and ninety-five steps per minute and on the ramp at seventy. Marked activity was noted during the rest of the conditions.

Posterior deltoid muscle. This muscle exhibited more activity than any other upper-extremity muscle during gait. Moderate activity was recorded during the slow cadence on both the 0 and 15-degree grades. Marked activity was shown during the remaining test conditions.

The posterior deltoid muscle acted during backward swing and during the last portion of forward swing.

Relationship of Activity to Cadence

Table 1 reveals the average percentage increase of muscular activity which occurred for each cadence when the incline was changed from the level surface to a 15-degree incline. The greatest increase occurred in the teres major muscle. It increased its activity 12 percent at 70 steps, 13.4 percent at 95 steps, and 11.6 percent at 120 steps. The other muscles exhibited very small changes, with the range being 0 to 4 percent.

Table 2 illustrates the average changes of activity which occurred as the cadences were increased on the same incline. Specifically, the percentages at 95 steps per minute represent the increases over 70 steps, and the percentages at 120 steps represent the increases over 95

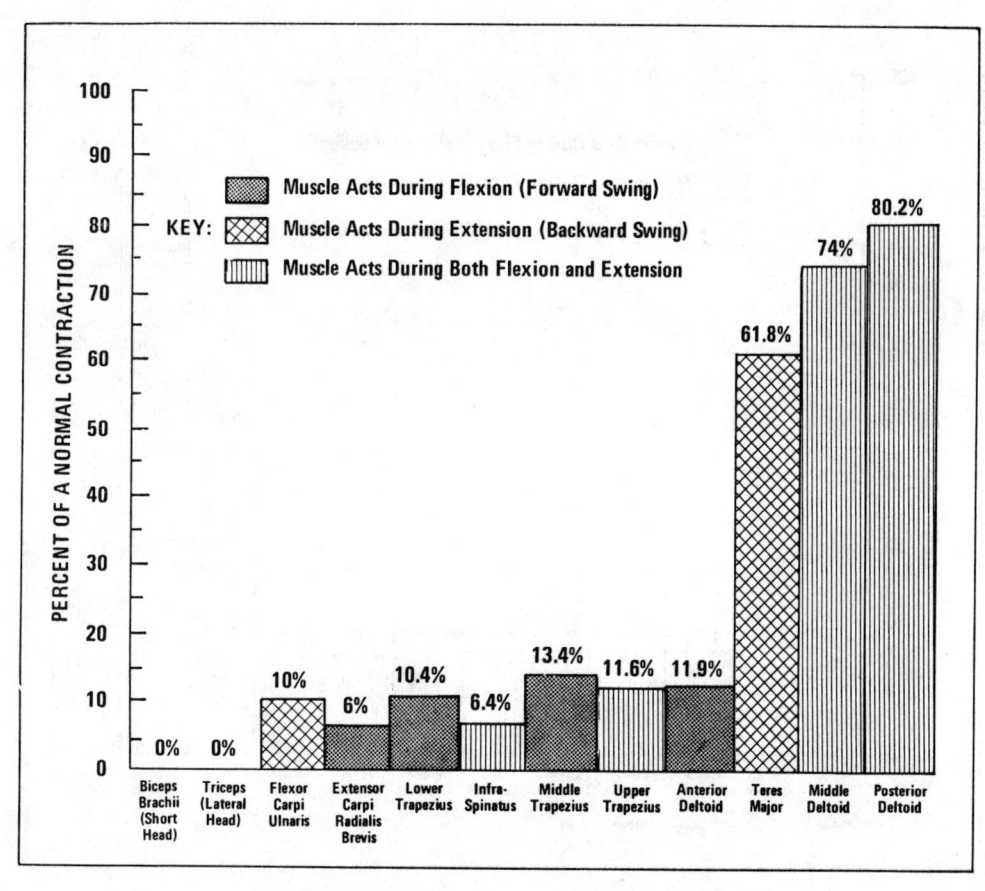

Fig. 7. Function of upper-extremity muscles at 95 steps per minute on incline of 15 degrees.

TABLE 2
MEAN PERCENTAGE INCREASE OF MUSCULAR ACTIVITY
ON LEVEL AND INCLINED SURFACES

Muscle	Level Surface			15-degree Incline		
	Steps per Minute 95	120	Total Increase from 70 to 120 Steps per Minute	Steps per Minute 95	120	Total Increase from 70 to 120 Steps per Minute
Biceps brachii	0.0	0.0	0.0	0.0	0.0	0.0
Triceps brachii	0.0	0.0	0.0	0.0	0.0	0.0
Flexor carpi ulnaris	6.0	4.0	10.0	8.2	5.0	8.7
Extensor carpi radialis brevis	5.0	3.0	8.0	5.0	4.0	9.0
Lower trapezius	0.4	2.4	2.8	5.0	1.4	6.4
Infraspinatus	1.8	4.4	6.2	1.8	6.4	7.2
Middle trapezius	6.6	2.8	9.4	4.8	4.6	9.4
Upper trapezius	3.6	1.8	5.4	0.2	2.4	2.6
Anterior deltoid	1.4	3.8	5.2	3.1	1.5	4.6
Teres major	11.4	5.8	17.2	12.8	4.0	13.2
Middle deltoid	12.0	10.8	22.8	18.0	7.6	25.6
Posterior deltoid	10.6	2.6	13.2	13.4	4.6	18.0

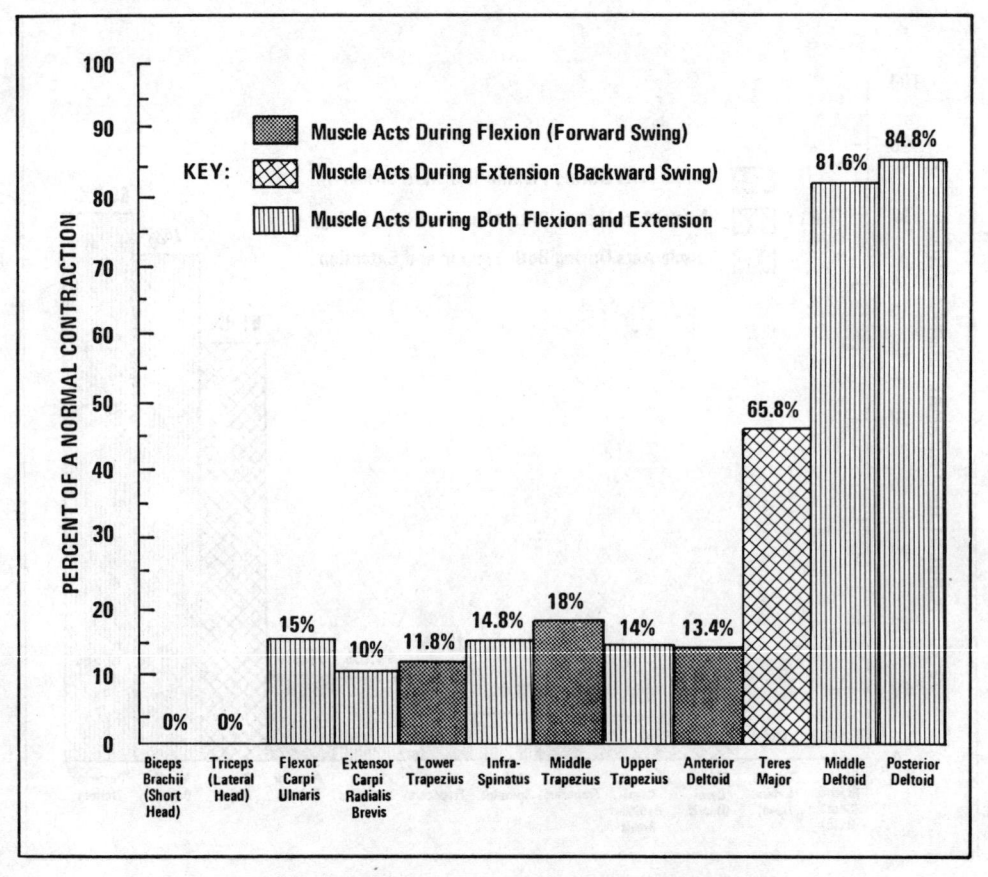

Fig. 8. Function of upper-extremity muscles at 120 steps per minute on incline of 15 degrees.

steps. The total increase is the sum of the two. The middle deltoid muscle showed the greatest increase in activity as the subjects walked faster. This was true on both the level surface and the 15-degree incline. The teres major and posterior deltoid muscles were the only other muscles which had a total increase of more than 10 percent.

DISCUSSION

The subjects had individual differences in muscular activity; therefore, there was a range at each test condition. For example, the anterior deltoid muscle ranged from 0 to 20 percent, and the teres major muscle ranged from 20 to 80 percent. The mean represents an approximation of the gross differences of upper-extremity muscular activity during gait.

After consultation with a statistician, it was decided that refined statistical manipulation of the data would be of no practical value. It is believed that this study, within its limitations, gives an accurate estimation of the upper-extremity muscular activity during human locomotion.

Major Muscles of Activity

The results indicate that three muscles provided most of the activity during arm swing. The teres major, middle deltoid, and posterior deltoid muscles contracted 50 percent, or more, of their normal contractions during gait. The activity came in short spurts.

The teres major muscle assisted moderately in backward swing of the arm. The deltoid muscles were more consistent in their activity. They appeared to be the major muscles which swung the upper extremity during normal locomotion. The middle deltoid muscle showed consistent activity during both flexion and extension of the arm at the shoulder joint because the arm had to abduct in order to clear the side of the body during both movements. The posterior deltoid muscle acted during backward swing and forward swing, not only extending the arm but also decelerating the arm during the last phase of flexion.

Sum Total of Other Muscles

Although the teres major, middle deltoid, and posterior deltoid muscles were the major muscles which swung the arm during locomotion, the functions of the other muscles should not be ignored. The sum total of their activity probably contributes to arm swing, although additional study is needed. The stabilizing action of the scapular muscles is just as important as actual movement. The middle trapezius muscle appeared to act with the posterior deltoid muscle to decelerate the forward swinging arm.

As a result of this study, the arbitrary division of the trapezius muscle into three or four parts and the traditional functions assigned to each are open to question.

Increased Activity during Gait

The results disclosed that as the subject walked faster, the muscular activity of the upper extremities increased. However, as he walked on a greater incline at the same cadence, no appreciable increase of activity was observed. The primary cause for the increased muscular activity in gait seemed to be the speed of the walk rather than the incline ascended. This increase was true of the three major muscles. The remaining muscles showed a minimal increase in activity.

Concurrent Lower-extremity Movements

Although the upper-extremity activity during gait is small when compared to that of the lower extremity, it is reasonable to assume that one upper extremity works with the contralateral leg to control excessive movement of the trunk at midstance. This balancing factor agrees with Steindler's conclusion that the function of locomotion was taken over by the lower extremities in the philogenetic development of human gait, while the upper extremities assisted in balancing the trunk over the pelvis.[5] The arm decelerators work with the opposite leg decelerators to help control the limits of trunk rotation. The upper-extremity activity may be another determinant of gait through its influence on the movement of the center of gravity.

Arms as Pendulums

According to the results of this study, the arms do act as pendulums during gait. However, this is not to say that gravity is solely responsible for arm swing. A pendulum swings freely from the combined action of gravity and momentum. Momentum is the product of mass times velocity. The velocity for arm swing is furnished by muscle action as well as by gravity; therefore, the upper-extremity muscles are not an incidental part of arm swing during gait. The terms *pendulum* and *muscle action* are not mutually exclusive.

SUMMARY

This study was designed to determine the muscular activity of the upper extremities during gait at different cadences and inclines.

The subjects were fifteen college students. Twelve different muscles were tested. The method of investigation was a multichannel electromyographic apparatus. An imbedded intramuscular copper wire served as the active electrode. The electrical activity was transmitted via telemetry. Specific electromyographic action potentials were obtained on one channel and integrated on another. The activity of each muscle during each test condition was interpreted as a certain percentage of a normal contraction. The subjects were tested at the cadences of 70, 95, and 120 steps per minute on a level surface and a 15-degree incline.

The posterior deltoid muscle exhibited the greatest amount of electrical activity during the experiment. It extended the backward swinging arm and decelerated the forward swinging arm. The other two muscles which showed significant activity were the middle deltoid and teres major. Their activity increased as the subjects walked faster. They did not show an appreciable increase in activity as the subjects ascended a steeper incline. Muscles below the shoulder region did not contribute to arm swing.

Although the activity of the scapular muscles was minimal, the sum total of their activity may play a part in swinging the arm during gait. It is reasonable to assume that upper-extremity muscle activity occurs concurrently with the opposite lower-extremity movements in order to decelerate the arm and leg, help control the limits of trunk rotation, and assist in maintaining balance in the frontal plane during gait. The upper extremities do act as pendulums, but the pendular action is due to muscular activity as well as to gravity.

Acknowledgment. Lee Kohler, R.N., of Cleveland, Ohio, supplied the wire electrodes used in this study.

REFERENCES

1. Elftman, H. The function of the arms in walking. Hum. Biol. 11:528–535, December 1939.
2. Murray, M. P., S. B. Sepic, and E. J. Bernard. Patterns of sagittal rotation of the upper limbs in walking: a study of normal men during free and fast speed walking. Phys. Ther. 47:272–284, April 1967.
3. Fernandez-Ballesteros, M. L., F. Buchthal, and P. Rosenflack. The pattern of muscular activity during the arm swing of natural walking. Acta Physiol. Scand. 63:296–310, March 1965.
4. Buchthal, F., and M. L. Fernandez-Ballesteros. Electromyographic study of the muscles of the upper arm and shoulder during walking in patients with Parkinson's disease. Brain 88:875–897, December 1965.
5. Steindler, A. Kinesiology of the Human Body. Springfield, Illinois: Charles C Thomas, Publisher, 1955.

SELECTED READINGS

1. Murray, M. P., A. B. Drought, and R. C. Kory. Walking patterns of normal men. J. Bone Joint Surg. 46A:335–360, March 1964.
2. Murray, M. P., R. C. Kory, B. H. Clarkson, and S. B. Sepic. A comparison of free and fast speed walking patterns of normal men. Amer. J. Phys. Med. 45:8–24, February 1966.
3. Gray, James. Muscular activity during locomotion. Brit. Med. Bull. 12:203–209, September 1956.

the author

Raymond E. Hogue, Ph.D., is director of the Physical Therapy Department and assistant professor of the Physical Therapy Curriculum at the University of Missouri, from which he earned his M.A. and Ph.D. degrees. Dr. Hogue has served as staff physical therapist at St. Mary's Hospital, Kansas City, Missouri; as physical therapy specialist at the U.S. Army Hospital, Ft. Gordon, Georgia; and as consultant to the Cerebral Palsy Development Center in Columbia, Missouri.

Basic Research

Development of Gait at Slow, Free, and Fast Speeds in 3- and 5-Year-Old Children

RUTH ROSE-JACOBS

The purpose of this study was to describe and statistically analyze 3- and 5-year-old children's gait at slow, free, and fast speeds in terms of stride length, step length (adjusted for leg length), stride width, included angle of feet, and cadence. The study also correlated gait factors and motor development. Gait patterns were recorded with a clinical, footprint method. In general, stride length and cadence were significantly different for age and speed, and step length and stride width were significantly different only for speed. Included angle of feet was not significant for age or speed. Motor ability as measured by the McCarthy Scales of Children's Abilities correlated only with stride length and cadence. It was concluded that gait patterns in 3- and 5-year-old children are not fully mature. Perhaps the interrelationship between gait factors, age, and speed, as well as the relationships among gait factors, present a more realistic analysis of gait in these children than if the variables were considered in isolation. Further research is needed to determine how variability of a child's gait decreases and which gait factors and conditions can be used appropriately to determine gait maturity.

Key Words: Child development, Gait, Research design.

The preschool years are a transitional period in the development of gait, and little quantitative research has been conducted in this area. The few quantitative preschool gait studies reported in the literature have examined few subjects, used expensive equipment, and lacked sufficient information regarding the relationships of gait speed, gait development, and general motor development.[1-6] Physical therapists have few quantitative guidelines that can be used to measure gait clinically and objectively during this developmental period.

Although a child may learn a motor skill, such as gait, at an early age, the development of control and elimination of extraneous movements occurs during the preschool years.[7] Only after the child has first attained a considerable level of skill does consistency of the motor activity occur. With maturation and opportunities for experience and motor learning, movements and postural adjustments become more automated to meet the new demands and situations in the child's life.[8, 9] Ultimately, the adult or mature level of gait is one of the most consistent human activities in which voluntary control plays a part.[10] Individuals with mature gait patterns are able to ambulate safely and comfortably within a wide variety of gait speeds[11] and maintain highly consistent details of each step at any speed.[11-13] The specific criteria to apply in determining whether gait is mature are highly controversial.[1-3, 14, 15]

Stride and step length, stride width, included angle of foot placement (the sum of a successive right and left foot angle), cadence, and variability of each of the above gait factors have all been used to measure children's gait. At free speed, stride and step length increase and cadence decreases until gait maturity. How and at what age gait maturity occurs is debated.[1, 16-18] The extent to which older children's step length is due to increased height and leg length is also unclear.[2] Scrutton[1] found that in 1- to 4-year-olds, there was a minimal change in stride width, but Burnett[3] found the narrowing of stride width to occur within weeks of attaining independent gait by the infant at 9 to 17 months. In infants, angle of foot placement decreases as the child progresses from supported to independent gait[3, 17]; however, little data describe angle of foot placement in the post-inde-

---

Ms. Rose-Jacobs was a graduate student, Sargent College of Allied Health Professions, Division of Physical Therapy, Boston University, when this study was conducted. She is now Assistant Professor, Department of Physical Therapy, Boston-Bouve College of Human Development Professions, Northeastern University, 360 Huntington Ave, 308 Robinson Hall, Boston, MA 02115 (USA).

This study was supported in part by an Allied Health Professions Traineeship from the Public Health Service Department of Health, Education and Welfare, Washington, D.C.

This article was adapted from a paper presented at the Fifty-seventh Annual Conference of the American Physical Therapy Association, Washington, DC, June 28-July 2, 1981.

This article was submitted May 24, 1982; was with the author for revision 25 weeks; and was accepted for publication February 18, 1983.

pendent stages of gait. Before age 3, patterns of foot movement, stance duration, and change of foot force vary with each step.[4] Three-year-olds have a variety of gait patterns ranging in similarity from children younger and older than themselves to adults. Others have found variability in gait in children even 8[19] and 10 years of age.[14]

The speed of ambulation must also be considered.[20] As the child's gait speed increases, the width of gait may become erratic and almost similar to the pattern prior to an infant's independent gait.[3] Gesell's concept of reciprocal interweaving is applicable to this phenomenon; the child attains a new level of developmental maturity in the area of speed but temporarily reverts to a somewhat less mature stride width.[9]

The purpose of this study is to describe and compare 3- and 5-year-old children's gait at slow, free, and fast speeds. Stride length, step length (adjusted for leg length), stride width, included angle of feet, and cadence are compared. This study will also attempt to relate children's level of gait maturity to gross motor development. The following hypotheses are presented: 1) As gait speed increases, the 3-year-old and 5-year-old groups will have longer strides and step lengths, wider stride width, smaller included angle of feet, and higher cadence; 2) as the speed increases, the 5-year-old group will have a longer stride and step length, narrower stride width, smaller included angle of feet, and lower cadence; 3) at each gait speed, the 5-year-old group, when compared with the 3-year-old group, will have a longer stride and step length, narrower stride width, smaller included angle of feet, and lower cadence; 4) at each gait speed, the 5-year-olds, compared with the 3-year-olds, will have less variability of stride length, step length, stride width, and included angle of feet; 5) at each age and at each speed, there will be a correlation between gait factors; and 6) at slow, free, and fast gait speeds, a correlation will exist between a child's scale index equivalent on the motor section of the McCarthy Scales of Children's Abilities[21] and the child's gait factors as measured in this study.

METHOD

Subjects

Fifteen normal (with no known neurological, orthopedic, or developmental problems) 3-year-old (38–44 months) and 16 normal 5-year-old (58–63 months) children were tested; mean ages were 40.67 (±1.91) and 59.44 (±1.46) months. Seven of the 3-year-olds and six of the 5-year-olds were girls. The sample was not totally random in that children were from middle to upper middle class socioeconomic backgrounds and were selected from four preschools in the greater Boston area. All children of appropriate ages and medical criteria, whose parental consent forms were returned, were included as subjects.

Procedure

Data were obtained by modifying the method used by Ogg.[22] The investigator tested the children at their preschools. Children wore their regular shoes.

Before the test, each child's functional leg length (distance from the palpation of the greater trochanter of the right femur to the point on the floor located just anterior and medial to the medial malleolus on the ipsilateral side) was measured in the standing position and recorded in centimeters. The subject was

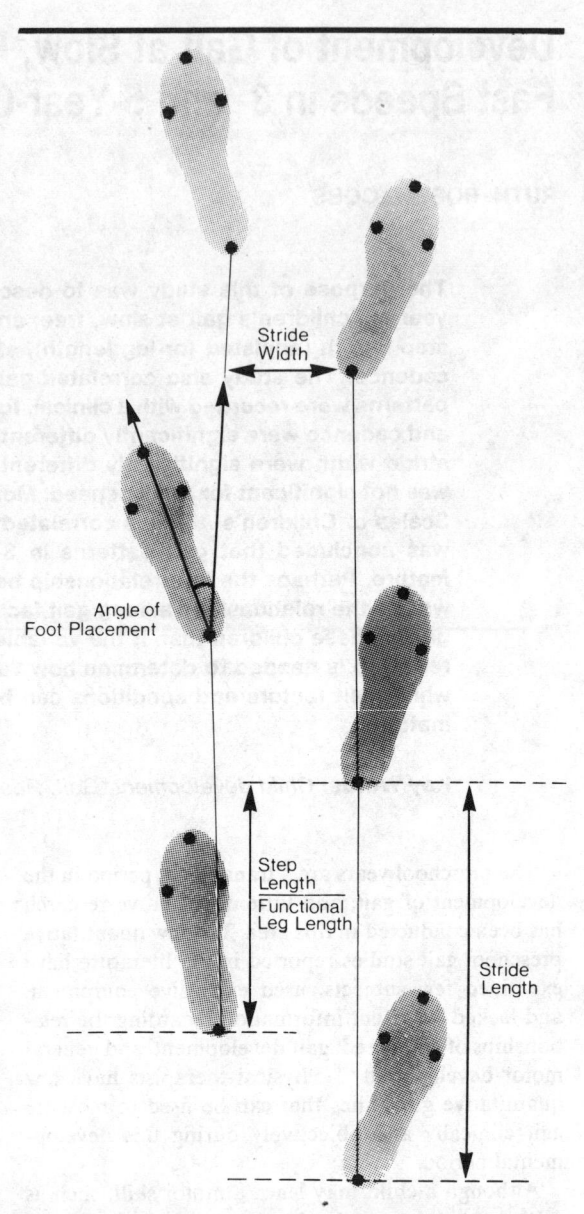

Figure. Measurements from ink footprints on the paper strip.

Basic Research

RESEARCH

TABLE 1
Means and Standard Deviations for Each Gait Factor at Each Age and Speed of Gait

Gait Factor and Age	n	Slow Gait		Free Gait		Fast Gait	
		$\bar{X}$	s	$\bar{X}$	s	$\bar{X}$	s
Stride length (cm)							
3 yr	15	59.85	8.08	70.51	10.85	84.41	13.81
5 yr	16	65.66	8.88	86.23	9.59	100.47	8.95
Step length (cm/cm leg length)							
3 yr	15	0.71	0.14	0.78	0.12	0.93	0.15
5 yr	16	0.64	0.11	0.81	0.09	0.95	0.10
Stride width (cm)							
3 yr	15	8.81	2.15	8.22	2.56	6.84	2.05
5 yr	16	8.63	3.43	8.00	2.41	6.21	2.23
Angle of feet (°)							
3 yr	15	4.31	13.30	5.05	15.38	7.56	14.55
5 yr	16	8.43	14.38	6.35	12.22	7.27	12.31
Cadence (steps/minute)							
3 yr	15	103.00	27.07	143.73	30.17	190.93	38.14
5 yr	16	87.56	19.39	125.63	32.10	155.43	24.14

then seated on a chair at the end of a paper measuring 5 ft by 20 in. Four round, thin, felt pads* 2 cm in diameter were secured to the sole of each shoe at the following landmarks identified by palpation: back center of the heel, bases of the first and fifth toes at the metatarsophalangeal joint and distal end of the second toe. The lateral landmarks were used for orientation purposes; the other landmarks were used at the completion of the study to measure the dependent variables. Markers of each foot were inked with a different color. Each child walked once up and down the 5-ft length of paper to become familiar with the testing procedure. The child was again seated, and markers were re-inked in preparation for the test. The child was then asked to walk normally to the chair at the end of a 26-ft length paper. A stopwatch timed the walk. Speed was controlled by asking the subjects to walk at the speed of the tester who had been trained to walk at 70 cm/sec and 120 cm/sec.[1,2,23,24] Order of slow and fast speeds were randomly presented to subjects. Immediately after the walking, the motor section of the McCarthy Scales of Children's Abilities was administered.[21]

* Dr. Scholl's® Felt Corn Pads, Scholl, Inc, Chicago, IL 60610.

Gait Variables

The gait variables were measured from imprints on the paper in the following manner. The first and last 3 ft of paper were discarded. *Stride length*, *step length*, and *stride width* were measured on the remaining 20 ft of paper (Figure). Stride length was not adjusted for leg length; however, step length was adjusted by the step factor ratio of step length/functional leg length.[1,23] This procedure determined whether an older child's longer stride was due solely to a longer leg length. *Included angle of feet* was determined from the angle of foot placement (Figure). The sum of a successive right and left foot angle equaled an included angle. *Cadence* was the number of step imprints per minute. Within-subject variability of gait factors was defined as the variance of each factor for each subject. Motor development level was measured by the raw score and scale index number of the motor section of the McCarthy Scales of Children's Abilities.[21]

Means for all gait factors were analyzed using two-way analysis of variance (speed by age) and specific differences were analyzed by the Newman-Keuls test.

TABLE 2
Analysis of Variance with Repeated Measures: Effects of Age and Gait Speed on the Mean Stride Length

Source of Variance	df	MS	F	p
Speed	2,58	6922.66	88.42	<.001
Age	1,29	3656.81	23.77	<.001
Speed × age	2,58	263.44	3.36	<.05

TABLE 3
Analysis of Variance with Repeated Measures: Effects of Age and Gait Speed on the Mean Step Length

Source of Variance	df	MS	F	p
Speed	2,58	.54	57.13	<.001
Age	1,29	.00	.03	NS
Speed × age	2,58	.03	2.69	NS

TABLE 4
Analysis of Variance with Repeated Measures: Effects of Age and Gait Speed on the Mean Stride Width

Source of Variance	df	MS	F	p
Speed	2,58	39.08	16.65	<.001
Age	1,29	3.02	.21	NS
Speed × age	2,58	.58	.25	NS

TABLE 5
Analysis of Variance with Repeated Measures: Effects of Age and Gait Speed on Cadence

Source of Variance	df	MS	F	p
Speed	2,56	46061.00	113.51	<.001
Age	1,28	13425.00	7.76	<.01
Speed × age	2,56	1107.00	2.73	NS

The F distribution was used to determine homogeneity of variance between the two ages for various gait factors. Variance for each group was calculated by summing each child's variance. Finally, the Pearson product-moment correlation was used to analyze the relationships between gait factors at each age and speed and between gait factors and raw motor scores on the McCarthy Scales for each gait speed of the combined ages.

Reliability

Two raters administered and scored the McCarthy Scales of Children's Abilities for two children and scored gait test papers for three children. Interrater reliability was 93 percent and 100 percent, respectively.

RESULTS

Table 1 presents the means and standard deviations of 3- and 5-year-olds for each gait factor at each speed. A preliminary data analysis determined that right and left stride and step lengths were not statistically different at each age group and speed and, therefore, only the right stride and step lengths were considered. Similarly, the set speed of gait of the 3- and 5-year-old groups at slow, free, and fast speeds was not significantly different. The gait speeds, therefore, were combined and mean speeds were as follows: slow, 51.9 cm/sec; free, 83.8 cm/sec; and fast, 128.6 cm/sec. Finally, when the Scale Index Scores of the McCarthy Scales of Children's Abilities were adjusted for age, the scores of the two groups were found to be equivalent.

Stride Length

There were significant differences in mean stride length due to speed of gait, age, and the interaction of speed and age (Table 2). Stride length for combined age groups at a fast speed was significantly longer ($p < .01$) than at slow or free speeds. Mean stride length at free speed was significantly longer ($p < .01$) than at slow speed. At both free and fast speeds, the 5-year-old group had significantly longer strides ($p < .01$) than the 3-year-old group at the same speeds; however, the difference was not significant between the age groups at slow speed.

Within subjects, the 3-year-old group, compared with the 5-year-old group, had significantly greater variance of stride length at slow ($F = 9.48$, $df = 14, 15$, $p < .001$) and free ($F = 3.04$, $df = 14, 15$, $p < .025$) speeds. No significant difference was present at a fast speed.

TABLE 6
Correlations Between Gait Factor at Each Age and Speed

Gait Factors	3-yr-olds				5-yr-olds			
	Step Length (adjusted)	Stride Width	Angle of Feet	Cadence	Step Length (adjusted)	Stride Width	Angle of Feet	Cadence
Slow Gait								
Stride length	.15	.27	−.63[a]	−.57[b]	.88[c]	−.64[d]	.55[b]	−.08
Step length (adjusted)		.21	−.09	−.14		−.48[e]	.54[b]	.02
Stride width			−.53[c]	−.23			−.72[c]	−.11
Angle of feet				.08				.24
Free Gait								
Stride length	.82[c]	−.23	−.30	.00	.69[d]	−.24	.34	−.47[e]
Step length (adjusted)		−.33	−.16	.18		−.48	.43[e]	−.35
Stride width			.13	−.49[e]			−.85[c]	.23
Angle of feet				.03				−.08
Fast Gait								
Stride length	.88[c]	−.12	−.05	.08	.74[c]	−.04	−.39	−.48[e]
Step length (adjusted)		−.22	.04	.18		.11	−.24	−.31
Stride width			−.53[b]	−.06			−.64[d]	−.15
Angle of feet				−.13				.06

[a] $p < .01$, [b] $p < .025$, [c] $p < .0005$, [d] $p < .005$, [e] $p < .05$.

Step Length (Adjusted for Leg Length)

Significant differences in mean step length were due only to speed of gait (Tab. 3). Mean step length at fast speed was significantly longer than at slow ($p < .01$) or free ($p < .01$) speeds. Mean step length at free speed was significantly longer ($p < .01$) than at slow. When mean step lengths of 3- and 5-year-olds were compared at each speed, no significant differences were present.

Within subjects, the 3-year-old group, compared with the 5-year-old group, had significantly greater variance of step length at slow ($F = 4.0$, $df = 14, 15$, $p < .01$) and free ($F = 5.54$, $df = 14, 15$, $p < .001$) speeds. At a fast speed, variability of step length of the 3-year-old group equaled the variability of the 5-year-old group.

Stride Width

Significant differences in mean stride width were due only to speed of gait (Tab. 4). Mean stride width was significantly narrower at a fast speed than at a free ($p < .01$) or slow ($p < .01$) speed. There was no significant difference in mean stride width between free and slow speeds. When mean stride width of 3-year-olds and 5-year-olds was compared at each speed, no significant differences were present.

Within subjects, the 5-year-old group, compared with the 3-year-old group, had significantly greater variance of stride width at a slow speed ($F = 6.8$, $df = 14, 15$, $p < .001$). Within-subject variability of the 3-year-old group approximately equaled that of the 5-year-old at free speed, and no significant difference was present at fast speed.

Included Angle of Feet

Included angle of feet was not significantly different even at the .10 level for speed, age, or the interaction of speed and age. When included angle of feet of 3- and 5-year-olds was compared at each speed, no significant differences were present.

Within-subject variability of included angle of feet of the younger group approximately equaled that of the older group at slow and fast speeds. No difference was present at a free speed.

Cadence

Differences in cadence were significant for speed and age but only reached the .10 level for the interaction of speed and age (Tab. 5). There was a significantly higher cadence at fast than at free ($p < .01$) or slow ($p < .01$) speeds. Also, there was a significantly higher cadence at free than at slow speed ($p < .01$).

When the cadence of the 3- and 5-year-old groups was compared at each gait speed, the cadence was not significantly different at slow or free speeds. Only at a fast speed did the older group have a significantly higher cadence than the younger group ($p < .01$).

TABLE 7
Correlations Between Gait Factors (Ages Combined) at Each Gait Speed, and Raw Motor Scores on the McCarthy Scales of Children's Abilities (df = 29)

Gait Factors	Slow	Free	Fast
Stride length	.42[a]	.51[b]	.60[c]
Step length (adjusted)	−.23	.19	.16
Stride width	.07	.01	.07
Angle of feet	−.03	−.13	−.14
Cadence	−.40[d]	−.31[e]	−.40[d]

[a] $p < .01$, [b] $p < .005$, [c] $p < .0005$, [d] $p < .025$, [e] $p < .05$.

Correlations of Gait Factors

Table 6 presents the correlations between gait factors at each age and speed. For the 5-year-old group, significant correlations existed between stride and step length at all speeds. In the 3-year-old group, significant correlations between stride and step length occurred only at free and fast speeds.

Significant correlations were also present between stride or step length and other gait factors. When stride length and width were correlated, only the older group at a slow speed demonstrated a significant inverse relationship. When step length and stride width were correlated, significant inverse relationships were present in the 5-year-old group at both slow and free gait speeds. Five-year-olds at a slow speed had similar positive correlations between stride or step length and included angle of feet, whereas 3-year-olds at the slow speed had an inverse correlation between stride length and included angle of feet and no significant correlation between step length and included angle of feet. At free or fast speeds, the only significant correlation between stride or step length and included angle of feet was a positive correlation between step length and included angle of feet for 5-year-olds at a free speed. At a slow speed, only the 3-year-olds with a longer stride length had a lower cadence. At free and fast speeds, only 5-year-olds with longer stride lengths had lower cadences. At neither age nor at any speed did step length significantly correlate with cadence.

At each age and speed, with the exception of 3-year-olds at a free speed, children with a larger stride width had a significantly narrower included angle of feet. Only 3-year-olds at a free speed had a significant relationship between stride width and cadence; those with a lower cadence had a wider stride. No significant correlations between included angle of feet and cadence were present at either age or any speed.

Table 7 presents correlations between gait factors and raw scores of the motor section of the McCarthy Scales of Children's Abilities. Only stride length and cadence significantly correlated with the raw motor score. The raw motor score increased as stride length increased at each gait speed. As the raw motor score increased, cadence decreased at each gait speed.

DISCUSSION

The results of this study emphasize the complexity of gait and gait development in 3- and 5-year-old children. Perhaps the interrelationships between gait factors, age, and speeds, as well as the relationships among gait factors, present a more realistic analysis of gait in these subjects than if the variables were considered in isolation.

Variability of Gait Factors

Because adults ambulate with a high consistency of gait details within a variety of speeds, one would expect children with a more mature gait to demonstrate less variability of gait factors within a variety of speeds. As expected, at slow and free speeds, the older children had less within-subject variability of stride and step length and, therefore, had fewer extraneous stride or step length movements than the younger children. At slow and free speeds, the older children were perhaps better able to control stride and step length despite the increased need for stability at slower gait speeds. At a fast speed with less need for medial-lateral stability,[2,15] and presumably less time spent in a unilateral support position, the 3-year-olds probably had less difficulty than at slower gait speeds and, therefore, variability of stride and step length at a fast speed was not statistically different from that of the 5-year-olds.

The 5-year-old group had a greater within-subject variance of stride width at slow speed than the 3-year-old group. These unexpected results help support the theories that the gait of 5-year-olds is not fully matured.[14,19] Perhaps stride-width consistency is established together with a cluster of other gait and skeletal factors, as will be discussed later, or perhaps Burnett's discussion of Gesell and Amatruda's concept of reciprocal interweaving can be applied.[3] The 5-year-old may have a more consistent stride or step length at a slow speed, but may temporarily revert to a more variable stride width in order to maintain lateral stability, especially at a slow speed.

Within-subject variance of included angle of feet of the two age groups was not significantly different. Variability of included angle of feet, therefore, may be established at an age prior to 3, or perhaps it is incorrect to consider included angle of feet without also considering other factors such as stride length and width.

In general, within-subject variability of gait factors as one means of assessing gait maturation has often been overlooked. Further research is needed to determine how variability of children's gait factors decreases, and which gait factors and conditions can be used appropriately to determine gait maturity. Consistency of gait factors in addition to the factors of increased age and varying speeds may better describe the development of preschool children's gait.

Stride and Step Length

As hypothesized, both stride and step length significantly increased with speed in both groups. Therefore, the mature relationship of increasing stride or step length with speed[12, 13, 23] is established before the age of three. Although it was expected that both stride and step length would increase with age, only stride length, not step length (adjusted for leg length) increased. This result is in agreement with the studies on adults[10] and older[1, 2] children. At free gait speed, an adult male's stride length is systematically related to height.[10] Sutherland[2] and Scrutton[1] found that at a free gait speed, minimal change in children's step length (adjusted for leg length) occurs subsequent to 3 years of age.

As expected, the 5-year-olds, compared with the 3-year-olds, had a significantly longer stride length at free and fast speeds. Either the difficulty both groups may have experienced walking at a slower gait speed, or lack of the medial-lateral and anterior-posterior stability in the supporting limb[2, 15] at a slow gait speed may have caused a similarity of 3- and 5-year-olds' stride lengths at a slow speed. The finding that the stride length of the 5-year-old group was not significantly longer than the 3-year-old group at a slow speed, as well as the lack of consistency of stride width at a slow speed, may indicate absence of a fully mature gait pattern in the 5-year-old group.

Because adult gait is a highly automated and consistent motor activity and step length is a part of stride length, one would expect to find a consistent relationship between stride and step length in both the adult[11] and preschooler gait. At both ages and at all speeds, except for 3-year-olds at a slow speed, the expected significant stride-step length correlation was present. The low stride-step length correlation of the 3-year-old group at a slow speed supports the previous observation that the slower speed was more difficult, especially for the younger children. Previous children's gait studies have not addressed this issue.

Stride Width and Included Angle of Feet

Because Morton and Fuller[25] suggest that the stride width and angle of foot placement together meet the adult's need for medial-lateral and anterior-posterior stability, stride width, and included angle of feet will, therefore, be discussed at the same time. The increased need for the lateral stability at a slower speed may have been the factor that caused the children's wider stride at a slower speed. The responses of the children on some of the tasks in the McCarthy Scales of Children's Abilities indicated that balance was not fully developed. In contrast to the adult,[13, 24] the included angle of feet of the 3- to 5-year-old children did not significantly decrease, a finding that again may be attributed to the need for lateral stability and balance in this group of children. As in the adult, the stride width and the angle of foot placement together help meet the children's gait stability needs.[13, 24] The method by which preschool children combine the two gait factors to accomplish gait stability with modified speed, however, differs from that of adults. Either stride width and included angle of feet are established at an early age, or the development of both factors are indeed more complex than previously thought. The complexity of gait development is demonstrated by the very different relationships in the 3- and 5-year-old groups when stride or step length and included angle of feet were correlated. Only the 3-year-old group at a free speed did not have a significant inverse relationship between stride width and included angle of feet (Tab. 2). The effect of age, speed, and within-subject variability on stride width and included angle of feet demonstrate that an adult pattern of gait with regard to these gait factors has not yet become established in these 3- and 5-year-old children. The varying relationships between stride width or included angle of feet (Tab. 2) and other gait factors at specific ages or speeds, as well as the unexpected direction of the relationship of stride width change due to speed, emphasizes the complex methods by which medial-lateral and anterior-posterior stability requirements are satisfied in a gait pattern that is not fully matured. Further study is required to understand the development of stride width and included angle of feet. Additional variables should be included such as angle of femoral neck, tibial femoral angle,[3] pelvic span as compared with ankle span, rotations of lower extremities and trunk, height of center of gravity, and duration of single stance during gait.

Cadence

Sutherland determined that a mature gait occurred at 3.4 to 4 years of age.[2] The criterion for gait maturity at free speed was a decrease in cadence with a concomitant increase in step length and walking velocity. In the present study, however, it was found that only at a fast gait speed did the 5-year-olds have a significantly lower cadence than the 3-year-olds. The difference in results may be because other studies either used fewer preschool children[5] or statistically analyzed data for a different purpose.[2] Other studies did not statistically analyze cadence for the interaction of both age and gait speed.[1, 4] Because the fast gait speed may be the most automatic of the speeds, the older children may have demonstrated the more mature gait pattern in regard to cadence.

When considering the variety of correlations between cadence and stride length (Tab. 2) of the 3- and 5-year-olds, it is apparent that the method by which the two age groups have attained a cadence at a designated speed has not been fully established. Leg length, the need for anterior-posterior and medial-lateral and single stance mobility, and the development of stride width and angle of feet are all factors that may determine the manner in which cadence is established.

Motor Development and Gait Factors

In this study, the initial assumption was that children with more mature gaits would have a higher motor development level; however, only stride length (unadjusted for leg length) and cadence correlated significantly with the raw motor scores. Because both stride length and cadence are dependent on leg length or height, the correlations may have reflected the longer leg lengths of the older children. A more in-depth motor evaluation needs to be used to determine whether there are more subtle motor developments related to gait factors and gait development. Older children accomplished some of the motor tasks on the McCarthy Scales with greater ease than the younger children although they did not always obtain a higher score on the task. The total possible points earned for these particular tasks, however, were often minimal, or perhaps the scoring system was not sufficiently graded to incorporate subtle developmental accomplishments. The use of a detailed motor test could shed more light on the developmental skills that relate to a more mature gait. The relationships of motor and gait development is an area that warrants further investigation.

Observations and Implications

Slow and fast gait speeds should have been controlled at 70 and 120 cm/sec respectively. The actual average slow speed (51.9 cm/sec) was slower than anticipated. The planned slow speed (70 cm/sec) was, therefore, more similar to the average free speed of 83.8 cm/sec than had originally been anticipated. Because the children were always asked to walk first at a free speed, they may have identified the proposed slow speed as too close to their normal free speed and, therefore, slowed their gait and lagged behind the investigator. The actual fast gait speed of 128.6 cm/sec was very similar to the intended speed, 120 cm/sec.

During the testing procedure for both age groups, the slow gait speed appeared to be the most difficult and the fast gait speed the least difficult. This hierarchy of relative difficulty was based on observations of facial expressions and upper extremity movements. The trend can also be seen in some of the objective gait factor measurements previously discussed.

Conclusions

The results and implications of this study can only be cautiously generalized beyond the group of children studied. Although analysis of variance has as one of its assumptions the presence of a random sample, this study did not have a totally random sample. Furthermore, the sample was relatively small. Also, although this study involved the linear and angular gait measurements that can be measured by the feet, important temporal and angular measurements of more proximal areas involved in gait were not included. Given the above cautions, however, this study has significance and implications for future study.

The gait of two normal 3- and 5-year-old groups has been described and compared at three different gait speeds using a clinical method that can be easily and inexpensively duplicated. The statistical testing of gait data collected for the two groups at three different speeds has emphasized the complexity of gait development between different aged children and in comparison to adults. To obtain more information about the influence of gait speed on gait factors, researchers in future gait studies of preschool children should continue to control speed of gait as they test gait factors. Future gait studies should also involve additional numbers of subjects and statistically test collected data to separate spurious findings attributable only to chance from those of statistical significance. Motor development and gait development should be studied together so that physical therapists working with handicapped preschool children will have a more knowledgeable basis for measuring and treating developmental disabilities.

Acknowledgments. The author wishes to thank Jane Coryell, PhD, and Cynthia Norkin for their advice in the research and writing of this article.

REFERENCES

1. Scrutton DR: Footprint sequences of normal children under five years old. Dev Med Child Neurol 11:44–53, 1969
2. Sutherland D, Olshen RA, Copper L, et al: The development of mature gait. Read at Orthopedic Research Society of the American Academy of Orthopedic Surgeons, Dallas, 1978, 1–19
3. Burnett CN, Johnson EW: Development of gait in childhood Part I and II. Dev Med Child Neurol 13:196–215, 1971
4. Endo B, Kimura T: External force of foot in infant walking. J Faculty of Science, The University of Tokyo, section 5, vol 4, part 2:103–117, March, 1972
5. Grieve DW, Gear RJ: The relationships between length of stride, step frequency, time of swing and speed of walking for children and adults. Ergonomics 5:379–399, 1966
6. Okamoto T, Kumamoto M: Electromyographic study of learning process of walking in infants. Electromyography 12:149–151, 1972
7. Horton ME: Development of movement in young children. Physiotherapy 57:148–158, 1971
8. Stott L: Child Development: An Individual Longitudinal Approach. New York, NY, Holt, Rinehart & Winston General Book, 1967, pp 20–144
9. Gesell A: Reciprocal interweaving in neuromotor development. J Comp Neurol 70:161–180, 1939
10. Murray MP, Drought AB, Kory RC: Walking patterns of normal men. J Bone Joint Surg [Am] 46:335–360, 1964
11. Murray MP: Gait as a total pattern of movement. Am J of Phys Med 46:290–333, 1967
12. Andriacci TP, Ogle JA, Galonte JO: Walking speed as a basis for normal and abnormal gait measurement. J Biomech 10:261–268, 1978
13. Murray MP, Kory RC, Clarkson BH: Comparison of free and fast speed walking patterns of normal men. Am J of Phys Med 45:8–24, 1966
14. Bernstein NA: Coordination and Regulation of Movements. New York, NY, Pergamon Press Inc, 1967, pp 6–95
15. Stratham L, Murray MP: Early walking patterns of normal children. Clin Orthop 79:8–22, 1971
16. McGraw MG: The Neuromuscular Maturation of the Human Infant. New York, NY, Hafner Publishing Co, 1963, pp 74–92
17. Shirley MM: The First Two Years. A Study of Twenty-five Babies I: Postural and Locomotor Development. Minneapolis, MN, University of Minnesota Press, 1931, pp 52–168
18. Popova T: Rudiments of erect forward movement before independent walking in children. In Bernstein NA (ed): Investigations on the Biodynamics of Walking, Running, Jumping. Central Scientific Institute of Physical Culture, Moscow, 1940, chap 4, pp 34–35
19. Tsurumi N: Electromyographic study on the gait of children. Journal of the Japanese Orthopedic Association 43:611–628, 1978
20. Connolly K: Response speed, temporal sequencing and information processing in children. In Connolly K (ed): Mechanisms of Motor Skill Development. London, England, Academic Press Inc, 1970
21. McCarthy D: Manual for McCarthy Scales of Children's Abilities. New York, NY, Psychological Corp, 1972
22. Ogg, HL: Measuring and evaluating the gait patterns of children. Phys Ther 43:717–720, 1963
23. Smidt GL: Hip motion and related factors in walking. Phys Ther 51:9–21, 1971
24. Murray MP, Kory RC: Walking patterns of normal women. Arch Phys Med Rehabil 51:637–650, 1970
25. Morton DJ, Fuller DD: Human Locomotion and Body Form. Baltimore, MD, Williams & Wilkins, 1952

Basic Research

Gait Cycle Duration in 3-Year-Old Children

DARLENE S. SLATON

The gait cycle duration of 11 children who were 3 years old was recorded using cinematography. The variation of cycle duration from cycle to cycle during free walking with each child and the variation of mean cycle duration among the children are reported. I analyzed individual subject characteristics including sex, height, weight, leg length, and age at onset of independent walking for strength of relationship to recorded cycle duration. Results suggest that cycle duration in 3-year-old children during free walking is shorter than that expected for adults. A minimum of variability exists in average cycle duration across subjects at 3 years of age; however, considerable variation of cycle duration from cycle to cycle within individual subjects is not uncommon. I identified no significant relationships between subject characteristics and gait cycle duration. The data reported are consistent with the results of other studies showing a trend of gradually increasing cycle duration and decreasing variability with increasing age in early childhood.

Key Words: *Child development, Gait, Research design.*

Researchers have studied human locomotion extensively over the past 40 years. Quantitative norms have been reported for temporal and spatial characteristics of gait, angular displacements, patterns of muscular activity, and many other characteristics of normal adult gait.[1-5]

Less information is available regarding gait during childhood, especially during the preschool years. A casual observer, even with an untrained eye, recognizes that a young child's gait is distinctly different from that of an adult in speed, coordination, and patterns of limb movement. Yet, few norms are available for gait in childhood.

This study reports descriptive information regarding timing characteristics of childhood gait. I chose cycle duration because it can be measured simply and accurately without awkward devices or markers attached to the child. Cycle duration is also a characteristic that can easily be recorded quantitatively by therapists in the clinical setting.

LITERATURE REVIEW

Human locomotion is an extraordinarily complex phenomenon. It has been studied extensively with a wide range of techniques and conceptual approaches. Saunders and associates defined normal locomotion as the "translation of the center of gravity through space along a pathway requiring the least expenditure of energy."[1] They described the following six determinants of normal human locomotion:

Ms. Slaton is Clinical Assistant Professor, Division of Physical Therapy, Department of Medical Allied Health Professions, School of Medicine, Medical School Wing E 222H, University of North Carolina at Chapel Hill, Chapel Hill, NC 27514 (USA).

This research was supported in part by Grant 149, Postgraduate Programs for Physical and Occupational Therapists in Maternal and Child Health Care, Department of Health and Human Services, Public Health Service, Health Services Administration, Bureau of Community Health Services, Rockville, MD 20852.

This article was submitted December 2, 1983; was with the author for revision 13 weeks; and was accepted July 11, 1984.

1. Pelvic rotation—the alternate movement of the pelvis to the right and left relative to the line of progression.
2. Pelvic tilt—the downward list of the pelvis relative to the horizontal plane on the side opposite to that of the weight-bearing limb.
3. Knee flexion at mid-stance.
4. and 5. Foot and knee mechanisms working together at heel and toe-off.
6. Lateral displacement of the pelvis over the weight-bearing limb.

Saunders and associates suggested that these determinants are necessary for minimizing the displacement of the center of gravity during forward progression and, thus, maximizing the efficiency of gait.

Milner and associates suggested that the normal cadence chosen by a free-walking adult requires a minimum of muscular activity.[2] Normalcy in gait, therefore, seems to be closely related to efficiency. Studies by Murray,[3] Tucker,[4] and Dubo and associates[5] implied a similarity of mean cycle duration among healthy, free-walking adults (Murray, 1.06 sec; Tucker, 1.0–1.1 sec; Dubo, 1.13 sec). If Milner and associates' theory is valid, then the results of these three studies, when averaged, suggest that a cycle duration of approximately 1.1 sec or a cadence of 109 steps/min represents the most efficient gait speed for most adults.[2]

Current literature suggests that gait cycle duration during free walking is shorter in the preschool child than in the adult and that cycle duration tends to lengthen as age increases (range: 0.68 sec at 1 year of age to 0.96 sec at 5 years of age).[6-11] Other characteristics of adult gait are well-established by the age of 3 years with refinement continuing on into later childhood.[7,9] Little quantitative information is available about the variability of gait characteristics within individual children or changes in the amount of variability with growth and maturation.

METHOD

Subjects

The subjects were 11 healthy white boys and girls, who were 3 years of age (36 to 47 months) and who attended a preschool program in Eastern North Carolina. The subjects were not randomly selected and, therefore, the sample may be biased in socioeconomic factors, education of parents, or other unmeasured variables.

Equipment

I recorded cycle duration by cinematography using a teledyne 16-mm movie camera* with a 25-mm lens. The camera was stationary on a stable surface located 40 ft† directly perpendicular to a smooth, black rubber walkway elevated 6 in‡ on a stage. I used black and white Kodak Tri-X Reversal film 7278 (ASA 164) at a speed of 100 frames per second. Two tungsten spotlights cross-lighted the walkway area. Distance multipliers were located at 1-ft intervals along the walkway with white adhesive tape, and a plumb bob for vertical orientation was hung in the background. Black curtains were hung behind the walkway to absorb excessive light and to provide adequate contrast. The camera's field captured a 15-ft length of the walkway, which extended 8 ft on each end to provide sufficient distance for warm-up and slow-down steps. The filmed portion of the walkway allowed the camera to capture three to five complete gait cycles for each trial. After processing, I used a Vanguard motion analyzer** with x- and y-coordinate markers and a computerized counter to analyze the film frame by frame.

Procedure

The protocol for the study was approved by the committee for the protection of the rights of human subjects and consent forms were signed by parents of all children who participated as subjects. During a standard interview with each potential subject's parent, I reviewed a screening questionnaire that included questions about age, health, and birth and developmental history. To assure that the child had no apparent developmental delays, I also administered a Denver Developmental Screening Test (DDST) for each child.[12] I completed both the parent interview and the DDST in the child's home within two weeks before filming.

The actual filming procedure required no more than 15 minutes. I filmed each child separately. After entering the room, the child was asked to remove his or her shoes and socks. I measured height, weight, and the distance from the right anterior superior iliac spine to the floor.

I then escorted the child to the beginning of the walkway and gave the following instructions:

> Start here. When I say go, walk to the other side of the stage, turn around and come back. Wait here until I say go. Go.

I gave no instructions regarding speed, arm swing, or other gait characteristics. Children who seemed hesitant or confused were asked to retrieve a 1-in red cube placed at the far end of the walkway. I attached no devices or markers to the child. At least three opportunities to practice were provided before filming began. For a pass across the walkway to be included in the data, the child had to have at least one foot in partial contact with the floor at all times to exclude the possibility of running and had to have no pause in forward progression. I filmed two passes across the walkway for each child and included data from both passes in the analysis. Subjective comments regarding the child's postural attitude or reluctance to participate were recorded by me.

Data Analysis

Because the time of toe-off could be identified more distinctly than the time of heel strike, gait cycle durations were recorded from toe-off to successive toe-off of the same foot. For the purposes of film analysis, I defined toe-off as the first upward or forward progression of the toe after initial toe contact. Slight increments of upward or forward progression were determined by using the x- and y-axes as reference indicators. I recorded the number of frames required for each complete cycle. At a film speed of 100 frames per second, each frame is equivalent to 0.01 sec. Reliability trials of my counting indicated differences in recorded cycle durations on three different days were not more than 0.01 sec (1 frame).

Step length was recorded in a similar manner. The distance from toe-off of one foot to toe-off of the opposite foot was recorded along the x-axis. I then calculated actual distance by using the distance multiplier marked on the walkway. Every individual step length and cycle duration included on the film was recorded for each child. After recording cycle durations and step lengths, I ran the film at full speed and recorded my subjective comments regarding general posture and upper extremity position.

I calculated Pearson product-moment correlation coefficients to test the strength of relationship between cycle duration and sex, age, age at onset of independent walking, number of months of experience with independent walking, height, weight, and leg length.

RESULTS

Table 1 presents individual characteristics and average cycle durations for each subject. The age at onset of independent walking ranged from 11 to 16 months. The range of independent walking experience was from 24 to 35 months. Height and weight ranged from the 10th to the 90th percentile compared with the anthropometric chart of the Children's Medical Center in Boston. Leg length varied from 44.4 to 50.8 cm.

Individual mean step lengths varied from 32.1 to 44.3 cm with an overall sample mean of 38.1 cm ± 3.9 cm. Individual mean cycle durations ranged from 0.69 to 0.86 sec with an overall sample mean cycle duration of 0.74 sec ± 0.06 sec.

Table 2 is a sample of the recordings for cycle duration of each subject and the subjective notations I made for each child during film analysis (Tab. 2). The average standard deviation for cycle duration within an individual subject is less than 0.06 sec; however, I recorded variations from the mean of up to 0.19 sec. No relationship seems to exist between the value of mean cycle duration and the amount of variations

---

* Teledyne Camera Systems, 131 N 5th Ave, Arcadia, CA 91006.
† 1 ft = 0.3048 m.
‡ 1 in = 2.54 cm.
** Vanguard Instruments Corp, 1860 Walt Whitman Rd, Melville, LI, NY, 11746.

TABLE 1
Subject Characteristics

Subject #	Sex	Age (mo)	Age at Walking Independently (mo)	Months of Walking Experience	Height (cm)	Weight (lb)[a]	Leg Length (cm)	Mean Cycle Duration (sec)
1	M	37	12	25	97.8	31	47.0	0.72
2	F	44	16	28	96.6	30	48.2	0.69
3	M	37	13	24	96.6	36	48.2	0.72
4	M	38	11	27	99.1	37	50.8	0.73
5	M	42	14	28	95.2	36	45.8	0.85
6	F	41	14	27	102.9	40	48.3	0.69
7	F	47	12	35	100.3	33	48.2	0.73
8	M	38	13	25	99.1	37	47.0	0.69
9	M	44	12	32	100.3	38	49.5	0.86
10	F	40	15	25	100.3	40	48.3	0.71
11	M	45	14	31	96.5	31	44.5	0.74
$\bar{X}$		41	13	28	98.6	35	47.8	0.74
s		3.49	1.49	3.45	2.29	3.58	1.70	0.06

[a] 1 lb = 0.4536 kg.

demonstrated by a subject. I found no significant relationships between cycle duration and subject characteristics (Tab. 3).

The subjective notations I made at the time of filming and during film analysis suggest that all of the children were uncomfortable during the procedure. Their facial expressions seemed to relay a cautious attitude, and most of the children held one hand to their mouth or looked at the camera during part of the filming. Nine of the children demonstrated a normal reciprocal arm swing, but only two of these children maintained the arm swing throughout the filming procedure. No relationship was identified between upper extremity position or postural attitude and the duration of corresponding gait cycles.

DISCUSSION

The mean cycle duration of 0.74 sec is consistent with the mean cycle duration of 3.5-year-olds reported by Sutherland and associates.[9] The reason for the shorter cycle duration or increased cadence of the children in comparison with that of adults is unclear. Statham and Murray suggested that increased cadence may be an attempt to decrease the lateral displacement of a high center of gravity.[6] In the adult, the center of gravity lies just anterior to the second sacral vertebra.[13] In the young child, it lies above the umbilicus.[13] Indeed, Murray found that in adults, faster walking speeds did decrease the lateral displacement of the trunk.[3] If increased cadence is an attempt to decrease the lateral displacement of a high center of gravity, then as the child grows older and the center of gravity gradually descends, the cycle duration should also gradually lengthen. The results reported here and those of Sutherland and associates[9] seem to support this hypothesis.

An alternate explanation is related to the acquisition of the six determinants of normal gait as defined by Saunders and associates.[1] If the shorter cycle duration in the child who has been walking a short period of time is caused by the lack of these characteristics, then the cycle duration should lengthen to approximately that of an adult by the time a mature gait is achieved. Sutherland and associates' data suggest that cycle duration rapidly lengthens during the initial period of inde-

Table 2
Individual Cycle Durations (Subject #5)

Cycle #	Cycle Duration (sec)	Subjective Comments
1	0.84	Head turned to the right
2	0.93	toward camera
3	0.81	(cycles 1, 5, 6)
4	0.84	
5	0.88	
6	0.88	
7	0.89	Left hand to mouth
8	0.89	(cycles 1–8)
9	0.79	
10	0.82	Normal arm swing,
11	0.86	head slightly flexed in
12	0.80	midline (cycles 9–12)
$\bar{X}$	0.85	
range	0.79–0.93	

pendent walking until 3 years of age.[9] From 3 years of age to 7 years of age, this lengthening continues but at a slower rate. This suggests that cycle duration lengthens in direct relation to the acquisition and refinement of mature gait characteristics.[7,9] A longer cycle duration would then indicate greater refinement of gait. I did not measure the characteristics of mature gait defined by Sutherland and associates[9] and by Saunders and associates.[1] The relationship between mature gait characteristics and cycle duration, therefore, could not be determined.

The gradually decreasing variability of cycle duration across subjects with increasing age reported by Sutherland and associates may be related to the variability of motor development rates across children that gradually converge in the later preschool years.[9] Children who develop motor skills earlier than expected may acquire the six determinants of mature gait before children who develop more slowly. These developmental differences result in a wide distribution of values across subjects from 1 to 3 years of age. The results of this

Table 3
Pearson Product-Moment Coefficients

	Cycle Duration	Sex	Age	Age at Walking Independently	Months of Walking Experience	Height	Weight	Leg Length
Cycle Duration	1.00	−.44	.32	−.18	.42	.05	.10	.06
		NS	NS	NS	NS	NS	NS	NS
Sex		1.00	.41	.47	.19	.44	.11	.18
			NS	NS	NS	NS	NS	NS
Age			1.00	−.08	.91	.08	−.36	−.23
				NS	NS	NS	NS	NS
Age at Walking Independently				1.00	.29	.16	.39	−.02
					NS	NS	NS	NS
Months of Walking Experience						.15	.27	.02
						NS	NS	NS
Height						1.00	.72	.33
							$p = .006$	NS
Weight							1.00	.47
								NS
Leg Length								1.00
								NS

study, however, demonstrated no relationship between the age at onset of independent walking and gait cycle duration at 3 years of age. The standard deviation of cycle duration for 3 year olds of 0.06 sec reported here is similar to that reported by Sutherland and associates for 3.5 year olds (0.07 sec).[9] Both of these are smaller than the standard deviation of 0.09 sec reported by Murray for cycle duration in adults.[3]

Values of variation in cycle duration within one individual subject for children or adults have not been reported. In this study, the average standard deviation in the 3-year-old child was 0.06 sec with ranges up to 0.33 sec. More information is needed regarding the normal range of cycle duration in an individual. The amount of variation may be as important an indicator of normalcy as mean cycle duration alone. Large variations may be indicative of gait pathology.

Although no significant relationships were found between cycle duration and the characteristics of individual subjects recorded in this study, relationships may still exist. The failure to identify relationships may be a result of the age of the sample chosen for this study. A relationship may be found between the number of months of independent walking and cycle duration if the sample includes children 18 to 35 months of age. A similar relationship may also exist between cycle duration and leg length. A larger sample must be studied to identify such relationships.

If strong relationships can be identified between cycle duration and the acquisition of the six determinants of mature gait or even between cycle duration and number of months of independent walking, cycle duration could possibly become a simple quantitative value that clinicians could use to assess further the maturity or abnormality of gait in childhood. Changes in recorded cycle duration could reflect improvements or regression in gait over time.

Collecting quantitative data regarding gait characteristics from young children who are by nature wary of strange and unusual circumstances is difficult. The subjective comments I recorded at the time of filming and during film analysis suggest that the data collected may be somewhat distorted by the emotional state of the subjects during filming. The cautious attitude of the children probably resulted in a decreased cadence and a slightly longer gait cycle duration than is normal for free walking.

Similar subjective data were not reported in the studies of other investigators. Many of these studies required the attachment of switches, straps, cables, dowels, and many other devices to the children during data collection. Interrupted light photography requires a darkened room with brightly flashing strobe lights throughout the recording session. These methods would appear to have an even greater psychological effect on the children than my methods. Quantitative measurements of the effect of these external variables on the emotional state of the child are not available. Their effect, however, must be considered when the information gathered by these methods is used for comparison with children in the clinical setting.

Further investigation is needed to establish norms for gait characteristics during the preschool years. Studies involving a much larger sample followed longitudinally as well as cross-sectionally could explore the relationship between the acquisition of mature gait characteristics and changes in cycle duration over time. Specific information regarding the variability of cycle duration within individual subjects at all ages also needs to be collected.

To capture the most natural and valid representation of gait, a method of data collection that eliminates the need for attachment of markers and recording devices to the child is desirable. Every effort needs to be made to record the characteristics of the child's gait in the most natural environment possible.

Ultimately, studies will need to be completed to determine the feasibility of using gait cycle duration or cadence as a quantitative measurement of gait maturity or abnormality in the clinical setting. Normative data must be gathered using methods that are practical for use by therapists in a variety of clinical situations. Perhaps, when collecting data, investigators should use a number of instruments concurrently with each child (eg, stopwatch, cinematography, or foot switches) to establish guidelines and expected cycle duration values for each method of recording.

Additional studies are needed to identify changes that occur with pathological conditions. Single-case studies could provide helpful information about changes in cycle duration that occur as a result of therapeutic intervention or the acquisition of developmental motor skills.

CONCLUSIONS

The cycle duration of 3-year-old children during free walking is shorter than that of adults. The reason for this shorter cycle duration is unclear, but a higher center of gravity in children and immature gait patterns may influence cycle duration. The variability across subjects is no greater than that expected for adults. The data reported here are consistent with the results of other studies showing a trend of gradually lengthening cycle duration and decreasing cycle variability across subjects with increasing age in early childhood.

REFERENCES

1. Saunders JB, Inman VT, Eberhart HD: The major determinants of normal and pathological gait. J Bone Joint Surg [Am] 35:543–558, 1953
2. Milner M, Basmajian JV, Quanbury AO: Multifactorial analysis of walking by electromyography and computer. Am J Phys Med 50:235–258, 1971
3. Murray MP: Gait as a total pattern of movement. Am J Phys Med 46:290–333, 1967
4. Tucker J: The Clinical Implications of Timing in Patients with Central Nervous System Disorders. Read at the Conference on Recovery of Motor Function Following Brain Damage, Columbia University, New York, NY, March 9–13, 1979
5. Dubo HIC, Peat M, Winter DA, et al: Electromyographic temporal analysis of gait: Normal human locomotion. Arch Phys Med Rehabil 57:415–420, 1976
6. Statham L, Murray MP: Early walking patterns of normal children. Clin Orthop 79:8–24, 1971
7. Burnett CN, Johnson EW: Development of gait in childhood: Part II. Dev Med Child Neurol 13:207–215, 1971
8. Rose-Jacobs R: Development of gait at slow, free, and fast speeds in 3- and 5-year-old children. Phys Ther 63:1251–1259, 1983
9. Sutherland DH, Olshen R, Cooper L, et al: The development of mature gait. J Bone Joint Surg [Am] 62:336–353, 1980
10. Aptekar RG: Light patterns as a means of assessing and recording gait: 1. Methods and results in normal children. Dev Med Child Neurol 18:31–36, 1976
11. Scrutton DR: Footprint sequences on normal children under five years old. Dev Med Child Neurol 11:44–53, 1969
12. Frankenburg WK, Dodds JB, Fandal AW: Denver Developmental Screening Test. Denver, CO, University of Colorado Medical Center, 1967
13. Watson EH, Lowrey GH: Growth and Development of Children. Chicago, IL, Year Book Medical Publishers Inc, 1967, pp 120–127

Basic Research

Comparison of Gait of Young Women and Elderly Women

PATRICIA A. HAGEMAN
and DANIEL J. BLANKE

> The purpose of our study was to describe and compare free-speed gait patterns of healthy young women with healthy elderly women. The evaluation was completed with high-speed cinematography using synchronized front and side views of 26 healthy volunteers. One group was composed of 13 subjects 20 to 35 years of age, and the other group was composed of 13 subjects 60 to 84 years of age. Each subject participated in one test session consisting of three filmed trials of free-speed ambulation down a 14-m walkway. The processed film was analyzed for 10 gait characteristics. Differences in gait characteristics between the two groups were examined using a correlated t test ($p < .01$). The elderly women demonstrated significantly smaller values of step length, stride length, ankle range of motion, pelvic obliquity, and velocity when compared with the younger women. The results of our study suggest that the physical therapist should not establish similar expectations for young women and elderly women during gait rehabilitation.
>
> **Key Words:** *Gait, Physical therapy.*

Understanding the effects of aging on movement and function is becoming increasingly important because of longer average life spans and a growing elderly population. Changes in stereotypic movements such as walking patterns have been reported as early as 60 years of age.[1,2] Because elderly people frequently utilize physical therapy services to achieve the maximal functional ability of motor activities such as gait, the collection of gait analysis data of healthy elderly subjects is essential to establish realistic rehabilitation expectations of the elderly population.

Although advances in technology have made objective methods of measuring gait more available, the physical therapy literature contains only a few studies of gait comparisons of healthy elderly women with healthy young women.[3-5] These studies report that elderly women demonstrate shorter step and stride lengths, lower average velocities,[3,5] and greater variability in stride width[4] when compared with young women. None of these studies, however, provides conclusive evidence of the effects of aging on the gait patterns of the elderly population because of small sample sizes and limitations in the number of specific gait characteristics examined.

Additional comparisons of the gait characteristics of healthy young women and healthy elderly women are necessary because gait training is a major portion of geriatric physical therapy rehabilitation and because women constitute a majority of the over-60-years age group. The purpose of this study was to describe and compare the free-speed gait characteristics of matched groups of healthy young women and healthy elderly women.

METHOD

Subjects and Selection Procedure

Twenty-six female volunteers, thirteen 20 to 35 years of age and thirteen 60 years of age or older, were accepted as subjects. Each subject provided informed consent in accordance with the procedures of the University of Nebraska Institutional Review Board.

The health status of each subject was evaluated on the basis of a medical review and an objective examination by a registered physical therapist, and all subjects were found to be free of disabling physical conditions or minor ailments that could affect or influence locomotion. Specifically, the subjects were without musculoskeletal or neurological involvement or medication for these conditions. Because leg length is an important determinant of stride length,[6] leg length was measured to ensure that each subject was without a leg-length discrepancy (± 1.9 cm), as defined by Subotnick.[7] The percentage of body fat of each subject was determined using skinfold measurements to ensure that no subjects who were extremely lean or obese would be included in the study. All subjects were within one standard deviation of the age-specific average percentage of body fat listed by Jackson et al.[8] For those subjects 20 to 35 years of age, the mean percentage of body fat was 22.7 ± 6.8; for those subjects 60 years of age and older, the mean percentage of body fat was 31.1 ± 7.5.

The elderly women meeting these criteria were tested first. Young women meeting these criteria were recruited to match the elderly women on the basis of right leg length. The matching of right leg lengths was within the same range suggested by Subotnick for leg-length discrepancies[7] to achieve

---

Mrs. Hageman is Instructor, Division of Physical Therapy Education, University of Nebraska Medical Center, 42nd and Dewey Ave, Omaha, NE 68105 (USA).

Dr. Blanke is Associate Professor, Department of Health, Physical Education, and Recreation, University of Nebraska at Omaha, Omaha, NE 68182.

This article was submitted June 4, 1985; was with the authors for revision 12 weeks; and was accepted January 29, 1986.

a close pairing of subjects between the young and elderly groups.

Instrumentation

High-speed cinematography was used to record the subjects' gait. The motion-recording system included two high-speed cameras positioned to orient their optical axes at 90 degrees with one camera parallel to and the other camera perpendicular to a 14-m walkway (Figure). The front camera, a Photec IV* fitted with a 50-mm Nikon[†] lens, was positioned 15.6 m from the center of the walkway. The side camera, a LoCam[‡] fitted with a 25-mm Cosimcar[‡] lens, was positioned 8 m from the center of the walkway. The cameras were positioned according to the procedure of Sutherland and Hagy.[9] Each camera was set to run at 100 frames a second. A 1-m reference scale was included in the field of view of both cameras. A lighting device also was placed in the view of both cameras to provide a common reference point so that frames could be matched later to analyze any phase of the walking cycle.

The processed film was displayed on a Lafayette Dataviewer[§] rear projection system. This system projects the film image of the subject onto a viewing screen and allows the film to be viewed frame by frame or advanced up to 24 frames a second, depending on the viewer's needs for each variable that is measured. The desired measurements were made directly from the projected image. A Numonics[‖] digitizer was used in conjunction with the projection system to assign separate X,Y-coordinate values for any landmark from both front- and side-view films. The coordinate values for the landmarks were stored in a computer[#] and were used for calculating the variables.

Procedure

Each subject was scheduled for one 45-minute testing session. All testing was performed at the Gait Analysis Laboratory at the University of Nebraska at Omaha. Appropriate shorts and sleeveless shirt were the required dress. A 1.9-cm white dot with a 0.6-cm blue center was placed on the following anatomical points in accordance with the procedures of Sutherland and Hagy[9] and Sutherland et al[10]: right and left anterior-superior iliac spines (ASISs), center of both right and left patellae with the knee flexed to 25 degrees, right and left malleoli, and the space between the second and third metatarsals of both the right and left feet. Other markers included a pelvic stick that consisted of a 15.5-cm dowel directed perpendicularly from an Orthoplast® base. The base was attached to a web belt with buckle closure. The belt was placed on the subject so that the stick projected anteriorly from a point midway between the ASISs. A tibial stick of similar construction was placed around the maximal circumference of the calf so that the stick projected anteriorly from the tibial crest.

The subjects then walked barefoot along the 14-m walkway. Each subject was advised to walk at the pace she normally would choose when walking on a clear sidewalk. The first

---

* Photomic Systems, Inc, 265 H Sobrante Way, Sunnyvale, CA 94086.
† Nikon, Inc, 623 Stewart Ave, Garden City, NY 11530.
‡ Redlake Corp, 1711 Dell Ave, Campbell, CA 95008.
§ Lafayette Instrument Co, PO Box 5729, Lafayette, IN 47903.
‖ Numonics Corp, 418 Pierce St, Lansdale, PA 19446.
Model 4052, Tektronix, Inc, PO Box 500, Beaverton, OR 97077.

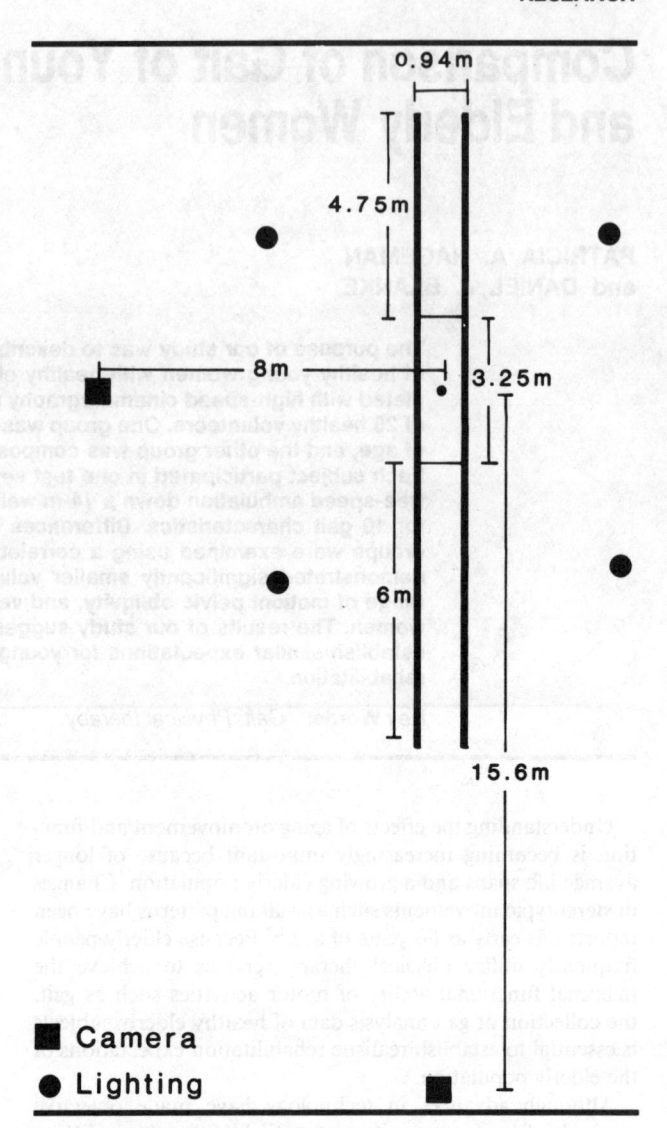

Figure. Walkway plan of Gait Analysis Laboratory.

4.75 m of the walkway allowed each subject to accelerate to her chosen walking speed before reaching the filmed area. The area from which measurements were taken was 3.25 m long, allowing one to two gait cycles, depending on the size of the subject and her walking speed. The last 6 m of the walkway ensured that each subject did not decelerate until she had left the filmed walkway area. Each subject performed three trials.

Measurements

We used the procedure that was described by Sutherland and Hagy[9] and validated in 1980 by Sutherland et al[10] to obtain the measurements with the processed film. Reliability of the measurements taken from our processed film was high when test-retest results were compared during a pilot study. The same observer (P.A.H.) made all of the test-retest measurements from the film, recording a maximum deviation of 2.5 degrees for rotational measurements and a maximum deviation of 2 cm for distance measurements.

Variables measured from the side view included: ankle plantar-flexion and dorsiflexion range of motion, average velocity of the center of gravity, step length, stride length, and vertical excursion of the center of gravity. Step length was measured as the distance in the line of travel between the right heel-strike and the following left heel-strike, beginning with the first full-body side view of the right heel-strike. A scale factor was calculated using the 1-m reference scale in the cameras' field of view. The measured film step length was multiplied by the scale factor to determine the actual step length.

Stride length was measured as the distance in the line of travel between successive points of foot-floor contact of the right foot, beginning with the first total-body view involving a right foot step. Actual stride length was calculated by multiplying the scale factor by the measured film stride length.

Average walking velocity was recorded as the total distance traveled by the subject's center of gravity during one gait cycle divided by the time elapsed during the movement, recorded in centimeters per second. Center of gravity was calculated at the initial point of the right foot-floor contact during the first total-body view and the successive points of right foot-floor contact. Center of gravity was determined according to the segmentation method.[11] We calculated cycle time by counting the number of frames for one gait cycle and dividing the number by the film speed.

Vertical center-of-gravity excursion was calculated by comparing the center of gravity of digitized frames at mid-stance and double-support phases of the gait cycle. Vertical center-of-gravity excursion was determined by subtracting the lowest vertical point from the highest vertical point.

A side view provided the method for determining the ankle plantar-flexion–dorsiflexion range of motion. The total degree of movement at the ankle formed by the line between the knee and ankle center and the line along the bottom of the foot was recorded in degrees. These measurements were obtained directly from the viewing screen with a protractor.

The front-view camera provided the data for determining stride width, lateral center-of-gravity excursion, pelvic obliquity, pelvic rotation, and tibial rotation. Stride width was measured as the horizontal distance between two consecutive steps measured from a point between the second and third metatarsals of each foot. Because the subject image became larger as the subject approached the camera, the appropriate scale factor was used for each point measured and confirmed by measurements from the side view. The actual distance between the metatarsal points of both feet was the stride width.

Lateral center-of-gravity excursion was considered to be the total lateral movement the body's center of gravity traveled during one gait cycle. The center-of-gravity (X,Y) coordinates for the mid-stance position of full weight bearing on the right foot and on the left foot were calculated. Using the appropriate scale factor for each center of gravity, the actual distance between the two center-of-gravity points was calculated.

Pelvic obliquity was the arc of upward and downward movement from the horizontal plane of the right ASIS during one gait cycle. This measurement was obtained directly from the viewing screen with a protractor. When the right ASIS was at the highest vertical point, the angle of upward movement was measured as the angle formed by the intersection of two lines. The first line was the segment between the right ASIS and the center point between the right and left ASISs located at the base of the pelvic-stick attachment. The second line was the segment between the base of the pelvic-stick attachment to the horizontal plane. The angle of downward movement was determined in a similar fashion and added to the angle of upward movement for the total arc of movement.

Pelvic rotation was the degree of rotation that the pelvis moved about a vertical axis. At 0 degrees of rotation, the 15.5-cm pelvic stick pointed directly ahead. At the point of greatest observed rotation to the right, the distance that the tip of the pelvic stick had rotated to the right from the neutral position was measured. This distance was converted into an actual distance using the appropriate scale factor. Considering this distance and the stick length to be two sides of a triangle, we used a trigonometric function to calculate the angle of rotation to the right. The same method was used for recording the greatest observed rotation to the left. The total rotation was the sum of the degrees of rotation to the right and left.

Tibial rotation was the degree the tibia rotated during foot-floor contact of the right foot. The rotation of the tibial stick about a vertical axis was calculated in the same manner as the pelvic-stick rotation.

Data Analysis

Descriptive statistics were calculated for each variable measured in both groups. An independent t test was used to compare the basic descriptive characteristics between the groups. Because the groups were nonrandom and matched for leg length, which may have affected the experimental variables, a correlated t test was used to compare gait characteristics between the groups. All comparisons were evaluated at the .01 level of significance.

RESULTS

The basic descriptive characteristics of both groups of women are presented in Table 1. The groups appeared to be well matched for leg length because no statistical differences were found between the two groups for either right or left leg-length comparisons. Although no differences were found in height or weight between the groups, the elderly women had a higher percentage of body fat than the younger women. Both groups, however, were within the normal range for percentage of body fat based on their age ranges.

Gait characteristics measured from the film of the side-view camera are reported in Table 2. A comparison of the means of these variables reveals significant differences of all variables except the vertical excursion of the center of gravity. The younger female group demonstrated a longer step length and stride length than the elderly women. Greater ankle movement was observed in the walking patterns of the young women than in those of the elderly women. The young women also ambulated at a significantly faster rate than the elderly women.

Table 3 lists the comparison of variables measured from the front-view camera. The only variable that revealed a significant difference between the two groups of women was pelvic obliquity because the young women demonstrated substantially greater pelvic obliquity compared with the elderly women. We found no significant differences between the groups for pelvic or tibial rotation or for lateral center-of-gravity excursion.

TABLE 1
Basic Descriptive Characteristics of the Groups

Variable	Elderly Women (n = 13)			Young Women (n = 13)			t
	X̄	s	Range	X̄	s	Range	
Age (yr)	66.85	7.60	(60.0–84.0)	23.92	3.57	(20.0–33.0)	
Height (cm)	161.00	9.16	(138.4–172.7)	165.10	8.15	(154.9–182.9)	−1.20[a]
Mass (kg)	61.43	17.04	(37.7–107.9)	60.43	8.20	(49.8–74.9)	0.19[a]
Body fat (%)	25.27	5.79	(17.2–35.1)	20.37	2.79	(16.0–25.4)	2.75[a,b]
Leg length (cm)							
Left	86.98	4.71	(78.5–95.6)	86.81	4.24	(80.0–94.1)	0.39[c]
Right	86.69	4.66	(78.3–96.9)	86.65	4.40	(80.9–95.6)	0.10[c]

[a] $df = 24$.
[b] $p < .01$.
[c] $df = 12$.

DISCUSSION

Our study resulted from a need to gain a better understanding of the gait characteristics of young women and elderly women. Whether physical therapists should expect elderly women to have the same rehabilitation potential as young women is not conclusive. Previous gait studies of healthy women either did not consider the specific matching of groups[3] or used a smaller sample size[5] than that used in our study. Previous studies comparing the gait characteristics of young and elderly women have emphasized gait intercycle variability[4] and the influences of heel height on gait.[5] This study of the linear, temporal, and rotational aspects of the gait patterns of healthy young women and healthy elderly women may be helpful to the physical therapist who uses gait characteristics to evaluate a patient's progress.

We adhered strictly to the criteria established for subject selection in this study. Matching the young group with the elderly group using leg-length measurements was considered crucial because of the influence of leg length on stride length.[6]

The results of our study are in close agreement with the findings of other gait studies involving adult women and those involving adult men. The step- and stride-length measurements of the elderly women in our study are similar to values published in a study by Murray et al[5] involving 30 women aged 20 to 70 years and in a study by Chao et al[12] involving 37 women aged 32 to 85 years. Finley et al[3] reported shorter step and stride lengths for the young women and elderly women in their study than those of our study. The shorter step and stride lengths of the subjects in the Finley et al study may have represented the effects of cumbersome equipment worn by their subjects during walking. The healthy elderly women in our study demonstrated similar step- and stride-length values to those reported by Sutherland and associates[10] for 15 healthy men aged 19 to 40 years.

The young women demonstrated significantly larger values for step length than the elderly women. The values of the young women in our study were similar to the values of 30 healthy men aged 20 to 65 years recorded during free-speed gait by Murray et al.[13] The mean step and stride lengths of the men were 78 cm and 156 cm, respectively. The young women in our study demonstrated a greater walking velocity than the women aged 20 to 36 years in a 1984 study by Murray et al.[14]

Our finding of larger means for step and stride lengths for the young women is not surprising because the mean walking velocity of the young group was significantly greater than the

TABLE 2
Comparison of Gait Characteristics Measured from the Side-View Camera

Variable	Elderly Women (n = 13)		Young Women (n = 13)		t[a]
	X̄	s	X̄	s	
Step length (cm)	66.34	6.77	80.68	5.43	7.10[b]
Stride length (cm)	134.92	14.71	162.70	10.84	6.47[b]
Ankle range of motion (°)	24.62	4.61	31.31	5.22	−3.93[b]
Velocity (cm/sec)	131.94	23.85	159.53	16.39	−4.90[b]
Vertical center-of-gravity excursion (cm)	2.87	1.34	3.51	1.77	−1.33

[a] $df = 12$.
[b] $p < .01$.

TABLE 3
Comparison of Gait Characteristics Measured from the Front-View Camera

Variable	Elderly Women (n = 13)		Young Women (n = 13)		t[a]
	X̄	s	X̄	s	
Lateral center-of-gravity excursion (cm)	3.03	2.13	2.39	1.50	0.10
Stride width (cm)	10.02	3.58	8.31	3.12	1.58
Pelvic obliquity (°)	6.77	2.05	9.86	2.38	−3.65[b]
Pelvic rotation (°)	11.77	4.30	11.77	4.44	0.00
Tibial rotation (°)	15.31	7.10	16.69	5.50	−0.51

[a] $df = 12$.
[b] $p < .01$.

mean walking velocity of the elderly group ($p < .01$). The ambulation rate of the elderly women, however, was not abnormally slow. Their mean walking speed of 131.9 ± 23.9 cm/sec was similar to values obtained by Murray et al[5] involving women aged 20 to 70 years whose mean free-speed velocity was 130.0 ± 15.0 cm/sec. The mean walking velocity of the elderly women also was similar to values obtained in a study by Sutherland et al involving men aged 19 to 40 years whose mean walking velocity was 121.6 cm/sec.[10]

A progressive increase in pelvic obliquity corresponding to increased walking speeds has been reported.[6,14] Pelvic obliquity was greater in the young women, as compared with the elderly women. The pelvic obliquity values of adult men reported in the literature ranged from five to eight degrees,[10] which is similar to the findings of our study.

Both groups of women maintained lateral and vertical center-of-gravity excursions within a 5-cm range reported in the literature.[6] Stride widths, however, were extremely variable among both groups. Gabell and Nayak reported similar findings.[4] Young subjects (21–47 years of age) and elderly subjects (66–84 years of age) in their study demonstrated variability within the gait cycle, primarily in stride width.

The elderly women demonstrated substantially less movement at the ankle during free-speed gait than the young women. In a study by Murray et al that compared the gait patterns of young men and elderly men, the elderly men (60–65 years of age) also showed a marked reduction of ankle movement during ankle plantar flexion at the end of ipsilateral stance.[13] The reason for the decrease in ankle movement is not clear, but it may have resulted from slower gait speeds, as suggested previously.[5]

Rotation about the thigh and tibia have been reported in phase with pelvic rotation. The rotary displacement increases progressively from the pelvis to the thigh to the tibia with values of 8 degrees of rotation documented at the pelvis to 19 degrees of rotation measured at the tibia.[15] This progressive increase in rotation from the pelvis to the tibia was demonstrated by both young and elderly subjects in our study, and no significant differences were found between groups. The values obtained in our study were similar to reported ranges for adult women during free-speed gait,[5] but were larger than the values reported for the free-speed gait patterns of men.[6,10]

Because pelvic rotation facilitates forward movement of the hip joint of the swinging leg, increased pelvic rotation would be expected during an increased stride length. A significant increase in pelvic rotation was not demonstrated by the young women, however, even though they demonstrated a significantly larger stride length when compared with the elderly women. This finding may be attributed to individual variation in the interaction between stride length, walking velocity, and pelvic rotation.[4]

Despite significant differences among several gait characteristics of both the young and elderly women, the elderly women demonstrated values of step length, stride length, walking velocity, pelvic rotation, and tibial rotation that were similar to or exceeded those values of healthy young men and women in other studies. These findings suggest that both the young women and the elderly women from our study walk faster today than their counterparts of 15 to 20 years ago.

The differences in the results of our study and those of previous studies may have been caused by differences in subject selection and measuring techniques. Subject cooperation and ability to follow directions may have influenced the results. Because we used a small subject sample, a true random sampling of the age groups may not have been represented. Some subjects may have had an undiagnosed or unrecognized pathological condition that affected their gait, despite our adherence to the guidelines established for subject selection.

Clinical Implications

The results of our study suggest that the clinician should not expect the same gait training rehabilitation potential for both young women and elderly women because differences exist between the gait characteristics of healthy young women and healthy elderly women. The degree to which a pathological condition may further affect the rehabilitation expectations of both groups during gait training is beyond the scope of this study.

Based on the results of our study, the clinician may expect an elderly woman to ambulate with a smaller step and stride length, a slower walking speed, less pelvic obliquity, and less ankle movement than a younger woman with a similar leg length. No differences between the young and elderly women would be expected in center-of-gravity excursion, pelvic and tibial rotation, or stride width.

Physical therapists are involved in the gait training of geriatric patients. Because of the frequency with which they treat elderly women, clinicians must know the effects of age on gait to understand the potential of gait training rehabilitation for this patient group.

Further study in this area is needed before definitive statements can be made about the effects of aging. Care must be taken when applying the results of this study to other populations. Further research could focus on the comparison of additional gait characteristics such as hip motion, hip rotation, angling of the feet, and upper extremity movement of young women and elderly women.

CONCLUSIONS

For the sample of subjects we examined, the following conclusions can be made:

1. The young women and the elderly women did not demonstrate significant differences in vertical center-of-gravity excursion, lateral center-of-gravity excursion, stride width, pelvic rotation, or tibial rotation.
2. The young women demonstrated significantly larger values than those of the elderly women in step length, stride length, ankle range of motion, pelvic obliquity, and walking velocity.
3. The values of the gait characteristics of both the young women and the elderly women in this study were larger than those of their counterparts of 15 years ago. Despite these apparent changes, the effects of aging were observed in these gait characteristics: step length, stride length, ankle range of motion, pelvic obliquity, and walking velocity.

REFERENCES

1. Fisher M, Birren J: Age and strength. J Appl Physiol 31:490–497, 1947
2. Berry G, Fisher R, Lang S: Detrimental incidents, including falls, in the elderly institutional population. J Am Geriatr Soc 29:322–324, 1981
3. Finley F, Cody F, Finizie R: Locomotive patterns in elderly women. Arch Phys Med Rehabil 50:140–146, 1969
4. Gabell A, Nayak V: The effect of age on variability in gait. J Gerontol 39:662–666, 1984
5. Murray M, Kory R, Sepic S: Walking patterns of normal women. Arch Phys Med Rehabil 51:637–650, 1970
6. Inman V, Ralston H, Todd R: Human Walking. Baltimore, MD, Williams & Wilkins, 1981
7. Subotnick S: The short leg syndrome. J Am Podiatr Med Assoc 66:720–723, 1976
8. Jackson A, Pollock M, Ward A: Generalized equations for predicting body density of women. Med Sci Sports Exerc 12:175–182, 1980
9. Sutherland D, Hagy J: Measurement of gait movements from motion picture film. J Bone Joint Surg [Am] 54:787–797, 1972
10. Sutherland D, Olsen R, Cooper L, et al: The development of mature gait. J Bone Joint Surg [Am] 62:336–353, 1980
11. Nutter J, Blanke D, Wang T: Microcomputers aid movement analysis. Collegiate Microcomputer 3:1–11, 1985
12. Chao E, Laughman R, Schneider E, et al: Normative data of knee joint motion and ground reaction forces in adult level walking. J Biomech 16:219–232, 1983
13. Murray M, Drought A, Kory R: Walking patterns of normal men. J Bone Joint Surg [Am] 46:335–360, 1964
14. Murray M, Mollinger L, Gardiner G, et al: Kinematic and EMG patterns during slow, free, and fast walking. J Orthop Res 2:272–280, 1984
15. Levens A, Inman V, Blosser J: Transverse rotations of the segments of the lower extremity in locomotion. J Bone Joint Surg [Am] 30:859, 1948

Basic Research

Comparison of Gait of Young Men and Elderly Men

The purpose of this study was to describe and compare the free-speed gait characteristics of healthy young men with those of healthy elderly men. Data collection consisted of high-speed cinematography resulting in synchronized front and side views of 24 healthy male volunteers, 12 between 20 and 32 years of age and 12 between 60 and 74 years of age. Young men were recruited to match the elderly men on the basis of right-leg length. Each subject participated in three filmed trials of free-speed ambulation down a 14-m walkway. The processed film was analyzed for eight gait characteristics. Differences in characteristics between the two groups were examined using a correlated t test ($p < .01$). No significant differences were observed between the groups for step and stride length, velocity, ankle range of motion, vertical and horizontal excursions of the center of gravity, and pelvic obliquity; however, the younger men demonstrated a significantly larger stride width than the elderly men ($p < .01$). The results suggest that the two populations of healthy adult men have similar gait characteristics. [Blanke DJ, Hageman PA: Comparison of gait of young men and elderly men. Phys Ther 69: 144–148, 1989.]

Key Words: *Aging; Gait; Kinesiology/biomechanics, gait analysis.*

Daniel J Blanke
Patricia A Hageman

Elderly people frequently use physical therapy services to achieve their maximal functional ability in motor activities such as gait. Clinicians are interested in the normal gait characteristics of all age groups, especially the elderly population.

Changes in walking patterns have been reported as early as 60 years of age.[1,2] Murray et al found that timing and stride dimensions were not systematically related to age in their gait study of 60 men aged 20 to 65 years; however, the subjects 60 to 65 years of age differed from the younger subjects in that they demonstrated shorter step lengths and stride lengths, decreased ankle extension, and decreased pelvic rotation.[2] A gait analysis of 64 men aged 20 to 87 years divided into eight age groups revealed differences in the gait characteristics of stride length, cadence, vertical oscillation of the head, and movements of the shoulders and ankles in the three groups of men over 65 years of age when compared with the younger groups of men during free-speed gait.[3]

Gabell and Nayak found no significant differences between 32 healthy elderly adults (aged 61–87 years) and 32 healthy adults (aged 21–47 years) in their gait analysis of intercycle variability of stride time, step length, and stride length.[4]

Several investigations and reports analyzed the gait characteristics of healthy men within a broad span of ages, including the elderly, to determine ranges of values for gait characteristics,[5,6] establish relationships of gait characteristics with speed,[7,8] or compare the gait characteristics of healthy men with male populations with pathological conditions.[9] These investigations, however, did not directly compare gait characteristics between the young and elderly men.

Differences observed in the gait characteristics of elderly men when compared with young men were similar to those found in the gait characteristics of elderly women when compared with young women.[10] Analysis of the gait patterns of men separate from our previous comparison of gait characteristics between matched groups of healthy elderly women and

---

D Blanke, PhD, is Associate Professor, Department of Health, Physical Education, and Recreation, University of Nebraska at Omaha, Sixtieth and Dodge St, Omaha, NE 68182.

P Hageman, MS, PT, is Assistant Professor, Division of Physical Therapy Education, School of Allied Health Professions, University of Nebraska Medical Center, 42nd and Dewey Ave, Omaha, NE 68105-1065 (USA). Address all correspondence to Mrs Hageman.

This article was submitted February 16, 1988; was with the authors for revision for 12 weeks; and was accepted August 23, 1988.

healthy young women is necessary because physiological differences have been documented between the sexes.[11]

Because few studies have investigated the differences between healthy elderly men and healthy young men, additional comparisons are needed. Previous studies attempted to compare groups of men who were similar in height and weight, although specific matching of subjects between groups was not completed.[2,3]

Based on the results of previous research, we hypothesized that a significant difference would exist between the healthy young and healthy elderly groups on eight gait characteristics. The purpose of this study was to describe and compare the free-speed gait characteristics of matched groups of healthy elderly men and healthy young men.

Method

Subjects and Selection Procedure

Twenty-four male volunteers, 12 between 20 and 32 years of age and 12 between 60 and 74 years of age, were accepted as subjects in this study. Each subject provided his informed consent in accordance with the procedures of the University of Nebraska Institutional Review Board.

All of the subjects were found to be free of disabling physical conditions or minor ailments that could affect or influence locomotion based on a medical review and an objective examination by a licensed physical therapist. Specifically, the subjects were without musculoskeletal or neurological involvement or medication for these conditions.

One tester (PAH) took all measurements. Height and mass were measured to provide descriptive characteristics for the groups. An average of three independent measurements was used to determine values for leg length and for skinfold thickness, which were used to determine the percentage of body fat. Leg lengths were measured to ensure that each subject was without a leg-length discrepancy ($\pm$ 1.9 cm) as defined by Subotnick.[12] This criterion is important because leg length is a major determinant of stride length.[13]

The percentage of body fat of each subject was determined using skinfold measurements to ensure that no subjects who were extremely lean or obese would be included in the study. Percentage of body fat for the young male subjects was calculated using the age-specific formula of Jackson and Pollock.[14] All young male subjects were within one standard deviation of the average percentage of body fat (13.4% $\pm$ 6.0%) for 18- to 24-year-old men as reported by Jackson and Pollock.[15] Subjects 60 years of age or older also were within one standard deviation of their age-specific average percentage of body fat (22.6% $\pm$ 4.1%).[16]

The elderly men meeting these criteria were tested first. Young men meeting these criteria were recruited to match the elderly men on the basis of right-leg length. The matching of right-leg lengths was within the same range suggested by Subotnick[12] for leg-length discrepancies to achieve a close pairing of subjects between the young and elderly groups.

Instrumentation

Data collection consisted of high-speed cinematography resulting in synchronized front and side views of each subject's free-speed gait down a 14-m walkway. The instrumentation, gait laboratory, and measurement methods for analysis have been described in detail previously.[10,17,18]

The front camera was a Photec IV* fitted with a 50-mm Nikon lens† set 15.6 m from the walkway. The side camera, a LoCam‡ fitted with a 25-mm Cosmicar lens,‡ was positioned 8 m from the center of the walkway. Both cameras were set to run at 100 frames per second. A 1-m reference scale and a lighting device were placed in the view of both cameras to provide a common reference for distance and for synchronizing front and side film frames during the film analysis.

The processed film was displayed on a Lafayette Dataviewer§ rear-projection system. The desired measurements were made directly from the projected image. A Numonics digitizer‖ was used in conjunction with the projection system to assign separate X,Y-coordinate values for any landmark from both front- and side-view films. The coordinate values for the landmarks were stored in a computer# and used for calculating the variables.

The procedure described by Sutherland and Hagy[17] and validated in 1980 by Sutherland et al[18] was followed in this study to obtain the measurements with the processed film. Reliability of the measurements taken from our processed film was high when test-retest results were compared during a pilot study. The same observer (PAH) made all of the test-retest measurements from the film, recording a maximum deviation of 2.5 degrees for rotational measurements and a maximum deviation of 2 cm for distance measurements.

Gait variables measured from the side view included ankle plantar-flexion and dorsiflexion range of motion, velocity, step length, stride length, and vertical excursion of the center of gravity. The front-view camera provided the data for determining stride width, lateral center-of-gravity excursion, and pelvic obliquity.

---

*Photomic Systems, Inc, 265 H Sobrante Way, Sunnyvale, CA 94086.

†Nikon, Inc, 623 Stewart Ave, Garden City, NY 11530.

‡Redlake Corp, 1711 Dell Ave, Campbell, CA 95008.

§Lafayette Instrument Co, 3700 Sagamore Pky N, PO Box 5729, Lafayette, IN 47903.

‖Numonics Corp, 418 Pierce St, Lansdale, PA 19446.

#Model 4052, Tektronix, Inc, PO Box 500, Beaverton, OR 97077.

Basic Research

Table 1. *Basic Descriptive Characteristics of Subject Groups (N = 24)*

Variable	Elderly Men (n = 12)			Young Men (n = 12)			df	t
	X̄	s	Range	X̄	s	Range		
Age (yr)	63.58	5.58	(60.0–74.0)	24.50	3.73	(20.0–33.0)		
Height (cm)	175.53	3.86	(168.2–179.7)	175.68	9.46	(151.1–185.4)	22	−0.05
Mass (kg)	80.11	11.01	(62.7–94.1)	77.32	7.45	(67.2–94.5)	22	0.72
Body fat (%)	21.08	4.80	(15.4–28.7)	11.01	3.31	(4.6–16.5)	22	5.94[a]
Leg length (cm)								
Right	93.53	2.79	(87.1–98.0)	92.98	2.84	(88.4–97.2)	11	1.96
Left	93.61	2.70	(87.4–98.0)	93.06	2.83	(88.8–97.3)	11	1.31

[a] $p < .01$.

Procedure

Each subject participated in one 45-minute testing session at the Gait Analysis Laboratory at the University of Nebraska at Omaha. The required dress included shorts and a sleeveless shirt. Tape markers were placed on anatomical points of each subject for easy reference on the processed film. The description and specific placement of the markers have been described previously.[10,12,13]

The subjects then walked barefoot along the 14-m walkway. The subjects were requested to walk at what they considered their natural pace when walking down a sidewalk without obstructions. The first 4.75 m of the walkway allowed each subject to accelerate to his chosen walking speed before reaching the filmed area. The area from which measurements were taken was 3.25 m long, allowing one to two gait cycles depending on the size of the subject and his walking speed. The last 6 m of the walkway ensured that each subject did not decelerate until he had left the filming area. Each subject performed three trials.

Data Analysis

Means and standard deviations were calculated for all of the variables. An independent t test was used to compare descriptive characteristics between the groups. Differences in the gait characteristics between the two groups were examined using a correlated t test because the groups were nonrandom and matched for leg length. Significance was accepted at the .01 level.

Results

No significant differences were found between the groups of men for either right or left leg-length comparisons, suggesting that the groups were well matched for leg length. The elderly men had a higher percentage of body fat than the younger men. Both groups, however, were within the normal range for percentage of body fat based on their age ranges. The basic descriptive characteristics of both groups are reported in Table 1.

No significant differences were found between the two groups for all variables measured from the side view, including step length, stride length, ankle ROM, velocity, and vertical center-of-gravity excursion. The gait characteristics from the side view are presented in Table 2.

The elderly men demonstrated a significantly smaller stride width ($p < .01$) compared with the young men. No significant differences were found between the groups for lateral excursion of the center of gravity or pelvic obliquity. The values obtained for the gait characteristics measured from the film of the front-view camera are shown in Table 3.

Discussion

This study resulted as a follow-up to a previous study comparing healthy

Table 2. *Comparison of Subjects' Gait Characteristics Measured from Side-view Camera (N = 24)*

Variable	Elderly Men (n = 12)		Young Men (n = 12)		t[a]
	X̄	s	X̄	s	
Step length (cm)	94.17	11.99	87.58	6.46	1.49
Stride length (cm)	189.58	23.39	192.58	18.03	−0.30
Ankle range of motion (°)	19.08	4.96	21.25	5.67	−0.85
Velocity (cm/sec)	138.93	23.41	131.32	17.52	0.77
Vertical center-of-gravity excursion (cm)	7.42	4.21	8.22	5.00	0.37

[a] $df = 11$.

elderly women with healthy young women. Data for the male subjects in this study were compared and analyzed separately from the previous study of women because of the documented physiological differences between the sexes.[11] We were interested in whether the elderly men demonstrated the same changes in gait that were demonstrated by the elderly women. We were also interested in whether the elderly men in our study demonstrated the same changes in gait that were found in elderly men by other authors.

Specific comparison of the results of many previous studies with data from this study is limited because normalization of gait measurements with respect to body size is an unresolved problem in human locomotion research.[13] Matching of the young group with the elderly group in our study using leg-length measurements was considered crucial because of the influence of leg length on stride length.[13]

The values of step length and stride length of the men in our study are much greater than values published in a study by Murray et al[2] involving 60 men aged 20 to 65 years and in a study by Kirtley et al[7] of 10 male subjects 18 to 63 years of age walking at their natural speed. Larsson et al examined the stride length of 32 male and female subjects aged 20 to 70 years during walking speeds classified as "very slow," "slow," "ordinary," "fast," and "very fast."[8] The mean values of stride length from both groups of men in our study are closest to the mean value of stride length in the very fast category (1.93 m) in the Larsson et al study.[8]

Compared with the results of other studies, both groups of men in our study demonstrated very long stride lengths; however, the mean values of velocity for both groups of men in this study were smaller than the mean values of velocity from other studies.[3,7,8] The conflicting results may be attributed to the differences in the population studied or methodologies used for testing. Velocity of walking may result from many combinations of stride length and cadence, explaining the differences found between the studies. High values of stride length reported from our subjects may be a result of longer leg lengths.

The mean velocities of the elderly men and the young men during free-speed ambulation are similar to values published by Katoh et al[9] of 32 male and female subjects ambulating at their chosen natural speed. The velocity values of the men in this study are also similar to the values of 534 men during functional ambulation at locations such as commercial, business, and residential areas.[5] The similarity of results suggests that the men in our study selected a functional walking speed as their natural free-speed gait.

Pelvic obliquity values of the men reported in the literature ranged from 5 to 8 degrees, which is similar to the findings in our study.[3,18] Both groups of men in this study maintained a lateral center-of-gravity excursion within the 5-cm range reported in the literature.[13] The vertical excursion of the center of gravity for both groups of men in this study exceeded this 5-cm range. The long stride lengths in relation to the moderate gait velocities may explain the larger values for vertical center-of-gravity excursions.

Significant differences were observed between the two groups of men in stride width; however, both groups demonstrated great variability in values of stride width. Gabell and Nayak[4] reported similar findings in their study of young subjects (21–47 years of age) and elderly subjects (66–84 years of age) during gait. Variability in values of stride width was also observed in our previous study of healthy elderly and healthy young women.[10] Although the younger men in our study demonstrated statistically larger values of stride width than the elderly men, the mean stride-width values from both groups are within the range of 2.5 to 12.7 cm reported in the literature.[3,6]

No significant differences were observed between the two groups in this study for values of ankle movement. Murray et al, however, found slightly less excursion during ankle movement in subjects 60 to 65 years of age.[2] Older men (aged 81–87 years) demonstrated significantly less ankle extension than younger men at the end of stance phase.[3] Research findings of elderly women also showed decreased ankle movement.[10]

Caution is advised when applying the results of this study to other populations because of the differences in subject selection and measuring techniques. A true random sampling of the age groups may not have been represented because of the small sample size. Subjects' motivation and ability to follow instructions may have influenced the results. Despite our adherence to the guidelines established for subject selection, some subjects may have had an undiagnosed or unrecognized pathological condition that affected their gait.

Table 3. *Comparison of Subjects' Gait Characteristics Measured from Front-view Camera (N =24)*

Variable	Elderly Men (n = 12)		Young Men (n = 12)		t^a
	X̄	s	X̄	s	
Lateral center-of-gravity excursion (cm)	2.28	1.20	1.70	.72	1.23
Stride width (cm)	8.25	5.09	10.80	3.94	−3.13[b]
Pelvic obliquity (°)	6.08	2.50	7.42	2.11	−1.55

$^a df = 11$.
$^b p < .01$.

Clinical Implications

This study of the linear and temporal aspects of the gait patterns of healthy young men and healthy elderly men may assist physical therapists who use gait characteristics to evaluate a patient's progress. The degree that a pathological condition would further affect the gait characteristics of both groups is beyond the scope of this study.

The results of this comparison of healthy elderly men with healthy young men contrast with the results of a previous study that used the same methodology and compared matched groups of healthy elderly women with healthy young women.[10] The gait characteristics between the two groups of men did not differ, whereas the healthy elderly women demonstrated a smaller step and stride length, a slower walking speed, less pelvic obliquity, and less ankle movement than the healthy young women. These conflicting results may be attributed to the documented physiological differences between the sexes,[11] suggesting that the aging process affects healthy women differently than healthy men. It is unclear whether physical activity affected the results in both studies. Although the studies of men and women involved healthy subjects, the elderly men were more likely to report participation in vigorous activities (eg, softball and hunting) than the elderly women who participated in an earlier study.

Further study of gait characteristics in the elderly population is needed before definitive conclusions may be made about the effects of aging on gait. Future research could focus on the relationship of physical activity to the gait characteristics observed.

Conclusions

The effects of aging were not observed. The young men and the elderly men did not demonstrate significant differences in seven of the eight gait characteristics examined (ie, step length, stride length, ankle ROM, velocity, vertical and horizontal excursions of the center of gravity, and pelvic obliquity). Although the young men demonstrated significantly larger values in stride width than the elderly men, the values of stride width for both groups were within the range published for the healthy population. The statistically different values were not clinically relevant.

References

1 Berry G, Fisher R, Lang S: Detrimental incidents, including falls, in the elderly institutional population. J Am Geriatr Soc 29:322–324, 1981

2 Murray M, Drought B, Ross C, et al: Walking patterns of normal men. J Bone Joint Surg [Am] 46:335–360, 1964

3 Murray M, Kory R, Clarkson B: Walking patterns in healthy old men. J Gerontol 24:169–178, 1969

4 Gabell A, Nayak V: The effect of age on variability in gait. J Gerontol 39:662–666, 1984

5 Finley F, Cody K: Locomotive characteristics of urban pedestrians. Arch Phys Med Rehabil 51:423–426, 1970

6 Bampton S: A Guide to the Visual Examination of Pathological Gait. Philadelphia, PA, Temple University Press, Rehabilitation Research and Training Center #8, 1979

7 Kirtley C, Whittle W, Jefferson R: Influence of walking speed on gait parameters. J Biomed Eng 7:282–288, 1985

8 Larsson L, Odenrick P, Sandlund B, et al: The phases of the stride and their interaction in human gait. Scand J Rehabil Med 12:107–112, 1980

9 Katoh Y, Chao Y, Laughman R, et al: Biomechanical analysis of foot function during gait and clinical applications. Clin Orthop 177:23–33, 1983

10 Hageman PA, Blanke DJ: Comparison of gait of young women and elderly women. Phys Ther 66:1382–1387, 1986

11 McArdle W, Katch K, Katch V: Exercise Physiology: Energy, Nutrition, and Human Performance, ed 3. Philadelphia, PA, Lea & Febiger, 1986

12 Subotnik S: The short leg syndrome. J Am Podiatr Med Assoc 66:720–723, 1976

13 Inman V, Ralston H, Todd R: Human Walking. Baltimore, MD, Williams & Wilkins, 1981

14 Jackson A, Pollock M: Generalized equations for predicting body density in men. Br J Nutr 40:497–504, 1978

15 Jackson A, Pollock M: Prediction accuracy of body density, lean body weight, and total body volume equations. Med Sci Sports 9:197–201, 1977

16 Latin R, Johnson S, Ruhling R: An anthropometric estimate of body composition of older men. J Gerontol 42:24–28, 1987

17 Sutherland D, Hagy J: Measurement of gait movements from motion picture film. J Bone Joint Surg [Am] 54:787–797, 1972

18 Sutherland D, Olsen R, Cooper L, et al: The development of mature gait. J Bone Joint Surg [Am] 62:336–353, 1980

Basic Research

Research Report

Biomechanical Walking Pattern Changes in the Fit and Healthy Elderly

David A Winter
Aftab E Patla
James S Frank
Sharon E Walt

A descriptive study of the biomechanical variables of the walking patterns of the fit and healthy elderly compared with those of young adults revealed several significant differences. The walking patterns of 15 elderly subjects, selected for their active life style and screened for any gait- or balance-related pathological conditions, were analyzed. Kinematic and kinetic data for a minimum of 10 repeat walking trials were collected using a video digitizing system and a force platform. Basic kinematic analyses and an inverse dynamics model yielded data based on the following variables: temporal and cadence measures, heel and toe trajectories, joint kinematics, joint moments of force, and joint mechanical power generation and absorption. Significant differences between these elderly subjects and a database of young adults revealed the following: the same cadence but a shorter step length, an increased double-support stance period, decreased push-off power, a more flat-footed landing, and a reduction in their "index of dynamic balance." All of these differences, except reduction in index of dynamic balance, indicate adaptation by the elderly toward a safer, more stable gait pattern. The reduction in index of dynamic balance suggests deterioration in the efficiency of the balance control system during gait. Because of these significant differences attributable to age alone, it is apparent that a separate gait database is needed in order to pinpoint falling disorders of the elderly. [Winter DA, Patla AE, Frank JS, et al. Biomechanical walking pattern changes in the fit and healthy elderly. Phys Ther. 1990;70:340–347]

Key Words: *Equilibrium; Geriatrics; Kinesiology/biomechanics, gait analysis; Posture, tests and measurements.*

The reduction of frequency of falls among the elderly is the goal of many researchers addressing the resultant injuries, death, and loss of mobility.[1]

Research has focused on epidemiological studies to provide a better description and assessment of the extent of the problem and on characterizing the changes in the standing balance control system that occur with age. The epidemiological data have implicated some aspects of locomotion (ie, initiation of walking, turning, walking over uneven surfaces, stopping) in almost all incidences of falls.[2-5]

Despite this strong evidence linking locomotion to falls, studies of changes in the balance control system have been limited mainly to tests that probe the integrity of the system during quiet standing. Performance on these tests does not correlate with incidence of falls and is a poor pre-

---

D Winter, PhD, PEng, is Professor, Department of Kinesiology, University of Waterloo, Waterloo, Ontario, Canada N2L 3G1. Address all correspondence to Dr Winter.

A Patla, PhD, is Associate Professor, Department of Kinesiology, University of Waterloo.

Frank, PhD, is Assistant Professor, Department of Kinesiology, University of Waterloo.

S Walt, MASc, is Research Assistant, Department of Kinesiology, University of Waterloo.

Financial support for this study was provided by the Medical Research Council of Canada (Grant MT4343) and Health and Welfare Canada (Grant 6606-3675-R).

This study was approved by the University of Waterloo's Office of Human Research.

This article was submitted July 19, 1989, and was accepted January 19, 1990.

dictor of fallers.[6] Even during perturbed standing tests,[7] the predictions have been no better than 30% (60% of fallers predicted, and 30% of nonfallers are false positives). This finding is hardly surprising because the balance challenges during walking are quite different from those involved in maintaining upright posture.

During standing, the goal is to maintain the body's center of gravity (CG) within the base of support. The initiation of gait, however, is an unstabilizing event whereby the body's CG is made to fall forward and outside of the stance foot.[8] By the time the selected cadence is achieved, the only stabilizing period is double-support stance, and even during that time period the one limb is pushing off with considerable force while the other limb is accepting the full weight of the body.[9] During natural cadence, 80% of the stride period is single-support stance, when the CG of the body has been shown to be outside the foot[10]; the closest it gets to the base of support is when it passes forward along the medial border of the foot. Even during the two 10% double-support stance periods, both feet are not flat on the ground. During the first half of double-support stance, or heel contact (HC), the weight-accepting foot is being lowered to the ground; during the latter half of double-support stance, the final stage of push-off has weight only under the toes. Thus, the body is in an inherent state of instability. Most of the findings from balance studies during standing, therefore, have very limited relevance to gait. The dynamic balance of the head, arms, and trunk (HAT) and the safe transit of the foot during the swing phase of gait (safe toe clearance and a gentle foot landing) present a challenge to the central nervous system during walking. The HAT constitutes two thirds of the body mass, and the HAT's center of mass (CM) is located about two thirds of the body height above ground level. The CM is the point where all the mass of the HAT can be considered to act in all three axes as compared with the CG, which is its location in the gravitational axis. In the sagittal plane, even in slow walking, the horizontal momentum of the HAT results in inherent instability. The role of the ankle muscles in standing balance is paramount, but in walking the role of the ankle plantar-flexor and dorsiflexor muscles for balance has not been seen to be important.[11] The moment of inertia of the HAT about the ankle is about eight times what it is about the hip.[12] Thus, during the first half of stance, for example, when a posterior acceleration at the hip is attempting to collapse the HAT in the forward direction, the ankle muscles do not act to intervene. If they did, they would require a plantar-flexor moment of about 300 N·m to control the huge inertial load. Instead, the ankle muscles produce a small dorsiflexor moment to lower the foot to the ground, followed by a small plantar-flexor moment to control the forward leg rotation. The hip extensor muscles, however, intervene to control the lesser inertial load in conjunction with a tight coupling with the knee muscles.[11,13] The tight coupling of these two motor patterns has been labeled an "index of dynamic balance."[14] This balance control of the large inertial load of the HAT acts primarily during single-support stance with a transfer of responsibility between limbs taking place during double-support stance.

The swing phase of gait has been shown to be executed with considerable precision[15] with average toe clearances of about 1 cm, and this clearance occurs while the horizontal velocity is maximal (3.6–4.5 m/sec). The heel velocity is also reduced drastically in both horizontal and vertical directions immediately prior to HC. Thus, any degeneration in this fine motor control of the foot may result in problems of stumbling during swing and in rebalancing immediately after HC.

Numerous studies have addressed the changes in the gait patterns of the elderly compared with those of the younger adult. The majority of these studies[16–19] have concentrated on basic outcome measures (ie, stride length, cadence, velocity) and the variability of those measures. Several of these studies have related these gait changes to falls,[18] mobility,[19] and post-fall anxiety.[20] All of these studies have made inferences about the reasons for the observed changes: lower cadence, shorter and more variable step length, increased head and torso flexion, and increased knee and elbow flexion. The suggested reasons imply a degeneration of balance control combined with a general loss of muscle strength. The measures reported, however, were outcome measures, which provide limited insight into the changes in the motor system for balance control and limit our ability to identify the mechanisms behind the observed changes.

With this background in mind, there is a need to document the motor pattern changes that occur in the gait of the elderly and to determine whether those changes are related to balance. Fit and healthy elderly individuals were chosen for this initial study to eliminate effects of a sedentary life style or pathological conditions on walking patterns. Of interest was the normal biological degeneration that takes place with age prior to the advent of any identifiable neural, muscular, or skeletal disorder. All kinematic and kinetic patterns were examined in detail in order to pinpoint major or subtle changes that would point to the degeneration or to compensations that reduce the chance of stumbling or losing balance. Simultaneously, a second major goal was achieved, that of developing a full database of kinematic and kinetic profiles against which to compare individual elderly patients with known or suspected balance or tripping disorders.

Method

Subjects

Fifteen elderly subjects were screened based on a life-style and medical questionnaire and examined by a geriatrician to eliminate any volunteers who had any pathological condition related to the human locomotor system. Informed consent forms were signed by each subject prior to the walking trials. These fit and healthy elderly individuals (10 men, 5 women) ranged in age from 62 to 78 years ($\bar{X}$ = 68 years).

Procedure

The protocol for the biomechanical gait analyses was identical to that reported previously[9,11,13,14] and is summarized as follows. Each subject was instrumented with reflective markers to define the following joint centers and segments: toe, fifth metatarsal, heel, lateral malleolus (ankle), head of the fibula, lateral epicondyle of the femur (knee), and greater trochanter (hip). Additional markers, not part of this link-segment analysis, were also attached to the trunk and head to define upper body kinematics: L4-L5, sternum, C1-C2, ear canal, and forehead. A standard link-segment model of the lower limb was developed for the foot, leg, and thigh segments in order to calculate the moments of force at the ankle, knee, and hip.[12,21] Each subject walked at his or her natural cadence on a level walkway a minimum of 10 times; the repeat trials were conducted over a period of about one hour (one trial every 5 or 6 minutes). Each subject walked over a force platform* while a Charge-Coupled Device (CCD) video camera† located 6 m to the side of the walkway recorded the marker trajectories over the stride period. The CCD camera was electronically shuttered at 1 msec with a field rate of 60 Hz. The video signal was stored on a Sony Motion Analyzer‡ and subsequently digitized using a specially designed video interface into an IBM PC-AT™ computer.§ The precision of the marker centroids was calculated to within 1 mm. The raw coordinate data were digitally filtered with a fourth-order zero-lag Butterworth filter with a cutoff at 6 Hz. The smoothed coordinates then became inputs to the standard link-segment model.

In addition to the joint moments of force, the mechanical power generated and absorbed at each joint was calculated[22] and the area under each power burst was integrated to determine the mechanical work performed during each of the generating and absorbing phases. The support moment, as defined a decade ago,[9] was calculated and is equal to the sum of the moments at the ankle, knee, and hip (extensor moments were set positive, and flexor moments were set negative). The support moment is the total motor pattern of the lower limb, which has been seen to be positive (extensor) during most of stance, negative (flexor) during late double-support and early swing, and positive (extensor) during late swing.[14] The ensemble average of the moment-of-force patterns over all the strides yielded a mean variance measure for the ankle, knee, and hip profiles, from which the hip-knee and knee-ankle covariances were readily calculated.[13] The kinematics of toe markers over the stride period yielded the toe clearance during midswing. *Toe clearance* was defined as the difference in the vertical displacement of the toe marker at its lowest point in stance (just before toe-off) and its lowest point in mid-swing.

Data Analysis

Identical measures were taken from our database on 12 young adults (7 men, 5 women), ranging in age from 21 to 28 years ($\bar{X}$ = 24.6 years). Because the population variances were not identical, a modified *t* test[23] was used to determine any significant differences between selected kinematic and kinetic variables that had potential impact on balance and falling during walking. These variables are presented in the Table.

Results and Discussion

The kinematic and kinetic patterns of one elderly subject are used in this section to illustrate the nature and format of the data. The mean cadence for this subject was 105 steps/min (s = 1.8), and the following ensemble-averaged waveforms were plotted at 2% intervals over the stride period (HC = 0%, next HC = 100%). The average toe-off for this subject was 65.7%, so it was set to the nearest 2% interval (66%). The following profiles are presented: ankle, knee, and hip angles (Fig 1); toe vertical displacement, vertical velocity, and horizontal velocity (Fig 2); ankle, knee, hip, and support moments (Fig 3); and ankle, knee, and hip powers (Fig 4). In all of these diagrams, the mean of the repeat trials is plotted as a solid line with one standard deviation plotted at each 2% interval over the stride period. The mean coefficient of variation (CV) is reported and represents the average variability over the stride period expressed as a percentage of the mean signal amplitude.[13] The CV measure is a single score that allows comparison of the percentage of variability of any waveform over any group of repeat walking trials.

Figure 1 shows the variability of this subject's ankle, knee, and hip joint angles to be quite low. The CV for the ankle, knee, and hip joints was 21%, 8%, and 8%, respectively. Similar low variabilities have been reported for intrasubject repeat trials performed across days as well as minutes apart on young adults.[13] These consistent results caution against any inferences about similar invariance in the motor patterns. The indeterminacy of the human motor system during stance is such that many combinations of moments of force at the ankle, knee, and hip can still result in the same lower limb kinematics, especially at the hip and knee, and this finding is supported by the data for this subject.

---

*Advanced Medical Technology Inc, 141 California St, Newton, MA 02158.

†Model TI-50ES, NEC America, 1255 Michael Dr, Wood Dale, IL 60191.

‡Model SVM-1010, Sony of Canada, 88 Horner Ave, Toronto, Ontario, Canada K2B 8K1.

§International Business Machines Corp, PO Box 1328-S, Boca Raton FL 33432.

The toe trajectory data (Fig 2) show the vertical displacement (upper trace), the vertical velocity (middle trace), and the horizontal velocity (lower trace). These trajectory plots all have low CVs, indicating a highly consistent control of the distal segment of the limb, the toe. The average toe clearance of 1.5 cm ($s = 0.5$) for this subject occurred at 80% of stride as the toe reached its peak horizontal velocity of 4.3 m/sec. The complex nature of this end-point control task needs to be recognized. The length of the link-segment chain is over 2 m, starting with the stance phase foot and continuing up to the hip, across the pelvis, and down the swing limb, and the chain involves at least 12 degrees of freedom at the joints and scores of muscles. The generation and execution of such a consistent toe trajectory is evidence of fine motor control.

The moment-of-force curves for this elderly subject are presented in Figure 3 with extensor moments plotted as positive, along with the support moment,[9] which is the algebraic sum of the three joint moments. The interpretation of the support-moment pattern has been discussed in detail previously.[9,13] In summary, the support moment quantifies the total limb synergy, which is extensor during most of stance, becomes flexor during late double-support and early swing, and returns to extensor during late swing. We have identified this support synergy in over 50 assessments on a wide variety of gait pathologies in healthy young ($n = 200$) and elderly ($n = 15$) subjects.

The variability of these moment patterns varies with the joint. This subject's CV was 9% at the ankle, 31% at the knee, 19% at the hip, and only 9% in the support moment. Because CV is a ratio of mean variance and mean signal, the low CV for support moment was partially due to increased mean signal as well as decreased mean variance. It has been shown that the variance in these motor patterns is not random, especially in the highly variable hip and knee patterns.[13] There is a tight neural and anatomical coupling between the knee and hip motor patterns. The covariance between the hip and knee moments can reach 89% in repeat strides assessed days apart and ranges from 60% to 70% for repeat assessments performed minutes apart.[14] This covariance is expressed as a percentage of the maximum possible and would reach 100% if the covariance were equal to the sum of the knee and hip variances. This coupling between the joint moments is revealed in the small CV for the summation of hip and knee moments, which was 14% for this set of repeat trials. The reason for these trade-offs between the hip and knee moments is related to a second limb synergy, that of dynamic balance.[11,14] This balance synergy is described as follows: On a stride-to-stride basis, the anterior-posterior balance of the HAT is controlled by the hip flexors and extensors during stance (mainly single-support stance). Each stride is somewhat different, and the regulation of this large mass (two thirds of body mass) requires a modified hip motor pattern on each stride. Thus, the high variance in the hip moment during stance is directly due to a continuously changing balance control task. The hip moment, however, is also part of the support synergy. To keep the support pattern nearly constant, there must be an opposite change in the knee moment, which is almost as variable, but in the opposite direction. Such a trade-off between

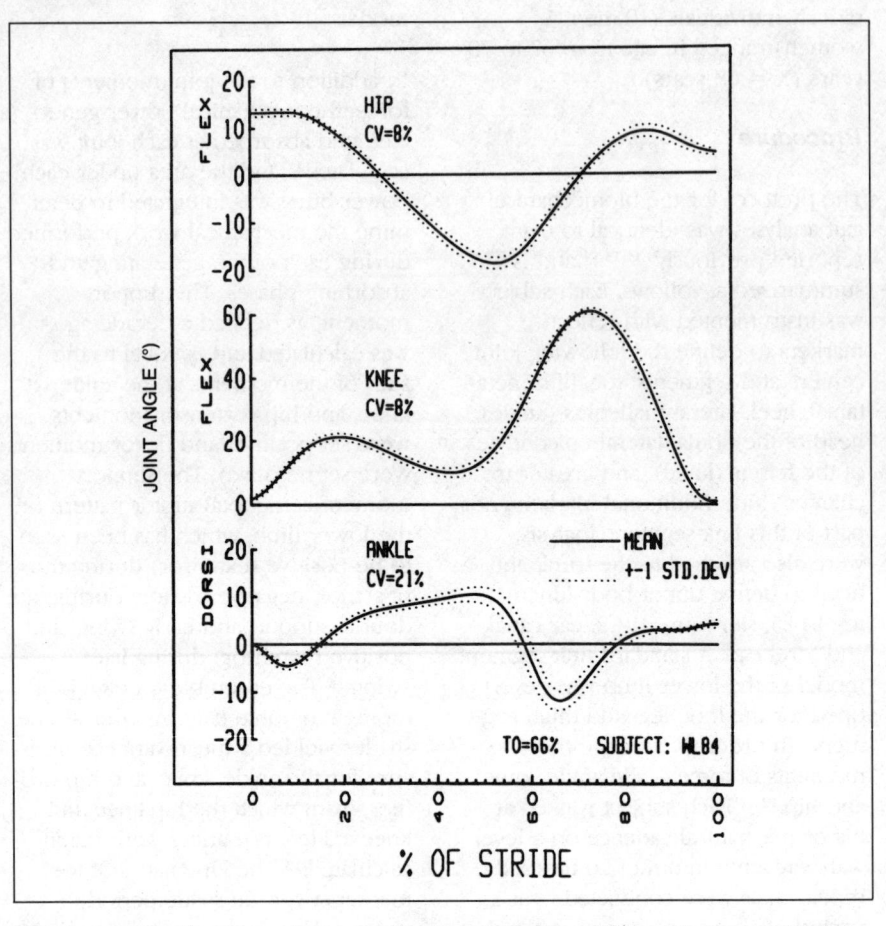

Fig 1. *Ensemble-averaged joint angles for 11 repeat walking trials of one elderly subject. Stride period is normalized to 100% from heel contact (HC) to HC, and for this subject the average toe-off (TO) was 66%. Solid lines plot average joint angle, and dotted lines represent one standard deviation at each 2% interval of stride period. As demonstrated by low coefficient of variation (CV) scores, lower limb kinematics remained very consistent. (DORSI = dorsiflexion; FLEX = flexion.)*

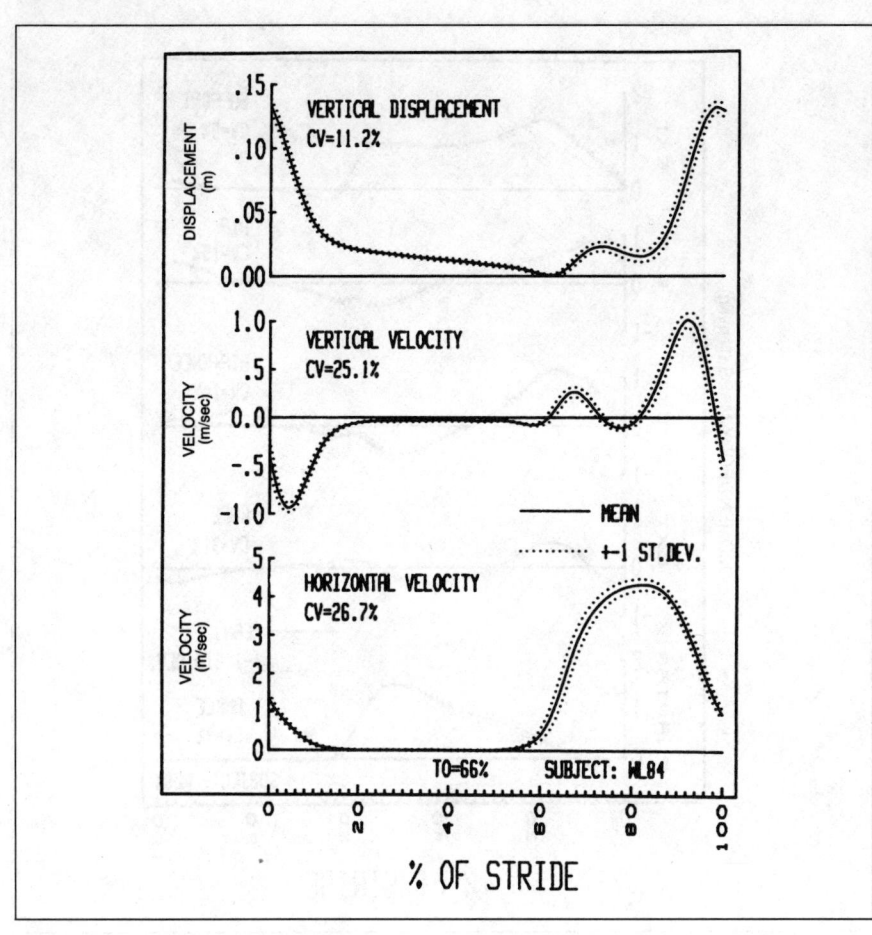

Fig 2. Ensemble-averaged toe trajectory plots for same subject as in Figure 1 over 11 repeat walking trials. Vertical displacement of toe (top trace) shows a minimum (set to 0) just prior to toe-off (TO) and minimum toe clearance during swing at about 80% of stride period when horizontal velocity (bottom trace) is near its maximum. (CV = coefficient of variation.)

the hip and knee moment patterns is almost one-for-one and is the reason for the low variance in the summation of the hip and knee moments.[11,14] The covariance between the hip and knee moments is a measure of this synergistic trade-off and has been labeled an "index of dynamic balance."[14] For this subject, it was 59.3%.

The comparison between the young adults' gait and that of the elderly subjects is presented in the Table. Nineteen gait variables, ranging from basic outcome measures (temporal, cadence), a key swing-phase kinematic variable (toe clearance), percentage covariances between the hip and knee and the knee and ankle, and key energetic variables (work performed during each power phase) are listed. Five of these gait variables showed significant differences ($p < .01$) between the two groups, and two were borderline ($p < .07$).

The natural cadence of these fit and healthy elderly adults was no different than that of the young adults, but the stride length was significantly shorter, independent of whether it was documented in meters or as a fraction of body height (statures). Previous studies of elderly gait all showed a reduction in both cadence and stride length.[16–18] The major possible explanation is that our subjects were screened carefully to eliminate the unfit and those with any gait-related pathological condition. All of our subjects were enrolled in a fitness program and had a generally active life style, and these factors appear to have kept their cadence up to normal. Associated with this shorter stride length was an increase in the stance time (elderly subjects, 65.5%; young adults, 62.3%), which was also statistically significant ($p < .01$). Although this increase appears small, it did result in a somewhat larger percentage of change in total double-support stance (elderly subjects, 31.0%; young adults, 24.6%). Toe clearance for the elderly subjects was not statistically different from that of the younger adults. This low toe clearance was achieved with less variability in the elderly subjects, despite the large number of degrees of freedom in the link chain (made up of stance and swing limb). This reduced variability appears to be a consequence of the shorter step lengths adopted by the elderly subjects.

The knee-hip covariance (% COV hip-knee) was marginally less for the elderly subjects (elderly subjects, 57.7%; young adults, 67.0%; $p < .07$). The interpretation of this score as an index of dynamic balance suggests that the elderly are less able to make the anterior-posterior shifts in the moment patterns on a stride-to-stride basis to dynamically control the balance of the HAT in the sagittal plane and at the same time maintain the extensor support moment. Currently, it is not possible to speculate whether the covariance reduction is functionally significant. Only after a large number of balance-impaired patients are analyzed will the safety threshold of this synergy be evident. Because of the somewhat higher variability in the hip-knee covariance score for the elderly subjects, these individual scores were examined and revealed that the elderly subjects had a biomodal distribution, with 10 of them falling within the same range as the young adults and 5 of them with quite low covariances. Our cautious interpretation of this finding is that some of our healthy elderly subjects may have a balance impairment that has not yet been detected by the current simple clinical tests.

The last three significant differences were seen in the mechanical power profiles at the three joints. The work performed (absorbed or generated) during each of these concentric and eccentric bursts is illustrated by the power curves shown in Figure 4 and is described in the Table. Figure 4 shows the average power plots for the 11 repeat trials for the same subject discussed previously. The time integral of each of these power phases (in watts per kilogram) yields the mechanical work (in joules per kilogram) performed by the muscles. The push-off generation (A4 work) by the elderly subjects was considerably reduced (elderly subjects, 0.191; young adults, .296 J/kg; $p < .01$) at the same time as the absorbed energy (K3 work) was increased (elderly subjects, −0.087; young adults, −0.047 J/kg; $p < .01$). Thus, the vigor of push-off by the elderly individual is drastically reduced. As stated previously, push-off normally starts at about 40% to 44% of the walking cycle, when the push-off leg is about 30 degrees forward of vertical and the contralateral limb has not yet reached HC.[22] Thus, a normal push-off is a "piston-like" thrust from the ankle, which acts upward and forward, and is destabilizing. The elderly subjects in this study appear to have recognized this fact and are reducing that potential for instability. Another possibility is that their plantar flexors may have reduced in strength, and, because of the overpowering gravitational load associated with push-off, a small reduction in strength resulted in a significant reduction in power generation. By-products of this weaker push-off were a shorter step length and the increased double-support time already discussed. Finally, because of the shorter step length, the angle of the foot relative to the ground at HC was reduced in the elderly subjects; thus, the need for absorption of energy by the dorsiflexors (A1 work) in lowering the foot to the ground would be reduced. This difference was borderline significant (elderly subjects, −0.0028; young adults, −0.0074 J/kg; $p < .08$).

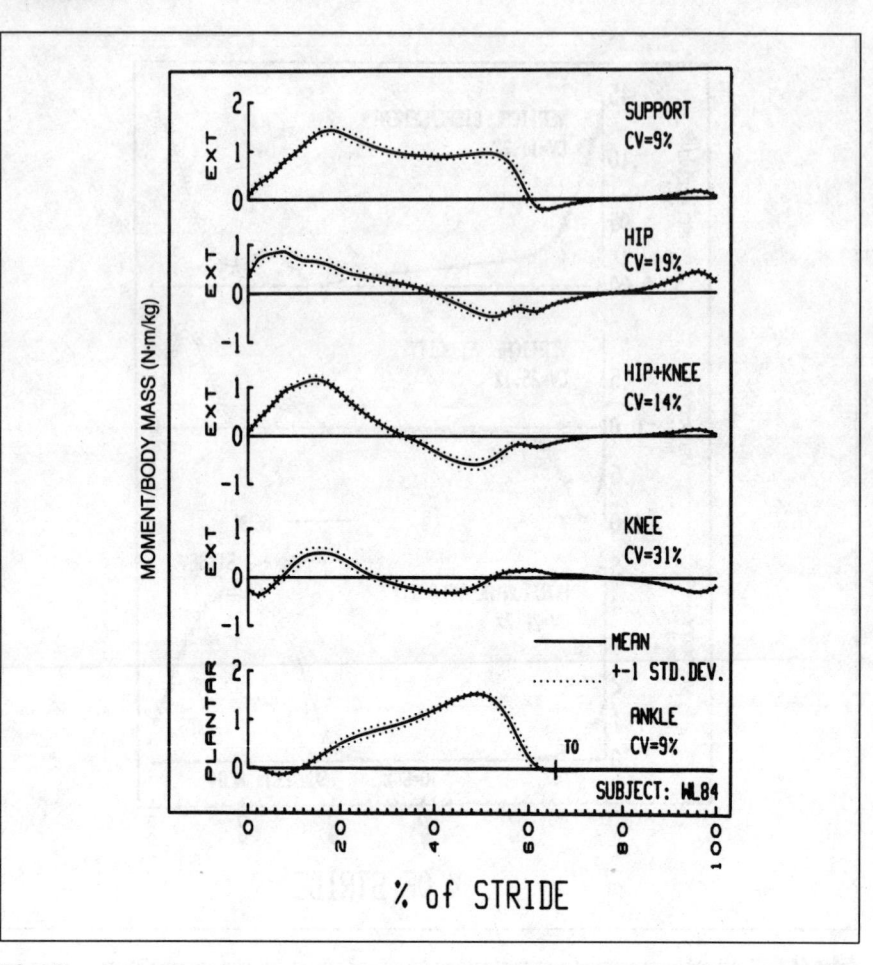

Fig 3. *Ensemble-averaged moment-of-force profiles for same subject as in Figure 1 over 11 repeat walking trials. Extensor moments at each joint are shown as positive. Variability of ankle moment for these repeat trials was low (9%) but considerably higher at the knee (31%) and hip (19%). (PLANTAR = plantar flexion; EXT = extension; CV = coefficient of variation; TO = toe-off.)*

Note that all the remaining variables that showed a significant difference were related and reflect functional changes in the gait pattern of the elderly subjects, as represented in the "circular" interrelationships presented in Figure 5. Three possible causes could equally account for all of the observed changes. First, the elderly subjects may have increased their double-support time and reduced the foot angle at HC to improve their restabilizing time. This adaptation would be accomplished with a shorter step length, which could be achieved at the motor level by a less vigorous push-off. A second cause could be that they felt more stable with a shorter step length or a lower velocity, with the associated more flat-footed landing achieved by a weaker push-off and with the longer double-support stance time being a natural consequence. Finally, the primary adaptation may have been a reduced push-off, caused either by muscle weakness or the inherent instability involved in that task, the consequence being a shorter step length and increased double-support stance time. With these three equally acceptable explanations, the exact primary cause of the adaptations may never be known. However, these age-related adaptations by the healthy elderly are important to recognize when researchers and therapists assess elderly individuals with balance disorders. This recognition will enable researchers and therapists to pinpoint

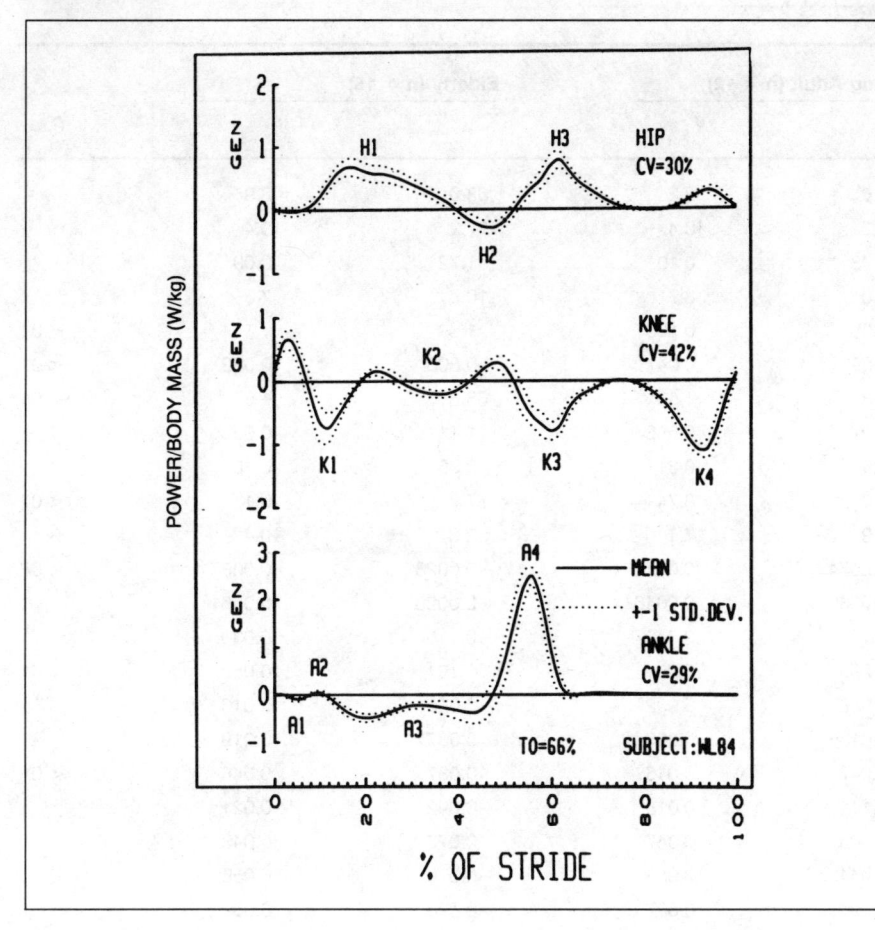

Fig 4. *Ensemble-averaged mechanical power curves for same subject as in Figure 1 over 11 repeat walking trials. Major focus is on reduced ankle push-off power (A4) by ankle plantar flexors and increased energy absorption by quadriceps femoris muscle (K4) during late stance and early swing. (See Table for definitions of work phase abbreviations.) (CV = coefficient of variation; TO = toe-off; GEN = generation.)*

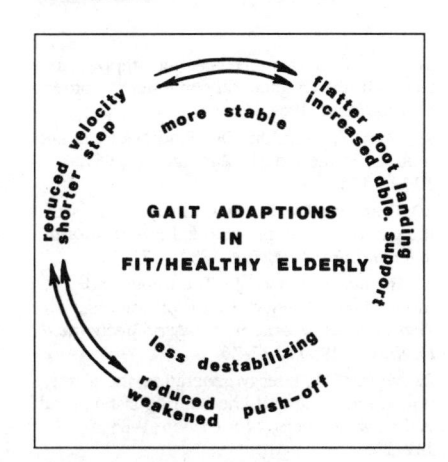

Fig 5. *Schematic "circular" argument showing possible explanations for major gait adaptations by the fit and healthy elderly group.*

changes attributable to the disorder and not to age.

Based on previous findings with young adults where no gait-related sex differences were evidenced, this study assumed that the mix of sexes in our elderly group would not alter our findings. In future work, we plan to expand the elderly subject pool to determine whether that assumption was correct.

Summary and Conclusions

This biomechanical study of the gait of young adult and fit and healthy elderly subjects revealed the following:

1. The natural walking velocity of the elderly subjects was significantly reduced; this reduction was not due to a decrease in cadence, but rather to a reduction in stride length. Accompanying this decrease was an increased double-support stance time.

2. Toe clearance in the elderly subjects was not significantly different from that of the younger adults.

3. The covariance between the hip and knee moments of force patterns, which has been identified as an "index of dynamic balance," was reduced slightly in the elderly subjects.

4. Significant differences, which were related to a less vigorous push-off and a more flat-footed landing, were noted in the mechanical power patterns.

5. The significant differences noted above are all attributable to an adaptation related to a safer (less destabilizing) gait stride.

6. Because of the significant differences attributable to age alone, it appears that a separate database is necessary in order to pinpoint falling disorders of the elderly.

Acknowledgment

We acknowledge the technical research assistance of Paul Guy.

References

1 Baker PS, Harvey H. Fall injuries in the elderly. In: Radebough TS, et al, eds. *Clinics in Geriatric Medicine*. Philadelphia, Pa: WB Saunders Co; 1985:501–508

2 Gabell A, Simons MA, Mayak USL. Falls in the healthy elderly: predisposing causes. *Ergonomics*. 1986;28:965–975

3 Gryfe CI, Ames A, Askley MJ. A longitudinal study of falls in an elderly population, I: incidence and morbidity. *Age Ageing*. 1977;6:201–211

4 Prudham D, Evans JG. Factors associated with falls in the elderly: a community study. *Age Ageing*. 1981;10:141–146

5 Sheldon JH. On the natural history of falls in old age. *Br Med J*. 1960;2:1685–1690

6 Overstall PW, Exton-Smith AN, Imms FI, et al. Falls in the elderly related to postural imbalance. *Br Med J*. 1977;1:261–264

Table. *Comparison of Young Adults and Elderly Subjects*

Variable[a]	Young Adult (n = 12)		Elderly (n = 15)		p
	X̄	s	X̄	s	
Age (yr)	24.6	2.2	68.0	3.9	
Weight (kg)	69.2	10.4	77.2	13.4	
Height (m)	1.73	0.10	1.72	0.09	
Cadence (steps/min)	111.0	8.7	110.5	7.3	
Stride length (m)	1.55	0.103	1.39	0.14	<.01
Stride length (statures)	0.895	0.047	0.808	0.082	<.01
Stance time (%)	62.3	1.55	65.5	1.67	<.01
Toe clearance (cm)	1.27	0.588	1.11	0.53	
Toe clearance variance (cm)	0.45	0.311	0.35	0.20	
% COV[b] (hip-knee)	67.0	8.74	57.7	15.1	<.07
% COV (knee-ankle)	50.9	17.1	52.6	10.4	
A1 work (J/kg)	−0.0074	0.0072	−0.0028	0.0037	<.08
A2 work (J/kg)	0.0036	0.0046	0.0029	0.0051	
A3 work (J/kg)	−0.111	0.042	−0.115	0.030	
A4 work (J/kg)	0.296	0.051	0.191	0.050	<.01
K1 work (J/kg)	−0.048	0.032	−0.040	0.016	
K2 work (J/kg)	0.0186	0.026	0.037	0.019	
K3 work (J/kg)	−0.047	0.015	−0.087	0.048	<.01
K4 work (J/kg)	−0.114	0.015	−0.092	0.021	
H1 work (J/kg)	0.103	0.047	0.072	0.043	
H2 work (J/kg)	−0.044	0.029	−0.070	0.056	
H3 work (J/kg)	0.090	0.027	0.098	0.032	

[a] Work phase: A1 = absorption by dorsiflexors after heel contact; A2 = generation by dorsiflexors to pull the leg forward over foot; A3 = absorption by plantar flexors as leg rotates forward over foot; A4 = generation of energy by plantar flexors at push-off; K1 = energy absorbed at knee by quadriceps femoris muscle during weight acceptance; K2 = energy generated by quadriceps femoris muscle as knee extends during mid-stance; K3 = energy absorbed by quadriceps femoris muscle as knee flexes during late stance and early swing; K4 = energy absorbed by knee flexors (hamstring muscles) as knee extends late in swing; H1 = energy generated by hip extensors as hip extends (hip flexion reduces) during weight acceptance; H2 = energy absorbed by hip flexors in mid-stance as backward-rotating thigh is decelerated; H3 = energy generated by hip during late stance and early swing to accelerate to lower limb upward and forward.

[b] % COV = percentage of covariance.

7 Wolfson LI, Whipple R, Amerman RN, et al: Stressing the postural response: a quantitative method for testing balance. *J Am Geriatr Soc.* 1986;34:845–850

8 Mann RA, Hagy JL, White V, et al. The initiation of gait. *J Bone Joint Surg Am.* 1979;61:232–239

9 Winter DA. Overall principle of lower limb support during stance phase of gait. *J Biomech.* 1980;13:923–927

10 Shimba T. An estimation of center of gravity from force platform data. *J Biomech.* 1984;17:53–60

11 Winter DA. Balance and posture in human walking. *Engineering in Medicine and Biology.* 1987;6:8–11

12 Winter DA. *Biomechanics of Human Movement.* New York, NY: John Wiley & Sons Inc; 1979

13 Winter DA. Kinematic and kinetic patterns in human gait: variability and compensating effects. *Human Movement Science.* 1984;3:51–76

14 Winter DA. Biomechanics of normal and pathological gait: implications for understanding human motor control. *Journal of Motor Behavior.* 1989;21:337–356

15 Winter DA. *Biomechanics and Motor Control of Human Gait.* Waterloo, Ontario, Canada: University of Waterloo Press; 1987:18

16 Finley FR, Cody KA, Finizie RV. Locomotion patterns in elderly women. *Arch Phys Med Rehabil.* 1969;50:140–146

17 Murray MP, Kory RC, Clarkson BH. Walking patterns in healthy old men. *J Gerontol.* 1969;24:169–178

18 Azar GJ, Lawton AH. Gait and stepping as factors in the frequent falls in elderly women. *Gerontologist.* 1964;4:83–84

19 Imms FJ, Edholdm OG. Studies of gait and mobility in the elderly. *Age Ageing.* 1981;10:147–156

20 Guimeres RM, Isaacs B. Characteristics of the gait of old people who fall. *International Rehabilitation Medicine.* 1980;2:177–180

21 Bresler B, Frankel JP. The forces and moments in the leg during level walking. *Transactions American Society of Mechanical Engineers.* 1950;72:27–36

22 Winter DA. Energy generation and absorption at the ankle and knee during fast, natural and slow cadences. *Clin Orthop.* 1983;197:147–154

23 Cochran WG. Approximate significance levels of the Behrens-Fisher test. *Biometrics.* 1964;20:191–195

Relationships Between Physical Activity and Temporal-Distance Characteristics of Walking in Elderly Women

Carol I Leiper
Rebecca L Craik

The purpose of this study was to investigate the relationships between physical activity and walking speed in women 64 years of age and over. Data were gathered from 81 nondisabled women ranging from 64.0 to 94.5 years of age. The women were categorized as sedentary, community active, or exercisers based on a combination of their living situation and level of daily activity. Subjects walked over a 3.84-m recording surface at five different paces, ranging from walking as slowly as possible to walking as quickly as possible. Actual walking speed and length of steps were measured. Stepping frequency and step length relative to leg length were derived measures. Mean walking speeds ranged from 0.43 m/s at the very slow pace to 1.42 m/s at the very fast pace. The walking speeds at the very slow pace were significantly different among the three physical activity groups. At the very slow pace, women who exercised were able to walk significantly more slowly than the other women. The groups were not significantly different at any other pace. Normal walking speeds for all groups were slower than those previously reported for younger women, with the walking speed of the fastest pace of the elderly women being closer to the normal walking speed of younger women. The results of this study indicate that physical therapists need to utilize age-appropriate values as the standard when evaluating performance. [Leiper CI, Craik RL. Relationships between physical activity and temporal-distance characteristics of walking in elderly women. Phys Ther. 1991;71:791–803.]

Key Words: Aging, Gait analysis.

Effects of Aging, Disuse, and Exercise on Functional Status

It is a common observation that as we age, performance of physical activities tends to improve during the first 2 to 3 decades of life, maintain or slightly decrease for another 1 or 2 decades, and then decline more consistently and at times rapidly. The decrements that occur in men with increasing age have been summarized for many physiological systems.[1] These data have been depicted schematically as linear projections, terminating at 80 years of age, using mean values for 20- to 35-year-old men as representative of optimal performance. Investigators have described naturally occurring age-related decreases in aerobic capacity,[2] bone mass,[3] joint flexibility,[4] and muscular strength.[5,6] Although such decreases have been associated with an increase in age, they are also similar to changes that occur with bed rest and in some instances inactivity.[7] For example, dur-

---

CI Leiper, PhD, PT, is Assistant Professor, Program in Physical Therapy, Beaver College, Glenside, PA 19038-3295 (USA). Address all correspondence to Dr Leiper.

RL Craik, PhD, PT, is Associate Professor, Program in Physical Therapy, Beaver College.

This study was completed in partial fulfillment of the requirements for Dr Leiper's doctoral degree at Temple University, Philadelphia, PA.

Dr Leiper's doctoral program was supported in part by the Foundation for Physical Therapy Inc.

This study was approved by the Temple University Research Review Committee for the Protection of Human Subjects.

This article was submitted December 3, 1990, and was accepted July 15, 1991.

ing periods of bed rest, bone mineral content decreases[8] and cardiovascular deconditioning occurs.[9] Encouragement of independence and maintenance of physical abilities may well have an increasingly important emphasis in programs for the prevention of aging-related problems.

Some investigators have demonstrated that certain aspects of physical ability are more strongly influenced by habitual activity than by age. In a longitudinal study to determine the aerobic capacity of master athletes who were between 50 and 82 years of age, Pollock et al[10] found that runners who maintained competitive levels of training maintained their aerobic capacity. Those who decreased training intensity showed significant declines in aerobic capacity.

The differential effects of age and habitual physical activity on arm reaction time have been studied in young (20–30 years of age) and elderly (60–70 years of age) men.[11] Each age group was divided into persons who were inactive and those who were active (ie, in either running or racket sports). The reaction times for older active men were faster than those for older sedentary men and were similar to those for much younger sedentary men. Performances in static balance, sit-and-reach flexibility, grip strength, and reaction times of index fingers have been investigated in both young ($\overline{X}$=22.2 years of age) and elderly ($\overline{X}$=68.7 years of age) women.[12] These groups of women were also divided into active or inactive subsamples. Except for grip strength, the performance of the older active women was significantly better than that of the older inactive women and more similar to that of the young women. These observations suggest that maintenance of regular physical activity may assist in prolonging active life and may possibly delay the onset of chronic or debilitating illness until close to the end of the life span.[13]

Gait Patterns

A variety of gait characteristics have been attributed to the aging process. These include visual observations of a slowed walking speed; flexion in the thoracic spinal area; a loss of the lordotic curve in the lumbar spine; a forward head position; short, shuffling steps; and a loss of normal arm action.[14–16] Limited information is available in these reports relating to the criteria used to select subjects for these observations, and it is possible that the samples included persons who had pathology. Other investigators reporting on samples with a declared pathology describe similar observations.[17,18]

In an attempt to describe and quantify normal adult patterns of walking and the relationships of pathology and aging, many investigators have examined selected kinetic and kinematic variables. Gait variables are often expressed in relation to a complete walking or gait cycle, which is defined as the time interval between sequential foot contacts for the same foot.[19,20] The simplest variables to observe are those that reflect the composite actions of the muscle activity and joint excursions—the resultant step length, cadence, and walking speed. Several studies in which comparisons between young and elderly women were made have been reported.[21–24] In each study, the authors found that the elderly women had a slower walking speed, a shorter step or stride length, and a longer stance time than did the young women.

Although walking is not essential for independent living, it is the form of locomotion, or moving from place to place, that is least encumbering and most efficient for function in daily life. Lundgren-Lindquist et al[25] have addressed the functional consequences of declining walking speed. They studied free-speed and maximal-speed walking performance in 226 Swedish men and women, all age 79 years. The average comfortable walking speeds were 1.03 m/s for men and 0.92 m/s for women. The mean maximal walking speeds were 1.43 m/s for men and 1.18 m/s for women. The authors noted that only the maximal walking speed for men was sufficient to achieve the 1.4-m/s walking speed necessary to safely negotiate traffic-light–controlled intersections in Sweden.

It is unknown whether habitual patterns of a physically active or sedentary lifestyle might be related to walking performance as one ages. The primary purposes of this investigation, therefore, were (1) to describe the range of walking speeds observed in a sample of women aged 64 years and over and (2) to investigate the relationships among physical activity, age, body mass index (BMI), and selected temporal-distance characteristics of walking. It was hypothesized that, when given a pace command, the actual walking speeds would not differ among the three physical activity groups. We also predicted that the variables of physical activity, age, and BMI would not result in a significant amount of variance in the actual walking speeds.

Method

Subjects

A sample of convenience of adult female volunteers was solicited from a number of locations.* At each location, institutional approval was granted prior to any solicitation of volunteers. Women 64 years of age and older were eligible for the study. Subjects were required to be in generally good health and free from impairments resulting in obvious gait deviations or asymmetry. Each subject was required to be able to walk a minimum of one city block, independently or with the maximal support of a cane. Subjects agreeing to participate in the study signed a statement of informed consent. Each subject was classified according to self-reported level of physical activity and then assigned to one of three groups defined

---

*Community Scholars Program, Beaver College, Glenside, PA 19038; Ft Washington Estates, Total Care Systems Inc, Springhouse, PA 19477; Heatherwood Retirement Community, Honeybrook, PA 19344; ActiveLife Program, Philadelphia Geriatric Center, Philadelphia, PA 19141.

Basic Research

Table 1. Age and Anthropometric Variables for Each Physical Activity Group and Total Sample

Variable	Group			
	Sedentary (n=25)	Community Active (n=27)	Exercisers (n=29)	Total (N=81)
Age (y)				
$\bar{X}$	81.34	76.30	71.12	76.00
SD	5.37	8.13	4.70	7.44
Range	72.0–90.5	65.5–94.5	64.0–81.5	64.0–94.5
BMI[a]				
$\bar{X}$	25.10	24.40	25.95	25.17
SD	3.48	3.95	4.07	3.86
Range	17.7–33.1	19.7–36.3	17.6–34.2	17.6–36.3
Standing height (m)				
$\bar{X}$	1.59	1.57	1.60	1.58
SD	0.06	0.06	0.07	0.06
Range	1.45–1.67	1.42–1.72	1.47–1.74	1.42–1.74
Weight (kg)				
$\bar{X}$	63.19	60.72	66.63	63.45
SD	8.98	10.32	11.26	10.59
Range	47.7–80.0	46.4–87.7	44.1–99.5	44.1–99.5
Leg length (cm)				
$\bar{X}$	77.96	76.87	78.21	77.68
SD	3.86	4.55	3.90	4.12
Range	70.0–89.0	68.0–87.0	73.0–85.7	68.0–89.0

[a]BMI=body mass index.

as sedentary, community active, or exercisers.

Sedentary group. Individuals in this group had lived in a retirement community for a minimum of 6 months and had limited physical activity. The women did not regularly prepare meals or housekeep, walked less than three city blocks per day, and climbed less than 30 stairs per week.

Community active group. These women regularly did their own shopping, meal preparation, and light housekeeping and went out socially. The women lived independently in the community, walked the equivalent of three city blocks or more per day, and climbed 30 or more stairs per week.

Exercisers group. The individuals in this group met the requirements described previously for the community active group and also participated in a regular, ongoing form of exercise or sport activity for a minimum of 4 months prior to testing. The minimum amount of exercise required for inclusion in this group was 30 minutes twice a week. Exercise consisted of any of the following activities: (1) free-speed or treadmill walking; (2) stationary bicycling; (3) swimming; (4) extremity flexibility, strength, or endurance exercise; (5) tennis or racquetball; (6) volleyball, basketball, or softball; (7) golf, if walking and pulling a golf cart; and (8) folk dancing or square dancing.

A total of 81 women, aged 64.0 to 94.5 years, were tested. The three physical activity groups were relatively equal in size, with 25 women in the sedentary group, 27 in the community active group, and 29 in the exercisers group. The descriptive data for age and the anthropometric variables are given in Table 1.

Instrumentation

Anthropometric measures. Weight and height measurements were obtained for each subject. Subjects were instructed to wear their usual walking shoes during these measurements, because the additional weight of the shoes and the resultant increase in leg length would affect the gait characteristics. Height was measured in the standing and sitting positions, based on methods described by Lohman and colleagues.[26] An anthropometer was devised that could be used for both the standing and sitting positions (Fig. 1). The standing apparatus consisted of a vertical piece of wood (182×8.5×1.75 cm) attached to a flat platform (44×47 cm). A small horizontal block of wood was attached to the vertical upright, with the

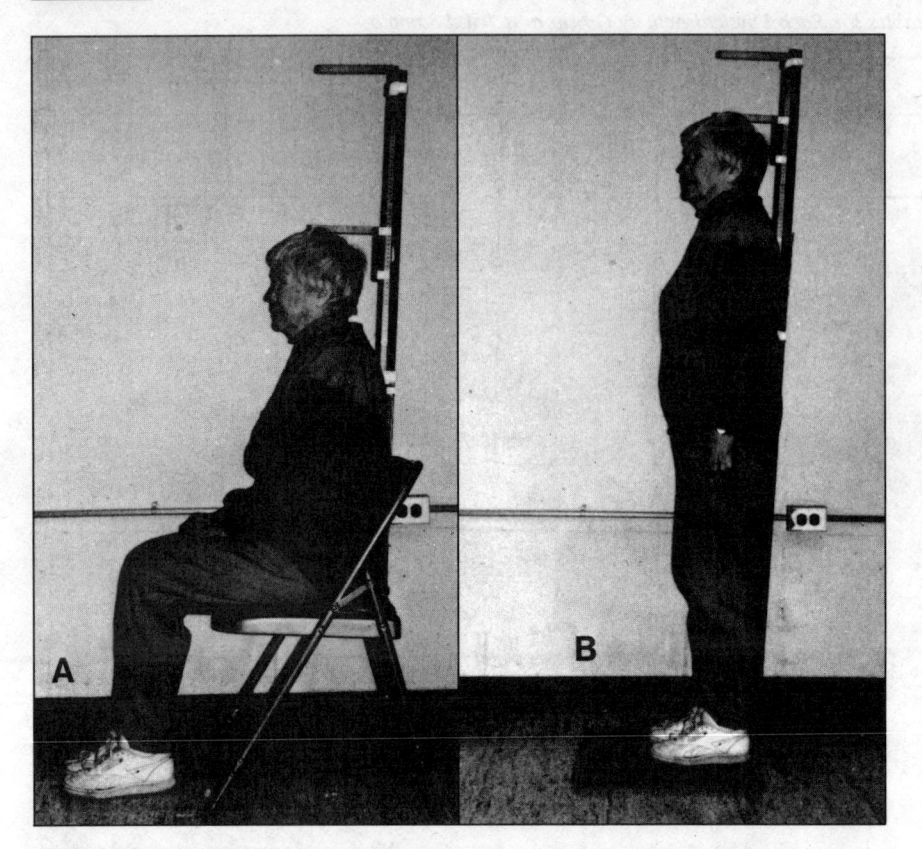

Figure 1. Measurement technique for sitting (A) and standing (B) heights.

top surface 104 cm above the platform. Two Velcro®[†] straps were attached to the upright, one near the top and one just above the horizontal block. A wooden caliper (79×1.5×1.5 cm, with two arms, each 17×1.5×1.5 cm) was placed vertically against the vertical upright, resting on the block and held in place by the Velcro® straps. A metal ruler (in centimeters) was attached to one side of the caliper. When a subject stood on the platform against the upright, the sliding arm of the caliper could be adjusted up or down to touch the top of the subject's head.

The same caliper was used to determine sitting height. For this measurement, a vertical piece of wood (122×3.5×1.75 cm) was attached to a flat platform (36×48 cm). A small horizontal block of wood was attached to the vertical upright, with the top surface 45 cm above the platform. The platform was placed on a folding chair. The caliper was used in the manner described previously to determine the subject's sitting height while she sat on the platform.

For both height measurements, each subject was asked to stand (sit) as erectly as possible while the caliper was adjusted. At the moment of recording, a second verbal command to stand (sit) "as tall as possible" was given. These two measurements were used to determine leg length via the following formula:

(1) Leg length (cm) = standing height (cm) − sitting height (cm) (1)

Body mass index was calculated from the formula proposed by Chumlea et al[27]:

(2) BMI = weight (kg)/standing height (m^2)

Gait measures. Measurements of the gait characteristics were collected by having each subject walk repeatedly over a portable instrumented walkway (Gait Mat[‡]) (Fig. 2). The Gait Mat (3.84 m × 70 cm) is divided into left and right sections and is constructed of 512 pressure switches, each 1.5×30.6 cm. The switches are oriented perpendicular to the direction of the subject's progression, with 256 switches under the straight path of each foot. The switches and circuitry of the walkway are covered by black rubber. The Gait Mat was connected to a Rockwell Aim 65 microprocessor.[§] As the subject walked over the walkway, each switch under the foot closed as weight was applied. The locations and times of foot contact and release were recorded along the length of the surface. Accuracy of the Gait Mat was determined electronically to be ±1 switch (1.5 cm) in location and ±20 milliseconds in timing. The microprocessor provided a printout specifying step length and step time bilaterally for each foot contact and the average walking speed for each pass over the walkway.

Procedure

When each subject arrived for testing, the procedure was explained and a health and activity questionnaire was administered. This questionnaire was used to record a description of the subject's living situation, any physical problems, an estimate of daily walking and stair climbing, and other specifics about daily activities and exercise. For this inventory, the investigator (CIL) asked the questions and entered the responses. The information was used to assign each individual to the appropriate group classification. The subject's standing and sitting height

---

[†]Velcro USA Inc, 406 Brown Ave, Manchester, NH 03103.

[‡]EQ Inc, 8427 Germantown Ave, Philadelphia, PA 19118.

[§]Rockwell International Corp, PO Box 3669, Anaheim, CA 92803.

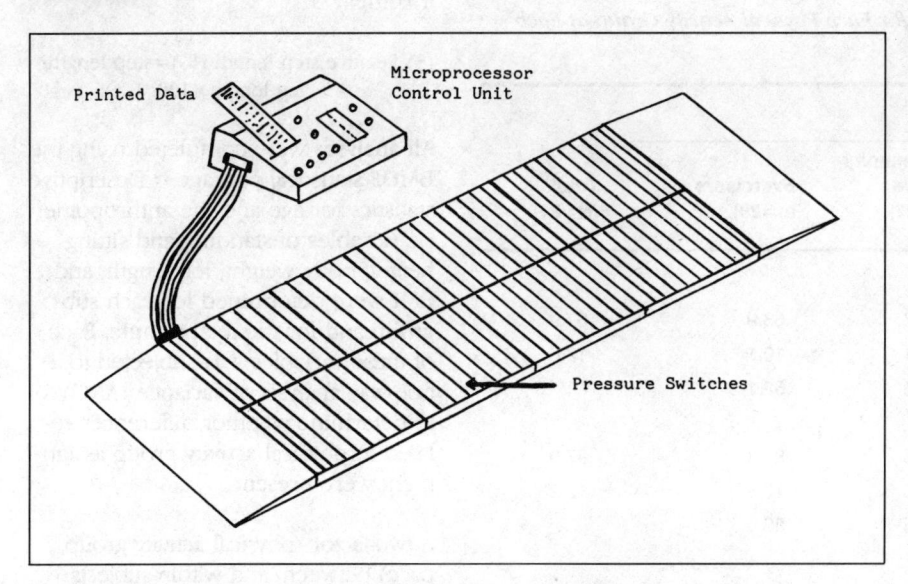

Figure 2. *Schematic of Gait Mat. Each side has 256 pressure switches, each 1.5×30.6 cm, which record the location and time of foot contact and release.*

and body weight were then measured and recorded.

Intrarater reliability of the anthropometric measurements for standing and sitting height was determined on a subsample of 6 of the 81 subjects. For these measurements, the investigator stood at the side of the caliper that did not have the ruler attached. She adjusted the height of the arm for the subject, and the value was recorded by an assistant. The caliper was then moved upward, and the subject moved away. The same procedure was then repeated. Intraclass correlation coefficients (ICC[3,1]) for fixed-effects models were used to determine the consistency of the measurements.[28] The ICCs were .99 for standing height and .98 for sitting height. The ICC (3,1) was selected, as we were primarily interested in our own intrarater accuracy pertaining to this investigation.

Each subject was then asked to walk across the Gait Mat at each of five different paces (the instruction given to each patient is shown in parentheses): very slow ("Walk as slowly as you can without stopping."); slow ("Walk slowly."); normal ("Walk at your preferred or normal speed."); fast ("Walk quickly."); and very fast ("Walk as quickly as you can without running.").

Each subject was asked to take several trial walks at the preferred pace over the walking surface to become comfortable with the procedure. When the testing sequence began, the subject walked across the mat at the requested pace and then paused. The data were stored in the memory of the microprocessor, the pace command was repeated, and the subject again walked over the pathway. Each subject repeated this sequence until 10 right and 10 left steps had been collected at each pace. The investigator then reminded the subject of the range of five paces and requested that she walk at the next slower or faster pace in the sequence. The sequence was the same for all subjects, starting with the preferred pace and moving

Table 2. *Walking Speed (in Meters Per Second) for Each Physical Activity Group at Each Pace*

Pace	Group			
	Sedentary (n=25)	Community Active (n=27)	Exercisers (n=29)	Total (N=81)
Very slow				
$\bar{X}$	0.38	0.47	0.43	0.43
SD	0.13	0.13	0.11	0.13
Age-adjusted $\bar{X}$	0.47	0.47	0.34	
Slow				
$\bar{X}$	0.62	0.73	0.74	0.68
SD	0.20	0.19	0.16	0.19
Age-adjusted $\bar{X}$	0.71	0.74	0.65	
Normal				
$\bar{X}$	0.89	0.99	1.03	0.96
SD	0.24	0.23	0.20	0.23
Age-adjusted $\bar{X}$	0.99	0.99	0.94	
Fast				
$\bar{X}$	1.07	1.17	1.24	1.15
SD	0.26	0.23	0.22	0.24
Age-adjusted $\bar{X}$	1.17	1.17	1.15	
Very fast				
$\bar{X}$	1.29	1.41	1.57	1.42
SD	0.30	0.23	0.29	0.30
Age-adjusted $\bar{X}$	1.39	1.41	1.49	

Table 3. Cadence (in Steps Per Minute) for Each Physical Activity Group at Each Pace

	Group			
Pace	Sedentary (n=25)	Community Active (n=27)	Exercisers (n=29)	Total (N=81)
Very slow				
$\bar{X}$	67.0	72.7	63.9	67.9
SD	13.0	13.1	10.1	12.6
Age-adjusted $\bar{X}$	69.0	72.5	62.1	
Slow				
$\bar{X}$	86.5	91.5	87.8	87.9
SD	13.6	10.3	10.8	12.2
Age-adjusted $\bar{X}$	88.5	91.5	86.0	
Normal				
$\bar{X}$	106.8	109.0	105.8	106.5
SD	14.5	11.1	9.3	11.8
Age-adjusted $\bar{X}$	108.8	109.1	104.0	
Fast				
$\bar{X}$	120.0	123.7	119.5	120.5
SD	16.7	11.6	12.3	13.7
Age-adjusted $\bar{X}$	122.0	123.8	117.7	
Very fast				
$\bar{X}$	136.4	140.5	141.6	139.3
SD	17.3	12.7	20.5	17.1
Age-adjusted $\bar{X}$	138.4	140.6	141.6	

to the slow and very slow paces. Following the trials at the very slow pace, each subject was asked to walk again at the preferred pace twice in preparation for the trials at the fast and very fast paces. Subjects were reminded that they could stop and rest at any time, if they desired. Total interview and testing time was approximately 40 minutes.

The set pace sequence, working progressively to the slowest and then to the fastest pace, was selected to examine the subjects' range of walking speeds and not for the purpose of determining whether a pace command could be produced independently from memory. The ability of the subjects to produce a consistent walking speed across the trials at each pace was determined by randomly selecting two subjects from each group. At each pace, an ICC (ICC[3,1]) was determined using the walking speeds for three trials of these six subjects. The following ICCs were obtained for the five paces: .90, very slow pace; .96, slow pace; .96, normal pace; .99, fast pace; and .99, very fast pace.

Data Analysis

Summary statistics for step length and step time, bilaterally and for walking speed, were calculated for each of the five paces for each subject. The independent variables in this investigation were (1) physical activity group, (2) age (in years), (3) BMI, and (4) the five paces. The dependent variables were the actual walking speed (in meters per second), cadence (in steps per minute), and relative step length produced at each pace. Relative step length was determined using the formula:

(3) Relative step length (%)=step length/ leg length×100

All analyses were completed using the BMDP statistical package.[29] Descriptive statistics for age and the anthropometric variables of standing and sitting height, body weight, leg length, and BMI were determined for each subgroup and for the total sample. Each of these variables was subjected to a one-way analysis of variance (ANOVA) to determine whether differences related to physical activity group assignment were present.

A two-factor (physical activity group, pace) between- and within-subjects analysis of covariance (ANCOVA) with age as a covariate was used to test the hypotheses relating to physical activity group effect on walking speed. An ANCOVA was also used to examine group effect on cadence and relative step length. Bivariate correlations were calculated for the pairs of actual walking speed, cadence, and relative step length measurements, and regression equations were developed to describe the relationships between each pair. A multiple-regression analysis was used to examine the relationships among physical activity, age, BMI, and actual walking speed at each pace. All hypotheses were tested against an alpha value of .05 for significance.

Results

Significant differences were present among the three physical activity groups for the variable of age only (F=18.09; df=2,78; P<.0001). A Tukey's *post hoc* test revealed significant intergroup age differences. Women in the sedentary group were the oldest ($\bar{X}$=81.3 years of age; SD=5.4), and those in the exercisers group were the youngest ($\bar{X}$=71.1 years of age; SD=4.7). Mean values for walking speed, cadence, and relative step length are shown for each physical activity group and for the total sample at each pace in Tables 2, 3, and 4, respectively. The age-adjusted means for walking speed ranged from 0.34 to 1.49 m/s across the five paces. The age-adjusted means for cadence ranged from 62.1 steps/min at the very slow pace to 141.6 steps/min at the very fast pace. Relative step length also increased across the five paces from a mean of 46% of leg length at the very slow pace to a mean of 80% of leg length at the very fast pace.

Effects of Physical Activity Group on Gait Variables

A two-factor (3×5) between- and within-subjects ANCOVA was used to test the hypotheses relating to group effect and pace for the dependent variable of walking speed. The demographic variable of age was determined to be a covariate because of age inequality among the three groups. Significant differences were present for the main effect of pace ($F=926.55$; $df=4,312$; $P<.0001$) and for the interaction ($F=4.74$; $df=8,312$; $P<.0001$). There were no significant differences relating to the main effect of physical activity group (Tab. 5).

Simple-effects tests were carried out to determine the nature of the interaction.[30] At the very slow pace, the walking speeds were significantly different among the three physical activity groups ($F=4.75$; $df=2,389$; $P<.025$). No other significant differences were apparent. Tukey's *post hoc* comparisons were conducted to further decompose the simple effects at the very slow pace. Significant differences were present between the sedentary group and the exercisers group and also between the community active group and the exercisers group. In each comparison, the exercising women walked more slowly than the others (Fig. 3).

Bivariate correlation tests were performed to determine the relationships among actual walking speed, cadence, and relative step length. Strong significant linear relationships were found for walking speed and cadence ($r=.92$, n=431, $P<.001$), velocity and relative step length ($r=.90$, n=431, $P<.001$), and relative step length and cadence ($r=.72$, n=431, $P<.001$). These relationships and associated regression equations are illustrated in Figures 4 through 6.

Predictors of Walking Speed

A multiple-regression analysis was used to determine the contribution of physical activity, age, and BMI to the actual walking speed at each pace. Regression equations were determined using an all-subsets regression, with the adjusted R^2 value as the factor to determine the best predictors.

Table 4. *Relative Step Length (in Percentage of Leg Length) for Each Physical Activity Group at Each Pace*

Pace	Group			
	Sedentary (n=25)	Community Active (n=27)	Exercisers (n=29)	Total (N=81)
Very slow				
$\bar{X}$	0.44	0.50	0.51	0.49
SD	0.12	0.10	0.08	0.11
Age-adjusted $\bar{X}$	0.49	0.50	0.46	
Slow				
$\bar{X}$	0.54	0.62	0.64	0.59
SD	0.13	0.12	0.08	0.12
Age-adjusted $\bar{X}$	0.59	0.62	0.59	
Normal				
$\bar{X}$	0.64	0.70	0.74	0.69
SD	0.12	0.12	0.09	0.12
Age-adjusted $\bar{X}$	0.69	0.70	0.69	
Fast				
$\bar{X}$	0.69	0.74	0.80	0.74
SD	0.12	0.11	0.08	0.11
Age-adjusted $\bar{X}$	0.74	0.74	0.75	
Very fast				
$\bar{X}$	0.73	0.78	0.85	0.79
SD	0.13	0.11	0.08	0.12
Age-adjusted $\bar{X}$	0.79	0.78	0.80	

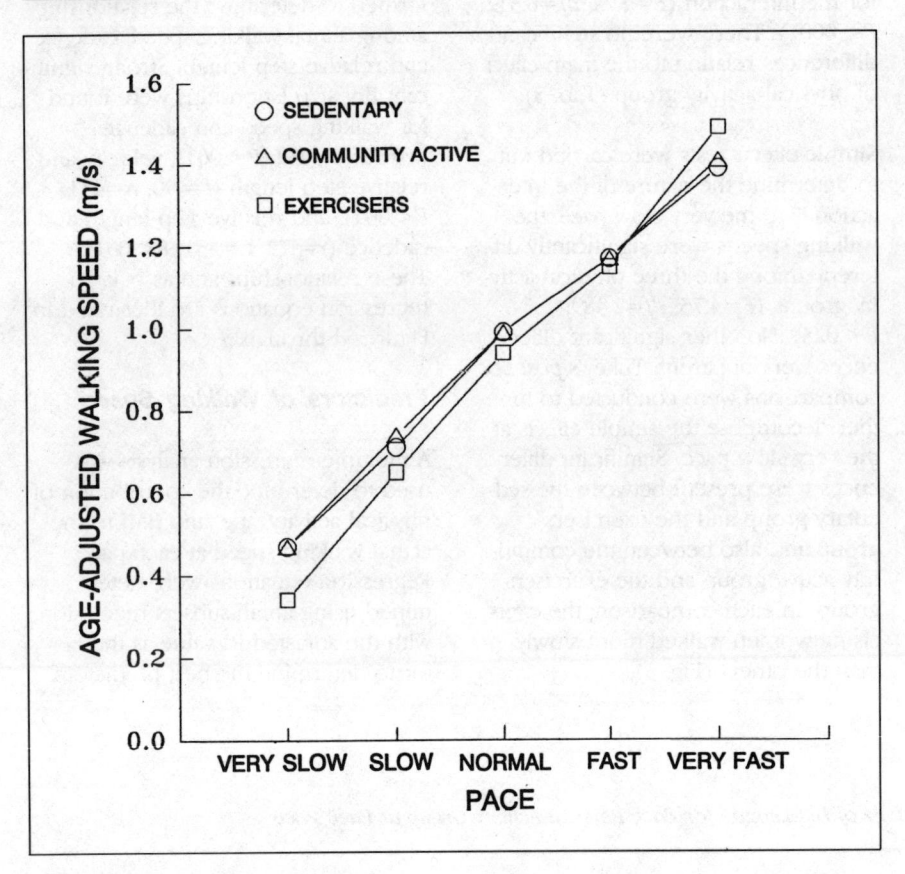

Figure 3. *Mean walking speed (in meters per second) for three physical activity groups at each of five paces illustrating interaction effect. Values adjusted for age covariate.*

The best-fit equations are given in Table 6. For all paces, age and BMI were the main predictors of walking speed. At each pace, except at the very slow pace, between 30% and 45% of the variability of walking speed was accounted for by age. In each case, BMI contributed only a maximum of 3% of the variability. The very slow pace presented a different situation. The criterion variables remained the same, but accounted for a total of only 5% of the variability in walking speed. The physical activity group to which the subjects were assigned appeared to have no role in prediction of walking speed. Thus, the null hypothesis of relationship between physical activity group and actual walking speed was accepted, and the null hypotheses relating to the relationship of age and BMI to walking speed were rejected.

Discussion and Implications

The procedure used in this study was designed to describe the spectrum of walking speeds used by our sample of elderly women. No external pace regulation or timer was used to produce specified cadences or walking speeds, because the natural characteristics were of interest. The subjects were informed of the five-phase procedure, but each individual selected the actual walking speed at each pace command. Every effort was made to ensure that each subject had actually performed at a complete range of her walking abilities. Subjects were repeatedly reminded of the current pace command and at the slowest and fastest commands were asked whether their performance was the best they could do.

The walking speeds produced at the slow, normal, and fast paces agree with those reported previously by Lundgren-Lindquist et al[25] and Himann et al.[31] In those studies, the ages of the subjects and the testing conditions were similar to those in our study (Tab. 7). The walking velocities produced in our study do not agree with those reported by other previous investigators.[21-23,32] In each of these investigations, however, either the sample or the experimental procedures were not comparable to those in our study. Murray et al[22] and Larsson et al[32] reported normal walking speeds at 1.30 and 1.28 m/s, respectively, somewhat higher than the average walking speed of 0.97 m/s recorded in this investigation. In both of those studies, care was taken to ensure representation of a wide age range of women (ie, 20–70 years), but

Table 5. *Two-Factor (3×5) Analysis-of-Covariance Results for Walking Speed Differences Among Physical Activity Groups with Repeated Measures on Pace and with Age as a Covariate*

Source	df	SS	MS	F	P
Between subjects					
Physical activity group	2	0.13	0.06	0.52	
Covariant (age)	1	4.84	4.84	39.97	.0001
Subjects within group	77	9.32	0.12		
Within subjects					
Pace	4	49.61	12.40	926.55	.0001
Physical activity group × pace	8	0.51	0.06	4.74	.0001
Pace × subjects within group	312	4.18	0.01		

the samples were very small and the data were reported as mean values for all women and were not stratified by age. The 13 women tested by Hageman and Blanke[23] were reported to have a comfortable walking speed of 1.31 m/s. The average age of their subjects was 66.8 years. These women also walked barefooted during the test. The slower preferred walking speeds for both elderly and young women (ie, 0.70 and 0.84 m/s, respectively) reported by Finley et al[21] may have been a result of the electrogoniometers attached to the major joints of the lower extremities.

Using our methods, we found that the elderly women in our study demonstrated a reduced capacity to walk quickly on command. When asked to walk at the very fast pace, the mean walking speed for the total sample of 81 women was 1.42 m/s. This maximal walking speed is only slightly different from that described as normal (ie, 1.29 m/s) for several younger and mixed samples of women.[22,23,32,33] Two other groups of investigators described asking subjects to "walk as quickly as possible" and reported speeds for that pace. Larsson et al[32] reported a walking speed of 2.4 m/s in their small, mixed-age (ie, 20–70 years) sample of women. In a sample more closely related to that of our investigation, however, Lundgren-Lindquist et al[25] reported a mean maximal walking speed of 1.18 m/s for 112 women, all 79 years of age. The necessity for producing very rapid walking speeds may be reduced in older women, depending on their lifestyle. In everyday living, the primary situation in which a rapid walking speed would be required is most likely to be when one is trying to cross a street. Perhaps even active persons are less likely to need to use rapid walking speeds if they customarily shop in a mall or shopping center.

Previous investigators have suggested that decrements in selected physical performance may be related to decreases in physical activity rather than

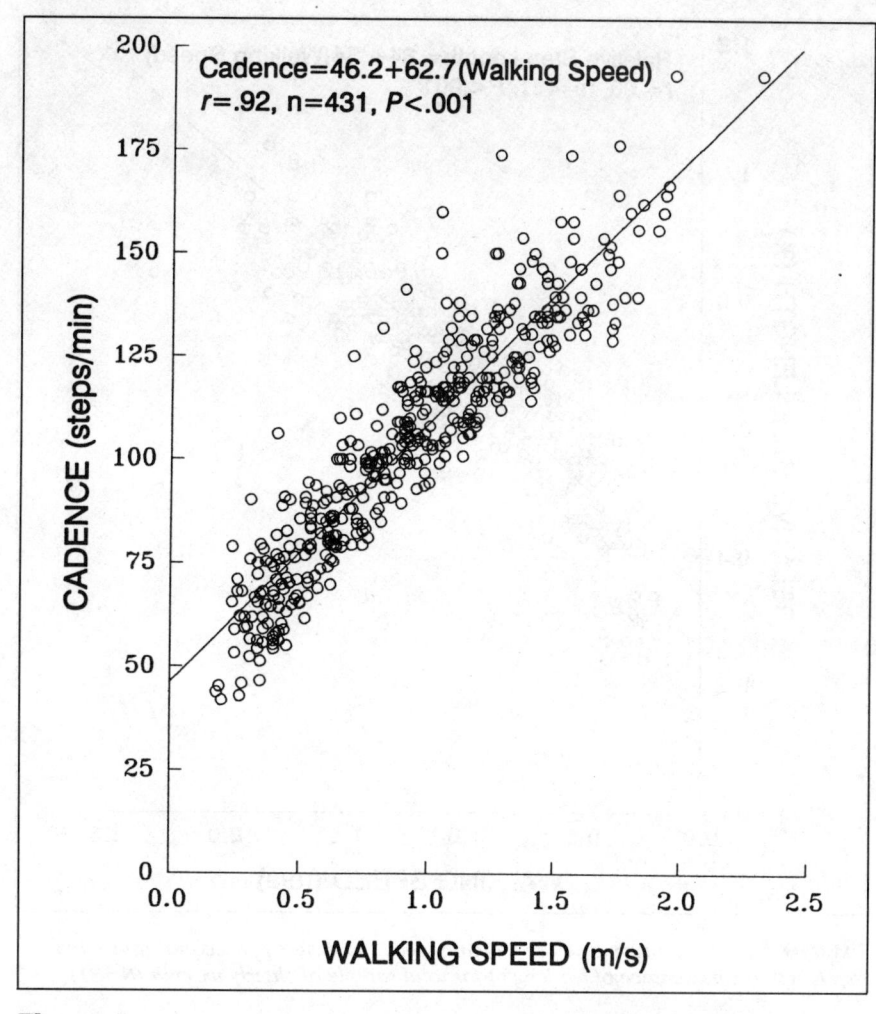

Figure 4. *Relationship between walking speed (in meters per second) and cadence (in steps per minute) for total sample of elderly women (N=81).*

Table 6. *Best-Fit Multiple-Regression Equations for Prediction of Walking Speed from Criterion Variables of Activity Group, Age, and Body Materials (BMI)*

Pace	Adjusted R^2	Equation	Contribution to R^2	
Very slow	.05	0.859−0.004(age)−0.005(BMI)	.06	(age)
			.02	(BMI)
Slow	.30	1.968−0.015(age)−0.006(BMI)	.31	(age)
			.01	(BMI)
Normal	.36	2.676−0.019(age)−0.009(BMI)	.37	(age)
			.02	(BMI)
Fast	.33	2.882−0.019(age)−0.010(BMI)	.34	(age)
			.03	(BMI)
Very fast	.45	3.832−0.028(age)−0.012(BMI)	.46	(age)
			.02	(BMI)

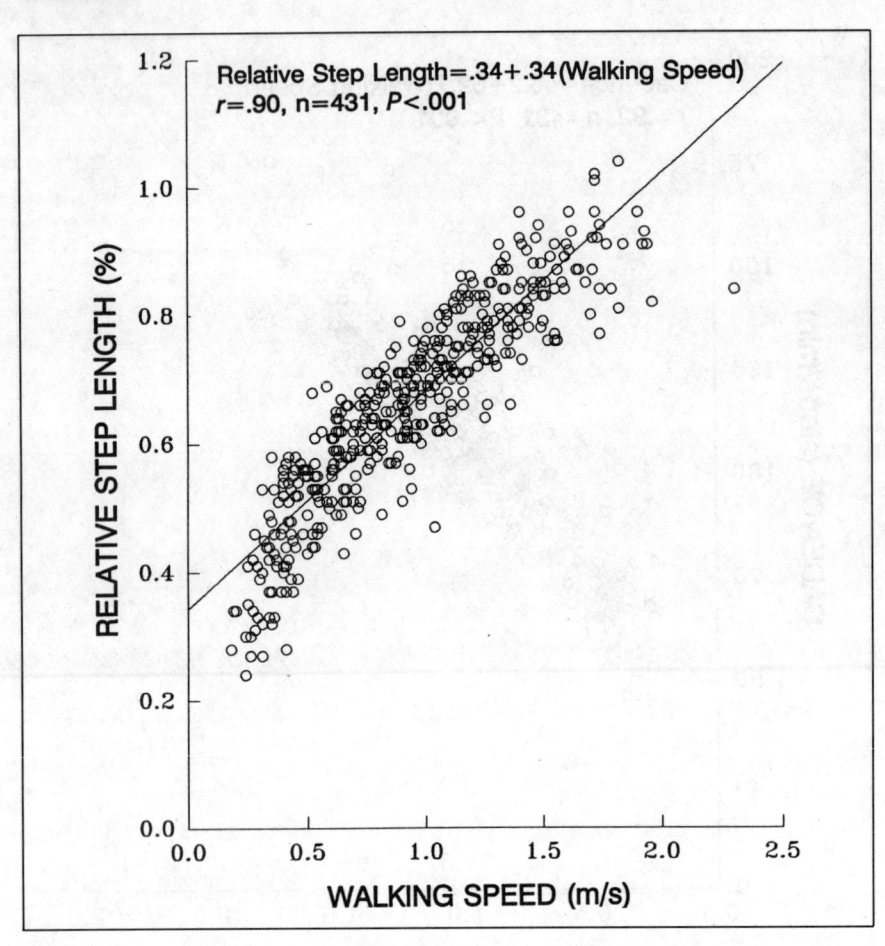

Figure 5. *Relationship between walking speed (in meters per second) and relative step length (in percentage of leg length) for total sample of elderly women (N=81).*

Plot shows: Relative Step Length = .34 + .34(Walking Speed), r=.90, n=431, P<.001

to an increase in age.[11,12] The results of our investigation indicate that physical activity, as defined for the purposes of our study, does not influence the speed of walking except at very slow paces. Perhaps because walking is such a fundamental and continuous motor task, changes attributed to disuse or lack of physical activity are less likely to be apparent than in tasks that require conscious coordinated movement or flexibility.

Although approximately one third of our subjects were described as exercisers, the type of exercise described seldom required quick or forceful activity. One half of the exercisers group participated in a structured program designed to promote cardiovascular fitness via graded treadmill and stationary bicycle activity. The remainder of the women in the exercisers group described a variety of activities and exercises performed with the intent on remaining physically active. Other investigators have suggested that age is not the primary predictor of walking speed. Cunningham et al[34] suggested that maximal aerobic power was the major determinant of walking speed, but their sample of men between the ages of 19 and 66 years was not comparable to our sample of women. In addition, their test protocol was designed for aerobic measurement. It should be noted that the walking test used in our study was not designed to determine the aerobic capacity of the women. Walking was not continuous during the test; thus, those women who regularly exercised aerobically may not have actually had an advantage over the other women on this particular test.

Another explanation for our findings could be a sampling problem. We attempted to recruit subjects with all levels of walking ability to participate in the testing. We noticed, however, that, within the retirement communities, women with observably limited walking ability (ie, those who were extremely slow or had obvious shortness of breath) were reluctant to volunteer, even after watching the simple procedure and talking with other volunteers. This factor created a bias that excluded the most inactive women. Related to the sampling problem were the actual definitions of the physical activity groups. Investigators of habitual physical activity patterns have been primarily interested in relationships with fitness and have reported a variety of direct and indirect methods to relate activity and fitness to health maintenance. Washburn and Montoye[35] have reviewed the questionnaires most commonly used to assess physical activity. They note that the majority of information relating to the validity and reliability of measurements obtained with these instruments has been limited to white middle-aged men, most of whom are still employed. The use of such instruments with a sedentary, elderly female sample is questionable. For this reason, we devised operational definitions for the three groups that were based on the kinds of activity to be expected within certain living situations. To date, no instrument for efficiently categorizing this type of daily physical activity level, as opposed to energy expenditure, has been determined to yield valid and reliable measurements.[36] By building the requirements of stair climbing, multiple trips outside the place of residence, and housekeeping activities into the operational definitions, we attempted to distinguish between truly sedentary women and those who were more physically active on a day-to-day basis. Clear demarcations between walking abilities may not have been realized using this classification scheme.

The walking speed data from this investigation were examined in two different ways. Both data sets seem to suggest that there is something distinctly different about the very slow walking pace as compared with the other walking paces. This impression can be seen as well in the scatterplots between cadence and walking speed and between relative step length and walking speed. It is known that alterations in walking speed are made by adjustments in step length or cadence, or both.[19] At walking speeds below 0.5 m/s, the linearity of the relationship seems to be lost. One interpretation of this observation is that walking at such a slow pace utilizes a different motor program than walking at the other paces. For example, although walking and running are both forms of locomotion, they are considered to be different activities. Perhaps this is also true of very slow walking. Craik and colleagues[37] have suggested that, at slow speeds, the nature of interlimb coordination of upper and lower extremities changes from an alternating reciprocal pattern to one of bilateral symmetrical activity of the upper extremities with each lower-limb step.

At a very slow pace, larger portions of the walking cycle are spent in the transition or double-support phases of the cycle. Winter et al[38] have suggested that, during the stance phase of walking, the hip joint is the major site of control for the large mass above consisting of the head, arms, and trunk. If a longer amount of time is spent in stance, and particularly in double-support stance, then the mass must be controlled or balanced for a longer period of time. Muscles may be required to contract in a coordinated manner more precisely at the slowest pace to regulate the movement. This element of fine or graded control may be lacking in the more sedentary older person. Many of the women spontaneously commented that they did not feel comfortable walking at the very slow pace. When asked why, the response frequently related in some way to a feeling of instability. At this pace, the feet of many women were observed to wobble in a mediolateral plane during stance. One might speculate that a different form of motor control is required to control the forward and lateral movement of the body at very slow paces.

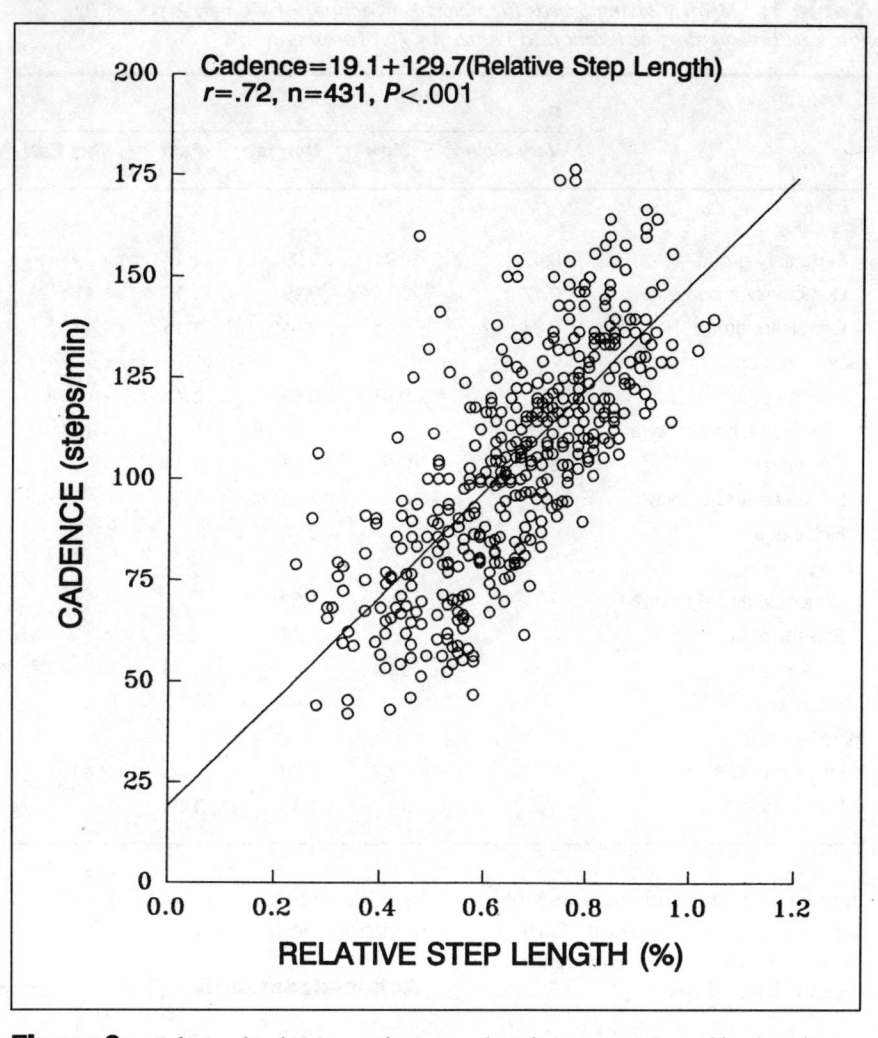

Figure 6. *Relationship between relative step length (in percentage of leg length) and cadence (in steps per minute) for total sample of elderly women (N=81).*

The unique characteristics of walking at an exceptionally slow pace require further investigation. It is seldom that nondisabled individuals naturally walk as slowly as the subjects in this study were asked to walk. It is not uncommon, however, for disabled persons or persons recovering from an accident or illness to produce walking speeds that are this slow or even slower. Frequently, physical therapists are observed supervising or assisting persons to walk at these exceptionally slow paces. One notices that, although the patient may appear to be in constant motion, the therapist is not. If very slow walking is a distinctly different task, then it should not be compared with a more normal or preferred pace, and the characteristics of preferred paces such as arm swing should not be imposed. The actions of body segments other than the legs may need to be described in order to determine what is a natural characteristic of the specific pace.

We believe that, although habitual physical activity appears to have little effect on short-distance walking speed, it may influence a variety of other activities that contribute to independence in living. The suggestion that women who exercise have better motor control than other women when walking at very slow speeds may have some implications for inves-

Table 7. Mean Walking Speeds (in Meters Per Second) of Women Reported by Previous Investigators in Relation to Values for This Investigation

	Pace				
	Very Slow	Slow	Normal	Fast	Very Fast
This study					
Sedentary group	0.47	0.71	0.99	1.17	1.39
Community active group	0.47	0.74	0.99	1.17	1.41
Exercisers group	0.34	0.65	0.94	1.15	1.49
Elderly women					
Himann et al[31]		0.67	0.89	1.14	
Lundgren-Lindquist et al[25]			0.92		1.18
Craik et al[37]	0.37	0.59	0.85	1.14	
Hageman and Blanke[23]			1.32		
Finley et al[21]			0.70		
Young women					
Hageman and Blanke[23]			1.59		
Blessey et al[33]			1.23		
Craik et al[37]	0.49	0.84	1.21	1.86	
Finley et al[21]			0.84		
Mixed sample					
Larsson et al[32]	0.46	0.87	1.28	1.70	2.40
Murray et al[22]			1.30	1.88	

tigators who are studying issues related to balance and falling. Further studies are needed to investigate these relationships.

Summary

This study was designed to investigate the relationships among physical activity, age, BMI, and the speed of walking in women 64 years of age and over. The specific purposes of the study were to describe the range of walking speeds observed in our sample of elderly women and to determine whether walking speed could be related to habitual patterns of physical activity, age, or BMI.

When the groups were adjusted for age inequality, the women walked in a similar manner, except when asked to walk at the very slow pace. At that pace command, the women who were in the exercisers group walked significantly slower than the women in the other groups. In addition, the primary determinant of walking speed was identified as age, with a small contribution from BMI.

Acknowledgments

We gratefully acknowledge the technical support of Oliver Woods, Barbara Hirai, and Albert Esquenazi, MD, of the Gait Analysis Laboratory, Moss Rehabilitation Hospital, Philadelphia, Pa, and the data-management assistance of Kim Gallo.

References

1 Shock NW. Systems integration. In: Finch CE, Hayflick L, eds. *Handbook of the Biology of Aging*. New York, NY: Van Nostrand Reinhold; 1977:639–665.

2 Plowman SA, Drinkwater BL, Horvath SM. Age and aerobic power in women: a longitudinal study. *J Gerontol.* 1979;34:512–520.

3 Smith EL, Rabb DM. Osteoporosis and physical activity. *Acta Med Scand [Suppl].* 1986;711:149–156.

4 Walker JM, Sue D, Miles-Elkousy N, et al. Active mobility of the extremities in older subjects. *Phys Ther.* 1984;64:919–923.

5 Murray MP, Duthie EH, Gambert SR, et al. Age-related differences in knee muscle strength in normal women. *J Gerontol.* 1985;40:275–280.

6 Aniansson A, Sperling L, Rundgren A, Lehnberg E. Muscle function in 75-year-old men and women: a longitudinal study. *Scand J Rehabil Med [Suppl].* 1983;9:92–102.

7 Bortz WM. Disuse and aging. *JAMA.* 1982;248:1203–1208.

8 Greenleaf JE, Kozlowski S. Physiological consequences of reduced physical activity during bed rest. *Exerc Sport Sci Rev.* 1982;10:84–119.

9 Saltin B, Blomquist G, Mitchell J. Response to exercise after bed rest and after training. *Circulation.* 1968;38(suppl 7):1–78.

10 Pollock ML, Foster C, Knapp D, et al. Effect of age and training on aerobic capacity and body composition of master athletes. *J Appl Physiol.* 1987;62:725–731.

11 Spirduso WW, Clifford P. Replication of age and physical activity effects on reaction and movement time. *J Gerontol.* 1978;33:26–30.

12 Rikli R, Busch S. Motor performance of women as a function of age and physical activity level. *J Gerontol.* 1986;41:645–649.

13 Fries JF. Aging, natural death and the compression of morbidity. *N Engl J Med.* 1980;303:130–135.

14 Koller WC, Wilson RS, Glatt SL, et al. Senile gait: correlation with computed tomographic scans. *Ann Neurol.* 1983;13:343–344.

15 Steinberg RU. Gait disorders in old age. *Geriatrics.* 1966;21:134–143.

16 Spielberg PI. Walking patterns of old people: cyclographic analysis. In: Bernstein NA, ed. *Investigations on the Biodynamics of Walking, Running, and Jumping, Part II*. Moscow, Union of Soviet Socialist Republics: Central Scientific Institute of Physical Culture; 1940:72–76.

17 Murray MP, Kory RC, Clarkson BH. Walking patterns in healthy old men. *J Gerontol.* 1969;24:169–178.

18 Imms FJ, Edholm OG. Studies of gait and mobility in the elderly. *Age Ageing.* 1981;10:147–156.

19 Inman VT, Ralston HJ, Todd F. *Human Walking*. Baltimore, Md: Williams & Wilkins; 1981.

20 Murray MP. Gait as a total pattern of movement. *Am J Phys Med.* 1967;46:290–333.

21 Finley FR, Cody KA, Finzie RV. Locomotion patterns in elderly women. *Arch Phys Med Rehabil.* 1969;50:140–146.

22 Murray MP, Kory RC, Sepic SB. Walking patterns of normal women. *Arch Phys Med Rehabil.* 1970;51:637–650.

23 Hageman PA, Blanke DJ. Comparison of gait of young women and elderly women. *Phys Ther.* 1986;66:1382–1387.

24 Drillis RJ. The influence of aging on the kinematics of gait. *The Geriatric Amputee.* 1961;919:134–145.

25 Lundgren-Lindquist B, Aniansson A, Rundgren A. Functional studies in 79-year-olds: walking performance and climbing capacity. *Scand J Rehabil Med.* 1983;15:125–131.

26 Lohman TG, Roche AF, Martorell R. *Anthropometric Standardization Reference Manual*. Champaign, Ill: Human Kinetics Publishers Inc; 1988.

27 Chumlea WC, Roche AF, Mukherjee D. Some anthropometric indices of body compo-

28 Lahey MW, Downey RG, Saal FE. Intraclass correlation: there's more than meets the eye. *Psychol Bull.* 1983;93:586–595.

29 Dixon WJ, Brown MB. *BMDP Biomedical Computer Programs P-Series.* Berkeley, Calif: University of California Press; 1985.

30 Kirk RE. Split-plot factorial designs. In: Kirk RE, ed. *Experimental Designs: Procedures for the Behavioral Sciences.* Monterey, Calif: Brooks/Cole Publishing Co; 1982:489–569.

31 Himann JE, Cunningham DA, Rechnitzer PA, Paterson DH. Age-related changes in sition for elderly adults. *J Gerontol.* 1986;41:36–39.

speeds of walking. *Med Sci Sports Exerc.* 1988;20:161–166.

32 Larsson L, Odenrick P, Sandlund B, et al. The phases of the stride and their interaction in human gait. *Scand J Rehabil Med.* 1980;12:107–112.

33 Blessey RL, Hislop HJ, Waters RL, Antonelli D. Metabolic energy cost of unrestrained walking. *Phys Ther.* 1976;56:1019–1024.

34 Cunningham DA, Rechnitzer DA, Pearce ME, Donner AP. Determinants of self-selected walking pace across ages 19 to 66. *J Gerontol.* 1982;37:560–564.

35 Washburn RA, Montoye HJ. The assessment of physical activity by questionnaire. *Am J Epidemiol.* 1986;123:563–576.

36 Montoye HJ, Taylor HL. Measurement of physical activity in population studies: a review. *Hum Biol.* 1984;56:195–216.

37 Craik RL, Herman RM, Finley FR. Human solutions for locomotion, II: interlimb coordination. In: Herman RM, Grillner S, Stein PSG, Stuart DG, eds. *Neural Control of Locomotion.* New York, NY: Plenum Publishing Corp; 1976:51–64.

38 Winter DA, Patla AE, Frank JS, Walt SE. Biomechanical walking pattern changes in the fit and healthy elderly. *Phys Ther.* 1990;70:340–347.

Basic Research

Weight-distribution Variables in the Use of Crutches and Canes

METTA L. BAXTER, M.A., RUTH O. ALLINGTON, B.A., and GEORGE H. KOEPKE, M.D.

▶ *In comparative studies of various gait patterns associated with the use of crutches and canes, the supporting forces are measured by pressure transducers attached to the weight-bearing points and connected to a Sanborn polygraph. The results indicated that when an individual with partial disability on one side walks with a cane, weight bearing is distributed most effectively if the cane is used on the side of the normal leg. During a two-point gait, the amount of weight shared by the assistive devices was relatively small. During a three-point gait with crutches, the amount of weight borne by the subject's leg varied considerably from the amount he had been trained to estimate.* ◀

DURING THE healing of leg fractures a risk of further damage always exists when the patient attempts to walk with the help of canes or crutches. The inappropriate use of an assistive device may result in excessive stress on a recent fracture, thus delaying healing and possibly disturbing the alignment of the bone segments. Proficiency in the use of such devices, however, not only enhances gait effectiveness but also facilitates union of the bone through the physiological effects of partial weight bearing and muscle action.[1]

A preliminary study carried out at the Veterans Administration Hospital, Ann Arbor, Michigan, has clarified some of the biomechanical principles involved in walking with canes or crutches. The purpose of the study was to determine the relative merits of various gait

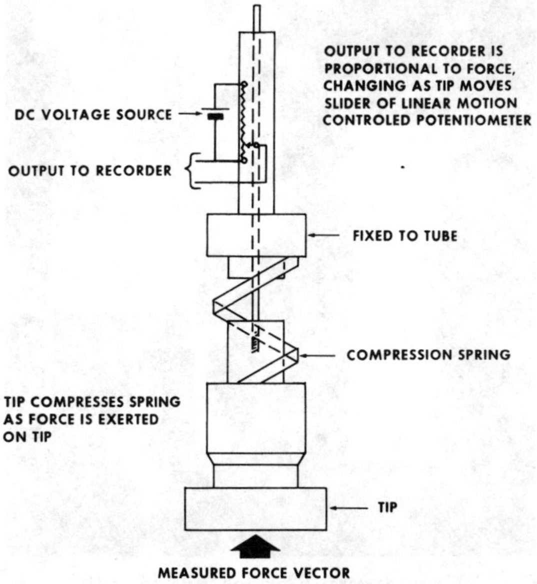

Miss Baxter: Chief, Physical Therapy Section, and Coordinator of Physical Medicine and Rehabilitation Service, Veterans Administration Hospital, Gainesville, Florida.

Mrs. Allington: Chief, Physical Therapy Section, Physical Medicine and Rehabilitation Service, Veterans Administration Hospital, Ann Arbor, Michigan.

Dr. Koepke: Consultant, Physical Medicine and Rehabilitation Service, Veterans Administration Hospital, and Professor, Department of Physical Medicine and Rehabilitation, University of Michigan Medical School, Ann Arbor, Michigan.

Fig. 1. Diagram of pressure transducer within tip of crutch or cane.

TABLE 1	
CRUTCHES, NONWEIGHT BEARING, RIGHT FOOT	
Subject	Percentage Right>Left[a]
A	3.32
B	2.08
C	8.09
D	11.25
E	4.12
F	5.38
G	3.17
H	4.07
I	3.28
J	1.98
Average	4.67

[a] Difference between body weight exerted on right and left crutches with complete nonweight bearing on right foot.

TABLE 2				
PERCENTAGE OF WEIGHT USING TWO CANES AND TWO-POINT GAIT				
Subject	Right Cane	Left Foot	Left Cane	Right Foot
A	23.2	76.8	19.9	80.1
B	16.7	83.3	16.0	84.0
C	37.6	62.4	30.4	69.6
D	33.1	66.9	29.4	70.6
E	32.3	67.7	32.5	67.5
F	30.8	69.2	27.0	73.0
G	27.8	72.2	23.8	76.2
H	31.4	68.6	27.9	72.1
I	27.9	72.1	32.8	67.2
J	32.3	67.7	20.9	79.1
Average	29.3	70.7	26.1	73.9
Median	29.4	70.9	26.1	73.9

patterns as indicated by the distribution of body weight on the foot and the tips of the assistive devices.

METHOD

The experimental subjects in this study were ten normal adults (seven female, three male), who ranged in age from twenty-two to fifty-nine. For the experiments they wore sandals equipped with switches on the soles that served as contact markers; the crutches and canes they used were equipped with pressure transducers added to the tips (Fig. 1). The switches and transducers were connected to a 4-channel direct-current recorder which was standardized.

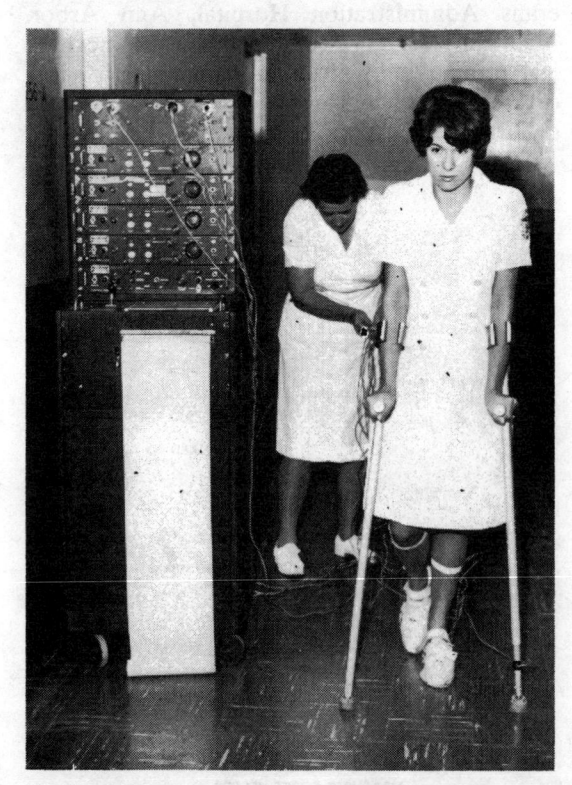

Fig. 2. Transducers in crutch tips to indicate the force exerted on the crutches and marking switches in sandals with leads to 4-channel recorder.

TABLE 3		
PERCENTAGE OF WEIGHT USING ONE CANE		
Cane in Right Hand (Ipsilateral)		
Subject	Cane	Right Foot
A	29.13	70.87
B	36.11	63.89
C	44.84	55.16
D	46.87	53.13
E	52.35	47.65
F	39.68	60.32
G	39.68	60.32
H	48.83	51.17
I	25.40	74.60
J	39.23	60.77
Average	40.21	59.79
Mean	39.68	60.32
Cane in Left Hand (Contralateral)		
Subject	Cane	Left Foot
A	26.49	73.51
B	25.00	75.00
C	35.15	64.85
D	40.00	60.00
E	42.35	57.65
F	36.15	63.85
G	36.50	63.50
H	34.30	65.70
I	28.68	71.32
J	36.92	63.08
Average	34.15	65.85
Mean	35.15–36.15	63.85–64.85

The calibration signal was checked during each of experiments to insure accuracy (Fig. 2). The error of calibration and interpretation with each recording was within 5 percent. Thus, the amount of weight shared by the assistive device could be recorded for the stance phase of each gait tested. Shear force was not measured.

After each subject demonstrated proficiency in protective gait patterns with canes and crutches, recordings were made during the two-point gait with minimal weight bearing on the feet. All experiments included ambulation for at least 25 feet, guided by a metronome at the rate of 58 steps per minute. The crutch was adjusted to provide suitable length for weight bearing.

Other crutch patterns included variables of the three-point gait: first nonweight bearing on the right foot; then with the right toe touching; and finally with the subject bearing an estimated 50 percent of his weight on the right foot. Care was taken to teach simultaneous placement of the crutches and right foot to insure appropriate distribution of the body weight. Measurements were taken at midstance phase to avoid the increased force at the moment of impact and to be certain that the left foot was off the floor. The subject was trained to bear an estimated 50 percent of his weight on the right foot by observing the appropriate weight recorded on a scale.

The value of training by this subjective method was assessed by repeating the ambulation experiment. After testing was completed with crutches, similar studies were used to determine the amount of weight borne by canes using a two-point gait. To protect the right foot, the cane was first carried in the left hand, and later in the right hand (Fig. 3). The cane

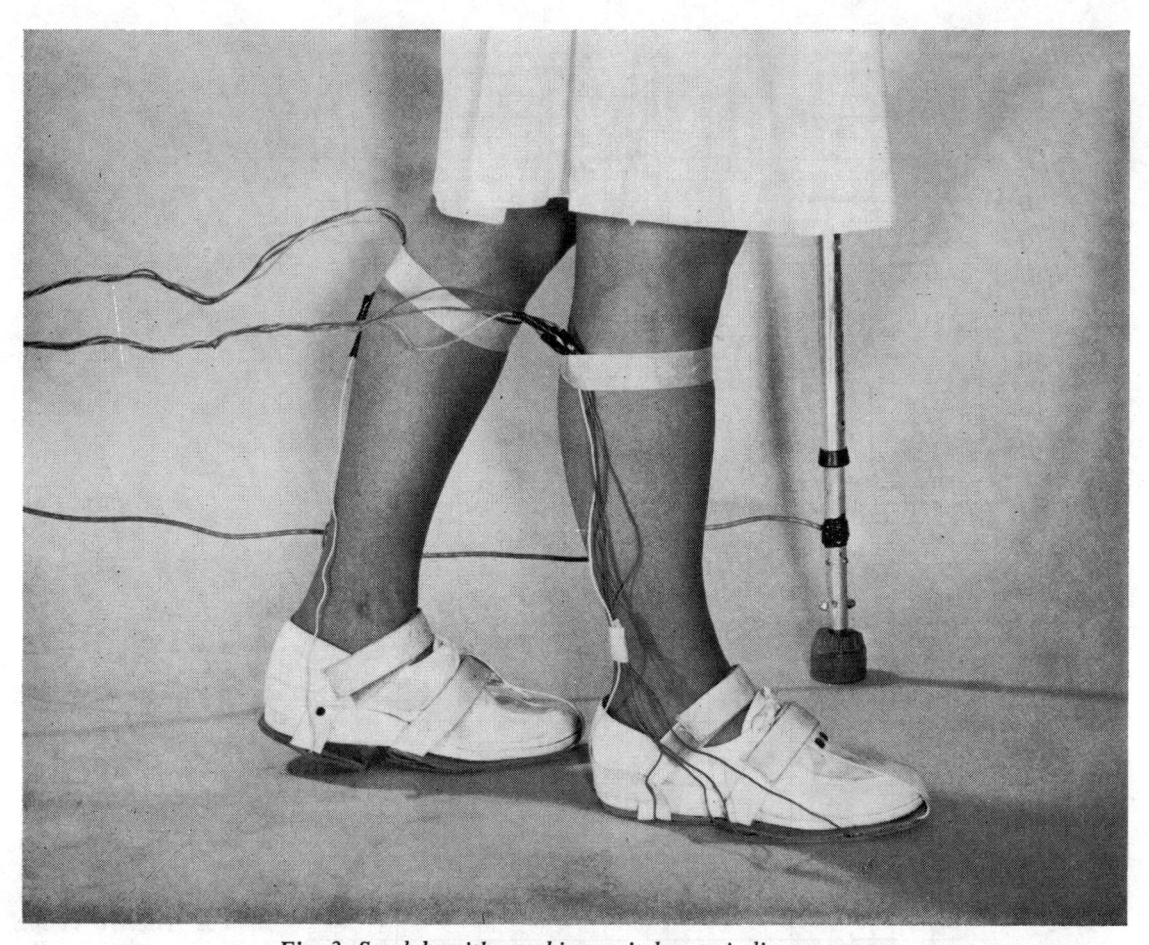

Fig. 3. Sandals with marking switches to indicate stance phase of walking. Transducer is within cane tip.

was adjusted to allow approximately 25 degrees of elbow flexion,[2,3,4] and the subject attempted to walk at the rate of 63 steps per minute.

The effect of one cane was repeated, walking at least 300 feet to determine any change that might occur with fatigue. An average of measured forces recorded during each experiment was taken as the representative value for each subject.

RESULTS

Figure 4 shows a sample recording during ambulation with crutches, using a three-point gait nonweight bearing on the right foot. As indicated in Table 1, all ten subjects exerted more force on the right crutch than on the left. This consistent increase of force on the right crutch ranged from approximately 2 percent to more than 11 percent over the force exerted on the left crutch.

It became apparent that verbal instruction to limit weight bearing to toe touching was ineffective (Fig. 5). Five subjects exerted essentially no weight on the protected right foot, but there was wide variation, and one subject exerted almost 40 percent of her body weight on the right foot (Fig. 5).

All the subjects experienced difficulty in estimating a force corresponding to 50 percent of the body weight. In preliminary trials, the amount varied among subjects from approximately 4 to 41 percent, with an average of approximately 22 percent. As Figure 5 shows, preliminary subjective training was of little value, inasmuch as weight bearing on the foot varied from 9 to 53 percent, with an average of approximately 27 percent after training.

The weight-sharing effect of canes during a two-point gait (two canes) varied among the subjects, resulting in approximtaely 62 to 84 percent of their body weights exerted on their

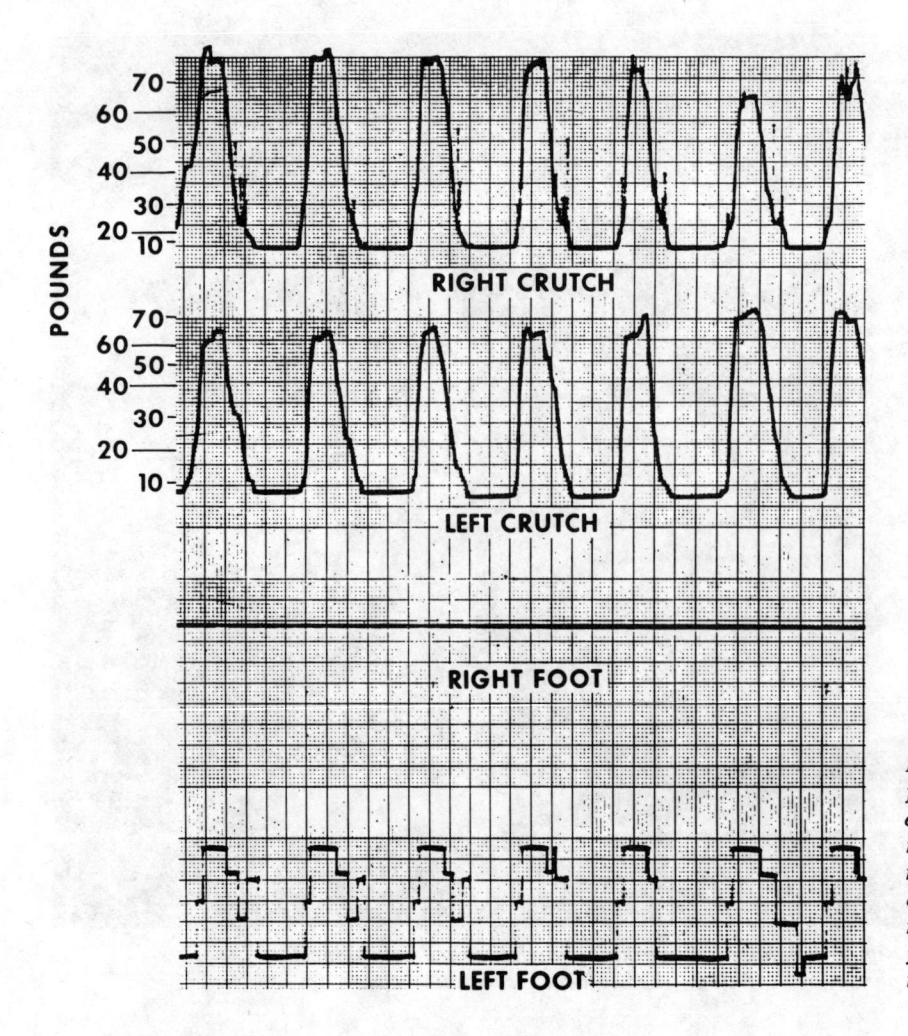

Fig. 4. Recording of forces (in pounds) exerted on crutch tips with nonweight bearing on right foot. The shift of body weight to the right is reflected in the force recorder of the right crutch.

feet with an average and mean of 72 percent (Table 2). One cane held in the right hand, i.e., ipsilateral side, resulted in weight bearing on the right foot ranging from approximately 47 to 75 percent of the total body weight. When the cane was used in the left hand (contralateral side), weight bearing on the right foot was about 57 to 75 percent of the body weight, or 6 percent more, on the average, than the value for the ipsilateral side (Table 3).

It was interesting to note that a cane on the ipsilateral side was uniformly more effective in sharing weight in all subjects except one, and that this individual was the only left-handed subject in the group. Another finding of special interest was related to the element of fatigue. A striking example is seen in Figure 6, which shows that at the outset of a test, twice as much force was exerted on the ipsilateral cane as on the contralateral cane, but after 100 steps the force exerted on the ipsilateral cane had dropped to one-half the amount of force exerted on the contralateral cane.

Thus it appeared that as fatigue increased, weight sharing by the ipsilateral cane gradually decreased, proportionately increasing weight borne by the protected foot. No significant change occurred when the cane was used on the contralateral side.

CONCLUSIONS

Within the limitations of the methods of investigation employed in this study, several inferences may be drawn.

- It appears that crutches are approximately 10 percent more effective than canes during a two-point gait.
- Despite practice, it is difficult to estimate the amount of weight bearing exerted on a foot during a three-point crutch gait.
- A toe-touching crutch gait may not prove

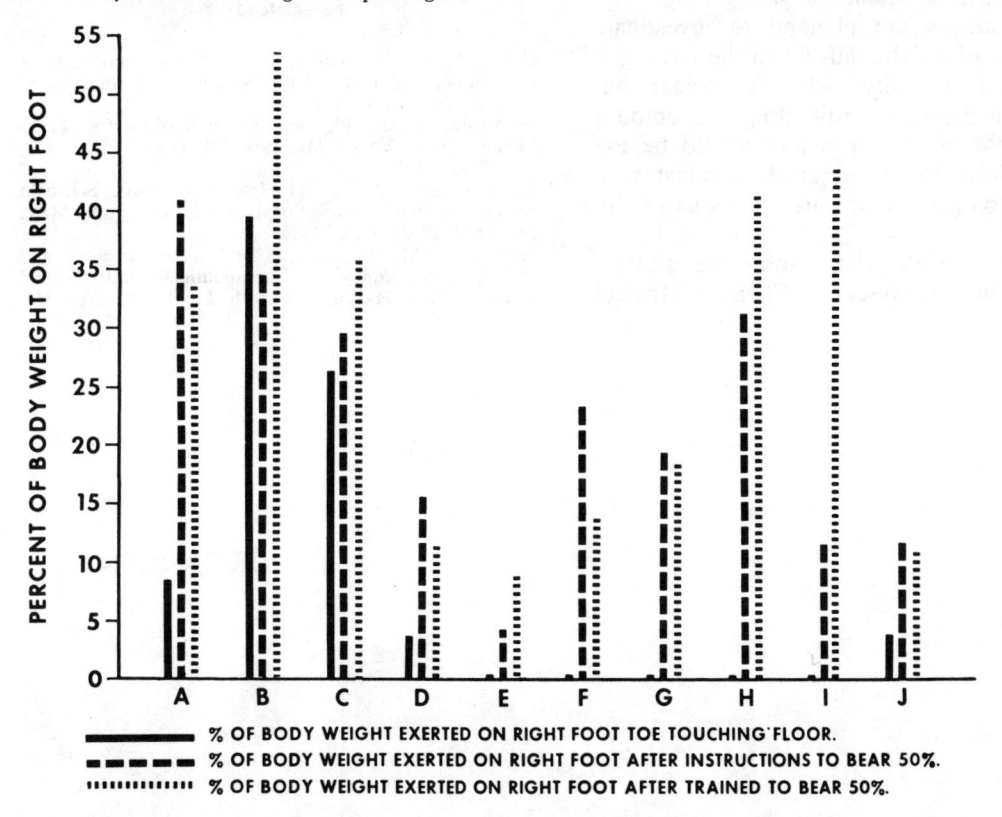

Fig. 5. Distribution of body weight with the use of crutches during three-point gait.

DISTRIBUTION OF BODY WEIGHT WITH THE USE OF CRUTCHES DURING 3 POINT GAIT

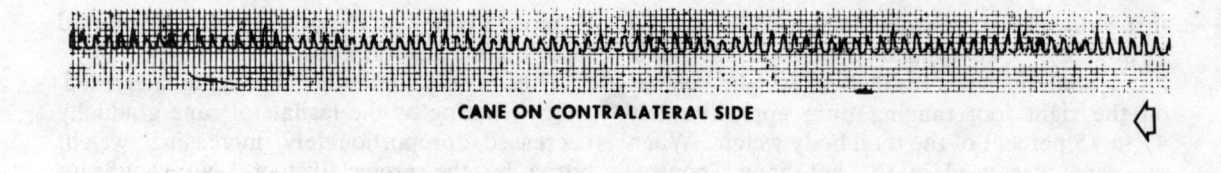

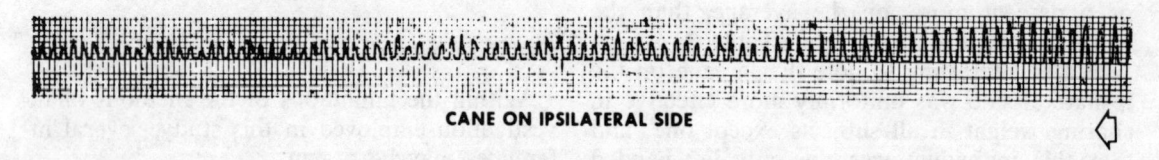

CANE TO PROTECT THE RIGHT FOOT

Fig. 6. Maximal force exerted on cane during walking approximately 250 feet. Only initially, large forces were exerted when cane was on ipsilateral side, but consistent force was exerted when cane was on contralateral side.

effective in avoiding weight bearing, but rather may result in the exertion of considerable force on the foot.

● When walking more than 50 feet, the weight-sharing effect of a cane is greater if it is used on the contralateral side.

Further studies are planned to investigate the influence of weight shift when the left leg is the protected extremity. Also, it is clear that an objective means of estimating the amount of weight placed on either foot would be extremely helpful to the physical therapist who must train patients in the use of assistive devices.

Acknowledgment. The authors are grateful to John Johnson, Research Chemist (Instrumentation), Veterans Administration Hospital, Ann Arbor, Michigan, for his technical assistance in the design and fabrication of the pressure transducer.

REFERENCES

1. Geiser, Max, and Joseph Trueta, Muscle action, bone rarefaction, and bone formation. J. Bone Joint Surg., 40B:282–311, May 1958.
2. Buchwald, Edith, Physical Rehabilitation for Daily Living. New York, McGraw-Hill Book Co., 1952, pp. 97 and 107.
3. Lowman, Edward W., and Howard A. Rusk, Self-help devices; crutch prescriptions: gaits. Postgrad. Med., 31, April 1962, pp. 392–394.
4. Hoherman, Morton, Crutch and cane exercises and use. In Licht, Sidney, ed., Therapeutic Exercise, Second Edition. New Haven, Elizabeth Licht, 1958, p. 361.

Basic Research

System of Reporting and Comparing Influence of Ambulatory Aids on Gait

GARY L. SMIDT, PhD,
and M. A. MOMMENS, MS

> The purposes of this study were to 1) present a standardized approach for describing gait when assistive devices are used, 2) report reference data for unassisted and assisted gait patterns for normal adults, and 3) discuss clinical implications for selected variables of gait. Using an automated gait system, measurements for temporal and distance factors and accelerometry were obtained for 25 normal young adults. In addition to the formulation of a new system for describing gait patterns when assistive devices are used, the results of the study were that 1) subjects walked slower with ambulatory aids than without them, 2) assisted gaits with the same number of counts per cycle tended to have similar measurements, 3) reciprocal swing times and stance times were symmetrical for all types of gait studied, 4) double stance times and step times were asymmetrical for three types of assisted gait, and 5) vertical accelerations were disproportionately elevated for most assisted gaits.
>
> **Key Words:** *Crutches, Gait, Physical therapy.*

Assistive devices such as canes, crutches, and walkers are commonly recommended for problems of pain, fatigue, equilibrium, joint instability, muscular weakness, excessive skeletal loading, and cosmesis. The physical therapist may evaluate the patient for abnormalities and identify the form of assistive device to be used and type of assisted gait to be learned. The clinician may judge the quality of the patient's walking performance using standards such as preconceived expectations of normality, information presented in scientific publications, or his clinical experience in dealing with manifestations of disorders.

Numerous reports on ambulatory aids and gait may be found in the literature.[1-7] A few articles have reported the effect of a cane or a crutch on patients' walking.[8-15] Data on the unassisted gait of normal subjects may be compared with patients who walk without an assistive device, in order to determine the extent of the patients' abnormality. No information is available, however, on normal gait for walking patterns associated with the variety of commonly employed types of assisted gait.

Inasmuch as humans walk by reciprocally placing each of their two feet in front of the other in forward progression, the sequence and number of contacts between the body and the walking surface are confined to any combination of two. When one or two ambulatory aids are added, the complexity of the sequencing and timing of floor contact is magnified. For assisted gait, the number of contact points is thus greater than two, and necessarily the possible order of contacts on the walking surface is exponentially increased. Walking techniques with canes and crutches have been described by Hoberman,[14] but explanations for assisted gait patterns could be improved to be more explicit and to convey a clear mental image of the walking pattern. One example of ambiguity is that a two-point gait may refer to the use of either one or two assistive devices, and these devices may be canes or crutches. A standardized method that accurately describes the type of assisted gait is needed.

The purposes of this paper are to 1) present a standardized approach for describing gait when assistive devices are used, 2) report reference data for

---

Dr. Smidt is Professor and Director of Programs in Physical Therapy Education, College of Medicine, The University of Iowa, Iowa City, IA 52242.

Ms. Mommens was a doctoral student in physical therapy, The University of Iowa, when this study was conducted. She is now self-employed and can be reached at 707 9th Ave, Coralville, IA 52241.

This paper was presented at the Fifty-second Annual Conference of the American Physical Therapy Association, New Orleans, June-July 1976.

This article was submitted October 3, 1978, and accepted November 7, 1979.

unassisted and assisted gait patterns, and 3) discuss clinical implications for selected variables of gait.

DEFINITIONS OF GAIT VARIABLES

For purposes of this study, the following definitions were used.

Walking Velocity: The rate of linear forward motion of the body; Product of distance and time; Measured in centimeters per second or meters per second.

Stance Time: The elapsed time that one foot is in contact with walking surface; Measured in seconds.

Swing Time: The elapsed time that one foot is not in contact with walking surface; Measured in seconds.

Gait Cycle: Usually considered the interval between successive ipsilateral foot contacts on the walking surface, but may be the interval between any recurring event during walking.

Cycle Time: The elapsed time during one gait cycle; Measured in seconds.

Double-Stance Time: The elapsed time during the gait cycle when both feet are simultaneously in contact with the walking surface; Measured in seconds.

Stride Length: The distance between two consecutive ipsilateral foot contacts on the walking surface; Measured in centimeters or meters.

Step Length: The distance in the direction of walking between the left and right feet; Measured in centimeters or meters.

Step Time: The elapsed time between consecutive foot-floor contacts; Measured in seconds.

Functional Lower Extremity Length: The distance between the center of the femoral head and a point on the floor located slightly anterior and medial to the medial malleolus on the ipsilateral side; Measured in centimeters or meters.

Ratios Reflecting Symmetry: Mathematically derived quotients obtained by dividing a measurement associated with one side of the body by a measurement for the same variable associated with the other side of the body.
a. Stride Length/Lower Extremity Length Ratio: Stride length divided by lower extremity length; Measurement unit is dimensionless.
b. Swing/Time Ratio: Right swing time divided by left swing time or, in a disabled individual, most involved side divided by least involved side; Measurement unit is dimensionless.
c. Swing/Stance Ratio: Swing time divided by stance time; Measurement unit is dimensionless.

Other ratios for which measurements were obtained during the course of this study were stance time, double-stance time, step time, step length, heel strike-foot flat time, foot flat-heel off time, and heel off-toe off time.

Body Acceleration: The rate of change of velocity of a point posterior to the sacrum; Measured in meters per second per second. Maximum values for acceleration in vertical, medial-lateral, and fore-aft directions were obtained in the study.

Harmonic Ratio: An index of smoothness of gait. A Fourier series analysis applied to the acceleration curves yields even- and odd-numbered coefficients. The harmonic ratio is the quotient derived from the sum of the coefficients for the even-numbered harmonics divided by the sum of the coefficients for the odd-numbered harmonics.

Cadence (stride or step frequency): The number of strides or steps per unit of time; Number of steps divided by time; Measured in strides or steps per minute or steps per second.

METHOD

A recently developed automated gait analysis system was used to obtain measurements for a large number of variables of gait.[15] The system included triaxial accelerometers, placed posterior to the sacrum, which were sensitive to changes in velocity in the anterior-posterior, medial-lateral, and vertical directions. The system also included pressure-sensitive foot switches attached to the heel and forefoot of each shoe, a signal-amplification unit, and a laboratory computer. Graduated strips of tape on the walkway permitted acquisition of step-distance measurements. After each walking sequence in the laboratory, measurement of temporal and distance factors (eg, walking velocity, step length, step time) and body acceleration could be viewed on the teletype. A detailed

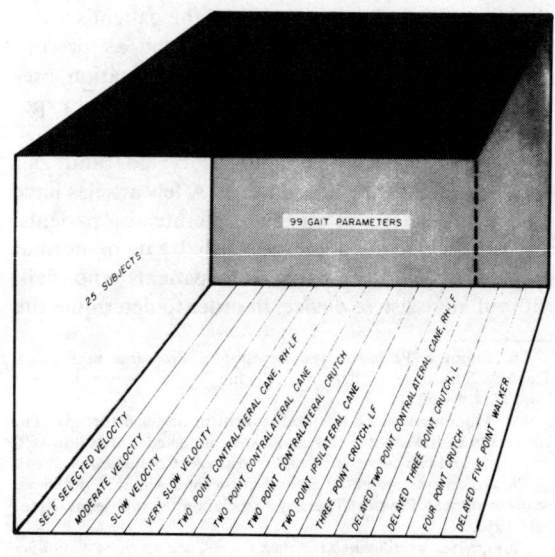

Fig. 1. Scope of study.

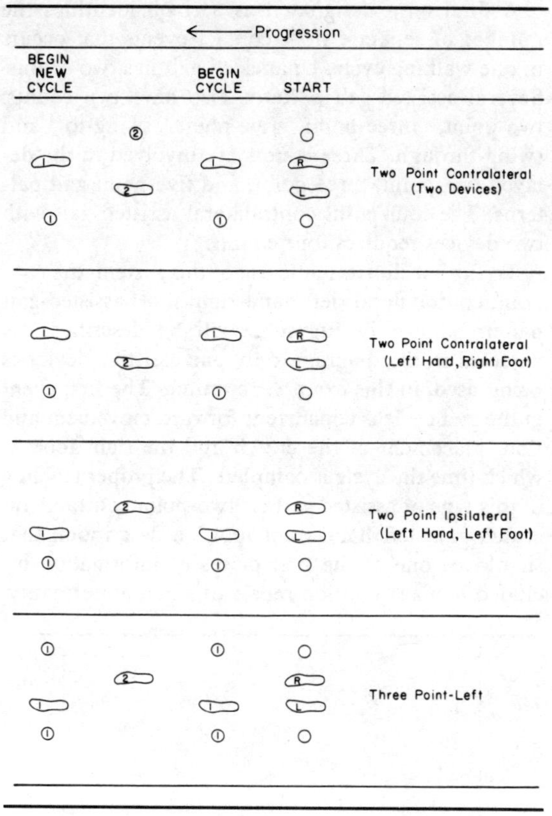

Basic Research

Fig. 2. Diagrammatic view of assisted gaits.

Approach for Labeling and Describing Assisted Gait Patterns

Specific terms are needed for describing assisted gait using ambulatory aids. The term *point* refers to the number of floor contacts on a line perpendicular to the direction of walking that simultaneously occurs during any part of stance phase for the lead foot. When the lead foot is clearly slower than the assistive device in making floor contact, the term *delayed* should be used. *Laterality* describes the associated placement on the walking surface of the side of the upper extremity holding the assistive device and the side of placement for the lead foot in the cycle. For example, laterality in the case of "ipsilateral left" indicates that both the assistive device held on the left and the left foot are concurrently in contact with the walking surface during a portion of the cycle. Another example is "contralateral, left hand-right foot" when the assistive device held on the left side and the right foot are concurrently in contact with the walking surface during a portion of the cycle. For completeness, the specific type of assistive device is identified. The recommended sequence for presenting descriptors for types of assisted gait is 1) delayed (when appropriate), 2) number of points, 3) laterality (when appropriate), and 4) type of assistive device. Common description of the system can be found in a previous publication.[15]

Twenty-five physical therapy students (12 men, 13 women) served as subjects for this study. The mean age was 22 years, mean height was 168 cm (5 ft, 6 in), and mean weight was 64.6 kg (142 lbs). After application of the equipment, the subjects were permitted to walk until they felt comfortable in the laboratory environment. During the first of two sessions, measurements of gait were randomly obtained for two walking sequences for each of the four categories of unassisted gait: a self-selected velocity, moderate velocity (71–110 cm/sec), slow velocity (31–70 cm/sec), and very slow velocity (30 cm/sec or less). At the end of the first session each subject was provided with properly adjusted ambulatory aids and received instruction to proficiency in nine different types of assisted gaits (Fig. 1). During the second testing session on another day, the foot switches and accelerometry apparatus were applied to the subject, and measurements of gait for two walking sequences of each assisted gait pattern were obtained. Before the test session, subjects were again permitted to walk with the measurement equipment attached and become acclimated to the laboratory environment.

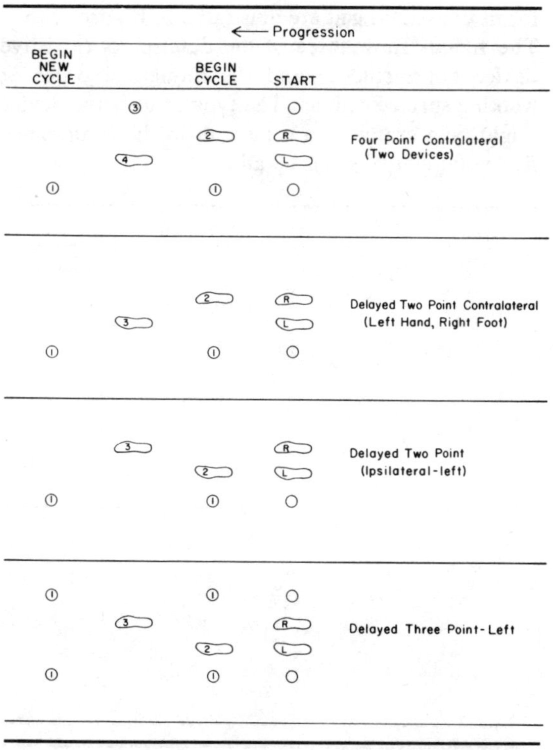

Fig. 3. Diagrammatic view of assisted gaits.

Fig. 4. Diagrammatic view of assisted gaits.

forms of assisted gait are illustrated in Figures 2 to 5. The figures show three of the descriptors (assistive device not included) and the numerical order of walking surface contact. The type of assistive device might be a crutch or a cane and, in the example of five-point gait (Fig. 5), a walker.

Fig. 5. Diagrammatic view of assisted gaits.

A final term designated as a *count* identifies the number of separate floor-contact events that occurs in one walking cycle. Unassisted gait has two counts. Several assisted-gait patterns also have two counts: two-point, three-point, five-point, swing-to, and swing-through. Three counts are involved in the delayed two-point, three-point, and five-point gait patterns. The four-point contralateral assisted gait with two devices requires four counts.

To further illustrate the use of this system, the two-point contralateral (left hand-right foot) assisted-gait pattern shown in Figure 2 will be described. As depicted in the diagram, only one assistive device is being used, in this example, a crutch. The first event in the gait cycle is concurrent forward movement and floor placement of the crutch and the right foot, at which time the cycle is complete. The proper labeling of this type of assisted gait is "two-point contralateral crutch gait, left hand-right foot," a description that should be one of the first pieces of information included in an evaluation report of a gait abnormality.

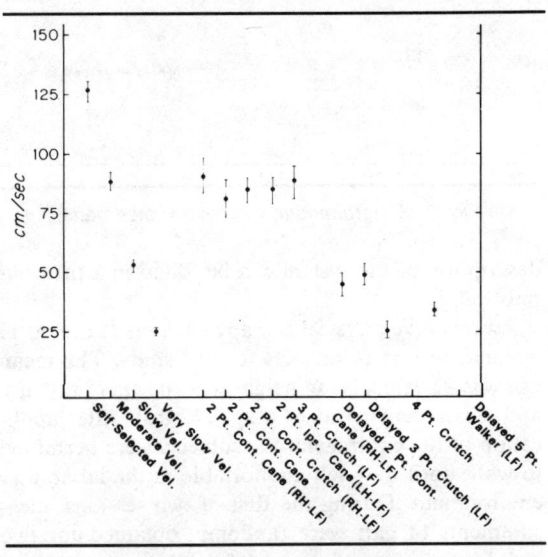

Fig. 6. Walking velocity.

RESULTS

Walking Velocity

Figure 6 shows, for walking velocity, the means and (95%) confidence intervals for the four unassisted gaits and nine assisted gaits studied. When using ambulatory aids, the subjects walked considerably slower than their customary self-selected velocity. Assisted gaits with two counts (two-point contralateral cane-right hand/left foot, two-point contralateral-two canes, two-point contralateral crutch-right hand/left foot, two-point ipsilateral cane-left, and

Basic Research

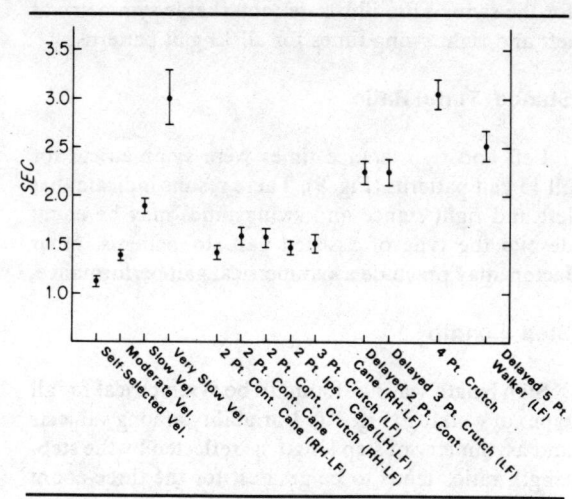

Fig. 7. Cycle time.

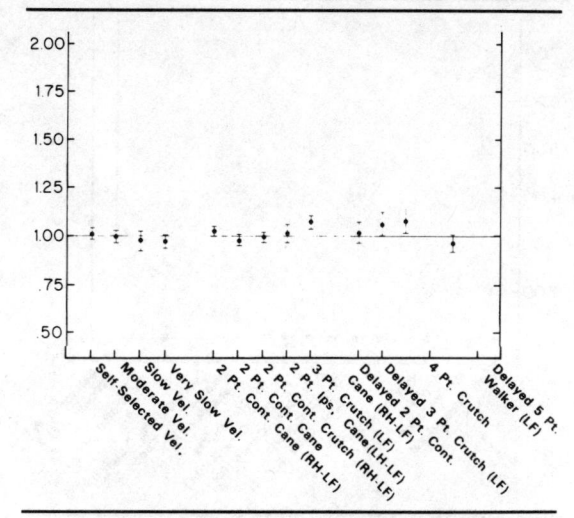

Fig. 8. Swing ratio L/R.

three-point crutch-left foot) were similar to the unassisted moderate velocity. The delayed three-point crutch gait and delayed two-point contralateral cane are similar to unassisted slow velocity; delayed five-point walker gait is similar to unassisted very slow velocity. The velocity for the four-point crutch is located between slow and very slow velocities.

Two clinical implications may be derived from these results. First, either the introduction of an assistive device to a patient or changes in the types of assisted-gait patterns will tend to alter the walking velocity. Secondly, types of assisted gaits with three and four counts may contain the upper limit for walking velocity. Patients with severe locomotor disorders may require a three- or four-count gait to permit walking; nevertheless, there appears to be a rather low upper limit for walking velocity.

Cycle Time

The cycle time, or time elapsed during one stride, is related to velocity and cadence (Fig. 7). The cycle

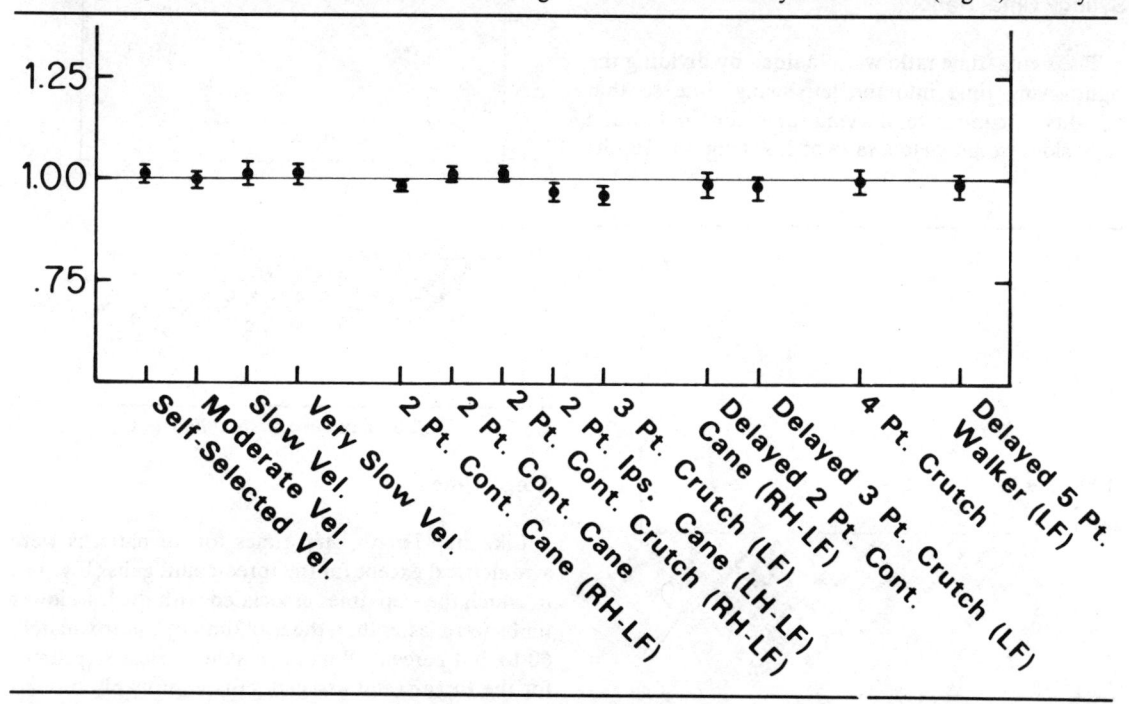

Fig. 9. Stance ratio L/R.

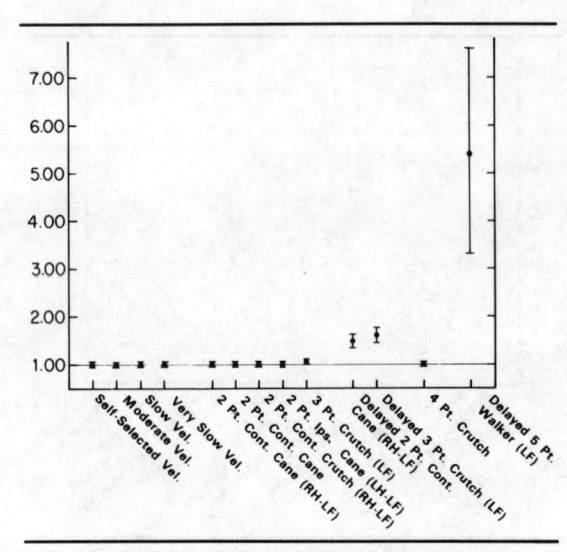

Fig. 10. Step-distance ratio RL/LR.

time for this study ranged from 1.13 sec for self-selected speed to slightly greater than 3 sec for the four-point crutch gait and the very slow unassisted gait, a three-fold difference. The results demonstrate that clinical isolation of movement abnormalities by observational gait analysis requires rapid scrutiny of body segments, particularly when walking is performed at a reasonable rate.

Swing/Time Ratio

The swing/time ratio was obtained by dividing the right swing time into the left swing time so that equality of contralateral swing times for the left and right sides would yield a ratio of 1.00 (Fig. 8). Results for the swing ratio illustrate remarkable symmetry of left and right swing times for all 13 gait patterns.

Stance/Time Ratio

Left and right stance times were symmetrical for all 13 gait patterns (Fig. 9). These results indicate that left and right stance and swing times may be equal despite the type of assisted gait. In patients, other factors may preclude a symmetrical gait performance.

Step Length

Step length was not found to be symmetrical for all types of walking (Fig. 10). Variability among subjects and asymmetry of step length as reflected by the step-length ratios tends to be greatest for the three-count assisted gaits (delayed two-point contralateral cane, delayed three-point crutch, and delayed five-point walker). The length of the lead step was largest. Therefore, a clinical objective of symmetrical step lengths for walking may be realistic, but will probably be most difficult to accomplish for the types of assisted gait that require three counts per walking cycle.

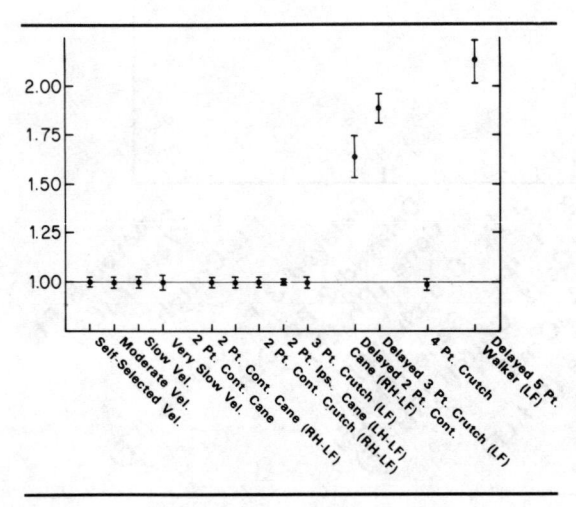

Fig. 11. Step-time ratio LR/RL.

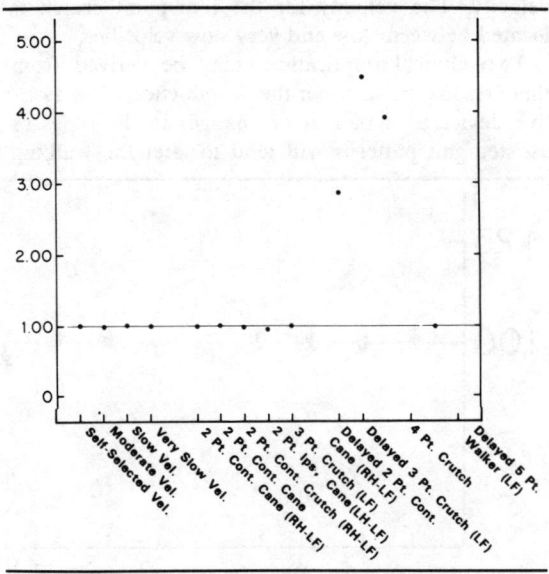

Fig. 12. Double-stance ratio RL/LR.

Step Time

Like step length, step times for all patterns were symmetrical except for the three-count gaits (Fig. 11), in which the step times associated with the lead lower limbs were faster than the trail limb by approximately 60 to 100 percent. Pursuit of symmetrical step times for the three-count assisted gaits is probably unrealistic, but can most closely be approximated by swing-

Basic Research

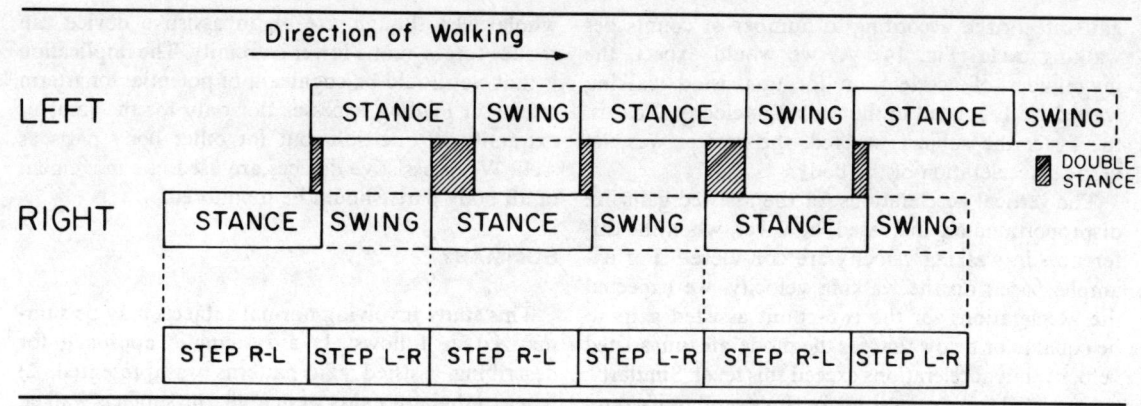

Fig. 13. Symmetrical-stance and swing; asymmetrical-step and double stance.

ing the assistive devices forward in a more rapid fashion. Step times for the remaining types of gaits were symmetrical and variability was small.

Double-Stance Time

The double-stance times tend to be symmetrical for all types of gait studied except the three-count assisted gaits, in which the right-left double stance was dramatically greater than the counterpart double-stance times associated with the left-right foot placement (Fig. 12). The results for three-count assisted gait patterns showed that stance and swing times were essentially symmetrical but that step times and double stance times were asymmetrical for the same types of assisted gait. This problem can be explained by the occurrence of a phase shift in the timing of the foot placement for either the right or left lower limb (Fig. 13).

Vertical Acceleration

The vertical acceleration near the center of gravity during walking tended to be similar for the assisted

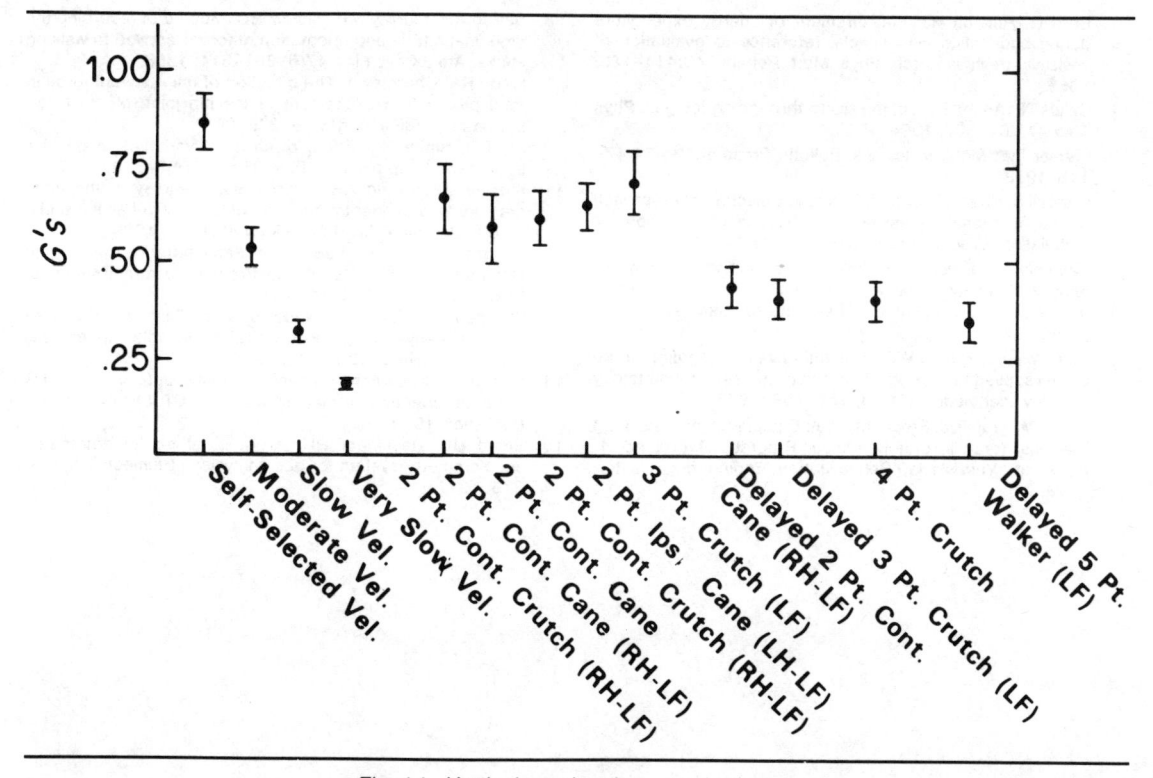

Fig. 14. Vertical acceleration—peak to peak.

gait categorized according to number of counts per walking cycle (Fig. 14). As we would expect, the magnitudes of acceleration for unassisted walking were directly related to the walking velocity. That is, the faster the subjects walked, the greater was the vertical acceleration of the body.

The vertical accelerations for the assisted gaits are disproportionately increased, however, when the differences in walking velocity are considered. For example, based on the walking velocity, we expected the accelerations for the two-count assisted gaits to be equal to or below that for the moderate, unassisted velocity, but accelerations exceed this level. Similarly, we expected the acceleration for three- and four-count assisted gait to be in the vicinity of the very slow velocity or less than the values for the slow velocity. However, the acceleration values for the three- and four-count assisted gaits are well above the level for the slow velocity.

Newton's second law (F = ma) states that acceleration is proportional to the force causing it. Application of this law to vertical acceleration results indicate that use of assistive devices tends to increase the vertical loading on the structures of the body as a whole even though use of an assistive device can reduce forces at one lower extremity. The implication is that we should be cognizant of potential long-term effects of loading excesses not only for the joint or extremity in question, but for other body parts as well. When assistive devices are used, the movement of all body parts should be monitored.

SUMMARY

This study involving normal subjects may be summarized as follows: 1) a systematic approach for describing assisted gait patterns was presented, 2) when ambulatory aids were used, the subjects walked slower than their unencumbered self-selected velocity, 3) measurements of gait tended to be similar for assisted gaits that require the same number of counts per cycle, 4) reciprocal swing times and stance times were symmetrical for all types of gait studied, 5) double-stance times and step times were asymmetrical for three types of assisted gait (delayed two-point, three-point, and five-point), and 6) vertical accelerations were disproportionately elevated for most assisted gaits.

REFERENCES

1. Bard G, Ralston HJ: Measurement of energy expenditure during ambulation with special reference to evaluation of assistive devices. Arch Phys Med Rehabil 40:414–420, 1959
2. Childs TF: An analysis of the swing-through crutch gait. Phys Ther 44:804–807, 1964
3. Farmer LW: Mobility devices. Bulletin Prosthet Res 14:47–118, 1978
4. Ganguli S, Bose KS, Dutta SR, et al: Biomechanical approach to the functional assessment of the use of crutches for ambulation. Ergonomics 17:365–374, 1974
5. Kauffman IB, Ridenour B: Influence of an infant walker on onset and quality of walking pattern of locomotion: An electromyographic investigation. Percept Mot Skills 45:1323–1329, 1977
6. Klenerman L, Hutton WC: A quantitative investigation of the forces applied to walking sticks and crutches. Rheumatology and Physical Medicine 12(3):152–158, 1973
7. Kljajic M, Krajnik J, Stopar M, et al: Equipment for measuring the axial force in crutch. Annual Progress Report no. 1. Ljubljana, Yugoslavia, Rehabilitation Engineering Center, 1978
8. Seireg AH, Murray MP, Scholz RC: Method of recording the time, magnitude and orientation of forces applied to walking sticks. Am J Phys Med 47(6):307–314, 1968
9. Elson RA, Charnley J: The direction of the resultant force in total prosthetic replacement of the hip joint. Medical and Biological Engineering 6:19–27, 1968
10. Ely DD, Smidt GL: Effect of cane on variables of gait for patients with hip disease. Phys Ther 57:507–512, 1977
11. Murray MP, Seirig AH, Scholz RC: A survey of the time, magnitude, and orientation of forces applied to walking sticks by disabled men. Am J Phys Med 48:1–13, 1969
12. Smidt GL, Wadsworth JB: Floor reaction forces during gait: Comparison of patients with hip disease and normal subjects. Phys Ther 53:1056–1062, 1973
13. Stauffer RN, Smidt GL, Wadsworth JB: Clinical and biomechanical analysis of gait following Charnley total hip replacement. Clin Orthop 99:70–74, 1974
14. Hoberman M: Crutch and cane exercises and use. In Licht S (ed): Therapeutic Exercise. New Haven, CT, Elizabeth Licht, Publisher, 1958
15. Smidt GL, Deusinger RH, Arora J, et al: An automated accelerometry system for gait analysis. J Biomech 10:367–375, 1977

Elbow Moment and Forces at the Hands During Swing-Through Axillary Crutch Gait

MARC REISMAN,
RAY G. BURDETT,
SHELDON R. SIMON,
and CYNTHIA NORKIN

We investigated swing-through axillary crutch gait (nonweight bearing on the left lower extremity) to determine the effects of gait speed, crutch length, and handle position on the forces exerted at the hands and on the moments exerted about the elbow joints. Ten healthy subjects, skilled in swing-through crutch gait, walked 1) at three speeds using fitted crtuches, 2) at a fixed speed with four different crutch lengths, and 3) at a fixed speed with four different handle positions. We collected ground reaction forces that exerted simultaneously on the right crutch and motion data with a force plate and three high-speed movie cameras. A biomechanical model was developed to calculate the forces exerted at the right hand and the moments exerted about the right elbow joint. Changing gait speed from slow to the normal gait of the subject showed statistically significant effects ($p < .05$) on the forces at the hand. When we changed crutch heights for the subjects, we found no significant effects on the forces at the subjects' hands. Changing handle position significantly affected the moment at the elbow. Increasing the elbow-flexion angle above 30 degrees by raising the crutch handle 1 to 2 in resulted in a 100 percent increase in elbow-extension moment. We found a correlation of .82 between actual average elbow-flexion angle and elbow-extension moment. Changing gait speed or crutch length did not affect elbow moment.

Key Words: Biomechanics, Crutches, Elbow joint, Gait.

Fitting axillary crutches for patients is a common procedure for physical therapists. The scientific basis for criteria used to adjust crutch length and handle position for a patient, however, has not been studied in detail. Standards proposed for measuring crutch length have included the following: 77 percent of the patient's height,[1] height minus 18 in,*[2] height minus 16 in,[3] and one and one-half or one to two fingers below the axillary fold to a point 4 in[4,5] or 8 in[6] from the side of the foot. Suggested standards for measuring hand-piece position have varied from designating specific joint placement of the elbow angle at 30 degrees of flexion[5,7] to vague descriptions that the elbow should be slightly bent.[8,9] None of these criteria has been based on scientific or biomechanical data.

Only recently have analytical studies of crutch gait been performed. McBeath et al compared energy requirements of crutch walking against normal gait.[10] They found that partial weight-bearing crutch gait required 33 percent more energy than normal gait, and that nonweight bearing crutch gait required 78 percent more energy. Peacock did a myographic analysis of swing-through crutch gait with still photographs.[11] He described an external flexion moment that acted around the elbow during crutch gait, which is balanced by an extension moment provided by the upper arm muscles, specifically the triceps brachii muscle. Wells found that the mechanical work during swing-through crutch walking was approximately the same as in normal walking but that more of this work was done by the upper extremities.[12] Because these muscles are not functionally designed for supporting the body's weight, they fatigued more rapidly than the lower extremity muscles. Shoup et al performed a biomechanical displacement analysis of a swing-through crutch gait to establish criteria for improving crutch design.[13] Shoup later used these criteria to develop a new type of forearm crutch for children.[14] Other investigators have measured the forces acting on crutches and other ambulatory aids.[15-17]

Although these studies have looked at energy consumption in crutch walking and at the forces acting on the crutches, the forces that acted on the joints of the body have not been adequately examined. The wrist and elbow joints are heavily relied on for support during the swing-through phase of axillary crutch gait. Although joint forces at the wrist[18] and torque at the elbow have been measured for isometric contractions,[19-21] these forces have not been measured during crutch walking. The purpose of this study was to show that variation of speed, crutch length, and handle position would positively affect the forces exerted by the hands and the moment exerted by the elbow joint during nonweight-bearing axillary crutch walking.

Mr. Reisman is a staff member, Washington University, Department of Physical Therapy, Irene Walter Johnson Institute of Rehabilitation, St. Louis, MO 63110 (USA).

Dr. Burdett is Assistant Professor, Program in Physical Therapy, School of Health Related Professions, University of Pittsburgh, Pittsburgh, PA 15261.

Dr. Simon is Director, Gait Analysis Laboratory, Children's Hospital, Boston, MA 02115.

Ms. Norkin is Assistant Professor, Sargent College of Allied Health Professions, Boston University, Boston, MA 02215.

This article was submitted August 2, 1983; was with the authors for revision 38 weeks; and was accepted November 11, 1984.

* 1 in = 2.54 cm.

METHOD

Subjects

Ten healthy women, randomly selected from a group of physical therapists and physical therapy students who had some skill in performing a swing-through crutch gait, were subjects in this study. Their average age was 27.6 years. To rule out height as a factor in measuring forces and moments, their heights were limited to a range of 63.5 to 66 in. None of the subjects reported any history of gait abnormalities. We used a form approved by Boston University and the Gait Analysis Laboratory of Children's Hospital to obtain informed consent from each subject. Approval for this study was received from the Human Subjects Committee of the University.

Procedure

For each gait session, the subjects wore shorts, t-shirts, and sneakers. To aid in identifying landmarks on film, small squares of black tape with white dots in the center were placed over the right acromion process, the right lateral epicondyle of the humerus, the dorsum of the right wrist midway between the ulnar and radial styloid processes, the right crutch tip, and the center of the axilla crossbar. The crutches used in this study were standard wooden axillary crutches with rubber-covered axilla crossbars, hand crossbars, and tips. Extra holes were drilled to allow for 1-in adjustments of the handles. A team of physical therapists at the Gait Analysis Laboratory decided the method for fitting the crutches based on commonly used criteria.[3,5,7] Crutch length was measured in the following manner: The subject stood with her arm abducted to 90 degrees. The crutches were measured from a point 2 in below the axilla to a point 12 in lateral to the midline of the body on a line along the tips of the shoes. The handle position was adjusted to provide 30 degrees of elbow flexion. Two physical therapists independently measured the elbow angle by centering a standard 7-in plastic goniometer over the lateral condyle of the humerus and by using the acromion and radial styloid processes as references.

Crutch gait may involve full, partial, or nonweight-bearing status on the part of either leg. For the purposes of this study, swing-through axillary crutch gait was defined as a nonweight-bearing gait with the left foot off the ground. The right foot, therefore, supported all of the force not carried by the crutches. The "normal" average speed of crutch gait for a subject with this height was determined in a pretest to be 0.73 m/sec. We considered that speeds of 0.46 m/sec and 1.12 m/sec deviated enough from the normal to be considered fast and slow speeds. The cadences and step lengths to achieve these speeds were also determined from the pretest. Table 1 gives the gait speeds, cadences, and step lengths from this study.

TABLE 1
Average Speed, Cadence, and Step Length of Crutch-Walking Trials

Speed	Cadence	Cadence	Step Length
Slow	0.46 m/sec	60 steps/min	1.5 ft/step
Normal	0.73 m/sec	72 steps/min	2.0 ft/step
Fast	1.12 m/sec	88 steps/min	3.5 ft/step

TABLE 2
Average Normalized Force on Hands During Crutch Stance of Different Trials

Speed	Slow (0.46 m/sec)		Normal (0.73 m/sec)		Fast (1.12 m/sec)
	.37		.41		.40
Crutch length	+2 in	+1 in	Fitted	−1 in	−2 in
	.41	.39	.41	.39	.40
Handle position	+2 in	+1 in	Fitted	−1 in	−2 in
	.39	.39	.41	.40	.41

To determine the effects of speed, crutch length, and handle position on the forces at the hands and moments at the elbow, we asked each subject to walk with the crutches under 11 different conditions: 1) at the three different speeds, using crutches fitted according to the standard selected; 2) at the normal speed, using four different crutch lengths (longer by 2 in and by 1 in; shorter by 2 in and by 1 in), but with the handle position set so that a 30-degree flexion angle existed at the elbow; and 3) at the normal speed, using four different handle positions (higher by 2 in and 1 in; lower by 2 in and 1 in), but with the crutch length the same as the fitted condition.

Speed was controlled by controlling both cadence and step length. Cadence was set by a metronome, and step length was controlled by marks placed along the floor. The subjects were asked to place each step of the foot or crutch near the markers. A practice session was taken for each walk so that the subject could become familiar with each speed, crutch length, and handle position. The walks were initiated from specific spots on the floor so that the right crutch would land in the middle of a force plate after three strides.

Force and Motion Analysis

The setup of the equipment, electronic systems, and computer programs for computing the elbow moment and the forces at the hands was developed at the Gait Analysis Laboratory and Children's Hospital, Medical Center, Boston. The body motion, in addition to ground reaction forces acting on the crutch tip in the vertical, anterior-posterior, and medial-lateral directions, were collected simultaneously by an Advanced Mechanical Technology force platform† and by three high-speed Photosonics 16-mm movie cameras‡ positioned orthogonally on three sides of the walkway. The right and left cameras were 12 ft§ away from the center of the force plate, and the front camera was 29 ft away. After the films were developed, they were analyzed on a Vanguard Motion Analyzer ‖ with a Graf Pen Sonic Digitizer# to collect the two-dimensional coordinates of specific points on each film. The three-dimensional coordinates of the wrist, elbow, and shoulder joints; crutch tip; and center of the axilla crosspiece were then calculated from these two-dimensional coordinates. Simon et al have described this force and motion analysis system in more detail.[22]

---

† Advanced Mechanical Technology, Inc, 141 California St, Newton, MA 02158.
‡ Instrumentation Marketing Corp, 820 S Mariposa St, Burbank, CA 91506.
§ 1 ft =.3048 m.
‖ Vanguard Instrument Co, Melville, NY 11747.
Science Accessories Corp, Southport, CT 06490.

Basic Research

Biomechanical Model

We developed a biomechanical model of crutch walking to calculate the forces acting on the hands and the moment exerted by the elbow extensors. The assumptions used in this model were 1) the mass and acceleration of the crutch, forearm, and hand could be neglected during crutch stance; 2) forces on the crutch occur only at the tip, handle, and crutch top; 3) the force at the crutch top is perpendicular to the sagittal plane; and 4) no twisting moments existed that were exerted on the crutch from the wrist (Figure).

By neglecting the mass and acceleration of the crutch during crutch stance, the forces acting on the crutch from the hand were calculated from Newton's laws of equilibrium. The direction of the axis of the elbow joint was assumed to be perpendicular to a plane formed by the shoulder-, elbow-, and wrist-joint centers. This axis is not, in general, perpendicular to the sagittal plane during crutch walking. Therefore, the component of the resultant force within this plane was calculated. This force was multiplied by the perpendicular distance between the elbow-joint center and the action line of this component to get the external moment exerted by the crutch about the flexion-extension axis of the joint. By neglecting the mass and acceleration of the forearm during crutch stance, the internal moment exerted by the elbow extensors was equal to this external moment exerted by the crutch. The resultant force was divided by body weight and was averaged over the crutch-stance time to give the average resultant normalized force at the hands. Elbow-extension moments were normalized by dividing by body weight and forearm length and were averaged over either the crutch-stance phase or the entire gait cycle. We thought average force and average moment were better indicators of muscle effort than peak force or moment, which may be exerted for only an instant. The actual elbow angle at each instant in time during crutch stance was also calculated and was used to obtain the average angle of the elbow during crutch stance.

Data Analysis

Six analyses of variance (ANOVAs) with repeated measures were performed to determine the effect of gait speed, crutch length, and handle position on the resultant normalized force at the hand and on the normalized moment at the elbow. The significance level was set at $p < .05$. Individual comparisons of means were made when appropriate by using the Neuman-Keuls multiple comparison test. A Pearson product-moment correlation coefficient was used to determine the correlation between elbow torque and elbow-flexion moment.

RESULTS

Table 2 shows the average resultant force exerted on the crutch by the hands during crutch stance as a function of speed, crutch length, and handle position. Analysis of variance showed the following results. 1) There was a statistically significant difference ($p < .05$) between the average force exerted at the hands during slow walking and normal walking. This force difference, however, is only 3.8 percent of body weight smaller than the force at normal speed. Although statistically significant, this difference may not represent any valuable clinical difference. 2) We found no significant difference ($p > .05$) among the average force values as a function of crutch length or handle position.

The forces at the hand create moments at the elbow joint that must be balanced internally by the triceps brachii muscle acting on its lever arm. These normalized moments are shown in Table 3. The average elbow moment changed very little with walking speed and indicated that elbow-extension muscle effort did not vary much with speed of walking. The average elbow moment exerted during crutch stance also did not vary significantly with crutch length. We found a significant variation, however, in elbow moment with changes in handle position. The two higher handle positions resulted in about twice as much moment as the fitted position or the two lower positions.

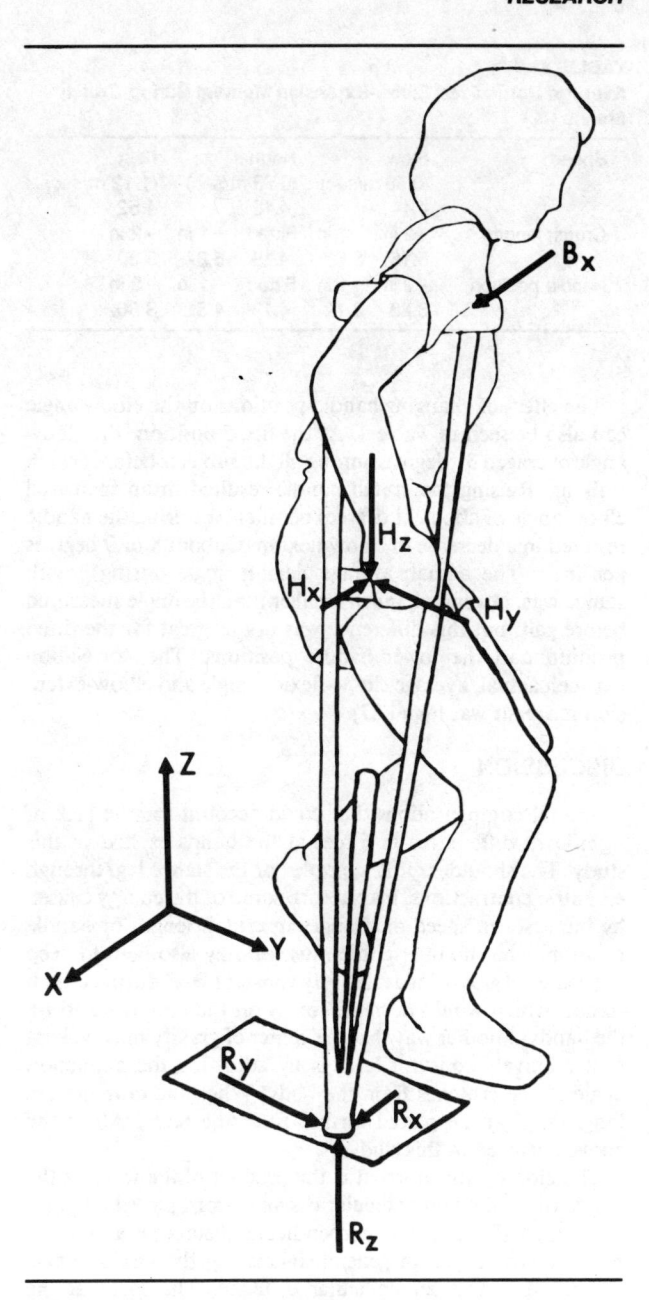

Figure. Forces exerted on the right crutch from the ground (R_x, R_y, R_z), the hand (H_x, H_y, H_z), and the body (B_x).

TABLE 3
Average Normalized Elbow-Extension Moment During Crutch Stance (%)

Speed	Slow (0.46 m/sec) 4.14		Normal (0.73 m/sec) 4.19		Fast (1.12 m/sec) 4.52
Crutch length	+2 in 5.15	+1 in 5.63	Fitted 4.19	−1 in 5.24	−2 in 5.30
Handle position	+2 in 8.20	+1 in 8.12	Fitted 4.19	−1 in 4.51	−2 in 3.86

The effect of changing handle positions on the elbow angle can also be seen in Table 4. At the fitted position, the elbow angle averaged 31 degrees among all the subjects before crutch walking. Raising the crutch handle resulted in an increased elbow angle of about 11 degrees per inch; lowering the handle resulted in a decrease in elbow flexion of about 8 or 9 degrees per inch. The actual, average flexion angle during crutch stance was, in general, much smaller than the angle measured before gait, but this difference was not as great for the fitted position and the lower handle positions. The correlation between actual, average elbow-flexion angle and elbow-extension moment was high (.82).

DISCUSSION

Several compensations that could account for the lack of significant differences in forces at the hands existed in this study. The shoulder girdle complex or the stance leg, through eccentric contractions, may absorb some of the energy caused by increases in speed or changes in crutch length or handle position. The shoulder girdle muscles may also help to keep the center of gravity at a relatively constant level during crutch stance, which would decrease forces on the crutches and on the hands. Another way that the center of gravity may be kept at a relatively constant level is by adjusting the abduction angle of the crutches from the body. When the crutches are long, they can be placed further from the body. All of the above occurred in this study.

The elbow-joint moment is the product of the force at the hands times the perpendicular distance from the elbow joint to the force (Figure). This perpendicular distance is a function of the elbow angle; in general, increasing the flexion angle will increase the perpendicular distance. The force at the hands was approximately the same for each crutch length and each gait speed, but the elbow angle was readjusted to 30 degrees of flexion after each length change. Therefore, the result of no significant difference in elbow moments with crutch-length changes and speed changes is not surprising. Energy consumption has been shown in other studies to increase with speed of crutch walking.[23] The present study indicates this increase in energy is probably caused by increased effort by muscles crossing other joints, possibly the shoulder girdle muscles and the muscles of the stance leg, rather than the elbow-joint muscles.

The difference in elbow moment that occurred with changes in handle position is almost entirely caused by an increase in the lever-arm distance between the elbow joint and the action line of the force on the hands. For the three, lower handle positions, the actual elbow angles were very similar, even though the initial angles were different. As a result, the average extension moments were also very similar. At the higher handle positions, the subjects used much larger elbow flexion angles during gait, and, therefore, the elbow-joint moments were much larger also. By using some relatively quick and inexpensive method of measuring elbow angle during gait, such as videotape or electrogoniometry, the elbow-joint moment could be estimated from the strong linear relationship between actual elbow angle and joint moment without the use of a force platform.

Clinical Implications

The results of this study indicate that the 30-degree resting elbow-flexion angle often used for fitting axillary crutches has a good biomechanical basis. If a therapist is going to deviate from this standard, it should be in the direction of lowering the handle slightly to decrease elbow flexion. Crutches should be made with fine enough adjustments at the handle to allow for proper fitting because increments of 1 in can make significant differences in elbow moment. An increase in elbow moment may result in a significant increase in the amount of force that the elbow extensors must exert. Muscle fatigue may, therefore, become one of the limiting factors in crutch walking.

CONCLUSIONS

The force exerted at the hands did not vary greatly with crutch length, handle position, or speed of walking. The moment exerted at the elbow joint also did not vary with crutch length or speed, but handle position did have a significant effect on elbow-extension moment. Increasing the resting flexion angle above 30 degrees by raising the crutch handle 1 or 2 in resulted in increasing the elbow moment by almost 100 percent, but decreasing the elbow-flexion angle by lowering the crutch handle did not significantly change the elbow-extension moment. We also found a high positive correlation between average elbow angle and elbow-extension moment.

Elbow-extensor muscle strength and endurance is not the only physical factor that affects the ability of someone to use axillary crutches effectively. Other factors, such as shoulder girdle muscle strength, may be more critical. Further examination of the biomechanics of the shoulder joint during crutch gait may elucidate how forces are kept constant at the elbow and hand during moderate changes of crutch height, handle height, and gait speed.

TABLE 4
Effect of Crutch Handle Position on Elbow Angle and Moment

Position	Flexion Angle (%)		Extension Moment (%) ($\bar{X}$)
	Pregait ($\bar{X}$)	During Gait ($\bar{X}$)	
2 in high	53	26	8.2
1 in high	42	23	8.12
Fitted	31	16	4.19
1 in low	22	13	4.51
2 in low	14	11	3.86

Basic Research

Effects of Selected Assistive Devices on Normal Distance Gait Characteristics

CHUKWUDUZIEM U. OPARA,
PAMELA K. LEVANGIE,
and DAVID L. NELSON

The purpose of this study was to investigate the effects of selected assistive devices on normal standards of gait. The gait characteristics of stride length, step length, step width, and foot angle were analyzed for 24 right-dominant, healthy men under four conditions: right ankle-foot orthosis (AFO), right hemiplegic arm sling (HAS), both devices (AFO+HAS), and no devices. The dependent variables were measured by a standard method from ink traces left by subjects walking on newsprint. Order of conditions was controlled, and cadence remained consistent across all four conditions for each subject. The AFO and AFO+HAS conditions produced statistically significant changes from normal gait characteristics. The HAS alone did not produce significant changes. Data from the study may be used as a basis for goal setting and as a guideline for the optimal level of function possible for a person wearing these devices. The extent of the patient's orthopedic and neurologic involvement should of course be considered.

Key Words: *Gait, Orthotic devices, Physical therapy.*

A primary objective of most physical therapy services is to maximize the patient's abilities. Physical therapists frequently use assistive devices, modalities, or interventions that effectively raise the patient's functioning levels. These devices, however, may simultaneously prevent the attainment of ideal function. Therefore, goal setting for the patient would be enhanced by a knowledge of the optimal level of function that can be expected under the conditions of the intervention. What level of function can reasonably be expected if the patient wears a specific assistive device? Continuing physical therapy services could be based on achieving a specified optimal level rather than on aiming for the ideal level of function.

In rehabilitating the hemiplegic patient, therapists concentrate on the patient's attainment of optimal ambulation. Goal setting in ambulation should take into account not only the patient's handicap but also the effects of any intervention. Two assistive devices sometimes used with hemiplegic patients are the ankle-foot orthosis (AFO) and the traditional hemiplegic arm sling (HAS). The purpose of this study is to identify the effects of these assistive devices on normal standards for gait. These devices might preclude perfectly normal gait in healthy people, although they partially correct patterns of gait used by patients with hemiplegia.

Researchers and clinicians have not agreed on the exact effects of the HAS in the rehabilitation of stroke patients. This sling has been used predominantly for preventing subluxation of the glenohumeral joint. Among many arguments against the HAS, some studies have shown that the sling interferes with the distribution of body weight and inhibits attaining or maintaining a normal walking pattern because the sling positions the arm in front of the body.[1-3] Delwaide et al monitored electromyographic (EMG) responses in healthy subjects and showed that the position of the upper limb induced lower limb reflexes even though the upper limb muscles had EMG quiescence.[4] Positioning the arm in an arm sling may cause a deviation from normal gait by inducing compensatory patterns. In healthy individuals, arm swinging appears to counteract excessive horizontal trunk rotation. A limitation of arm swing by use of the HAS may result in a lack of the counter effect, which, in turn, affects gait.

The AFO is widely accepted by clinicians and researchers.[5-7] Friedland found that an AFO, such as the double upright, improved the subject's gait considerably and was also cosmetically acceptable.[2] The brace facilitates safe, effective ambulation with minimum energy expenditure, especially for patients with marked weakness around the ankle and foot.[8] Magora et al, however, warned that the standard rigid lower limb brace produces changes in the contralateral lower extremity that may explain the early degenerative osteoarthritic changes, discomfort, and fatigue felt by many patients in the unbraced limb.[9] Smidt and Mommens supported this line of thought when in a study of the influences of some ambulatory aids on gait, they concluded that the use of assistive devices tends to increase the vertical loading on the body structures.[10]

Given the use in rehabilitation of the HAS and the AFO, clinicians need to know the optimal level of ambulatory function that can be expected of patients using one or both of these devices. Any changes produced in normal gait by these assistive devices should be anticipated in a patient's gait. This "modified" gait represents the best that the average

---

Mr. Opara is a doctoral student in the Department of Health Sciences, Sargent College of Allied Health Professions, Boston University, Boston, MA 02215 (USA).

Ms. Levangie is Assistant Professor of Physical Therapy, Sargent College of Allied Health Professions.

Dr. Nelson is Associate Professor of Occupational Therapy, Western Michigan University, Kalamazoo, MI 49008.

This study is based, in part, on Mr. Opara's thesis completed in partial fulfillment of the Master of Science degree at Sargent College of Allied Health Professions, Boston University.

This article was submitted August 16, 1983; was with the authors for revision 55 weeks; and was accepted January 31, 1985.

patient should be able to achieve; expectations should be further modified by the extent of the handicap.

Distance gait factors have been identified as one class of variables of importance in quantitative gait evaluation.[11] We chose the gait characteristics of stride length, step length, step width, and foot angle for our study. These characteristics are measurable by clinicians without access to sophisticated instrumentation.

METHOD

Subjects

Twenty-four men between the ages of 20 and 55 years were recruited from the university community. We deemed the wide age range acceptable because we compared the subjects with themselves in a Latin Square design using repeated measures. Only those whose reported height, weight, and age fell within the optimal range on a height-weight chart as listed by the Metropolitan Life Insurance Company were accepted for the study. The subjects were allowed ample time to read and sign an approved informed consent form. Right dominance was determined by the choice of hand used to sign the consent form and by kicking accuracy. Subjects wore their own shoes.

Procedure

Each subject's comfortable free-walking speed was established by letting him walk twice back and forth along a 30-ft* walkway. The light of a metronome was synchronized to his steps. This cadence was maintained for all testing conditions. All cadences fell within norms established for men.[11]

Subjects were randomly assigned to four groups of equal size. Subjects from each group experienced all four experimental conditions. Each group, however, experienced the conditions in a different order (in accordance with the Latin Square design). The four conditions were no devices, HAS alone, AFO alone, and AFO and HAS (AFO+HAS) simultaneously.

The HAS (Fig. 1) had a sliding buckle and metal loops for adjustment and a thumb-loop in the wrist-hand support. One-hundred-degree elbow flexion was maintained. The AFO was the double

---

* 1 ft = .3048 m.

upright, universal short leg brace without stops.

Assistive devices were worn on the right limbs only. Two 8.00- × 0.76-m walkways of newsprint were secured to the floor for each subject. Before starting the test, we affixed two strips of moleskin to the soles of both shoes in the middle of the widest point of the forefoot and at the midpoint of the heel (Fig. 2). The moleskin strips were soaked in water-based ink, and the subject walked the length of the walkway to the beat of the metronome. Each walkway recorded two of a subject's four trials. The particular experimental condition was noted on the walkway. The two trials were differentiated by ink colors and direction of footprints. The subject rested for 10 minutes between trials while the shoe pads were resoaked and the appropriate assistive devices were put on or taken off.

All measurements were based on the mean of the four strides after the first two steps. Stride lengths were measured as the linear distances between two consecutive heel-pad prints of the same foot. Step lengths were measured as the linear distances between one heel-pad print and the subsequent contralateral heel-pad print. Step width was measured as the distance between one heel-pad print and the opposite line of progression. (The line of progression is a line joining two consecutive heel-pad prints of the same foot.) Foot angle was measured as the angle formed by the line of progression and the line joining the midpoints of the heel and the forefoot pad prints of the same foot.[11-15]

Data Analysis

To test for differences between the four experimental conditions experienced in four different orders, we planned a two-way analysis of variance (ANOVA) with one repeated measure for each dependent variable (conditions × orders). If main effects for conditions were found, we planned a Newman-Keuls *post hoc* analysis.

RESULTS

Table 1 gives a descriptive summary of results. These scores are consistent with values of normal gait as obtained by other investigators.[11,15,16] Pearson product-moment correlations indicated insignificant correlations between cadence and the dependent variables;

Fig. 1. Right upper limb in HAS.

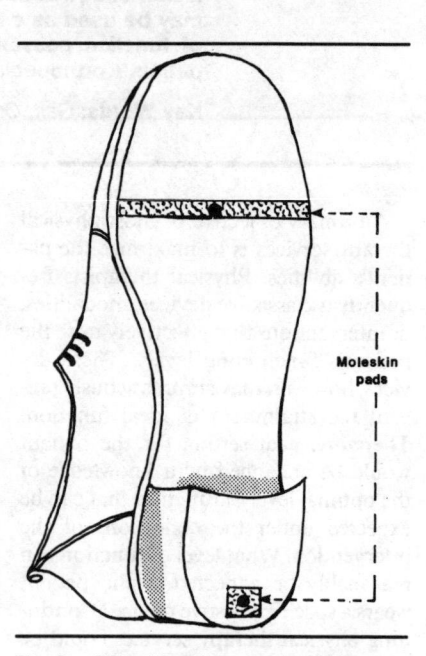

Fig. 2. Moleskin strips on sole of shoe.

therefore, cadence was not a confounding variable.

The main research questions posed by this study are answered in Tables 2 to 5. In this type of analysis, the main effect for conditions of gait tests whether the different types of assistive devices made a significant difference. Note that the main effect for conditions is significant in all analyses. In other words, the type of assistive device caused significant differences for right stride length, left stride length, right step length, left step length, step width, right foot angle, and left foot angle.

Order showed no significant main effects. Small but significant interactions existed between the order of presenta-

Basic Research

TABLE 1
Summary of Means and Standard Deviations of Gait Characteristics (N = 24)

Gait Characteristics		No Devices $\bar{X}$	s	HAS $\bar{X}$	s	AFO $\bar{X}$	s	AFO + HAS $\bar{X}$	s
Stride (cm)	left	153.9	(11.3)	153.2	(11.8)	149.6	(13.1)	149.8	(11.8)
	right	153.9	(11.7)	153.2	(11.4)	149.9	(13.3)	149.4	(11.7)
Step (cm)	left	77.5	(6.8)	77.1	(6.8)	75.1	(7.3)	74.9	(7.2)
	right	76.6	(5.3)	76.2	(5.4)	74.6	(6.4)	74.8	(5.2)
Width (cm)		8.9	(4.0)	8.6	(4.0)	7.9	(3.3)	7.4	(3.3)
Angle (°s)	left	7.9	(4.0)	7.7	(3.8)	10.2	(3.2)	11.3	(3.8)
	right	8.1	(4.0)	7.8	(3.9)	10.2	(3.2)	11.5	(3.5)

TABLE 2
Two-way Analyses of Variance for Right and Left Stride Length

Gait Characteristic	Source	df	SS	MS	F
Right stride length	orders	3	215	71.7	0.13
	error (between)	20	10843	542.2	
	conditions	3	367	122.3	4.25[a]
	orders × conditions	9	556	61.8	2.15[a]
	error (within)	60	1725	28.8	
Left stride length	orders	3	225	75	0.14
	error (between)	20	10870	543.5	
	conditions	3	360	120	4.24[a]
	orders × conditions	9	546	60.7	2.14[a]
	error (within)	60	1700	28.3	

[a] Significant at $p < .05$.

TABLE 3
Two-way Analyses of Variance for Right and Left Step Length

Gait Characteristic	Source	df	SS	MS	F
Right step length	orders	3	46.5	15.5	0.14
	error (between)	20	2285.7	114.3	
	conditions	3	72.4	24.1	3.64[a]
	orders × conditions	9	137.8	15.3	2.31[a]
	error (within)	60	397.9	6.6	
Left step length	orders	3	109.8	36.6	0.20
	error (between)	20	3587.4	179.4	
	conditions	3	127.8	42.6	3.94[a]
	orders × conditions	9	172.4	19.2	1.77
	error (within)	60	648.62	10.8	

[a] Significant at $p < .05$.

tion and conditions for stride, step lengths, and step width. This significance means that performance in one condition varied somewhat depending on the conditions following or preceding it.

Newman-Keuls *post hoc* analyses identified the specific conditions that were different from each other. They showed that wearing both the HAS and the AFO at the same time caused significant decreases ($p < .05$) in left and right stride length and in left step length. This combination of assistive devices also caused an increase in step width and both foot angles in comparison with the values obtained when no devices were worn ($p < .05$). The most significant changes were noted in foot angles ($p < .01$). The *post hoc* analyses also showed that wearing the AFO alone caused similar changes in comparison with wearing no devices: a significant decrease in left stride and right step lengths and an increase in step width and both foot angles.

Wearing the HAS alone did not cause any statistically significant changes from normal gait values in the characteristics measured. A trend was noted, however, toward a small deviation in the direction of a decrease from normal values for stride and step lengths and an increase for step width and foot angles.

DISCUSSION

Statistical significance does not necessarily imply clinical significance of great magnitude. The means presented in Table 1 indicate that the four conditions did not differ profoundly from each other. This study found statistically significant but not profound differences between gait in the AFO and AFO+HAS conditions and gait in the condition in which no devices were used. Although the AFO and the AFO+HAS conditions were essentially similar, the shift in significance from the left step for the AFO+HAS to the right step for the AFO alone was unaccountable. The HAS did not contribute to the significance of change, although some changes in values were noted with its use. The changes in gait characteristics were apparently induced predominantly by the AFO. We point out, however, that the results cannot automatically be generalized to all other types of short leg assistive devices.

This study was designed to identify the changes made by the AFO and HAS in selected gait characteristics. From the data gathered, a researcher might also look at the reason these changes occurred. In our study, cadence was held constant, but velocity was not. With a constant cadence, a decrease in step length yields a decreased velocity. Other investigators have shown that an interaction exists between velocity and the gait characteristics of step length, step width, and foot angle.[10,12,13] As velocity increases from the subject's customary gait, step length increases while step width and foot angle decrease. Decreases in velocity from customary gait might be expected in step width and foot angle. From this study alone, we cannot determine whether the AFO caused a change in velocity, which then affected the gait characteristics, or whether the initial effect of the AFO was on the gait characteristics themselves.

The observed changes may also be attributable to changes other than in velocity. The "push" of the braced limb may have had a restriction. Simkin et al identified this restriction as an important factor in forward motion.[17] Similarly, the AFO may have limited the description of the two intersecting arcs of foot and ankle motion. Other investigators have shown these arcs to be important components of the foot-knee mechanism—a major determinant of normal gait.[7,11,14,18] Possibly, wearing the AFO on one side required a mutual compensatory effort by both lower limbs. This explanation of the role of compensation lends support to the theory that early osteoarthritic changes in the unbraced limb may be linked to bracing.[9] Such compensation may have led to changes in distance gait characteristics. Changes in foot angle may be an attempt by the body to maintain comfortable balance while walking. Other possible ways of accounting for these changes might be a consideration of the ranges of joint motion, the muscles involved, the loading factors, and energy expenditure. Ultimately accounting for the causes of change was not within the scope of the study.

CONCLUSION

Our study found that the commonly recommended AFO has significant effects on the normal distance gait characteristics of right-dominant male subjects. These effects were most pronounced when the AFO was used in conjunction with the HAS. The HAS alone had little, if any, effect on distance gait characteristics. The AFO significantly reduced stride and step lengths and caused significant widening of the step width and foot angles in healthy subjects. The study provided documentation of the optimal level of function that can be achieved in terms of distance gait characteristics when the universal double-upright short leg brace is used in conjunction with the traditional hemiplegic arm sling. Consequently, the data may serve as a basis for goal setting when these devices are used in the clinic, considering, of course, the extent of the patient's orthopedic and neurologic deficit. In future studies, further information may be obtained by controlling the velocity of walking and by performing the test on non-right-dominant subjects and subjects of different age groups.

The study was not intended to discredit the use of assistive devices for patients, and the results cannot be generalized to all other assistive devices. Rather, it indicated that the aim of rehabilitation efforts should be to achieve an appropriate optimal functioning level.

TABLE 4
Two-way Analyses of Variance for Step Width

Gait Characteristic	Source	df	SS	MS	F
Step width	orders	3	96.8	32.3	0.71
	error (between)	20	903.3	45.2	
	conditions	3	31.9	10.7	3.86[a]
	orders × conditions	9	64.9	7.2	2.62[a]
	error (within)	60	165.5	2.8	

[a] Significant at $p < .05$.

TABLE 5
Two-way Analyses of Variance for Right and Left Foot Angle

Gait Characteristic	Source	df	SS	MS	F
Right foot angle	orders	3	153.8	51.3	1.14
	error (between)	20	900.1	45.0	
	conditions	3	218.3	72.8	24.57[a]
	orders × conditions	9	36.9	4.1	1.38
	error (within)	60	177.7	2.9	
Left foot angle	orders	3	162.2	54	1.20
	error (between)	20	901.2	45	
	conditions	3	222	74	26.73[a]
	orders × conditions	9	42.2	4	1.69
	error (within)	60	166.1	2	

[a] Significant at $p < .01$.

REFERENCES

1. Hurd MM, Farrell KH, Wayloni GW: Shoulder sling for hemiplegia: Friend or foe? Arch Phys Med Rehabil 55:519–522, 1974
2. Friedland F: Physical therapy. In Licht S (ed): Stroke and Its Rehabilitation. Baltimore, MD. Williams & Wilkins, 1975
3. Licht S: Stroke rehabilitation program. In Licht S (ed): Stroke and Its Rehabilitation. Baltimore, MD, Williams & Wilkins, 1975
4. Delwaide PJ, Fijiel C, Richele C: Effects of postural changes of the upper limb on reflex transmission in the lower limb: Cervicolumbar reflex interactions in man. J Neurol Neurosurg Psychiatry 40:616–621, 1977
5. Perry J: The mechanics of walking in hemiplegia. Clin Orthop 63:23–31, 1969
6. Perry J: Lower extremity bracing. Clin Orthop 63:32–38, 1969
7. Saunders JB, Inman VT, Ebenhart HD: The major determinants in normal and pathological gait. J Bone Joint Surg [Am] 35:543–548, 1953
8. Lehman JF: Lower limb orthotics. In Licht S (ed): Orthotics, Etcetera. New Haven, CT, Elizabeth Licht, Publisher, 1966
9. Magora A, Robin GC, Rozin R, et al: Investigations of gait 5: Effect of a below knee brace on the contralateral unbraced leg. Electromyogr Clin Neurophysiol 13:355–361, 1973
10. Smidt GL, Mommens MA: System of reporting and comparing influence of ambulatory aids on gait. Phys Ther 60:551–558, 1980
11. Murray MP, Drought BA, Kory RC: Walking patterns of normal men. J Bone Joint Surg [Am] 46:335–360, 1964
12. Norkin CC, Levangie PK: Joint Structure and Function: A Comprehensive Analysis. Philadelphia, PA, F A Davis Co, 1982
13. Andriachi TP, Ogle JA, Galente JO: Walking speed as a basis for normal and abnormal gait measurements. J Biomech 10:261–268, 1977
14. Ogg HL: Measuring and evaluating the gait patterns of children. J Amer Phys Ther Assoc 43:717–720, 1963
15. Boenig DD: Evaluation of a clinical method of gait analysis. Phys Ther 57:795–798, 1977
16. Perry J: Mechanics of walking: A clinical interpretation. Phys Ther 47:778–801, 1967
17. Simkin A, Magora A, Saltiel J, et al: Relationship between muscle action and mechanical stress in below knee braces. Electromyogr Clin Neurophysiol 13:495–503, 1973
18. Morton DJ, Fuller DD: Human Locomotion and Body Form: A Study of Gravity and Man. Baltimore, MD, Williams & Wilkins, 1952

Energy Expenditure of Ambulation Using the Sure-Gait® Crutch and the Standard Axillary Crutch

Annette L Annesley
Monica Almada-Norfleet
David A Arnall
Mark W Cornwall

Energy expenditure is increased for ambulation with various assistive devices such as canes, walkers, and crutches compared with unassisted ambulation. The purpose of the present investigation was to determine whether a significant difference in oxygen consumption and heart rate existed during ambulation with two different types of crutches. Ten healthy male subjects between the ages of 40 and 60 years participated in this study. Each subject ambulated at 1.5 mph on a treadmill using two different types of crutches—the standard axillary crutch and the Sure-Gait® crutch. After walking on the treadmill without an assistive device, subjects ambulated using a three-point, swing-to gait pattern with one of the two types of crutches. This procedure was repeated using the other type of crutch. Oxygen consumption and heart rate were analyzed using an analysis of variance for repeated measures design. The results of the study showed a significant difference (p < .01) between ambulation with crutches and unassisted ambulation for oxygen consumption and heart rate. No difference, however, was found between the two crutch types. [Annesley AL, Almada-Norfleet M, Arnall DA, et al: Energy expenditure of ambulation using the Sure-Gait® crutch and the standard axillary crutch. Phys Ther 70:18–23, 1990]

Key Words: *Crutches; Energy expenditure; Equipment, general; Gait; Orthotics/splints/casts, general.*

For many patients who must ambulate non-weight bearing (NWB) on one lower extremity, the energy expenditure during gait may be a limiting aspect to activity. Patients who use crutches or canes may have physical limitations that impair their ability to meet the metabolic demands of assisted gait. In elderly patients, who generally are less strong and have poorer endurance than younger patients, prolonged crutch walking may be impossible.

Many investigators have examined energy expenditure during ambulation with assistive devices.[1–6] It is well documented that ambulation using assistive gait devices increases energy expenditure[1–8] compared with normal walking. This finding suggests that walking with assistive devices is a significant physiological stressor that should be considered in any therapeutic plan.[7] Ganguli and associates reported that use of axillary crutches imposed a greater metabolic cost for lower extremity amputees during activities of daily living compared with the energy expenditure of healthy subjects or amputees using prostheses.[1] Waters and associates reported a significantly greater energy expenditure during NWB swing-through crutch gait when compared with the energy demands of walking with a weight-bearing cast.[7] These

---

A Annesley, BSPT, is Staff Physical Therapist, St Joseph's Medical Center, Physical Medicine and Rehabilitation, 350 W Thomas Rd, Phoenix, AZ 85013.

M Almada-Norfleet, BSPT, is Staff Physical Therapist, Northwest Hospital, 6200 N La Cholla Blvd, Tucson, AZ 85741.

Ms Annesley and Ms Almada-Norfleet, were students in the Bachelor of Science in Physical Therapy Program, Northern Arizona University, when this study was conducted in partial fulfillment of their degree requirements.

D Arnall, PhD, PT, is Assistant Professor, Department of Physical Therapy, Northern Arizona University, NAU Box 15105, Flagstaff, AZ 86011 (USA). Address all correspondence to Dr Arnall.

M Cornwall, PhD, PT, is Assistant Professor, Department of Physical Therapy, Northern Arizona University.

This article was submitted June 6, 1988; was with the authors for revision for 27 weeks; and was accepted May 26, 1989.

findings show that the energy demand is relatively high, and as a result, persons with low physical work capacity may not be able to ambulate with crutches.

Hinton and Cullen determined the energy cost of ambulation using the Ortho® crutch* and the standard axillary (SA) crutch in comparison with unassisted gait in healthy male college students.[2] Although a significant increase in energy cost was shown between crutch walking and unassisted gait, there was no significant difference between the Ortho® crutch and the SA crutch. They reported, however, that during NWB ambulation for short periods of time, the Ortho® crutch was less taxing than the SA crutch in energy costs and heart rate (HR) demands.[2]

The Sure-Gait® (SG) crutch* and the SA crutch are very similar in design. The major structural difference is the weight-bearing "tip" of the SG crutch (Fig. 1). The manufacturer of the SG claims that the device offers superior stability, superior gait, and a revolutionary crutch design. We believe that if the SG does provide increased stability and does represent a new, revolutionary crutch design, those features might be manifested in a gait that is less costly in terms of energy expenditure. To date, no published data comparing the energy demands placed on patients using the SA and SG crutches have been found. The purpose of this study, therefore, was to compare the oxygen consumption ($\dot{V}O_2$) and HR responses in subjects using the SA crutch compared with subjects using the SG crutch. Comparisons were made to determine whether walking with one crutch type was less demanding in terms of energy expenditure than walking with the other crutch type. The null hypothesis was that there is no significant difference in $\dot{V}O_2$ or HR during ambulation with the SA crutch compared with the SG crutch.

Method

Subjects

The subject selection procedure and subject test conditions met the criteria for acceptable research practice as determined by the Institutional Review Board of Northern Arizona University. Standard advertising methods were used to gather the subject sample from the community of Flagstaff, Ariz.

Ten healthy male volunteers, ranging in age from 40 to 60 years, participated in this study. All subjects completed an extensive medical questionnaire to ensure no history of cardiovascular, respiratory, or orthopedic disorders. After reviewing the medical history and interviewing all volunteers, the subjects were given final clearance by their personal physician to participate in the study. The subjects were fully informed of the potential risks and benefits of the study, and each subject signed a consent form.

To ensure that cardiovascular responses to treadmill walking were not artificially elevated (that is, to reduce test anxiety), all subjects performed a nonassisted normal gait at a constant speed of 1.5 mph for five minutes. Because the volunteers were unfamiliar with crutch walking, all subjects underwent a gait-training instruction period. They were taught a three-point, swing-to, crutch-walking gait pattern that was NWB on the side of their choice. Subjects practiced with both crutch types until they could ambulate freely without the necessity of contact guarding on level ground and on the treadmill. This instruction period familiarized the subjects with the two crutch types and ensured that all of them used the same gait pattern and that they could demonstrate reasonable expertise while using the crutches. In addition, the width of the treadmill belt was 24 in,[†] which allowed easy ambulation with assistive devices.

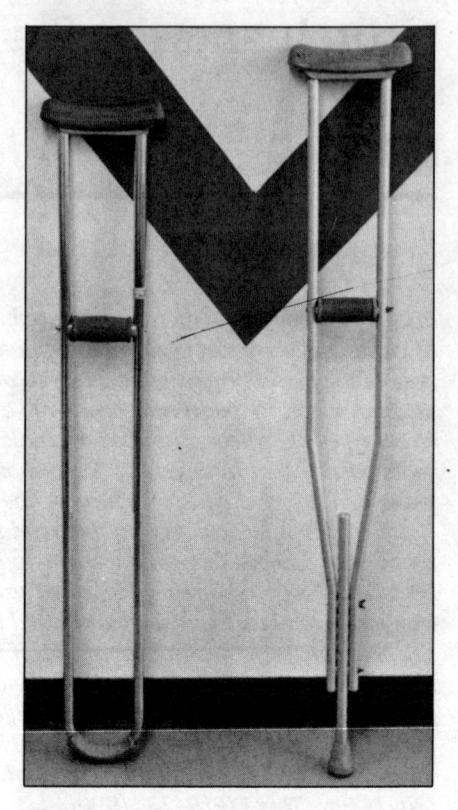

Fig. 1. *Sure-Gait® (left) and standard axillary (right) crutches.*

Instrumentation

The subjects were attached to a three-lead electrocardiograph[‡] using a modified lead II to monitor HR continuously throughout the exercise period. Oxygen uptake was determined by analyzing the expired gas volumes using a procedure outlined by Rasmussen.[9] Briefly, this procedure involved the following measurement techniques. During the treadmill-walking trials, the subjects' $\dot{V}O_2$ was continuously measured by placing noseclips on the subjects and having them breath through a mouthpiece connected to a Hans-Rudolph Model 2700 one-way valve.[§] During inspira-

---

*Lumex, Inc, 100 Spence St, Bay Shore, NY 11706.

[†]1 in = 2.54 cm.

[‡]Gilson Medical Electronics, 3000 W Beltline Ave, Middleton, WI 53562.

[§]Hans-Rudolph, Inc, 7200 Wyandotte St, Kansas City, MO 64114.

Basic Research

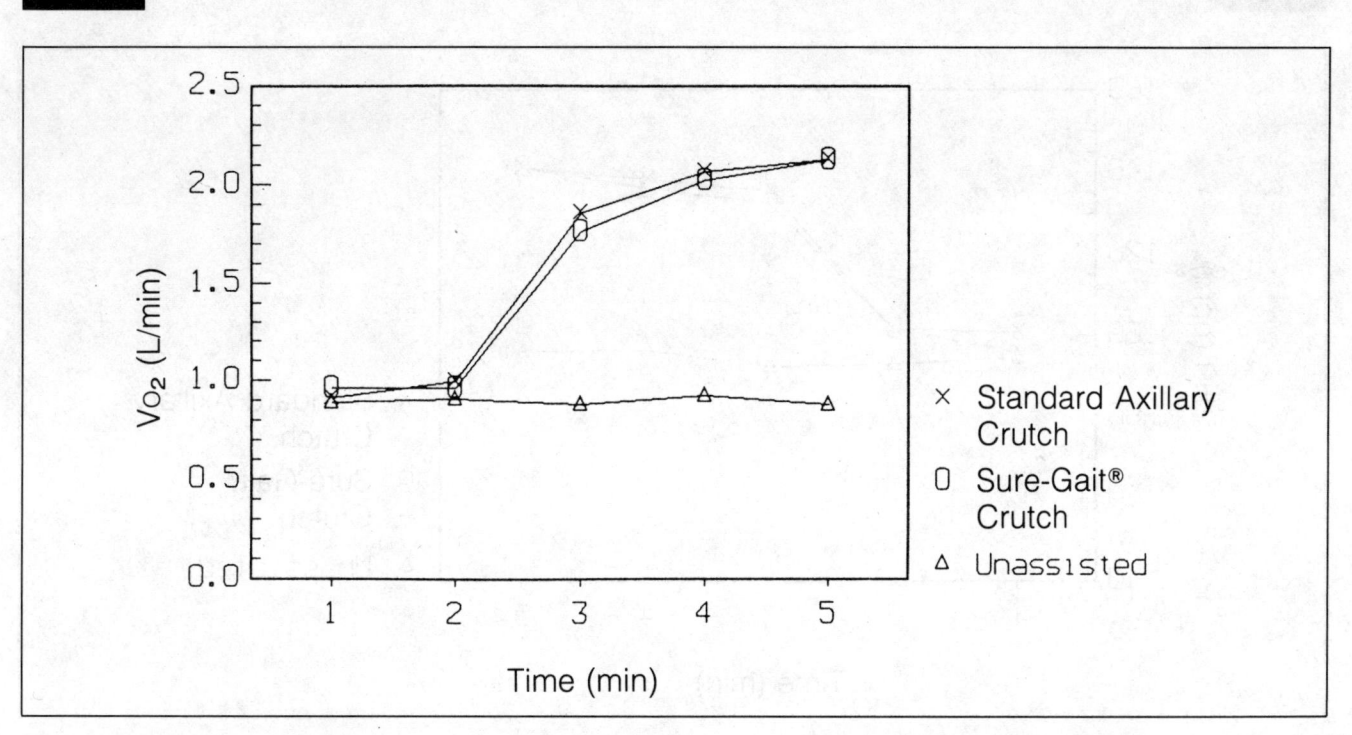

Fig. 2. *Effect of crutch type on oxygen consumption ($\dot{V}O_2$) during level ambulation. Crutch walking was initiated at end of minute 2.*

tion, air was drawn in through a Parkinson-Cowan flowmeter.[||] Flowmeter readings were recorded on a Gilson Model 5/6 medical recorder.[‡] Respiratory minute volumes were then computed from these readings. Air was exhaled through the mouthpiece into a mixing chamber. From the mixing chamber, a small sample of the expired air was pumped through an Applied Electrochemistry Model S-3A oxygen analyzer[#] and Model CD-3A carbon dioxide analyzer,[#] which had previously been calibrated against gas standards of known concentrations. The expired oxygen values obtained from the oxygen analyzer were subtracted from the ambient oxygen concentration to obtain the percentage of oxygen assimilated during exercise. Oxygen uptake was calculated by multiplying the assimilated oxygen concentration by the respiratory minute volumes, which were corrected for standard temperature and pressure-dry (STPD) conditions. In a similar fashion, carbon dioxide (CO_2) production was calculated by multiplying the exhaled CO_2 fractions by the respiratory minute volumes, corrected to STPD conditions. All data reduction was done manually, and the results were analyzed using a microcomputer.

Procedure

The SA and SG crutches were adjusted for each subject in a standing position so that a 1-in space was left under the axilla. The handgrips were positioned to accommodate roughly 15 degrees of elbow flexion.

The subjects were tested under three separate exercise conditions comprising one bout of unassisted treadmill walking, one bout of ambulation using the SG crutches, and one bout of ambulation with SA crutches. The order of unassisted gait and crutch-ambulation trials for each subject was randomly determined. Each treadmill test lasted five minutes. A rest period was provided between conditions so that each subject's HR returned to its initial baseline value. During the crutch-walking trials, subjects were brought to an increased exercise HR on the treadmill by ambulating without assistive devices at a 2% grade for the first two minutes. The two-minute 2% grade was used to rapidly increase the subject's HR and prevent excessive upper extremity muscle fatigue, which had occurred in earlier pilot trials when subjects ambulated on the treadmill with crutches for the entire five minutes. Once HR had increased, the moving treadmill was lowered to a 0% grade and the subjects performed three minutes of crutch walking at 1.5 mph. We observed that a steady-state HR was achieved by the fifth minute of treadmill walking (Figs. 2, 3). At this time, the trials were terminated.

Data Analysis

The data for minutes 3 through 5 were analyzed using a two-way analy-

---

[||]Parkinson-Cowan, Stretford, Manchester, England.

[#]Applied Electrochemistry, 735 N Astoria Ave, Sunnyvale, CA 94086.

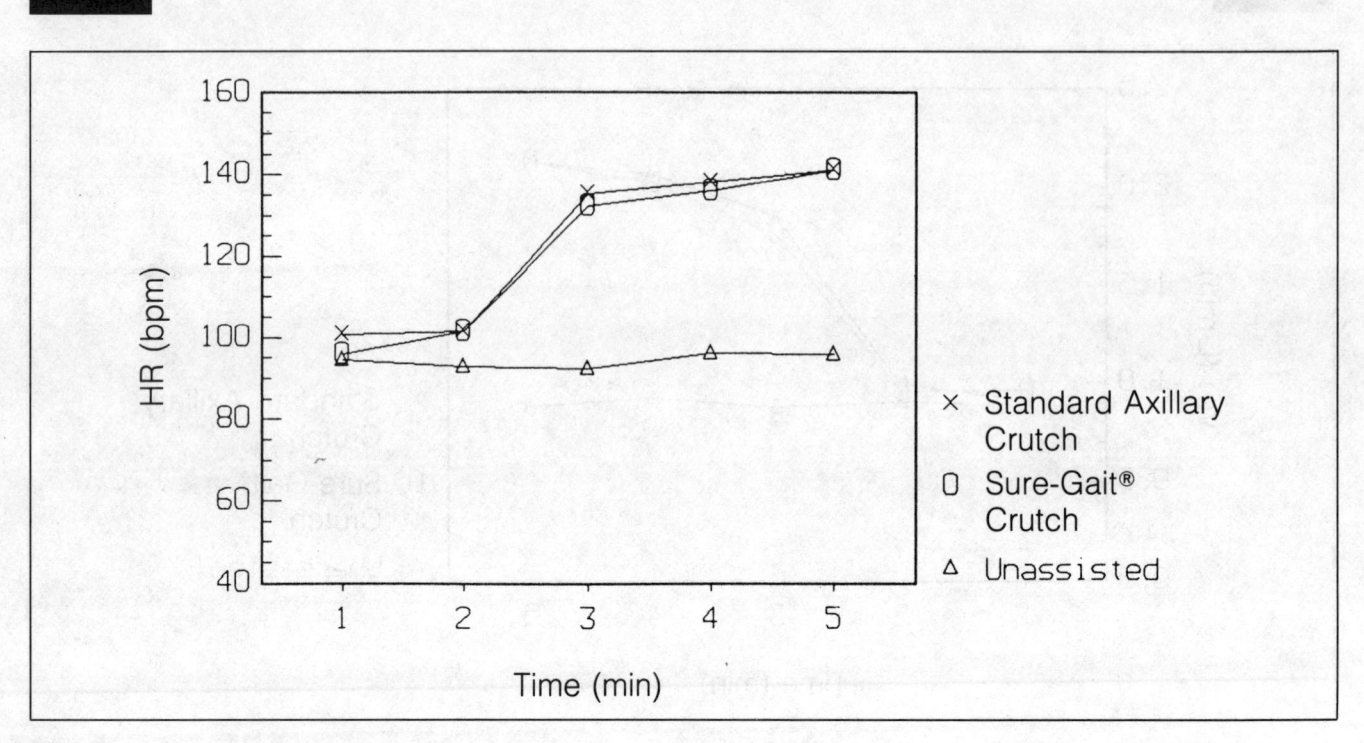

Fig. 3. *Effect of crutch type on heart rate (HR) during level ambulation. Crutch walking was initiated at end of minute 2.*

sis of variance (ANOVA) for repeated measures on both factors (crutch and time). A Newman-Keuls multiple comparison *post hoc* test for significance between means was performed following the determination of a significant *F*-ratio value. All tests were conducted at the .01 alpha level.

Results

Figures 2 and 3 show the mean $\dot{V}o_2$ and HR values, respectively, for the three experimental conditions. Significant differences ($p < .01$) were found between the three experimental conditions with respect to both $\dot{V}o_2$ and HR (Tabs. 1, 2). *Post hoc* comparisons of each of the dependent variables showed a significant ($p < .01$) difference between unassisted gait and either crutch gait. During the first minute of crutch walking (minute 3 of the test), there was a significant difference ($p < .01$) in the energy demands of the two crutch designs (Figs. 2, 3). The SA crutch was found to be metabolically more costly than the SG crutch. That difference, however, was transient and disappeared by the second minute of crutch walking (minute 4 of the test). There was at least a 141% increase in $\dot{V}o_2$ during either crutch-walking condition compared with unassisted gait (Fig. 2). In addition, the subjects experienced a 47% increase in HR during crutch walking compared with unassisted ambulation (Fig. 3).

Discussion

Corcoran and Brengelmann have criticized the use of treadmills for determining $\dot{V}o_2$ during assisted gait in handicapped individuals.[10] They believed that the treadmill belts were too narrow to accommodate awkward gait, especially when the use of canes or crutches was necessary. In addition, Corcoran and Brengelmann believed that handicapped persons feel unstable on a moving treadmill belt and that the anxiety produces artificial increases in their metabolic expenditure.[10] We believe we were able to overcome these difficulties in part by providing a wider treadmill surface (24 in) on which to test our subjects. This adaptation provided an appropriately wide surface moving at a constant speed during the test procedure. None of the subjects related feelings of anxiety or inhibition during ambulation on the treadmill. More than likely, this outcome was a result of the extensive familiarization procedures used in our study. We also believe that by familiarizing our subjects with the test conditions, much test anxiety was minimized, thus reducing the metabolic expenditure. We suggest a thorough familiarization process be considered a standard part of the testing procedure for all future treadmill gait analysis studies.

We also disagree with Corcoran and Brengelmann's[10] conclusion that a continuously moving treadmill acts to destabilize awkward gaits in handicapped individuals and, therefore, is not an appropriate tool to measure the energy consumption of walking. Although treadmill walking does not mimic the variable speed of normal, unassisted gait, it does offer a completely flat surface without the standard pitfalls of walking on uneven ground. We also believe that gait speed is an important variable to control because of the relationship between velocity and energy expenditure.[2] Because the treadmill belt moves at a constant speed, it allows the subjects to achieve steady-state work, ensuring an accurate measure of true $\dot{V}O_2$ for a known quantity of work. From our experience, the treadmill is a valuable and appropriate tool to measure $\dot{V}O_2$ during crutch-assisted gait.

It appears from our metabolic data that the upper extremity work demands during steady-state exercise are similar for walking with the two crutch types during much of the test. The significant increase in $\dot{V}O_2$ and HR during ambulation with crutches compared with unassisted gait is consistent with the findings of many other investigators.[1-8] Our findings at the various time periods during the walking trials are similar to those of Hinton and Cullen.[2] They reported that the metabolic cost of walking with either the SA crutch or the Ortho® crutch produced different energy demands for the first 2.5 minutes of walking. They suggested that differences in $\dot{V}O_2$ between walking with the two crutch types during the initial 2.5 minutes could be accounted for by the different designs in the two crutches. They did not find differences in $\dot{V}O_2$ with either crutch, however, from minutes 2.5 to 11.5.

Although our data are similar to those reported by Hinton and Cullen,[2] we would suggest an alternative explanation for the differences in $\dot{V}O_2$ between walking with the different types of crutches. We find it difficult to support the contention that differences in crutch design exert a significant oxygen demand in the first one to two minutes of crutch walking but do not exert that same demand with prolonged gait. If a design feature does exist that would make assisted gait more efficient, then that design feature would provide that efficiency throughout the period of crutch walking. If it did not, its benefits would be negligible. The differences in $\dot{V}O_2$ and HR during the initial portion of crutch walking is likely the result of differences among subjects in the rate at which they attain a steady state and not of differences in crutch design.

Table 1. *Two-way Analysis-of-Variance Summary for Oxygen Consumption During Minutes 3 Through 5 of Treadmill Testing*

Source	df	SS	MS	F
Crutch type	2	24.15	12.08	98.34[a]
Error 1	18	2.21	0.12	
Time	2	0.76	0.38	27.04[a]
Error 2	18	0.25	0.01	
Crutch type × time	4	0.37	0.09	10.43[a]
Error 3	36	0.32	0.01	
TOTAL	89	37.36		

[a] $p < .01$.

Table 2. *Two-way Analysis-of-Variance Summary for Heart Rate During Minutes 3 Through 5 of Treadmill Testing*

Source	df	SS	MS	F
Crutch type	2	36306.82	18153.41	92.44[a]
Error 1	18	3534.73	196.37	
Time	2	556.42	278.21	12.94[a]
Error 2	18	387.13	21.51	
Crutch type × time	4	84.71	21.18	1.77
Error 3	36	431.07	11.97	
TOTAL	89	48557.66		

[a] $p < .01$.

Because of the small sample sizes of this study and of Hinton and Cullen's study,[2] additional research of the metabolic demands of different gait devices is warranted. A larger sample would minimize the influence of subject variability and strengthen any conclusions made. Additional research might also focus on the use of a patient population sample, nontreadmill walking, or other dependent variables that may also reflect efficiency of gait (eg, balance, electromyographic activity).

Summary

Within the limitations of the present investigation, we conclude that ambulation with an assistive device results in increased metabolic demand, as represented by increases in $\dot{V}O_2$ and HR. The SA crutch is metabolically more demanding than the SG crutch during the initial minute of use. This difference, however, was not found after two minutes of ambulation. We believe that this initial difference was caused by subject variability rather than differences in crutch design.

References

1 Ganguli S, Bose KS, Datta SR, et al: Biomechanical approach to the functional assessment of the use of crutches for ambulation. Ergonomics 17:365–374, 1974

2 Hinton CA, Cullen KE: Energy expenditure during ambulation with Ortho crutches and axillary crutches. Phys Ther 62:813–819, 1982

3 Cordrey LJ, Ford AB, Ferrer MT: Energy expenditure in assisted ambulation. J Chronic Dis 7:228–233, 1958

4 Bard G, Ralston HJ: Measurement of energy expenditure during ambulation, with special reference to evaluation of assistive devices. Arch Phys Med Rehabil 40:415–420, 1959

5 Fisher SV, Patterson RP: Energy cost of ambulation with crutches. Arch Phys Med Rehabil 62:250–256, 1981

6 McBeath AA, Bahrke M, Balke B: Efficiency of assisted ambulation determined by oxygen consumption measurement. J Bone Joint Surg [Am] 56:994–1000, 1974

7 Waters RL, Campbell J, Thomas L, et al: Energy costs of walking in lower-extremity plaster casts. J Bone Joint Surg [Am] 64:896–899, 1982

8 Patterson RP, Fisher SV: Cardiovascular stress of crutch walking. Arch Phys Med Rehabil 62:257–260, 1981

9 Rasmussen S: Characteristics of oxygen intake. J Sports Med Phys Fitness 15:105–111, 1975

10 Corcoran PJ, Brengelmann GL: Oxygen uptake in normal and handicapped subjects, in relation to speed of walking beside a velocity-controlled cart. Arch Phys Med Rehabil 51:78–87, 1970

Basic Research

Research Report

Energy Cost, Exercise Intensity, and Gait Efficiency of Standard Versus Rocker-Bottom Axillary Crutch Walking

The purpose of this study was to investigate differences in selected biomechanical and physiological measurements and subjective preferences for ambulation with the standard single-tip axillary crutch versus the rocker-bottom–type axillary crutch. Self-selected walking velocities (S-SWVs) and stride length for each crutch type were determined for a two-point, non-weight-bearing, swing-through gait in 24 healthy volunteers. Relative exercise intensity, oxygen uptake ($\dot{V}o_2$), and gait efficiency were assessed for each crutch type at both S-SWVs. Subjects negotiated two architectural barriers (stairs and ramp) and completed a subjective questionnaire concerning crutch preferences. Walking with either crutch type resulted in slower S-SWVs, greater $\dot{V}o_2$, higher relative exercise intensity, and reduced gait efficiency compared with values for normal unassisted ambulation. An analysis of variance for these variables revealed nonsignificant between-crutch differences. Based on the subjective data, a preference for the standard single-tip crutch was evident. Within the scope of the study, the results supported no apparent advantage relative to energy expenditure to using the rocker-bottom crutch. [Nielsen DH, Harris JM, Minton YM, Motley NS, et al. Energy cost, exercise intensity, and gait efficiency of standard versus rocker-bottom axillary crutch walking. Phys Ther. 1990;70:487–493.]

David H Nielsen
Joan M Harris
Yvonne M Minton
Nancy S Motley
Jeri L Rowley
Carolyn T Wadsworth

Key Words: *Ambulation aids, crutches; Energy expenditure; Exercise, general; Kinesiology/biomechanics, gait analysis.*

To assess physiological variation during ambulation with assistive devices (eg, canes, walkers, crutches), energy cost, heart rate (HR), relative exercise intensity, and gait efficiency have commonly been used.[1-10] Blessey et al[11] determined values for these variables during unassisted ambulation. These values provide the basis of comparison for gait studies involving the use of assistive devices. Walking speed and step- or stride-length measurements are considered good clinical biomechanical indexes of overall walking performance.[12] Research indicates that energy cost and relative exercise intensity increase in a curvilinear manner with increases in walking velocity.[13-16] Step length and stride length also usually increase systematically with increases in walking speed.[12] Research on gait efficiency of normal and pathologic gaits has shown that individuals usually self-select the most efficient walking speed.[14-17] In this context, measurement of the energy cost per meter traveled (in milliliters of oxygen per kilogram-meter) has been used as a criterion measure of gait efficiency. Accordingly, the graphic plot of energy cost per meter traveled versus walking speed usually results in a minimum value at an individual's self-

---

D Nielsen, PhD, PT, is Associate Professor, Graduate Program in Physical Therapy, College of Medicine, The University of Iowa, 2600 Steindler Bldg, Iowa City, IA 52242 (USA). Address all correspondence to Dr Nielsen.

C Wadsworth, MS, PT, is Lecturer, Graduate Program in Physical Therapy, College of Medicine, The University of Iowa.

Ms Harris, Ms Minton, Ms Motley, and Ms Rowley were students in the Master of Physical Therapy Program, College of Medicine, The University of Iowa, when this study was completed in partial fulfillment of their degree requirements.

This study was approved by the Human Subjects Review Committee, College of Medicine, The University of Iowa.

This article was submitted August 29, 1989, and was accepted March 15, 1990.

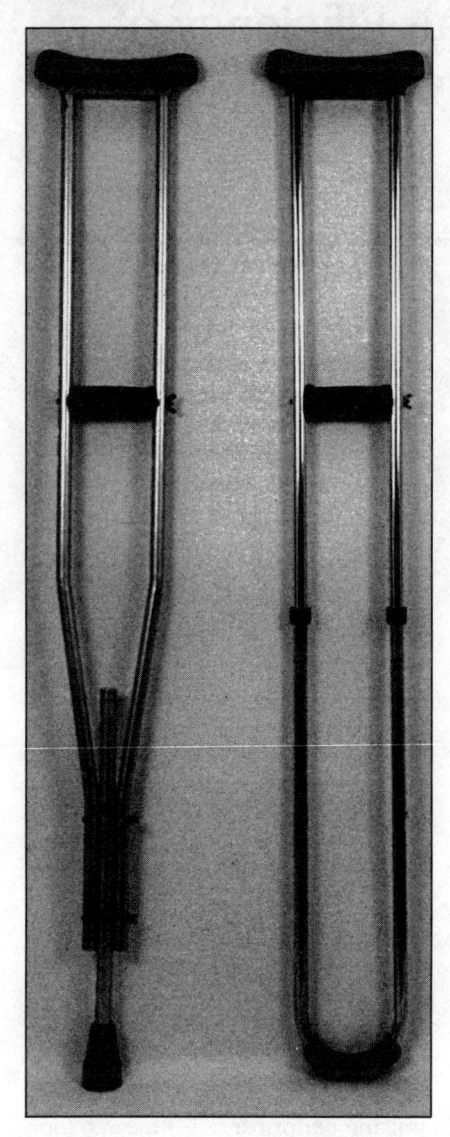

Figure 1. *Standard single-tip (left) and rocker-bottom (right) axillary crutches.*

selected walking velocity (S-SWV). From an energy-conservation point of view, the S-SWV at this minimum value would be the most efficient walking speed. Devices designed to assist ambulation should keep energy expenditure and its associated exercise intensity to a minimum while still permitting as near normal walking speeds as possible.[12,15] In this context, speed has been considered a good clinical index of general walking ability.[12]

Several types of axillary crutches are available to patients. The standard single-tip axillary crutch is currently widely used in clinics. Recently, an aluminum rocker-bottom–type axillary crutch with two uprights and a curved base has been introduced (Sure-Gait® axillary crutch*) (Fig. 1). The rocker-bottom crutch design, however, is not entirely new. In 1917, Joll[18] reported on GE Healing's invention of a roller-bottom crutch made of wood, which purportedly allowed for a faster ambulation pace compared with the ordinary axillary crutch of that time. Although crutch designs throughout history have changed in an attempt to increase comfort, safety, and ease of use,[19,20] few data have been collected that objectively indicate the actual effectiveness of these design alterations.

Limited information is available comparing the rocker-bottom–type crutch with the standard single-tip crutch. Gillespie et al[21] assessed physiological differences between a wooden rolling-bottom crutch and a standard single-tip crutch. Significant between-crutch variation and walking speed, however, confounded their experimental design, invalidating the final energy cost comparison. A recent report by Annesley et al[1] is the only published study to date that specifically compares the Sure-Gait® crutch with the standard single-tip axillary crutch. Only 10 subjects participated in that study, however, and testing was performed during ambulation on a motor-driven treadmill at one arbitrarily specified walking speed. The results indicated no significant between-crutch differences in oxygen uptake ($\dot{V}o_2$) and HR. The authors recommended additional research with a larger subject sample and overground walking before any definitive conclusions could be made.

The purpose of our study was to investigate differences in ambulation with the standard axillary crutch versus the rocker-bottom crutch (Sure-Gait® crutch). The specific objectives were to 1) test for differences in S-SWV and stride length; 2) test for differences in energy cost, gait efficiency, and relative exercise intensity; and 3) evaluate subjective preferences for crutch type in ambulation during stair climbing and ramp walking. Because of the limited amount of and somewhat conflicting information available on this topic, no experimental hypotheses were formulated.

Method

Research Design

We used a 2-×-2, two-factor (crutch type and walking speed), repeated-measures research design in this study.[22] Four testing orders were used to randomize crutch type and walking speed.

Subjects

The subjects in this study were 24 healthy female volunteers with a mean age of 23.8 years (SD = 2.2) and a mean weight of 62.5 kg (SD = 6.4). Participants were sedentary to moderately active. None were highly trained athletes. Written informed consent was obtained from each subject prior to participation in the study.

From a clinical perspective, the subjects were considered fairly representative of individuals seen for acute foot, ankle, or leg injuries requiring temporary use of crutches for ambulation. As often is the case in the clinic, none of the subjects had prior experience with crutch walking. For practical reasons, specifically the availability of subject volunteers, only female subjects were used. From a physiological point of view, we would expect the results to be comparable with those of male subjects of similar ages, except that the HR measurements could be slightly higher depending on sex differences in levels of cardiorespiratory fitness.

---

*Lumex Inc, 100 Spence St, Bay Shore, NY 11706.

Procedure

We collected data during two 1-hour test sessions scheduled at least 48 hours apart. Testing was conducted on a 60-m–long, flat, tiled hallway and an adjacent walking ramp and stairwell. Self-selected walking velocity and stride length were determined for each crutch type. Steady-state HR and $\dot{V}o_2$ measurements were obtained for each of four 5-minute walking tests identified according to crutch-type ambulation and S-SWV. Biomechanical (S-SWV and stride length) and physiological (HR and $\dot{V}o_2$) measurements were not taken during ramp walking or stair climbing.

During the initial test session, all necessary paperwork was completed and the subjects were fitted for crutches. After a demonstration and adequate time to practice walking with each crutch type, S-SWV and stride length were determined for one crutch type. These measurements were followed by HR and $\dot{V}o_2$ measurements during ambulation with both crutch types at the S-SWV of the first crutch type. Ten-minute rest periods were provided between each test trial. The second test session consisted of determining S-SWV and stride length of the second crutch type, followed by HR and $\dot{V}o_2$ measurements for both crutches at the second determined S-SWV. At the conclusion of this test session, the subjects ambulated with each crutch type up and down stairs and a ramp. They then completed a subjective questionnaire concerning crutch preferences.

Crutch fitting was performed according to generally accepted guidelines.[23] Care was taken to standardize crutch height and degree of elbow flexion between the two crutch types to ensure biomechanical equality. Both the standard single-tip and rocker-bottom crutches weighed 1.98 kg per pair.

Each subject was instructed in the performance of a two-point, non-weight-bearing, swing-through gait pattern. This gait pattern was selected for standardization purposes because it eliminated the need to control a partial weight-bearing gait. Further standardization was attained by having the subjects hold up (via partial knee flexion) their dominant leg during ambulation. Limb dominance was determined by the leg the subject preferred to use when kicking.

The S-SWV tests were conducted based on a previously established protocol.[17] The subjects were instructed to walk at a comfortable speed that they could maintain for at least 5 minutes. The hallway in which they walked was marked off into four consecutive 10-m segments. During the fifth minute of ambulation, the investigators used stopwatches to measure the subject's walking time for each of the four segments and counted the strides taken over the 40-m walkway. Self-selected walking velocity and stride length were calculated according to the following equations:

$$S\text{-}SWV = 10 \text{ m/mean time for four 10-m walking segments} \quad (1)$$

$$\text{Stride length} = 40 \text{ m/total number of strides} \quad (2)$$

Various techniques have been used to determine S-SWV. Multiple timed trials (ie, three-five repeated time measurements for back-and-forth walking over a calibrated distance) is probably the most common approach.[24] However, we have found this technique to be less reliable than a 5-minute steady-state test protocol.[17] Test-retest analysis of the data collected on 17 cerebral-palsied children for the 5-minute test protocol produced a nonsignificant 2.4 sec/min between-day difference in S-SWV and a significant Pearson product-moment correlation coefficient of .84. Based on these findings, we adopted the 5-minute steady-state method as our standard protocol for this as well as other gait studies.[15,25]

Walking velocity during HR and $\dot{V}o_2$ tests was controlled with an instrumented speedometer cane,[24] which one of the investigators used while walking in front of the subject. Heart rate was measured with a hand-held, digital pulse-rate monitor[†] immediately following each walking test. Oxygen uptake was determined through indirect calorimetry by the open-circuit method. Timed samples (40-60 L) of the subject's expired air were collected in plastic Douglas bags during the last minute of each walking test (Fig. 2). Subsequent air volume measurements and gas analyses were performed on a semiautomated on-line computer system. The system consisted of a 120-L–capacity Collins gasometer and mixing chamber[‡] and Beckman O_2 and CO_2 electronic gas analyzers,[§] all connected on-line to a preprogrammed XT-IBM[||]–compatible laboratory computer.

Oxygen uptake was used as the criterion measure of the energy cost of walking. Gait efficiency was calculated from the ratio of $\dot{V}o_2$ and S-SWV (in meters per minute):

$$\text{Gait efficiency} = \dot{V}o_2/S\text{-}SWC = \text{mL } O_2/kg \cdot m^{16} \quad (3)$$

Percentage of age-predicted maximum heart rate (%APMHR) was used as an index of the relative exercise intensity of walking:

$$\%APMHR = (\text{exercise HR}/220 - \text{age}) \times 100^{26} \quad (4)$$

The reliability of our $\dot{V}o_2$ and HR measurements has been established previously. In an earlier study,[27] we investigated the changes in $\dot{V}o_2$ and HR under standardized walking conditions (treadmill and continuous overground walking) for a functional range of walking velocities from 26.8

[†] 1-2-3 Heart Rate Monitor, Heart Rate Inc, 3188-E Airway Ave, Costa Mesa, CA 92626.

[‡] International Medical Equipment Co, 11950 Riverwood Dr, Burnsville, MN 55378.

[§] Beckman Instruments Inc, 3900 River Rd, Schiller Park, IL 60176.

[||] International Business Machines Corp, Old Orchard Rd, Armonk, NY 10504.

m/min (1.0 mph) to 107.3 m/min (4.0 mph) with 13.4-m/min (0.5 mph) increments. Between-day measurement analysis showed no significant differences in $\dot{V}O_2$ or HR. The range in mean differences was 0.03 to 0.43 mL $O_2 \cdot kg^{-1} \cdot min^{-1}$ for $\dot{V}O_2$ and 0 to 3 bpm for HR. The standardized walking conditions enhanced the measurement precision but also reduced the total measurement variability, resulting in small within- and between-subject variances. Between-test correlation coefficients were subsequently not computed.

Data Analysis

The statistical analysis was performed with the general linear model of the Statistical Analysis System (SAS) library program at the Weeg Computer Center of The University of Iowa (Iowa City, Iowa). The .05 level was adopted as the level of significance for this study.

Descriptive statistics (means, standard deviations, and within-subject standard errors) were calculated on all variables. Student's paired t tests were used to test for differences in S-SWV and stride length. Oxygen uptake, gait efficiency, and %APMHR were tested by an analysis of variance (ANOVA). A frequency analysis and a Wilcoxon matched-pairs signed-rank test were performed on the questionnaire data.[28]

Results

Figure 3 shows the mean S-SWV and stride-length values. Compared with unassisted ambulation,[11] the S-SWVs for crutch walking were appreciably slower. Visual inspection suggested negligible between-crutch differences, which were verified by nonsignificant t-test values ($P > .05$). As shown in Figure 4, $\dot{V}O_2$ and %APMHR were elevated compared with unassisted ambulation.[11] The ANOVA revealed nonsignificant F-test results ($P > .05$) for crutch-×-speed interaction and for the main effects for between-crutch differences. Similarly, the crutch-×-speed interaction and between-crutch differences in gait efficiency were also

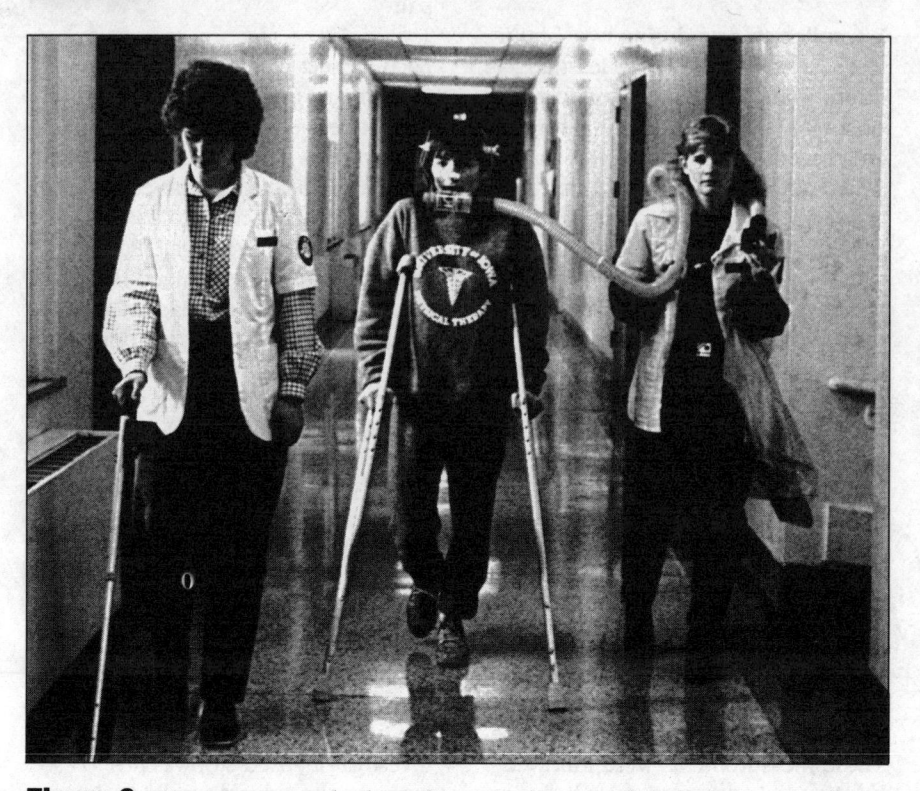

Figure 2. *Investigators and subject demonstrating procedure for oxygen uptake measurement.*

nonsignificant ($P > .05$) (Fig. 5). Analysis of the questionnaire data showed statistically significant ($P < .05$) preferences for the standard single-tip axillary crutch for stair-climbing, overall safety, and predicted long-term use. The rocker-bottom crutch was preferred for walking up a ramp. There were no preferences for walking down a ramp or for general assessment of efficiency.

Discussion

The research literature suggests that people spontaneously self-select an optimally efficient walking speed, referred to as the free-paced walking velocity or S-SWV.[15] This finding has been observed in individuals with normal gait as well as in selected patient groups with pathological gaits. Pathological gait usually results in shorter step lengths or stride lengths and slower S-SWVs, with the magnitude of the decreases being directly related to the degree of impairment. Self-selected walking velocity has sub-

sequently been considered an acceptable clinical index of general walking ability. In our study, S-SWVs for both crutches were slower than S-SWV for unassisted ambulation reported in healthy women.[11] Ambulation with either crutch type decreased S-SWV by a similar magnitude, approximately 23% as previously reported.[4,5,14] Of particular concern to our experimental questions, S-SWV for the rocker-bottom crutch was not different from S-SWV for the standard axillary crutch, indicating that neither crutch afforded any advantage in allowing faster absolute S-SWVs. Because we found no significant difference in S-SWV, we did not expect to find any appreciable between-crutch differences regarding the other variables investigated in this study.

Energy cost for both types of crutches was 60% greater than that reported for S-SWV during unassisted ambulation.[11] Our subjects' $\dot{V}O_2$ values are similar to those reported by Dounis et al[4] and by Fisher and Patterson[5]

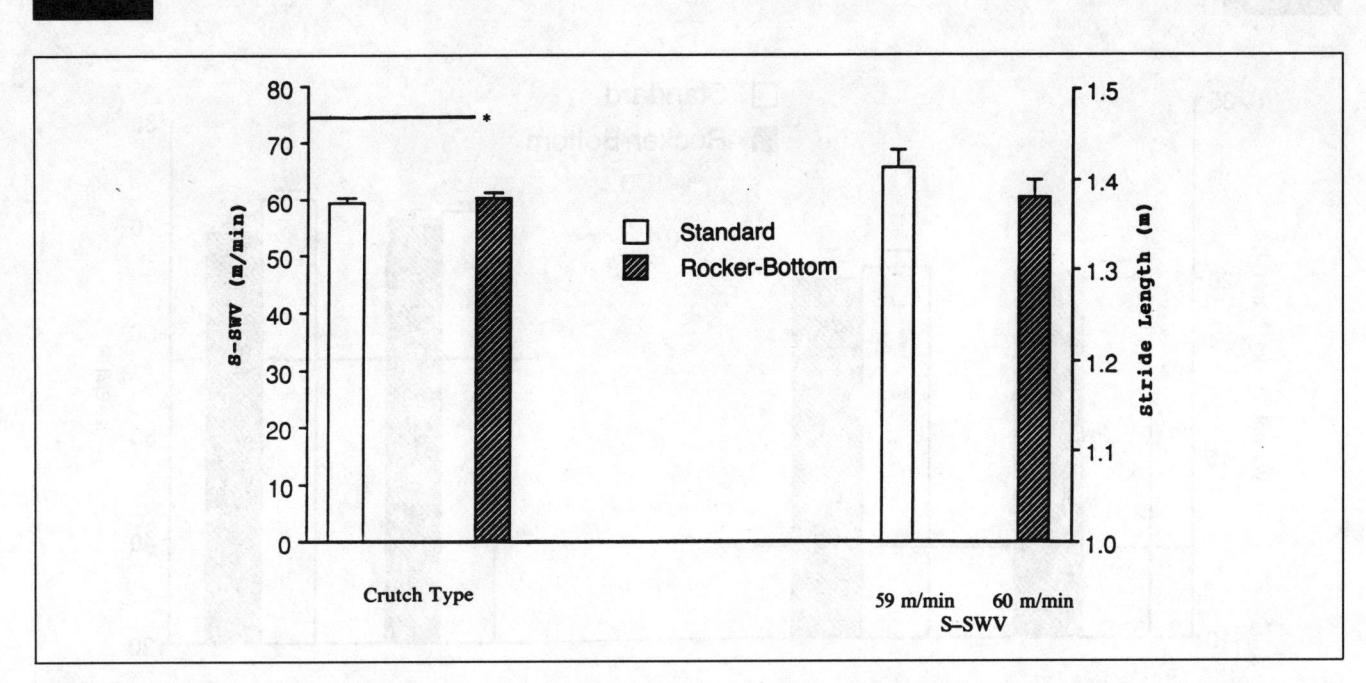

Figure 3. Means and standard errors for self-selected walking velocity (S-SWV) and stride length. Between-crutch differences were nonsignificant (P > .05). (Asterisk indicates mean S-SWV for healthy adult female subjects during unassisted ambulation.[11])

for the standard axillary crutch using a two-point, non-weight-bearing, swing-through gait. Direct comparison of our subjects' $\dot{V}O_2$ values with those reported by Annesley et al[1] for Sure-Gait® crutch walking is difficult because of differences in gait pattern (three-point, swing-to vs swing-through), walking speed (1.5 vs 2.2 mph), and mode of walking (treadmill vs overground). From an experimental point of view, however, the results corroborate each other because no between-crutch differences were found in either study.

Because the energy cost of crutch walking was increased, our criterion measure of gait efficiency—energy cost per meter traveled—correspondingly increased. The overall mean value (0.34 mL O_2/kg·m) is similar to the value reported by Fisher and Patterson.[5] This value is approximately two times greater than for unassisted ambulation,[11] which translates into an approximate 100% reduction in general gait efficiency.

From a clinical perspective, the relative exercise intensity of ambulation with assistive devices is often a primary concern. Based on the degree of walking impairment and the specific cardiorespiratory fitness status of the patient, the exercise stress of walking may be intolerable. In this context, the %APMHR has been used as a criterion measure of relative exercise intensity. Extended walking at values greater than 85% of APMHR is usually intolerable to most patients. Heart rate values during crutch walking may be slightly elevated because of the upper extremity component of exercise. The %APMHR, however, is still an acceptable index of relative exercise intensity. Our observed range of 69% to 72% of APMHR was approximately 45% greater than normal, unassisted ambulation as documented by Blessey et al.[11] Based on the average age of 50 years for the subjects in the study of Annesley et al,[1] the computed 83% of APMHR for treadmill walking exceeded our values, but the reported 47% increase over unassisted ambulation was comparable. As expected, our values better agreed with those reported by Fisher and Patterson[5] and Pagliarulo et al[8] (ie, 70%-75%) for similar crutch-walking conditions. Our values would also appear to be within the exercise tolerance of most healthy individuals. The lack of any significant difference in HR reported by Annesley et al[1] and no difference in %APMHR in our study suggest no difference in relative exercise intensity between standard axillary and rocker-bottom crutch walking.

Studies indicate that stride length mirrors walking speed and is directly related to general gait function.[12] We found no significant difference in stride length between the two crutch types. This finding suggests little biomechanical difference in walking performance between the crutches.

Prior to the recent study of Annesley and colleagues,[1] only one study[21] had investigated similar type crutches. Contrary to the results of our study, Gillespie et al[21] found the energy cost for the rolling-bottom crutch to be significantly lower than for the standard axillary crutch. However, subjects ambulated at a significantly slower velocity with the rolling-bottom crutch as opposed to the standard crutch. Consequently, the measured $\dot{V}O_2$ values would be expected to be lower.

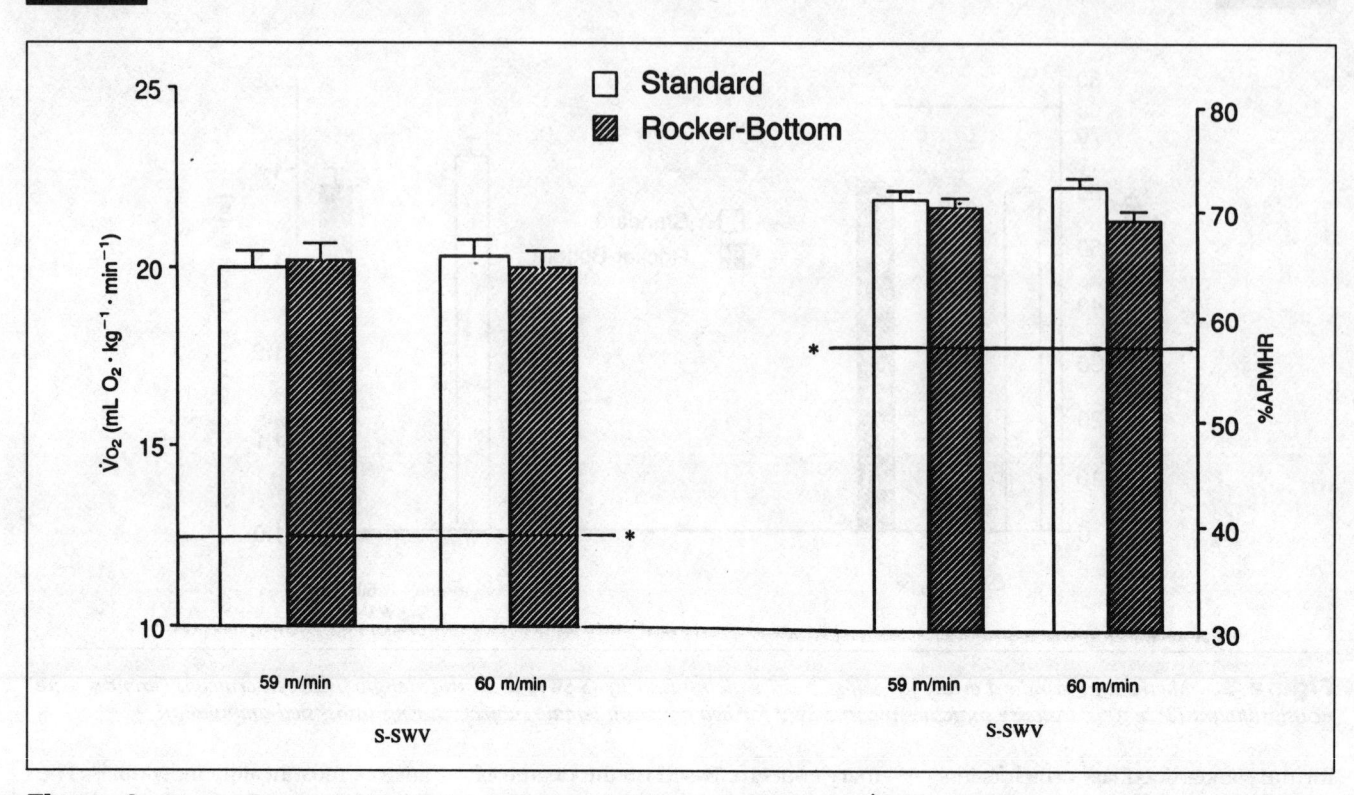

Figure 4. *Means and standard errors for energy cost of walking (oxygen uptake [$\dot{V}O_2$]) and relative exercise intensity of walking (percentage of age-predicted maximum heart rate [%APMHR]). Between-crutch differences were nonsignificant ($P > .05$). (Asterisks indicate mean $\dot{V}O_2$ and %APMHR values for healthy adult female subjects at self-selected walking velocity [S-SWV] of 74 m/min.[11])*

The manufacturer's product literature suggests that the rocker-bottom crutch is more comfortable, more convenient, and safer than conventional crutches. Our questionnaire data, however, suggest a preference for the standard single-tip crutch for stair-climbing and for safety and as the overall crutch of choice. A hypothesis is that the standard single-tip crutch may provide a sense of increased stability because of its fixed one-point contact and the cushioning and suctioning effects of the cup-shaped rubber crutch tip. Interestingly, the rocker-bottom crutch was preferred for ascending ramps. An explanation may be that the rocker-bottom crutch's curvature decreases the perception of "vaulting" uphill on the ramped surface. From a fiscal standpoint, there is no appreciable difference in cost between the two types of crutches. In our area, the retail price for the Sure-Gait® crutch and for the standard axillary crutch is $43.50 and $44, respectively.

Clinical Implications and Recommended Research

Based on the results of our study, there appears to be no apparent advantage in terms of energy consumption to using the rocker-bottom crutch over the standard single-tip crutch. Additional research considering various walking conditions, however, may be helpful. Specifically, we suggest similar investigations but testing during ambulation at different walking speeds, with different gait patterns on different grades and walking surfaces. The ability to negotiate different architectural barriers would also be of interest. The inclusion of male subjects and subjects of different ages would be advantageous.

Summary

Within the limitations of this investigation, crutch walking with a two-point, non-weight-bearing, swing-through gait pattern produced a 20% decrease in S-SWV, a 60% increase in energy cost, a 45% increase in relative exercise intensity, and a 100% reduction in gait efficiency compared with values reported for unassisted ambulation.[11] We found no difference in walking performance, however, between the standard single-tip and rocker-bottom axillary crutches. A subjective preference for the standard single-tip crutch was evident for ascending and descending stairs, overall safety, and long-term use.

References

1 Annesley AL, Almada-Norfleet M, Arnall DA, Cornwall MW. Energy expenditure of ambulation using the Sure-Gait® crutch and the standard axillary crutch. *Phys Ther.* 1990;70:18–23.

2 Bard G, Ralston HJ. Measurement of energy expenditure during ambulation, with special reference to evaluation of assistive devices. *Arch Phys Med Rehabil.* 1959;40:415–420.

3 Cordrey LJ, Ford AB, Ferrer MT. Energy expenditure in assisted ambulation. *J Chronic Dis.* 1958;7:228–233.

4 Dounis E, Rose GK, Wilson RSE, et al. A comparison of efficiency of three types of crutches using oxygen consumption. *Rheumatol Rehabil.* 1980;19:252–255.

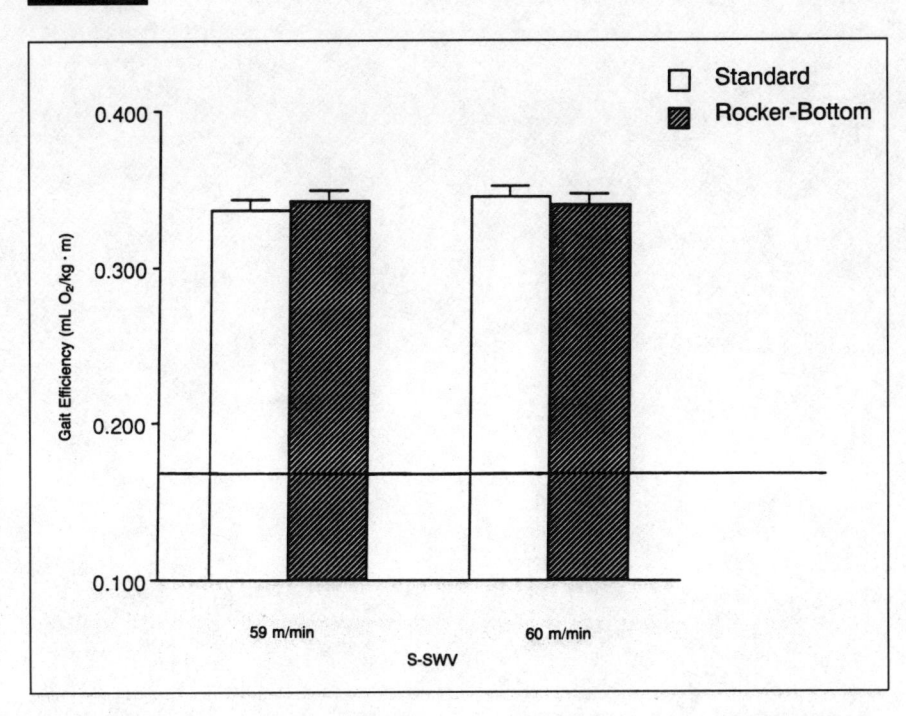

Figure 5. *Means and standard errors for gait efficiency. Between-crutch differences were nonsignificant ($P > .05$). (Asterisk indicates mean gait efficiency value for healthy adult female subjects at self-selected walking velocity [S=SWV] of 74 m/min.[11])*

5 Fisher SV, Patterson RP. Energy cost of ambulation with crutches. *Arch Phys Med Rehabil.* 1981;62:250–256.

6 Ganguli S, Bose KS, Datta SR, et al. Biomechanical approach to the functional assessment of the use of crutches for ambulation. *Ergonomics.* 1974;17:365–374.

7 Hinton CA, Cullen KE. Energy expenditure during ambulation with Ortho crutches and axillary crutches. *Phys Ther.* 1982;62:813–819.

8 Pagliarulo MA, Waters RL, Hislop HJ. Energy cost of walking of below-knee amputees having no vascular disease. *Phys Ther.* 1979;59:538–542.

9 Sankarankutty M, Stallard J, Rose GK. The relative efficiency of "swing-through" gait on axillary, elbow and Canadian crutches compared to normal walking. *J Biomed Eng.* 1979;1:55–57.

10 Stallard J, Sankarankutty M, Rose GK. A comparison of axillary, elbow and Canadian crutches. *Rheumatol Rehabil.* 1978;17:237–239.

11 Blessey RL, Hislop HJ, Waters RL, Antonelli D. Metabolic energy cost of unrestrained walking. *Phys Ther.* 1976;56:1019–1024.

12 Skinner HB, Effeney DJ. Special review: gait analysis in amputees. *Am J Phys Med.* 1985;64:82–89.

13 Corcoran PJ, Brengelmann GL. Oxygen uptake in normal and handicapped subjects in relation to speed of walking beside a velocity-controlled cart. *Arch Phys Med Rehabil.* 1970;51:78–87.

14 McBeath AA, Bahrke M, Balke B. Efficiency of assisted ambulation determined by oxygen consumption measurement. *J Bone Joint Surg Am.* 1974;56:994–1000.

15 Nielsen DH, Shurr DG, Golden JC, et al. Comparison of energy cost and gait efficiency during ambulation in below-knee amputees using different prosthetic feet: a preliminary report. *Journal of Prosthetics and Orthotics.* 1988;1:23–30.

16 Ralston HJ. Energy speed relation and optimal speed during level walking. *Int Z Angew Physiol.* 1958;17:277–283.

17 Smyntek DJ. *Efficiency and Assessment of Gait in Cerebral-Palsied Children.* Iowa City, Iowa: The University of Iowa; 1978. Master's thesis.

18 Joll CA. An improved crutch. *Lancet.* 1917;1:583.

19 Epstein S. Art, history and the crutch. *Ann Med Hist.* 1937;9:304–313.

20 Hall RG. A rolling crutch. *JAMA.* 1918;70:666–668.

21 Gillespie FC, Fisher J, Williams CS, et al. A physiological assessment of the rolling crutch. *Ergonomics.* 1983;26:341–347.

22 Lundquist EF. *Design and Analysis of Experiments in Psychology and Education.* Boston, Mass: Houghton Mifflin Co; 1953:17.

23 Minor MAD, Minor S. *Patient Care Skills.* Reston, Va: Reston Publishing Co Inc; 1984:162.

24 Nielsen DH, Gerleman DG, Amundsen LR, Hoeper DA. Clinical determination of energy cost and walking velocity via stopwatch or speedometer cane and conversion graphs. *Phys Ther.* 1982;62:591–596.

25 Lough LK, Nielsen DH. Ambulation of children with myelomeningocele: parapodium versus parapodium with ORLAU swivel modification. *Dev Med Child Neurol.* 1986;28:489–497.

26 Fox SM, Naughton JP, Haskell WL. Physical activity and the prevention of coronary heart disease. *Ann Clin Res.* 1971;3:404–432.

27 Rohrig W. *Submaximal Exercise Testing: Treadmill and Floor Walking.* Iowa City, Iowa: The University of Iowa; 1978. Master's thesis.

28 Sokal RR, Rohlf FJ. *Biometry.* San Francisco, Calif: WH Freeman & Co Publishers; 1969:400–403.

Effect of Load and Carrying Position on the Electromyographic Activity of the Gluteus Medius Muscle During Walking

DONALD A. NEUMANN
THOMAS M. COOK

> Physical therapists often teach people with hip osteoarthritis ways to decrease gluteus medius muscle activity of the stance limb during gait. The rationale for decreasing this muscle activity is that hip muscle contraction needed for frontal plane hip stabilization is responsible for a large component of the hip joint compressive forces during stance. The magnitude and carrying position of external loads during walking are both variables that influence requirements of gluteus medius muscle force. Therefore, the purpose of this study was to determine through EMG, the relative amounts of gluteus medius muscle electrical activity produced during the stance phase of gait when subjects used varying combinations of load size (10 and 20% of body weight) and carrying position of the hands (contralateral or ipsilateral to a given hip, or anterior or posterior to the chest). We studied 24 healthy subjects and used their EMG activity during the stance phase of gait as an indication of the relative amount of myogenic hip compression force. Results indicated statistical differences in EMG according to carrying condition with the contralateral position (with loads of 10 and 20% of body weight) producing the highest levels of EMG. We discuss the kinesiologic reasons for results and the prevention of hip osteoarthritis in occupational settings.
>
> **Key Words:** Electromyography, Hip, Muscles, Osteoarthritis.

Physical therapists often instruct people with unilateral osteoarthritis how to protect the hip joint by decreasing strong gluteus medius muscle contraction during the stance phase of gait. Because hip joint reaction forces are highly dependent on tensions developed by the hip musculature, decreasing the need for gluteus medius muscle contraction during gait will substantially decrease the joint reaction forces.[1,2] During the single-limb support (SLS) phase of gait, the gluteus medius muscle on the stance hip provides a force that contributes to the stability of the pelvis in the frontal plane. This muscular force, acting through a moment arm, creates an internal moment about the stance hip that neutralizes and counteracts an opposing external moment created by the person's body weight.

For simplicity, if a person ignores forces resulting from frontal plane angular accelerations of the body about the stance hip during gait, then a static model of equilibrium can be used to approximate forces needed for frontal plane equilibrium (Fig. 1). The gluteus medius muscle is obligated to generate large forces during the SLS phase of gait because this muscle must act through a relatively short internal moment arm. Studies show that during the stance phase of gait, joint reaction forces at the hip approximate 3.5 to over 6 times body weight.[2,3] Superimposed body weight over the single-support limb contributes to these joint forces, but the body weight component is small compared with the compression forces caused by the gluteus medius and other hip musculature. Bipedal locomotion necessitates a strong contraction of the hip musculature for frontal plane stability during SLS; therefore, human hip joints repeatedly receive high compressive forces that may predispose hyaline cartilage deterioration.

People who bear weight on a painful arthritic hip often lean to the side of the antalgic hip during weight bearing to pass the body weight through the stance hip to minimize the opposing internal moment produced from the overlying gluteus medius muscle. For similar reasons, patients are instructed that they should carry any load during walking on the side of the arthritic hip. Carrying a load in the hand contralateral to the affected hip theoretically increases the amount of gluteus medius muscle contraction needed to stabilize the affected hip during SLS.[4] The contralateral position of the external load increases the moment arm of the load from the diseased hip, which places increased demand on the gluteus medius muscle to provide greater force to effect a balancing internal moment.

The primary purpose of this study was to test whether the EMG activity of the gluteus medius muscle during the stance phase of gait increases with carrying a load in the hand contralateral to the muscle compared with carrying a load in the hand ipsilateral to the muscle. We expected the contralateral carrying positions would result in higher EMG from the gluteus medius muscle during the stance phase of gait than would the ipsilateral carrying positions. The secondary purpose of this study was to evaluate the gluteus medius muscle

---

Mr. Neumann is a physical therapist and a graduate student, University of Iowa, Physical Therapy Education, Iowa City, IA 52242 (USA).

Mr. Cook is a physical therapist and Lecturer, Physical Therapy Education, College of Medicine, University of Iowa.

This article was submitted February 3, 1984; was with the authors for revision six weeks; and was accepted June 22, 1984.

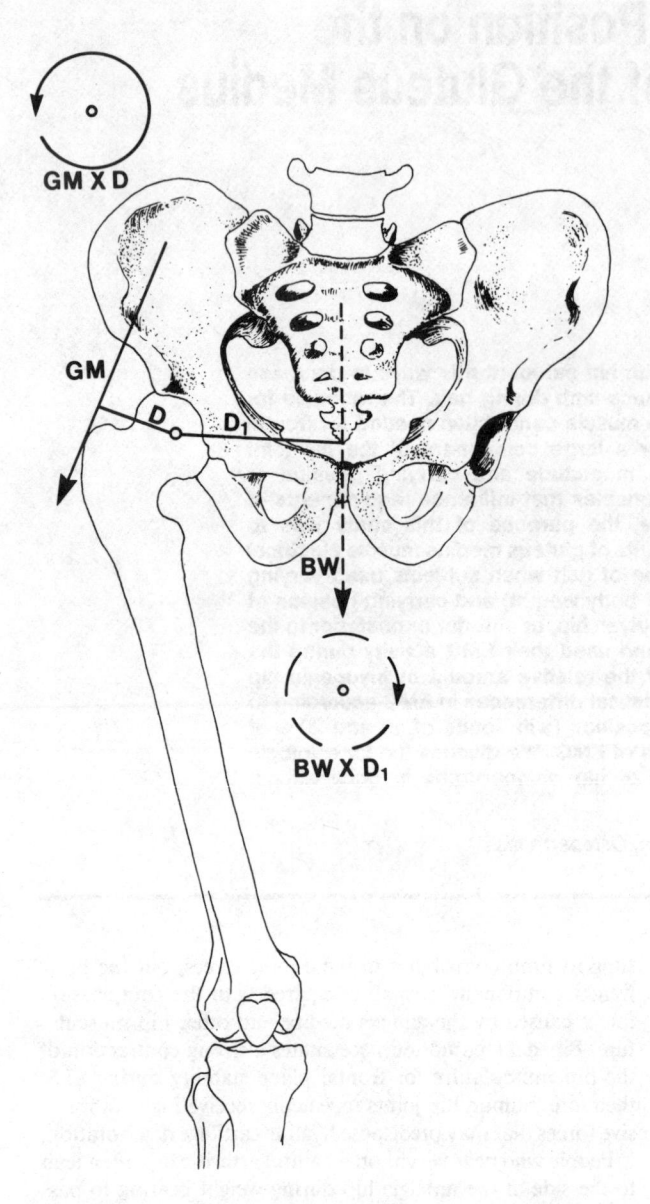

Fig. 1. Diagram representing the moments that are present when static rotatory equilibrium about the right stance hip is assumed during walking. The gluteus medius (GM) muscle force times its moment arm (D) produces an internal moment (counterclockwise moment; solid arrows) that neutralizes the external moment (clockwise moment; broken arrows) produced by body weight (BW) times its moment arm (D_1).

EMG activity during the stance phase of gait when loads were carried in a position anterior and posterior to the thorax. These carrying positions are very common alternatives to one-hand side carrying and, therefore, warrant study. We hypothesized we would find no differences in the EMG from the gluteus medius muscle for these anterior and posterior carrying positions.

We used the levels of gluteus medius muscle EMG that resulted from the various carrying conditions as a qualitative index for comparing the presumed myogenic joint reaction forces that accompany the muscle contraction. To date, controversy exists regarding the mathematical relationship between EMG and isometric tension. Some studies show a linear relationship, but others report a curvilinear relationship.[5-11] Both relationships, however, show a monotonic relationship (ie, as tension increases, EMG increases). Therefore, we believe using EMG as a tool was justifiable to indicate relative changes in tensions within the same individual. No attempts were made to quantify joint reaction forces based on EMG; only increases or decreases were addressed.

METHOD

Healthy subjects walked while carrying loads of various weights in four different carrying positions. We recorded surface EMGs of the gluteus medius muscle during the stance phase of each gait cycle. The average EMG amplitude (expressed as a percentage of maximum voluntary isometric contraction) during the stance phase of gait served as the dependent measure. The two independent variables were 1) carrying position (four positions: contralateral, ipsilateral, anterior, and posterior) and 2) load (two levels: 10 and 20% BW). Subjects also performed a no-load walk (with no external load) that served as a baseline measure of the influence of external load on the activity of the muscle.

Subjects

Twenty-four healthy subjects (12 men and 12 women) volunteered to participate in the study. All subjects signed a consent form before testing. The subjects were all considered naïve because we did not fully explain the specific nature and purpose of the study. The women had a mean age of 25.1 years with a range from 21 to 35 years. The men had a mean age of 26.8 years with a range from 22 to 36 years. Both sexes combined had a mean age of 25.6 years, weight of 68.7 kg, and height of 1.73 m. We screened all subjects for flatfeet, structural scoliosis, leg-length discrepancy greater than 1.3 cm (0.5 in), postural asymmetry, and chronic low back pain. All but one subject reported right-hand dominance.

Instrumentation

Electromyographic instrumentation included one surface electrode assembly, a small amplifier to wear at the waist and signal conditioner unit* (with attached battery power source), and a common ground electrode. The electrode assembly, measuring 33 × 17 × 10 mm contained circuitry appropriate for on-site preamplification with a gain of 35. Each electrode assembly contained two silver-silver chloride electrodes that were each 8 mm in diameter and separated by a fixed distance of 20 mm. The combined amplification of the amplifier and preamplifier was 100 to 10,000 with a bandwidth of 7 Hz to 6 kHz. The signal conditioner component of the amplifier produced a linear envelope signal by providing full-wave rectification and low-pass filtering with a low frequency cutoff of 8 Hz. Input impedance was less than 15 pF in parallel with 2 mΩ.

We used an electrical strip type of footswitch (two per shoe) to determine the stance phase of gait, and the footswitch served as a trigger for sampling EMG data by a laboratory minicomputer.[12,13] The EMG, footswitch, and ground reference signals were conducted by means of a 21.3-m umbilical

---

* Rehabilitation Engineering Center, Moss Rehabilitation Hospital, 12th St and Tabor Rd, Philadephia, PA 19141.

cable to an on-line Interdata 7/16 digital computer† that sampled the data at 100 samples a second. Both the EMG signal and the footswitch signal were monitored on a dual channel oscilloscope. We used a separate amplifier channel to evaluate the raw EMG signal for artifact.

Procedure

We applied an EMG electrode assembly over the subject's gluteus medius muscle and randomly chose the side of application (right or left). The side with the electrode and footswitch placement was considered the test hip and remained constant throughout the entire experiment.

Application of the electrode assembly was preceded by thorough cleaning of the skin with alcohol. Holes in double-sided adhesive tape coincided with the two electrodes on the assembly and were filled with conductive gel. High input impedance afforded by the amplifier and the on-site preamplification made skin resistance measurement unnecessary. We applied the electrode assembly perpendicular to the muscle-fiber direction at a point that equally divided a vertical line on a frontal plane connecting the apex of the greater trochanter and lateral iliac crest. Electrical activity resulting from active, frontal-plane hip abduction verified that our electrode placement was over the gluteus medius muscle. The common ground electrode was placed over the upper anterior-medial shaft of the tibia contralateral to the test hip. A heel and forefoot switch were taped to the sole of each subject's shoe. We placed both heel and forefoot switches on similar oblique angles roughly corresponding to the lateral-to-medial path of dynamic forefoot weight bearing.

All EMG data were normalized to a percentage of the individual's maximum voluntary isometric contraction (MVIC). The EMG activity associated with the MVIC was determined with the subject in a side-lying position and the hip abducted to 30 degrees while manual resistance was applied over the distal-lateral thigh. After two submaximal trial contractions, subjects performed three 4-second MVICs with a 60-second rest period between each contraction. The computer was manually triggered to collect EMG data during the last 3 seconds of each of the three MVICs as the investigator commanded "hold, hold, hold, hold." The computer determined the average amplitude of the smoothed and rectified EMG signal for each 3-second period and the mean of the three test periods was determined to be the subject's MVIC. The maximum contractions also allowed for the setting of the gains on the amplifier so that the signals remained within the voltage range of the computer's analog-to-digital converter (with a resolution of 0.05% of full scale). Once the gain was set, it remained unchanged for that subject.

A wooden carrying box, measuring 27 × 27 × 19 cm, was constructed with a handle so that the subjects could walk and carry loads in different carrying positions. The carrying box could also be placed in a standard aluminum-framed backpack designed for carrying children. To adjust loads, weights were added to the box by opening a hinged door and sliding barbell weights over a mounted threaded pipe. The weight of the unloaded box, including backpack assembly, was 3.6 kg.

Four different carrying positions were used during walking. In the first two positions, the box was held in the hand either

† Interdata, 2 Crescent Pl, Oceanport, NJ 07757.

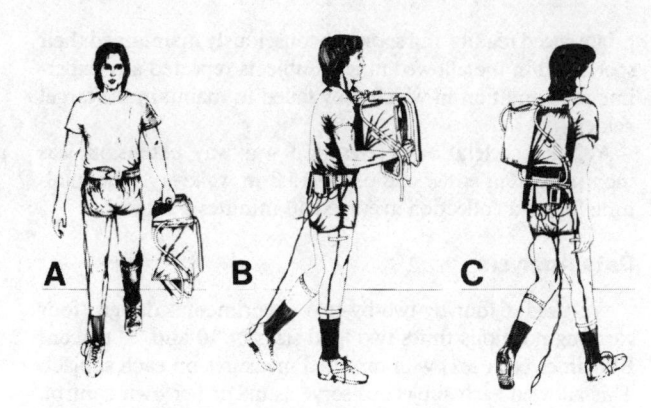

Fig. 2. Carrying positions when right hip is the test-hip: A—Contralateral carry position; B—Anterior carry position; and C—Posterior carry position.

contralateral or ipsilateral to the test hip. Figure 2A demonstrates the contralateral carrying position when the subject's right hip was used as the test hip. In the third position (anterior carry), each subject held the box anterior to the chest with both hands (Fig. 2B). In the fourth position (posterior carry), each subject carried the box on his or her back, using the padded straps provided by the backpack assembly (Fig. 2C). We adjusted the straps so that equal pressure was felt on each shoulder. For the first block of experimental conditions, the load carried in the four positions was equivalent to 10 percent of BW; a load of 20 percent of BW was carried in the second block of experimental conditions. We arranged all four positions within each block so that each of the 24 subjects used a unique sequence of carrying positions. Two control conditions were used where subjects walked without load. The first of these conditions (no-load 1) was performed before the 10 percent BW block, and the second (no-load 2) before the 20 percent BW block. We used the measurements recorded during these two no-load conditions as a baseline measure of EMG activity for comparing the experimental conditions. A baseline EMG measure could, thus, be established for both blocks of experimental conditions to minimize any confounding effects associated with fatigue. Fatigue was not anticipated, however, because the subjects were allowed to rest between carrying conditions.

Each of the 10 control and experimental conditions (two no-loads and four carrying positions for each of the two load sizes) were imposed on each subject as they made two passes on the 15.3-m walkway for a total walking distance of approximately 30 m for each condition. Pilot work helped us determine that the 30.6-m walk required approximately 24 seconds and resulted in EMG data from approximately 24 stance phases of gait. Considering all 24 subjects, data were collected from approximately 576 stance phases (24 subjects × 24 stance phases) for each experimental condition.

Before actual data collection, six practice walks were performed (two no-load and a block of 10% BW conditions) that allowed subjects to familiarize themselves with the instrumentation, ask any questions regarding procedure, and reduce anxiety. During these practice trials, subjects were encouraged to maintain a walking speed that did not deviate more than plus or minus 10 percent from 1.3 m/sec. Feedback was given to subjects after each trial on whether their walking speed was appropriate, too slow, or too fast. Subjects learned this appropriate speed readily and soon subconsciously maintained their speed within the allowed range. Subjects repeated any experimental condition in which they failed to maintain the target velocity.

A 2-m acceleration-deceleration walkway extension was incorporated at either end of the 15.3-m walkway. The total-time for data collection averaged 40 minutes a subject.

Data Analysis

We used a four-by-two-by-two experimental design (four carrying positions times two load sizes of 10 and 20 percent BW times each sex) with repeated measures on each subject. This allowed each subject to serve as his or her own control. The EMG data (expressed as %MVIC) were normalized by subtracting the no-load 1 EMG values from those obtained from the 10-percent BW load size conditions. The no-load 2 EMG values were subtracted from the 20-percent BW load size conditions following the same procedure. These differences from the no-load values (referred to as delta EMG values) were fitted to an analysis of variance (ANOVA) model.[14,15] The means of the experimental conditions were statistically compared with each other and with zero (ie, the no-load condition) by using multiple a posteriori t-test comparisons with Bonferroni adjustments. These adjustments maintain an a priori alpha level of .05 by dividing .05 by the number of comparisons made.

RESULTS

The ANOVA, summarized in Table 1, demonstrated statistical significance for the main effects of load size and position. Statistical significance was also demonstrated for the load size times position interaction. We found no main effect by sex. When considering load size, averaged across all four carrying positions, the 20-percent BW conditions produced an overall mean delta value of 5.6-percent MVIC, and the 10-percent BW conditions produced an overall mean delta value of 1.2-percent MVIC. These delta values were statistically different from one another and from the no-load values. When considering position, averaged across both load sizes, the following EMG delta values were produced: contralateral, 9.7-percent MVIC; anterior, 4.2-percent MVIC; posterior, 1.5-percent MVIC; ipsilateral, −1.8-percent MVIC (Fig. 3). All eight load size by position conditions were statistically different from the no-load position except for the posterior 10 percent BW and the ipsilateral 20-percent BW conditions. Table 2 reports the results of the EMG delta values for the two control and eight experimental conditions. Statistical comparisons for differences between all 28 possible combinations must be considered because of the significant interaction that existed between load size and position (Tab. 3).

DISCUSSION

A significant interaction existed between load size and position. Figure 4 graphically displays the plot of the EMG responses for all eight experimental conditions with reference to the no-load conditions. This figure demonstrates that from left to right on the horizontal axis (ie, from ipsilateral through contralateral), the differences in the delta EMG between the two load sizes, within positions, progressively increase. If the two lines were parallel, indicating no interaction, we could assume that doubling the load caused a similar increase in EMG across all four positions. In reality, however, the lines diverge quite strongly at the anterior and, especially, at the contralateral positions. This interaction prohibits logical discussion of the main effect of load size (without regard to position) and position (without regard to load size) on the delta EMG from the gluteus medius muscle. The effect of load size on the EMG is dependent on the position the load

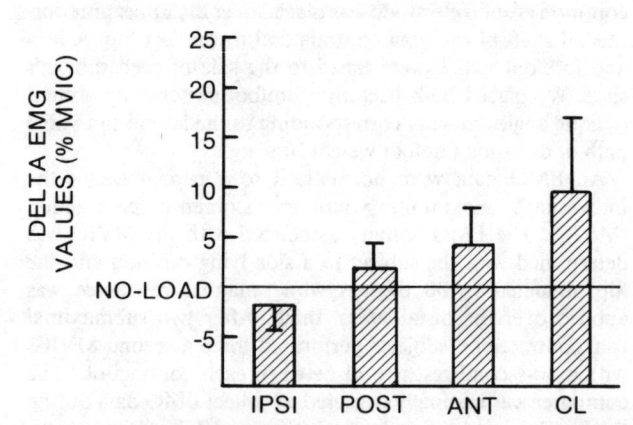

Fig. 3. Gluteus medius muscle EMG activity during the stance phase of gait for different carrying positions (averaged across both load sizes). Refer to text for the magnitude of the delta EMG values. (IPSI = ipsilateral, POST = posterior, ANT = anterior, CL = contralateral).

TABLE 1
Summary for Analysis of Variance of Delta EMG Values

Source of Variation	df	SS	MS	F
Sex	1	32.92	32.92	1.60[a]
ID (sex)	22	451.41	20.52	1.63
Load size	1	910.46	910.46	72.32[b]
Position	3	3378.10	1126.03	89.44[b]
Load size × position	3	527.07	175.69	13.96[b]
Error	161	2026.86	12.59	...

[a] Type I MS for ID (Sex) used as error term.
[b] Statistically significant ($p < .05$).

TABLE 2
Mean Delta EMG Values For All Ten Experimental Conditions[a]

Condition		Delta EMG Values (%MVIC)	
Position	Load Size (%)	$\bar{X}$	s
Contralateral	20	14.6	7.2
Anterior	20	6.5	3.9
Contralateral	10	4.8	2.7
Posterior	20	2.4	2.9
Anterior	10	2.0	1.9
Posterior	10	0.5[b]	2.5
No-load 1	...	0.0	...
No-load 2	...	0.0	...
Ipsilateral	20	−1.1[b]	3.4
Ipsilateral	10	−2.4	2.5

[a] Presented in descending order of delta EMG values.
[b] Not statistically different from no-load.

Basic Research

is carried, and the effect of the position is dependent on the magnitude of the load size.

The following sections allow an organized discussion of the results obtained from comparing the experimental conditions and the presumed effect on the myogenic hip joint reaction forces. Because we used healthy subjects in our study, we assume that the responses of the gluteus medius muscle to the experimental carrying conditions would be similar with patients with osteoarthritis of the hip.

Contralateral versus Ipsilateral Carrying Conditions

Table 3 shows the ranking of delta EMG according to condition as follows: contralateral 20 percent BW > contralateral 10 percent BW > ipsilateral 20 percent BW = ipsilateral 10 percent BW. The contralateral carrying conditions demonstrated the highest delta EMG values because in this position, the gluteus medius muscle has the responsibility of balancing an additional external moment created by the load times its long moment arm (Fig. 5A). Furthermore, doubling the magnitude of the load carried in the contralateral position from 10 to 20 percent BW tripled the EMG from the gluteus medius muscle. Both the magnitude of the load and the nature of contralateral position are significant factors in the external moment production that must be balanced by tension from the gluteus medius muscle. The contralateral position, in general, and specifically the 20-percent BW load, should be avoided by anyone interested in minimizing high myogenic joint reaction forces. In contrast to the contralateral positions, the weight of the ipsilateral conditions fell lateral to the hip axis and, therefore, produced a counterclockwise moment that assisted the muscle in counterbalancing the clockwise moment produced by the subject's BW (Fig. 5B). Therefore, results from this study suggest that a person with unilateral hip osteoarthritis should carry external loads on the *same* side as the most severe hip degeneration to protect the joint from additional joint forces. In fact, carrying a 10-percent BW load in the ipsilateral position demanded less from the gluteus medius muscle during the stance phase of gait than did walking without any external load at all. Fortunately, this carrying position frees the contralateral hand for the use of a cane, which can add additional protection to the joint.[3, 16, 17]

Note that not only were the ipsilateral 10-percent BW and ipsilateral 20-percent BW conditions statistically equal, but the ipsilateral 10-percent BW condition was significantly less than no-load and the ipsilateral 20-percent BW condition was not. The simple static biomechanical model described in Figure 5B serves as a basis for hypothesizing that the EMG delta values produced by an ipsilateral 10-percent BW condition would be less than that produced by the no-load position. The model would also predict that the ipsilateral 20-percent BW load would require even less out of the gluteus medius muscle than the ipsilateral 10-percent BW condition because a larger counterclockwise moment would result during stance. As the results showed, however, the ipsilateral 20-percent BW condition was statistically equal to (and not less than) the no-load. The simple static model does not apparently reflect the more complex real life situation. The additional counterclockwise moment produced by the ipsilateral 20-percent BW condition was evidently offset by other variables not measured or accounted for in the static model (eg, angular accelerations about the stance hip or changes in the center of gravity produced by the subject). Nevertheless, we find it interesting that a relatively large load such as 20-percent BW, if carried in the ipsilateral hand, required no more delta EMG than did the no-load carry.

Posterior Versus Anterior Carrying Conditions

Table 3 shows the following ranking of delta EMG according to condition: anterior 20 percent BW > anterior 10 percent BW = posterior 20 percent BW = posterior 10 percent BW. We hypothesize that we would find no difference in EMG between the anterior and posterior positions. Not considering the anterior 20-percent BW condition, all other comparisons

TABLE 3
Comparison Matrix of All 28 Comparisons

Column	Ipsi 10	Post 10	Ant 10	Cl 10	Ipsi 20	Post 20	Ant 20	Cl 20
Ipsi 10	...							
Post 10	O	...						
Ant 10	←	O	...					
Cl 10	←	←	O	...				
Ipsi 20	O	O	O	↑	...			
Post 20	←	O	O	O	←	...		
Ant 20	←	←	←	O	←	←	...	
Cl 20	←	←	←	←	←	←	←	...

Key:
← = column experimental condition > rows.
↑ = rows experimental condition > column.
O = no significant difference.
All differences are significant at $p \leq .002$ ($\alpha = .05$ divided by 28 comparisons).

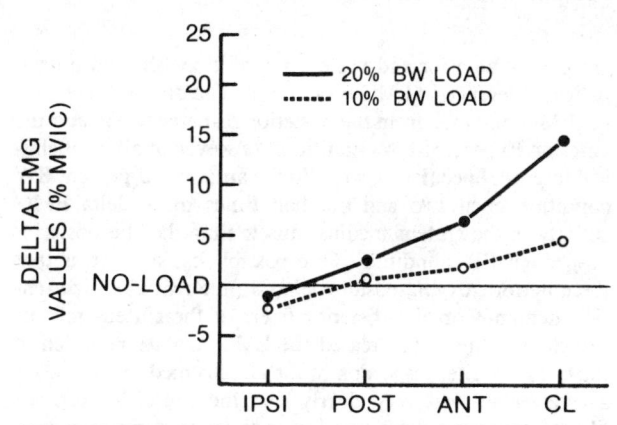

Fig. 4. Interaction plot between load size and carrying position. (Refer to Table 2 for the magnitude of the delta EMG values for each carrying condition.)

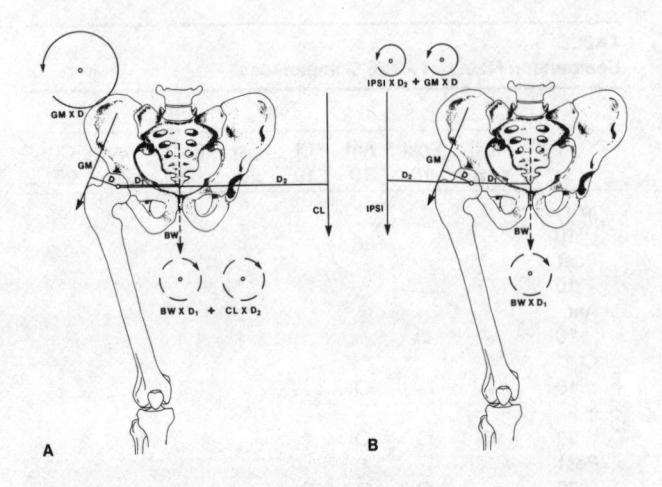

Fig. 5. Diagram representing the moments that contributed to static rotatory equilibrium about the right stance hip when a load was carried (assuming negligible frontal plane accelerations about the stance hip).
A—Contralateral carry: The gluteus medius (GM) muscle force times its moment arm (D) must produce an internal moment (counterclockwise moment; solid arrows) large enough to neutralize the combined external moments (clockwise moments; broken arrows) of BW times its moment arm (D_1) plus the contralateral load (CL) times its moment arm (D_2).
B—Ipsilateral carry: The ipsilateral load (IPSI), by passing lateral to the stance hip and thus acting through its moment arm (D_2), produced a counterclockwise moment (solid arrows) that was additive to that produced by the gluteus medius (GM) muscle. Fewer demands were therefore placed on the gluteus medius (GM) muscle to neutralize the clockwise moment produced by BW times D_1.

among the anterior and posterior conditions showed no statistical difference in EMG production. The disparity between the EMG obtained from the posterior 20-percent BW and the anterior 20-percent BW conditions, however small, is of clinical interest. Specifically, why did the anterior 20-percent BW condition elicit two and one-half times more delta EMG activity in the gluteus medius muscle than did the posterior 20-percent BW condition? One possible explanation is that the anterior carrying position placed preferentially high tension demands on the posterior fibers of the gluteus medius muscle and thereby increased the EMG activity recorded at the electrode site. Inspection of the gluteus medius muscle on a cadaver showed very clearly that the muscle's posterior fibers were in excellent position to provide a posterior moment about the mediolateral axis of the stance hip. This action of the posterior fibers of the gluteus medius muscle has been demonstrated by an earlier EMG study.[18] At least two factors associated with the anterior carrying position would have specifically caused increased posterior fiber electrical activity. First, as the load was held anterior to the chest, the subject's center of gravity shifted anteriorly. This created an exaggerated anterior moment (ie, increased tendency to flex at the trunk) at early stance by increasing the combined external load and BW's mass moment of inertia about the stance hip's mediolateral axis. The moment, created as the segment was accelerated, was equal to the mass moment of inertia of the segment times its angular acceleration. The mass moment of inertia can be thought of roughly as a quantity that takes into consideration the mass of a segment and the length of all the moment arms of all the particles within the mass. This result-

ing increased anterior moment called for additional activity in the posterior musculature, including the posterior fibers of the gluteus medius muscle. In contrast, the posterior carrying position because its weight passed posterior to the stance hip mediolateral axis decreased the tendency to flex at the trunk at early stance. Decreasing this tendency, therefore, would minimize the need for posterior muscle activity and presumed associated myogenic joint reaction forces.

A second reason for increased posterior muscular activity may have been inherent in how the anterior load was carried. Specifically, even though equal loads were used in both anterior 20-percent BW and posterior 20-percent BW conditions, the anterior load was actually 4.4 percent BW heavier because of the weight of the forwardly placed forearm-hand segment needed to carry the box.[19]

Finally, if individuals with hip osteoarthritis must carry a load in the anterior position, then the 10-percent BW load is recommended because this carrying condition produced statistically less EMG than did the 20-percent BW load.

Advantages of the Posterior Carrying Position

If patients with unilateral osteoarthritis must carry a load, then the ipsilateral carrying position is recommended. A comment, however, is in order regarding the benefit of the posterior carrying position. The posterior 20-percent BW condition produced a delta EMG value of only 2.4 percent MVIC and did not possess the undesired "trade-off" inherent to the ipsilateral carrying position. Protecting one hip at the expense of creating a contralateral carrying position for the opposite stance hip is not necessarily recommended. Therefore, for persons with bilateral osteoarthritis of the hips or for healthy individuals who need protection, the posterior carrying position serves as a logical alternative to the ipsilateral carrying position we have advocated. Furthermore, the posterior carrying position should be considered in occupational settings where loads are routinely carried during many walking cycles. Besides the relatively lower myogenic hip reaction forces that theoretically accompany the posterior carrying position, employees would have free use of their hands, facilitating safety and productivity. Work efficiency may diminish, however, because of the work required to mount a posterior load.

As Figure 4 shows, the delta EMG values obtained from the contralateral and anterior positions for the 20-percent BW load size were approximately three times greater than the same positions for the 10-percent BW load size. This difference was expected because as the external load was increased, the external moment about hip axes was also increased and must be neutralized by contraction of hip musculature. In contrast, no statistical difference existed between the ipsilateral 20-percent BW and the ipsilateral 10-percent BW conditions, or between the posterior 10-percent BW and the posterior 20-percent BW conditions. As the magnitude of the posterior load doubled from 10- to 20-percent BW, the delta EMG from the gluteus medius muscle did not significantly increase. Note also that the posterior 10-percent BW condition produced statistically the same delta EMG as did the no-load condition. Because the posterior carrying position does not theoretically predispose either stance hip to high myogenic joint reaction forces, these benefits of the posterior carrying position are enhanced.

Table 3 was designed to allow clinicians, employers, and patients to choose scientifically alternatives to carrying conditions when desiring to minimize gluteus medius muscle tension that may add to high joint reaction forces.

Limitations of the Study

Because the data in this study were collected from healthy people, only inferences may be made regarding the effect of carrying condition on the gluteus medius muscle of patients with osteoarthritis of the hip. The EMG responses obtained from healthy persons, however, can be used to educate the public about those carrying conditions that may predispose hip joints to excessive myogenic forces.

CONCLUSIONS

The following are the major conclusions drawn from our study.
1. The delta EMG values from the gluteus medius muscle during stance phase of gait could be ranked according to carrying condition as follows: contralateral 20-percent BW > contralateral 10-percent BW > ipsilateral 20-percent BW = ipsilateral 10 percent. If clinicians are interested in protecting a patient's hip from high myogenic joint reaction forces, then the ipsilateral carrying position is a favorable choice. The ipsilateral 10-percent BW condition actually "shuts-down" the gluteus medius muscle activity below that of the no-load walks. The ipsilateral 20-percent BW condition also produced no more EMG than did the no-load conditions. The ipsilateral carrying position also allows the contralateral upper extremity the use of a cane providing additional protection to the hip.

2. The delta EMG values from the gluteus medius muscle during the stance phase of gait could be ranked according to carrying condition as follows: anterior 20-percent BW > anterior 10-percent BW = posterior 20-percent BW = posterior 10-percent BW. The anterior 20-percent BW condition should be avoided because it produced high EMG activity most likely from the posterior fibers of the gluteus medius muscle. Decreasing the load of the anterior 20-percent BW condition by one half, however, statistically decreased the EMG activity by threefold.

3. Because an ipsilateral carrying position may protect an ipsilateral hip during the stance phase of gait, by necessity, the same load is carried contralaterally for the opposite stance hip. The clinician should be aware, therefore, of the preventive benefits of the posterior carrying position. Specifically, the posterior carrying position does not predispose either stance hip to high myogenic joint reaction forces that presumably accompany carrying positions with the hand. Also, a rather large load of 20-percent BW, if carried posteriorly, only produced gluteus medius muscle EMG values 2.4 percent larger than the no-load conditions. And finally, doubling the magnitude of a posterior load from 10- to 20-percent BW did not statistically increase the delta EMG from the gluteus medius muscle.

Acknowledgments. We thank Debbie Neumann, Gary Soderberg, Judy Biderman, Carl Kukulka, and Chanatip Suksang for their time and effort in helping with the preparation of this manuscript.

REFERENCES

1. Brand RA, Crowninshield RD, Johnston RC, et al: Forces on the femoral head during activities of daily living. Iowa Orthopaedic Journal 2:43–50, 1982
2. Crowninshield RD, Brand RA, Johnston RC: The effects of walking velocity and age on hip kinematics and kinetics. Clin Orthop 132:140–144, 1978
3. Brand RA, Crowninshield RD: The effect of cane on hip contact force. Clin Orthop 147:181–184, 1980
4. Bombelli R: Osteoarthritis of the Hip, ed 2. New York, NY, Springer-Verlag New York Inc, 1983, pp 372–373
5. deVries HA: Efficiency of electrical activity as a physiological measure of the functional state of muscle tissue. Am J Phys Med 47:10–22, 1968
6. Lippold OCJ: The relationship between integrated action potentials in a muscle and its isometric tension. J Physiol (Lond) 117:492–499, 1952
7. Moritani T, deVries HA: Reexamination of the relationship between the surface integrated electromyogram (IEMG) and force of isometric contraction. Am J Phys Med 57:263–276, 1978
8. Komi RV, Buskirk ER: Reproducibility of electromyographic measurements with inserted electrodes and surface electrodes. Electromyography 4:357–367, 1970
9. Kuroda E, Klissouras V, Milsum JH: Electrical and metabolic activities and fatigue in human isometric contraction. J Appl Physiol 29:358–367, 1970
10. Bigland-Ritchie B: EMG/Force relations and fatigue of human voluntary contractions. In Miller DI (ed): Exercise and Sport Science Reviews. Philadelphia, PA, The Franklin Institute Press, 1982, vol 9, pp 75–117
11. Woods JJ, Bigland-Ritchie B: Linear and non-linear surface EMG/force relationships in human muscles. Am J Phys Med 62:287–299, 1983
12. Smidt GL: Hip motion and related factors in walking. Phys Ther 51:9–21, 1971
13. Johnson RC, Smidt GL: Measurement of hip-joint motion during walking. J Bone Joint Surg [Am] 51:1083–1094, 1969
14. SAS Institute Inc: SAS User's Guide: Basics, 1982 Edition. Cary, NC, SAS Institute Inc, 1982, p 923
15. SAS Institute Inc: SAS User's Guide: Statistics, 1982 Edition. Cary, NC, SAS Institute Inc, 1982, p 584
16. Ely DD, Smidt GL: Effect of cane on variables of gait for patients with hip disorders. Phys Ther 57:507–512, 1977
17. Blount WP: Don't throw away the cane. J Bone Joint Surg [Am] 38:695–708, 1956
18. Soderberg GL, Dostal WF: Electromyographic study of three parts of the gluteus medius muscle during functional activities. Phys Ther 58:691–696, 1978
19. Dempster WT: Space Requirements of the Seated Operator. Washington, DC, US Dept of Commerce, Office of Technical Services, Wright Air Development Center Technical Report, 1955

Basic Research

Research Report

Influence of Body Weight Support on Normal Human Gait: Development of a Gait Retraining Strategy

Lois Finch
Hugues Barbeau
Bertrand Arsenault

The recovery of locomotion, following interactive training with graded weight support, in the adult spinal cat has led to the proposal that removal of body weight may be a therapeutic tool in human gait retraining. There would be benefits, however, in knowing normal responses of humans to partial weight bearing before applying this strategy to patients. In this study, 10 nondisabled male subjects walked on a treadmill while 0%, 30%, 50%, and 70% of their body weight was supported by a modified climbing harness. To dissociate the changes attributable to walking speed from those attributable to body weight, each subject walked at the specified body-weight-support (BWS) levels and at full weight bearing (FWB) at the same speed. Simultaneously, electromyographic data from the right leg muscles, footswitch signals, and video recording of joint motion were collected. The FWB and BWS gaits appeared similar, except at the highest level of BWS studied (ie, 70% of BWS). Significant differences among other BWS and FWB trials at comparable speeds included decreases in percentage of stance, percentage of total double-limb support time, and maximum hip and knee flexor swing angle. Other adaptations to BWS were a reduction in the mean burst amplitude of the muscles that are active during stance and an increase in the mean burst amplitude of the tibialis anterior muscle. The possible implications of this new gait retraining strategy for patients with neurological impairment are discussed. [Finch L, Barbeau H, Arsenault B. Influence of body weight support on normal human gait: development of a gait retraining strategy. Phys Ther. 1991;71:842–856.]

Key Words: *Body weight support, Electromyography, Gait, Kinematics.*

The locomotion of patients with neurological deficits is often impaired by poor muscular activation,[1] poor weight-bearing capacities,[2–4] and poor balance.[5] Although conventional treatment regimens usually focus on retraining of these components,[1,5] gait deviations often persist, despite improvement in muscle activation pattern[1] and patients' underlying abilities.[6]

A spinal animal model, however, demonstrates that recovery of locomotor function is possible. Rossignol and colleagues[7,8] have shown that the adult spinal cat can recover a locomotor pattern similar in many aspects to normal, with proper foot placement and weight support of the hindquarters, after a complete transection of the spinal cord. Their results are mainly due to an interactive locomotor training program that consists of appropriately graded weight support, provided by supporting the cat's tail and allowing the animal to bear only the amount of weight with which it can cope during treadmill locomotion.

L Finch, MSc(Rehab), is Neuro-coordinator, Department of Physiotherapy, Montreal Neurological Institute.

H Barbeau, PhD, PT, is Associate Professor, School of Physical and Occupational Therapy, McGill University, 3654 Drummond St, Montreal, Quebec, Canada H3G 1Y5. He is also Chercheur Boursier of the Fonds de la Recherche en Santé du Québec. Address all correspondence to Dr Barbeau.

B Arsenault, PhD, PT, is Associate Professor, Ecole de Réadaption, Faculté de Médicine, Université de Montréal, and Research Center, Montreal Rehabilitation Institute, 6300 Darlington Ave, Montreal, Quebec, Canada H3S 2J4.

This work was supported by the Medical Research Council of Canada and the Fonds de la Recherche en Santé du Québec.

This article was submitted May 14, 1991, and was accepted July 15, 1991.

Such a training approach, which allows for simultaneous retraining of different components of locomotion, is needed for patients with neurological disorders. Finch and Barbeau[9] have proposed that removal of body weight may facilitate the expression of gait patterns and therefore could be considered a therapeutic tool in gait retraining in patients who have neurological impairment. Although the effects of increased loading have been studied,[2,10-12] the effects of decreased loading on the electromyographic (EMG) characteristics of normal gait remain to be investigated. Hewes et al[13] studied the effects of replicated lunar gravity on the kinematics of walking and running. Lunar gravity was simulated by using a walkway tilted laterally to 9.5 degrees from the vertical. The subjects walked while being supported in slings. The three subjects studied walked and ran 60% slower than under normal conditions. The results showed that there was a decrease in the amplitude of the hip, knee, and ankle angular movements and an increase in the forward inclination of the body. The experimental paradigm, however, did not isolate the effects attributable to decreased loading from those attributable to slower walking speeds assumed by the subjects.

The rationale for our study, which led to the development of a new gait retraining strategy, was based on findings both from the spinal animal model and from clinical gait observations. Our rationale was that supporting a percentage of body weight during gait retraining may facilitate the expression of a more normal gait pattern. Reduction in load through the lower extremities during locomotion can be achieved by the use of a body-weight-support (BWS) system* that was developed in our laboratory.[14]

We believe that before these findings can be extrapolated to patient populations, normal gait studies on the effects of decreased loads in conjunction with walking speed changes during treadmill walking are needed. Thus, the purpose of this study was to investigate the effect of BWS on the EMG, kinematic, and temporal-distance characteristics of normal human locomotion during treadmill walking. Preliminary results demonstrated that progressive BWS facilitates the expression of a more normal gait pattern in patients with spinal cord injuries[15] and in patients with hemiplegia.[16]

Method

Preliminary Trials

Seven nondisabled volunteers participated in preliminary trials that allowed determination of the BWS levels used in this study. After walking at different BWS levels, only three of the seven subjects could walk with heel contact at 80% of BWS. Therefore, the upper BWS limit was set at 70%; 30%, and 50% of BWS were arbitrarily defined as the middle and lower limits, respectively. None of the seven subjects were able to walk at their natural, comfortable full-weight-bearing (FWB) speed at any BWS level. From these trials, a range of walking speeds for each BWS level was found: for 0% of BWS, 1.20–1.50 m·s^{-1}; for 30% of BWS, 0.90–1.00 m·s^{-1}; for 50% of BWS, 0.79–0.89 m·s^{-1}; and for 70% of BWS, 0.65–0.75 m·s^{-1}. Thus, specific walking speeds were predetermined by averaging, after 15 minutes of walking, the comfortable walking speed at each BWS level for the subjects who participated in our preliminary testing.

Experimental Protocol

Ten nondisabled male subjects, with a mean age of 31 years (SD=3.68), a mean height of 1.76 m (SD=0.04), and a mean weight of 72.7 kg (SD=7.00), walked on a Collins treadmill with 0%, 30%, 50%, and 70% of their body weight supported by a modified overhead harness. All subjects wore shorts and running shoes and were novice treadmill walkers with less than 2 hours of experience. All procedures were explained to the subjects, and each subject gave informed consent prior to experimentation. Prior to data collection, each unsupported subject was habituated to the experimental protocol by walking for 20 minutes on a motor-driven treadmill. Throughout all trials, simultaneous EMG and kinematic data from the right lower limb as well as the bilateral footswitches[†] were collected. Equipment information and specifications are presented in the Appendix.

Electromyographic and Footswitch Data

Electromyographic activity was detected by surface electrodes (Medi-Trace pellet electrodes[‡]) applied on the skin, 2 cm apart (center to center) longitudinal to the direction of muscle fibers. The electrodes were centered over the muscle belly following conventional skin preparation. Recordings were obtained for the following muscles on the subjects' right side: erector spinae (electrodes placed 2 cm lateral to the L4–5 interdiskal space), gluteus medius (4 cm posterior to the anterior superior iliac spine), vastus lateralis (10–12 cm superior to the upper edge of the patella and 5–6 cm lateral to the superior midline of the thigh), medial hamstrings (posterior one third of thigh), tibialis anterior (2 cm lateral and 4 cm below the tibial tubercle), and medial gastrocnemius (2 cm superior to the lower edge of the muscle and 5 cm medial to the midline of the calf). A surface electrode was placed medially on the right leg over the bony surface of the tibia to serve as a ground electrode. Electromyographic signals, after passing through preamplifiers[§] to eliminate movement artifacts,

---

*Industries Auteca Ltd, 3125 Bernard Pilon, St Mathieu de Balocil, Quebec, Canada J3G 4F5.

[†]Nortel Manufacturing Ltd, 2000 Ellesmere Rd, Scarborough, Ontario, Canada N1H 2W4.

[‡]Graphic Controls Canada Ltd, 215 Herbert St, PO Box 5500, Gananoque, Ontario, Canada K7G 2X1.

[§]Centre de Recherche en Sciences Neurologiques de l'Université de Montréal, Montreal, Quebec, Canada H3C 3J7.

Basic Research

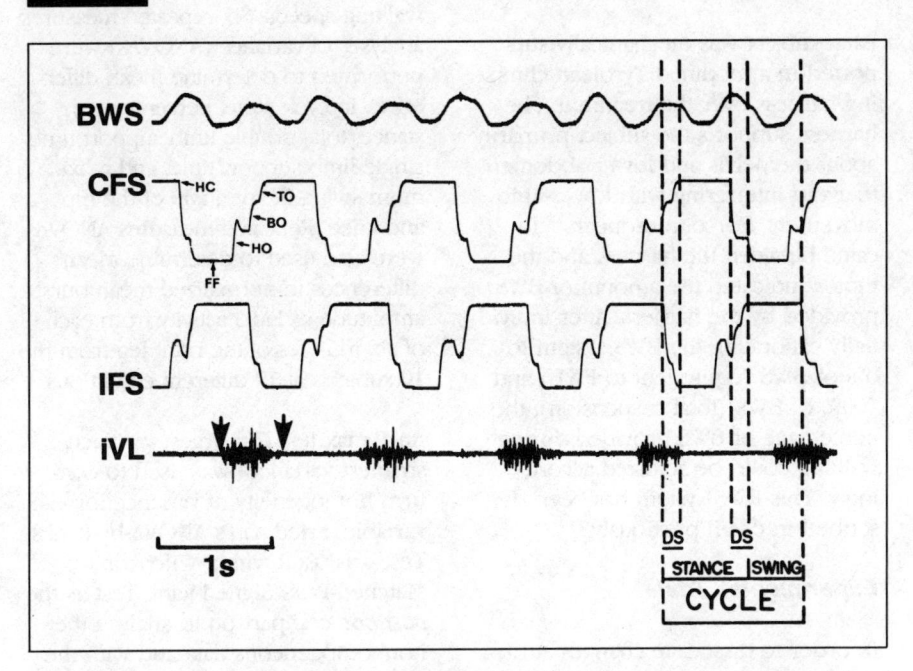

Figure 1. *Representative footswitch and electromyographic recordings of one muscle of one subject walking with 30% of body weight support (BWS) at 0.97 m·s⁻¹. Top trace is transducer readout at 30% of BWS, followed by left contralateral footswitch (CFS), right ipsilateral footswitch (IFS), and right ipsilateral vastus lateralis muscle (iVL) readouts. Arrows in the iVL trace demonstrate on and off times. (DS=total double-limb support times, HC=heel contact, FF=foot-flat, HO=heel-off, BO=ball-off, TO=toe-off.)*

were band-pass filtered (10–1,000 Hz), differentially amplified[§] (common mode rejection rate=80 dB), and recorded[||] together with footswitch signals, time code, and auditory signals on a 14-channel magnetic tape at 9.5 cm/s (3.75 in/s) (frequency response=2,500 Hz).

Footswitches were placed bilaterally beneath the heel, the head of the fifth metatarsal, and the big toe. Each footswitch produced a distinct voltage (Fig. 1), allowing for determination of temporal-distance characteristics. The time between one heel contact to the next heel contact of the same limb was considered as a 100% cycle. *Percentage of stance* was defined as the stance time divided by the cycle time. The two double-limb support times were summed to yield a total double-limb support time, which was then normalized to cycle time for subsequent analysis.

A representative sequence of artifact-free EMG signals was chosen for analysis from recorded data after having been played back on a polygraph. After passing through an anti-aliasing filter (450 Hz), the EMG signals were then digitized at 1 kHz for off-line computer analysis (PDP 11/34A[#]) using interactive programs.[17]

All data were displayed on a high-resolution terminal in 10-second sections. A minimum of 10 cycles were chosen for averaging for each BWS and FWS walking speed trial. Figure 1 represents the raw EMG activity recorded from the vastus lateralis muscle of one subject walking at 30% of BWS and at 0.97 m·s⁻¹. Our interactive computer programs allowed placement of arrows manually to define onset and termination of each EMG burst.[17] In Figure 1, the first arrow indicates the "on" time and the second arrow indicates the "off" time of the bursts of vastus lateralis muscle activity. The on-off timing for the muscle was normalized as a percentage of the gait cycle. Thus, the normalized on time of a muscle is equal to the percentage of the time from the immediately preceding right heel-strike to the start of the muscle burst divided by the cycle time. The off time was determined in a similar manner for the termination of muscle activity. Because of the wide intersubject variability, the on-off EMG timing was analyzed using descriptive statistics only.

The mean amplitude of each EMG burst, from on time to off time, was determined by the computer as the integrated area under the full-wave-rectified EMG signal divided by the burst duration. This has been a widely used method in quantifying EMG data in both animals[17] and humans.[18] The mean burst amplitude of a specific muscle, for each BWS session and for each control trial, was then normalized, within subjects, using the FWB mean burst amplitude at the speed of 1.36 m·s⁻¹ as a reference denominator:

$$\text{Normalized mean burst amplitude} = \frac{\text{Mean burst amplitude at BWS}}{\text{Mean burst amplitude at FWB} (1.36 \text{ m·s}^{-1})} \times 100\%$$

This normalized mean burst amplitude calculated for each muscle allowed for between-subject and between-trial analysis. Of the investigated muscles that have two bursts during the gait cycle, only the medial hamstring muscle burst occurring during stance and the tibialis anterior muscle burst occurring during swing were analyzed.

Joint Angle Data

To collect joint angle data, subjects were videotaped using a shutter video camera[**] placed 4 m from the center of and perpendicular to the treadmill.

[||] Honeywell, Test Instruments Div, 740 Ellesmere Rd, Scarborough, Ontario, Canada N1H 2W4.

[#] Digital Equipment Corp, 100 Herzberg Rd, PO Box 13000, Kanata, Ontario, Canada K2K 2A6.

[**] Panasonic, 8270 Mayrand St, Montreal, Quebec, Canada H4P 2C5.

Reflective joint markers were placed at the greater tuberosity of the right shoulder, the midline of the rib cage half way between the iliac crest and the shoulder, the greater trochanter, the lateral femoral epicondyle, the lateral malleolus of the fibula, 2 cm above ground in line with the heel of the shoe, the fifth metatarsophalangeal joint, and 2 cm above the sole in line with the toe of the shoe. Trials were recorded on videotape** at a speed of 60 frames per second. A remote search controller** was used for frame-by-frame viewing. The sagittal angular displacement, one cycle per subject, was manually measured from the monitor screen** at each 2% to 5% of the gait cycle, depending on cycle duration. This technique has been used in other studies,[13] and the experimental error of the angular measurements has been reported to be ±5 degrees. Because the camera was set in line with the BWS system and perpendicular to the sagittal plane of the treadmill, parallax error was minimized. The hip and knee angles were calculated with respect to the vertical, with the neutral position in standing being taken as 0 degrees of displacement of the hip, flexion being positive and extension being negative. The hip and knee angular positions attained at the critical events of heel-strike, foot-flat, mid-stance, heel-off, and toe-off and the maximum knee swing angle (MSA) were plotted on a relative time scale with the gait cycle normalized to 100%. Only the MSAs are reported.

At each BWS level, the distance from the greater trochanter to the floor was measured in a subsample of five subjects. This was done in order to determine whether the height of the greater trochanter was altered by the support system during the experiments. In addition, the distance between the toes of the left foot and the heel of the right foot was measured to record any change in "contact distance."

Body-Weight-Support System

Each subject was mechanically supported in a modified Tyrolean climbing harness over the treadmill. The harness supports the subject primarily about the pelvis and lower abdomen to avoid interfering with lower-limb movement. Force transducers†† located between the harness and the motor indicated the amount of BWS provided by the harness. After individually calibrating the BWS system to 0% of BWS (equivalent to FWB) and 100% of BWS (total suspension), the percentage of BWS provided during walking could be adjusted accordingly. This BWS system has been described in detail previously.[14]

Experimental Trials

In order to dissociate changes attributable to slower walking speeds from those attributable to the level of BWS provided, each subject participated in two experimental trials performed within a single session. During the first trial, each subject walked at four different, randomly ordered levels of BWS and at a specific speed (1.36 m·s^{-1} at 0% of BWS, 0.97 m·s^{-1} at 30% of BWS, 0.85 m·s^{-1} at 50% of BWS, and 0.70 m·s^{-1} at 70% of BWS). During the second trial, each subject, acting as his own control, walked FWB at the four previously assigned speeds. Therefore, the interaction of BWS with walking speed was determined by comparing results across the BWS trials. The effect of walking speed was determined by comparing results across the FWB control trials. The effect of BWS was determined by comparing each BWS trial with the control FWB trial at the specific speed. To minimize fatigue, a 10-minute rest period was provided after each walking trial.

Data Analysis

Different analyses were performed using kinematic, EMG amplitude, and footswitch data from the 10 subjects walking at four BWS levels and four walking speeds. Six repeated-measures analyses of variance (ANOVAs) were performed to determine mean differences in cycle time, percentage of stance, total double-limb support time, single-limb support time, and maximum swing flexor angle of the hip and knee. Repeated-measures ANOVAs were also used to determine mean differences in normalized mean burst amplitude of EMG activity from each of six muscles of the right leg from the 10 subjects at 17 different conditions.

An F max test (F=largest variance/smallest variance) was used to confirm homogeneity of variance for each variable. Friedman's ANOVA-by-Ranks Test was used with a Wilcoxon Matched-Pairs Signed-Rank Test as the *post hoc* comparison to analyze the nonhomogeneous data and with the Scheffé Multiple-Comparison Test as the *post hoc* comparison to analyze the homogeneous data.[19] All differences reported were statistically significant at .01.

Results

Temporal Data

The mean temporal components are listed in Table 1. The mean cycle time at 1.36 m·s^{-1} (the FWB control speed, which is equivalent to 0% of BWS) was 1,084 milliseconds. As determined by the Friedman's ANOVA-by-Ranks (χ^2=26.6) and Wilcoxon Signed-Rank tests, there was a statistically significant increase in cycle time with decreasing walking speed during the control FWB trials as well as a significant increase in cycle time across the BWS trials (Fig. 2A). Nevertheless, no significant difference was found between control FWB cycle time and equivalent BWS trials at a given walking speed. Hence, there was no main BWS effect on cycle time beyond that attributable to walking speed.

There was a trend toward increased percentage of stance with decreasing walking speed at FWB. This increase, however, was not significant across walking speeds at FWB. The percent-

---

††Intertechnology Inc, 3675 Boul de Sources, DDO, Quebec, Canada H9B 2T6.

Basic Research

Table 1. Temporal Results for Body-Weight-Support (BWS) and Full-Weight-Bearing (FWB) Trials (N=10)[a]

	Trials[b]						
	BWS (%)				FWB		
	0 (1.36 m·s^{-1})	30 (0.97 m·s^{-1})	50 (0.85 m·s^{-1})	70 (0.70 m·s^{-1})	0.97 m·s^{-1}	0.85 m·s^{-1}	0.70 m·s^{-1}
Cycle time (ms)							
$\bar{X}$	1,084.5	1,238.3	1,396.1	1,680.4	1,268.9	1,361.6	1,527.9
SD	62.1	97.6	163.8	249.5	40.2	82.7	126.4
CV (%)	5.7	7.8	11.7	14.8	3.1	6.0	8.2
Stance (%)							
$\bar{X}$	59.9	56.9	55.6	51.7	62.3	63.1	63.5
SD	2.9	2.3	2.2	5.1	2.6	2.4	2.5
CV (%)	4.8	4.0	3.9	9.8	4.1	3.8	3.9
TDST (%)							
$\bar{X}$	21.7	17.1	13.4	8.6	27.4	27.7	29.0
SD	4.4	4.4	4.2	5.7	3.8	3.0	3.4
CV (%)	20.2	25.7	31.3	66.2	13.8	10.8	11.7
SLST (%)							
$\bar{X}$	38.2	39.8	42.2	43.1	34.9	35.4	34.5
SD	7.3	6.7	6.4	10.8	6.4	5.4	5.9
CV (%)	19.7	16.8	15.1	25.6	18.3	15.2	17.1

[a]Stance, total double-limb support time (TDST), and single-limb support time (SLST) were recorded as percentage of cycle time. (Full weight bearing equivalent to 0% of BWS.)

[b]Walking speed at each level of BWS trials shown in parentheses. During control FWB trials, subjects walked at same speeds as during BWS trials at 30%, 50%, and 70% of BWS.

age of stance decreased significantly (repeated-measures ANOVA, $F=32.56$) across BWS trials. The percentage of stance values with BWS were all significantly decreased from the FWB control values at a given walking speed (from 60% at 0% of BWS to 52% at 70% of BWS). Thus, the decrease in percentage of stance values was affected mainly by BWS (Fig. 2B).

The total double-limb support time revealed a similar trend to that of percentage of stance. With decreasing walking speed at FWB, there was minimal change in percentage of total double-limb support time. A marked and significant decrease (repeated-measures ANOVA, $F=42.86$) in total double-limb support time, however, was observed across BWS trials. The total double-limb support time decreased from 22% at 0% of BWS to 9% at 70% of BWS (Fig. 2C). In addition, the percentage of single-limb support time, calculated by subtracting the percentage of total double-limb support time from the percentage of stance, was unaffected by walking speed in the FWB session, but increased slightly (from 38.2% to 43.1%) across BWS trials. The standard deviations were calculated by obtaining the square root of the sum of the total double-limb support time and percentage of stance variances.

Hip and Knee Angle Displacement

Figure 3 shows the hip and knee average angular displacement curves for the BWS and FWB control trials. Curves were connected through the points to denote trends. Note the open circles represent the BWS condition. These points represent the mean joint angles of the 10 subjects for each trial. The 0% of BWS plot is included in both sets of curves as a reference line. All curves were plotted normalized to the critical gait events of the 0% of BWS level, with the cycle normalized to 100%. The hip and knee joints demonstrated a similar pattern across all trials with the exception of amplitude of movement.

The angular displacements of the hip and knee at each critical event across the FWB control trials appeared to be similar, with minimal walking speed effect (top half of Figs. 3A and 3B). The MSA values of the hip and knee were not significantly different across FWB control trials (Tab. 2). With BWS, the largest differences in hip angular displacement were at heel-strike, foot-flat, and MSA (Fig. 3A), whereas the largest differences for the knee with BWS were at foot-flat, toe-off, and MSA (Fig. 3B). The ANOVAs and Scheffé comparisons (hip, $F=16.95$; knee, $F=24.66$) revealed that, for both the hip and the knee, the MSA of all BWS trials differed significantly from that of the 0% of BWS condition at 1.36 m·s^{-1}, and the MSA of the 70% of

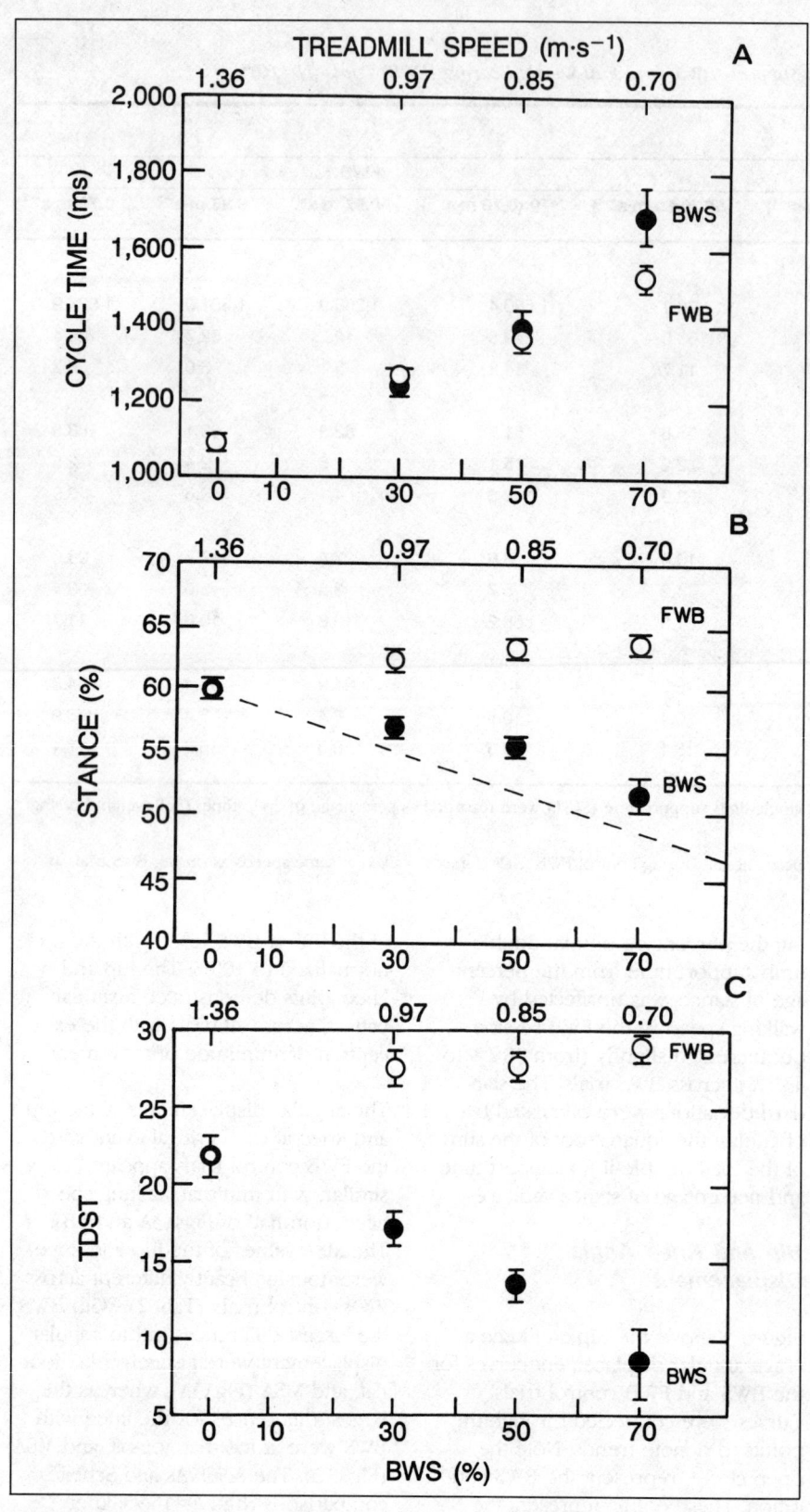

Figure 2. *Mean cycle time (A), percentage of stance (B), and total double-limb support time (TDST) (C) with one standard error of the mean for 10 nondisabled male subjects. Dashed line represents true percentage of stance attributable to body weight support (BWS). Numbers on upper x-axis indicate treadmill speed for both the control full-weight-bearing (FWB) and BWS conditions; numbers on lower x-axis represent percentage of BWS removed for BWS condition only. Control FWB is represented by open circles; BWS is represented by filled circles.*

BWS condition was significantly less than that of the FWB condition at any control walking speed (Tab. 2).

The measurement from the greater trochanter to the floor revealed an average increase of 1.5 cm at 70% of BWS across subjects. The contact distance decreased sequentially by 15% at each BWS level from 73.5 cm at 0% of BWS to 22.5 cm at 70% of BWS.

Electromyographic Timing

Results of EMG on-off timing for the muscles investigated are illustrated in Figure 4. In the control FWB trials, walking speed had a minimal effect on timing, except for a delay in both the vastus lateralis and gluteus medius muscles' off timing (Figs. 4B and 4C). The major BWS effects observed were an earlier onset in the first erector spinae muscle burst and delayed offset in the second erector spinae muscle burst (Fig. 4A). These effects led to an increased burst duration, similarly observed in the tibialis anterior and gastrocnemius muscles, as well as a delayed on-off time for the medial hamstring muscles when compared with the FWB controls.

Normalized Mean Burst Amplitude

The normalized mean burst amplitudes are presented in Table 3. Using Friedman's ANOVA-by-Ranks Test, it was found that the mean burst amplitudes of the erector spinae, gluteus medius, gastrocnemius, and tibialis anterior muscles (Fig. 5) were not significantly affected by decreasing walking speed during the FWB control trials. It is noteworthy that, with decreasing walking speed, the amplitudes of the first and second erector spinae muscle bursts were similar, except at the lowest walking speed. Moreover, there was a trend toward a decrease in the amplitude of the medial hamstring ($F=5.06$) (Fig. 5E) and vastus lateralis muscles with decreasing walking speed.

Neither the first nor the second erector spinae muscle burst amplitude was significantly affected by BWS, ex-

Basic Research

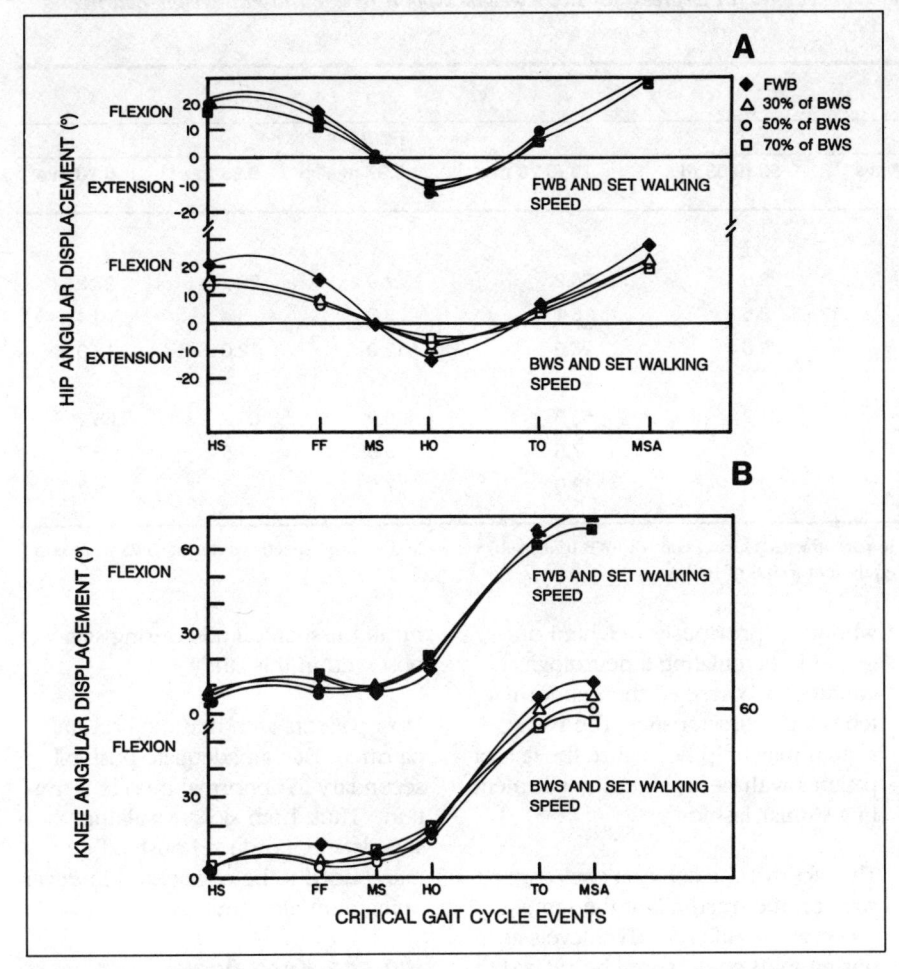

Figure 3. Mean hip (A) and knee (B) angle displacement (in degrees) for 10 nondisabled male subjects as a function of gait events. The full-weight-bearing (FWB) condition, symbolized by filled diamonds at 1.36-$m·s^{-1}$ walking speed, is included in all sets of curves as a reference line. All curves are plotted normalized to the critical gait events of the 0% of body weight support (BWS) trial with a cycle normalized to 100%. Heelstrike (HS) occurred at 0% of BWS, toe-off (TO) at 60% of BWS, maximum swing angle (MSA) for the knee at 65% of BWS, and MSA for the hip at 85% of BWS. The control FWB condition is represented by the filled symbols; the BWS condition (except for 0% of BWS) is represented by open symbols. (FF=foot-flat, MS=mid-stance, HO=heel-off.)

cept for the 70% of BWS condition, when the amplitude of both bursts decreased significantly ($\chi^2=21.89$). Throughout all BWS trials, the second erector spinae muscle burst amplitude was greater than the first burst amplitude (Fig. 5A). The gluteus medius muscle mean burst amplitude decreased significantly with BWS ($F=12.68$), to such an extent that, at the 50% and 70% of BWS trials, the amplitudes were significantly smaller than those at any FWB control trials (Fig. 5C). Vastus lateralis muscle burst amplitude was not significantly affected by BWS ($F=1.18$), because of the large between-subject variability (Tab. 3). The medial hamstring muscle burst amplitude, although remaining stable at 30% and 50% of BWS, decreased to 56.1% at 70% of BWS (Fig. 5D). The medial gastrocnemius muscle burst amplitude decreased significantly ($\chi^2=23.2$) with increasing BWS (as much as 50% of its FWB value at the 70% of BWS level), whereas the tibialis anterior muscle burst amplitude increased ($\chi^2=21.2$) (Figs. 5F and 5E, respectively).

Discussion

Temporal Data

The progressive decrease found for percentage of stance and percentage of total double-limb support time across BWS trials may have resulted from a pure BWS effect or from a possible interaction between BWS levels and the set walking speed. The results of the combined intervention of increasing BWS and decreasing walking speed suggest that these body weight effects may have been underestimated. For instance, in the case of percentage of stance, assuming the effects of decreasing walking speed and increasing body weight are additive, the dashed line in Figure 2B would represent the true BWS effect without the effect of walking speed (ie, the increase in percentage of stance attributable to FWB control sessions subtracted from the BWS percentage of stance values). Because our subjects were assigned a specific walking speed per BWS level, the walking speeds may not have matched the amount of body weight support removed for each subject, resulting in a higher-than-normal variability (Tab. 1). The relationship between walking speed and BWS remains to be further investigated.

The BWS effect is evident on the percentage of total double-limb support time; however, the percentage of single-limb support time increased only from 39.8% to 43.1% (Tab. 1). Thus, it would appear that the BWS effects on the overall percentage of stance are greater for the percentage of total double-limb support time, only minimally increasing the percentage of single-limb support time.

The percentage of total double-limb support time for the FWB control sessions (ie, 27%) is not only longer than that normally reported for overground walking (ie, 20%), but is also longer than any values obtained with BWS. The longer total double-limb support time observed in the FWB control sessions during treadmill locomotion is in agreement with previously reported results and is thought

Table 2. Hip and Knee Flexor Maximum Swing Angles (in Degrees) for Body-Weight-Support (BWS) and Full-Weight-Bearing (FWB) Trials (N=10)

	Trials[a]						
	BWS (%)				FWB		
	0 (1.36 m·s⁻¹)	30 (0.97 m·s⁻¹)	50 (0.85 m·s⁻¹)	70 (0.70 m·s⁻¹)	0.97 m·s⁻¹	0.85 m·s⁻¹	0.70 m·s⁻¹
Hip flexion							
$\bar{X}$	29.9	23.8	22.5	20.2	27.9	29.3	27.2
SD	3.0	7.5	6.5	5.1	3.3	3.4	3.4
CV (%)	10.0	32.0	29.0	25.0	12.0	12.0	13.0
Knee flexion							
$\bar{X}$	72.4	65.0	61.2	55.9	70.9	69.9	66.2
SD	3.9	7.3	5.6	7.5	5.0	4.5	5.7
CV (%)	5.0	11.0	9.0	13.0	7.0	6.0	9.0

[a]Walking speed at each level of BWS trials shown in parentheses. During control FWB trials, subjects walked at same speeds as during BWS trials and 30%, 50%, and 70% of BWS. (Full weight bearing equivalent to 0% of BWS.)

to reflect a greater need for balance on a moving surface.[20] Eke-Okoro and Larsson[2] also found an increased total double-limb support time, perhaps resulting from a need to compensate for the uneven loading, and subsequently a greater balance demand. With BWS, the subjects required less balance control to walk supported on a treadmill, as reflected in the reduced total double-limb support time. Further investigation of this relationship is needed.

The decreased total double-limb support time and increased single-limb support time could have implications for balance training, as a major problem in neurological gait is the lack of balance control.[5] During the BWS trials, the subjects were forced to support their body weight (albeit less body weight) on a single limb for longer periods of time. This finding stresses the balance component of gait and thus can be incorporated during training.

By providing BWS during treadmill walking, both balance and locomotion are simultaneously being retrained, rather than separately addressed as in conventional physical therapy practice.[5] Furthermore, Eke-Okoro and Larsson[2] showed that the removal of weight from nondisabled subjects, who were previously weighted on one side (simulating a neurological condition), corrected their abnormal temporal characteristics. The BWS system may help normalize the gait of patients with neurological impairment in a similar fashion.

The fact that the subjects could not walk on the treadmill at the same speed at the different BWS levels as during FWB could partly be related to the progressively raised greater trochanter. During normal walking at a comfortable speed, the center of gravity describes a smooth vertical displacement of 5 cm, the lowest point occurring when both feet are on the ground. The harness could limit the vertical displacement of the body, which is related to the decreased contact distance. As a result, the walking speed is perceived as being faster and the maximal comfortable treadmill speed must be reduced. Hewes et al[13] also reported that their subjects, in a simulated lunar gravity situation without any change in their center-of-gravity movement, walked 60% slower than normal. Margaria and Cavagna[21] hypothesized that subjects walking under reduced gravity conditions would walk slower because of a decrease in energy expenditure at push-off. This hypothesis is supported by the decreased gastrocnemius muscle mean burst amplitude during stance observed in this study.

Most patients with neurological impairment lack an adequate push-off secondary to abnormal muscle activation.[6] Thus, both slower walking speeds and a reduced push-off demand need to be considered in developing a training strategy.

Hip and Knee Angle Displacement

Although the angular displacement pattern for the hip and knee remained similar for both FWB and BWS trials, the amplitude of movement decreased with BWS. The consistent pattern and amplitude of movement demonstrated during the FWB trials compare favorably with the results of Smidt,[22] Murray et al,[23] and Winter.[24] Contrary to our results, however, Smidt[22] observed a decrease in the total hip movement, with a reduction in walking speed from 1.34 to 0.91 m·s⁻¹ in subjects walking over ground. Murray et al[23] also reported a decrease in hip displacement at heel-off for subjects walking at a comfortable speed on a treadmill, as compared with over-ground walking.

With BWS, the decrease in the hip and knee angles at foot-flat and at

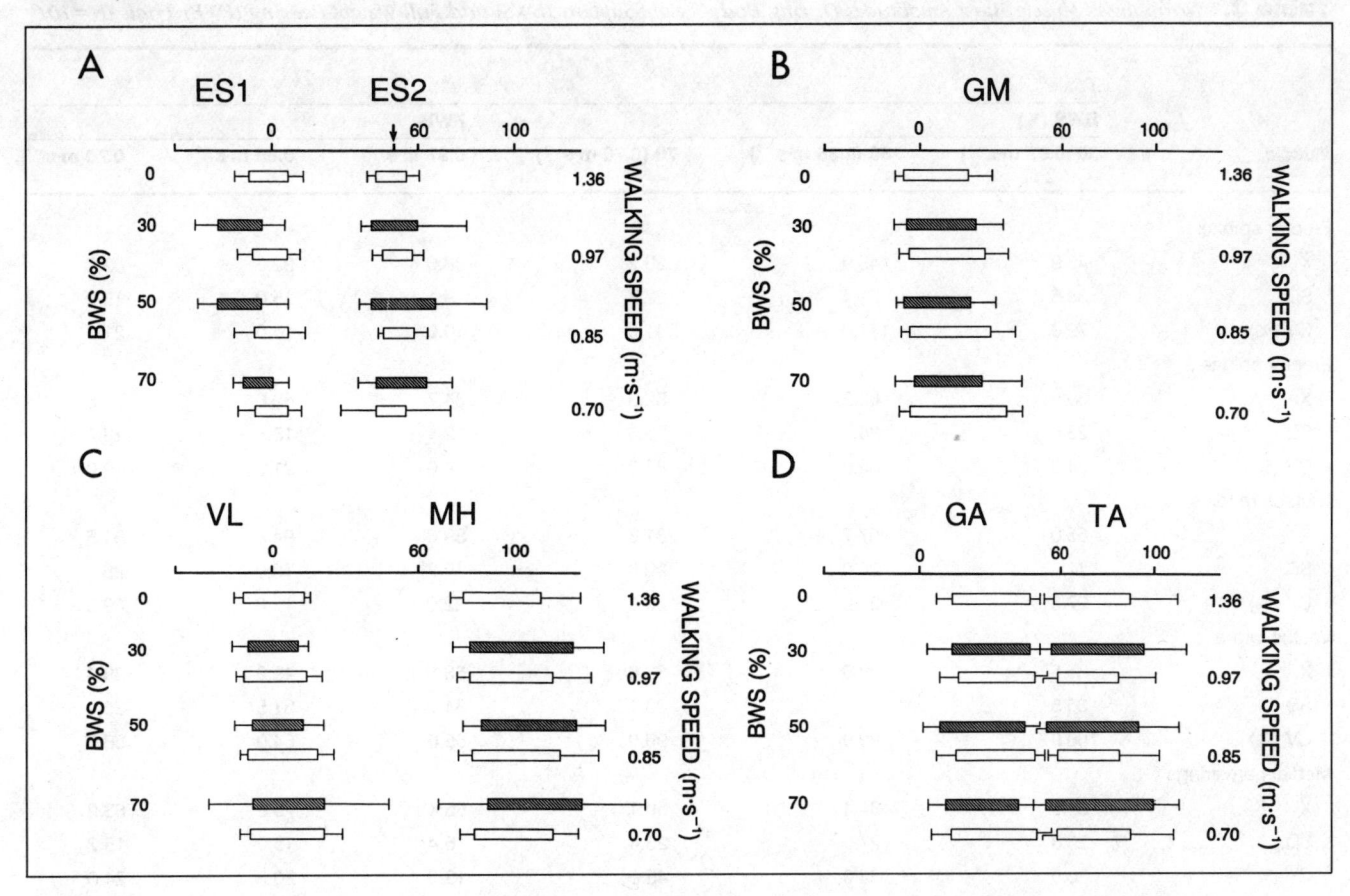

Figure 4. *Normalized "on" and "off" timing of (A) erector spinae (ES1 and ES2), (B) gluteus medius (GM), (C) vastus lateralis (VL) and medial hamstring (MH), and (D) medial gastrocnemius (GA) and tibialis anterior (TA) muscles. Timing was normalized to the gait cycle, with 0% representing right heel contact, 60% right toe-off, and 100% the subsequent right heel contact. Arrow in illustration A indicates left heel contact. Control full-weight-bearing walking speed conditions (ie, 0% of body weight support [BWS]) are represented by unhatched rectangles, and BWS walking speed conditions are represented by hatched rectangles. Walking speeds (in meters per second) are applicable to both conditions. One standard deviation for each on-off timing is also indicated.*

each respective MSA may be the result of harness constraints on the center of gravity. As previously discussed, the excursion of the center of gravity was limited, and each subject's body under BWS conditions (particularly at 50% and 70% of BWS) was prevented from moving downward at foot-flat by the harness. The resultant decrease in knee flexion at foot-flat, however, reinforces the findings of Hewes et al.[13] Their subjects showed a decrease in knee flexion at foot-flat, which the authors attributed to the decreased amount of weight supported and not to the harness constraints. The subjects in this experiment not only supported less weight, but their bodies were also prevented from descending at foot-flat. Thus, knee flexion at foot-flat was probably reduced by both factors. Another contributing factor should be considered. There may be a decrease in the transfer of kinetic energy from the reduced push-off and thus a decreased energy transfer and subsequent decrease in swing momentum and displacement. Patients with neurological impairment are often unable to control the energy transfer from the unaffected limb to affected limb.[25] These patients, however, would be able to control this transfer if it was graded to their ability using the BWS system. In addition, because most patients with acute neurological conditions lack the necessary amplitude of active joint movement, a system that progressively stimulates an increase in movement could be considered beneficial in gait retraining.

Electromyographic Timing

Despite variation, the EMG timing of flexor and extensor muscles appeared similar across walking speeds in FWB control trials. This timing demonstrated a link to the events of the gait cycle that is similar to that reported for nondisabled subjects at a comfortable walking speed[26,27] or a slower cadence.[28] The phase relationship between heel-strike and the onset of EMG activity for the subjects walking on a treadmill seems as consistent as that previously reported by Medeiros.[11]

A relationship between cycle time and lower-limb extensor muscle duration has been reported.[11,29–31] Because the cycle time across the BWS trials in our study was not significantly differ-

Table 3. Normalized Muscle Burst Amplitudes During Body-Weight-Support (BWS) and Full-Weight-Bearing (FWB) Trials (N=10)[a]

Muscle	Trials[b]					
	BWS (%)			FWB		
	30 (0.97 m·s^{-1})	50 (0.85 m·s^{-1})	70 (0.70 m·s^{-1})	0.97 m·s^{-1}	0.85 m·s^{-1}	0.70 m·s^{-1}
Erector spinae 1						
$\bar{X}$	48.9	45.9	23.2	88.6	92.3	89.2
SD	38.6	51.1	32.7	9.1	15.0	19.0
CV (%)	79.0	111.0	141.0	10.0	16.0	21.0
Erector spinae 2						
$\bar{X}$	69.5	63.2	32.3	84.7	80.6	70.6
SD	23.6	35.5	26.3	12.1	16.8	13.2
CV (%)	34.0	56.0	81.0	14.0	21.0	19.0
Gluteus medius						
$\bar{X}$	65.0	57.7	31.8	89.6	94.8	91.5
SD	17.6	20.0	20.0	19.3	18.0	26.3
CV (%)	27.0	35.0	65.0	22.0	19.0	29.0
Vastus lateralis						
$\bar{X}$	35.5	28.0	33.6	53.5	49.2	39.4
SD	35.5	27.6	33.1	34.7	31.5	26.1
CV (%)	100.0	99.0	99.0	65.0	64.0	66.0
Medial hamstrings						
$\bar{X}$	84.5	84.4	56.1	85.4	75.6	63.3
SD	25.3	27.0	25.4	16.4	15.0	15.2
CV (%)	30.0	32.0	45.0	19.0	20.0	24.0
Tibialis anterior						
$\bar{X}$	97.0	104.0	112.2	74.3	74.4	78.1
SD	19.2	35.0	45.4	12.4	14.3	21.9
CV (%)	20.0	34.0	40.0	17.0	19.0	28.0
Gastrocnemius						
$\bar{X}$	76.2	67.0	40.5	97.1	93.8	92.0
SD	16.6	16.9	14.8	4.7	8.8	8.5
CV (%)	22.0	25.0	37.0	5.0	9.0	9.0

[a]The mean burst amplitude at each BWS and FWB walking speed was normalized to 0% of BWS 1.36-m·s^{-1} mean burst amplitude before the data were pooled across subjects.

[b]Walking speed at each level of BWS trials shown in parentheses. During control FWB trials, subject walked at same speeds as during BWS trials at 30%, 50%, and 70% of BWS. (Full weight bearing equivalent to 0% of BWS.)

ent from that for the FWB control trials, it is conceivable that the EMG timing would remain similar between the two experimental conditions. The minor increases in vastus lateralis and gluteus medius muscle off timing under both conditions may be related to the longer positive total double-limb support moment required with the increased cycle time observed with slower walking speeds.

Body weight support did have some effect on the timing of activation of the back muscles studied. Thorstensson et al[32] demonstrated that small changes in trunk flexion (2°–4°) could cause a change in erector spinae muscle activity. Thus, the changes in timing of the erector spinae muscles may be related to modifications required to control trunk movement while supported in a harness over the treadmill. The erector spinae muscles appear sensitive to equilibrium changes caused by the removal of body weight or the stabilizing support of the harness. Furthermore, a relationship between swing duration and muscle timing may account for increased medial hamstring muscle burst duration.

Timing of events, especially that of muscle activation, is an important factor to consider in gait retraining. Examination of EMG activity in the mus-

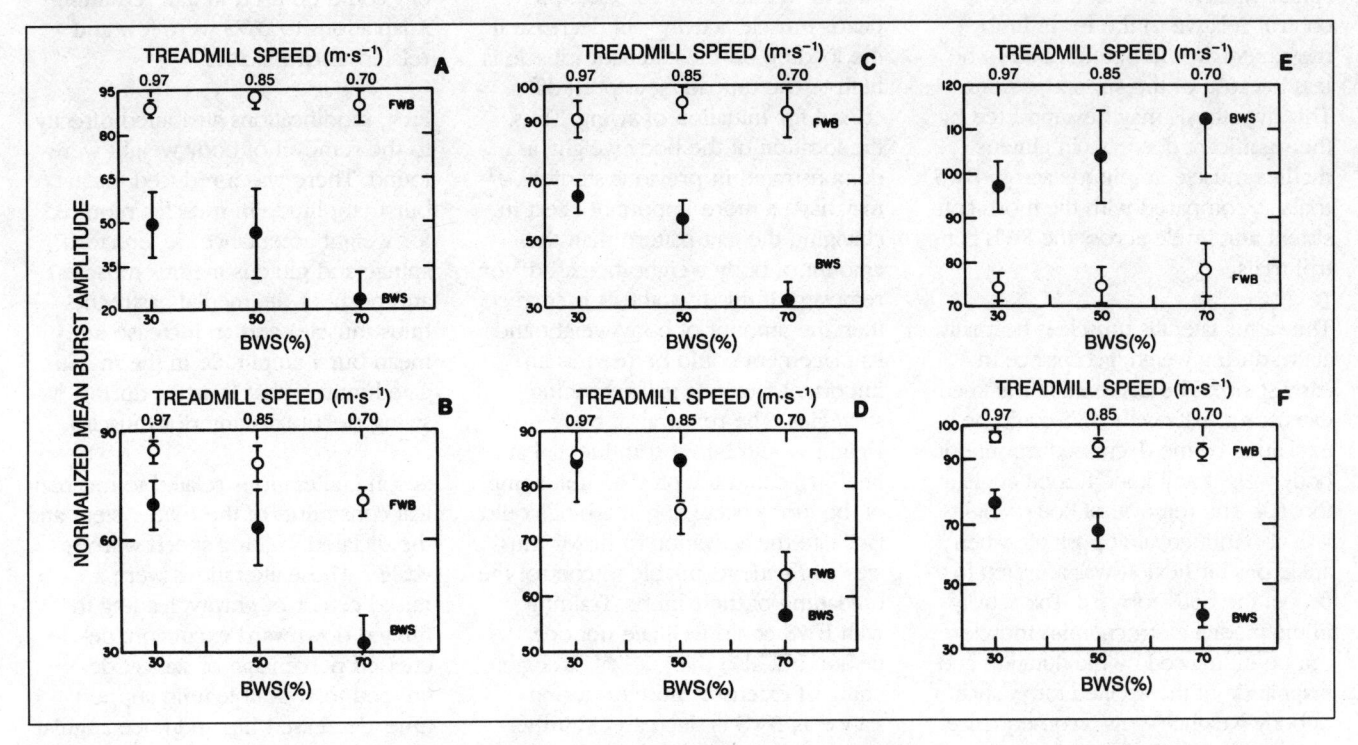

Figure 5. *Electromyographic normalized mean burst amplitudes (with one standard error of the mean) of first burst of erector spinae (A), second burst of erector spinae (B), gluteus medius (C), medial hamstring (D), tibialis anterior (E), and medial gastrocnemius (F) muscles. The y-axis represents amplitude in dimensionless units. Upper x-axis represents treadmill speed for both full-weight-bearing control (FWB) condition and body-weight-support (BWS) condition; lower x-axis represents the percentage of BWS removed for BWS conditions only. Control FWB condition is represented by open circles; BWS condition is represented by filled circles.*

cles of patients with hemiplegia revealed a mixed, unpredictable pattern of timing. In most instances, abnormalities appeared related to loading on the limb, especially early gastrocnemius muscle activation.[1] Electromyographic timing was not altered with BWS, however, possibly because of a decreased load (BWS), rate of application (slow treadmill speed), and duration of load (decreased percentage of total double-limb support time). The proper training of muscles (eg, gastrocnemius muscles) may be facilitated with BWS training, but further investigation of this hypothesis is needed.

Mean Electromyographic Burst Amplitudes

The changes in EMG amplitude attributable to the effect of walking speed in the FWB control sessions in our study are comparable with those reported by Yang and Winter.[28] The effects of walking speed on EMG amplitude were more evident at the hip than at the ankle muscles. The proportional changes in joint acceleration and deceleration with decreasing walking speed reported by Yang and Winter would thus be expected to modify the hip and knee muscle EMG amplitudes more than those at the ankle.

The normalized mean EMG burst amplitudes of the erector spinae, gluteus medius, vastus lateralis, and gastrocnemius muscles decreased with increasing BWS. As the body weight supported by the lower limbs was reduced, the muscles active during stance, especially during the heel-strike and push-off phases, demonstrated a decrease in burst amplitude.

It has previously been proposed that the erector spinae muscle decreases the forward and lateral motion of the trunk during the heel-strike and push-off phases of the gait cycle, respectively.[27,32] Thus, a smaller forward braking reaction would be required with decreasing body weight, as illustrated by the gradual decrease in the first EMG burst amplitude. The second burst amplitude, which was consistently larger than the first burst amplitude, however, was less affected by BWS in our study. Thorstensson and colleagues[27,32] have suggested that, when the second EMG burst amplitude is larger than the first burst amplitude, this is usually the result of a change from a frontal to a more sagittal plane of movement. Consequently, it seems that the removal of body weight did not decrease the need for control of lateral trunk movement. It was observed that the harness, in combination with body weight removal, may have been related to an increase in lateral trunk movement consistent with the continued second burst of erector spinae muscle activity

The gluteus medius muscle, a postural muscle needed for stance stability at heel-strike, demonstrated a similar modification to that of the erector spinae muscle. The need for pelvic control relative to the lower-limb movement would be expected to be less because of the support system. This hypothesis may be supported by the significant decrease in gluteus medius muscle amplitude across BWS trials as compared with the more consistent amplitude across the FWB control trials.

The vastus lateralis muscle is normally active during weight acceptance in early stance. The decrease in the knee extensor activity with BWS could be explained by the decreased amount of body weight and knee flexion angle at foot-flat. The removal of body weight was also influential at push-off, when ankle plantar flexion was activated to propel the body forward. The activity in the medial gastrocnemius muscle can be influenced by the duration and magnitude of the applied force, such as body weight.[30,33] As a consequence of increasing BWS, gastrocnemius muscle EMG amplitude decreased, as seen in animal experiments.[33,34] In addition, increasing BWS may decrease the stretch on the gastrocnemius muscle during stance, which may facilitate the proper phasing of this muscle for patients involved in a gait retraining program. This effect would be important for patients with neurological impairment, because early stretch activation in the triceps surae muscle is a common problem interfering with locomotion. The removal of weight also seems to affect muscles during the swing phase, as can be seen by the increased amplitude in the tibialis anterior muscle during the swing phase. This effect would also be important, because lack of dorsiflexion is another common problem in the locomotor pattern of most patients with neurological impairment.

It has been reported that the amount and timing of loading can have a strong influence on the locomotor pattern in animals.[34] For example, flexor muscle activity will decrease if the loading on the extensor muscle is high.[34] The unloading at push-off is critical for initiation of swing. Thus, the location of the body weight, as demonstrated in previous studies,[12,34] may have a more important effect in changing the gait pattern than the amount of body weight increased[11] or removed. If this hypothesis is correct, then the amount of body weight and its placement could be used as an important factor in gait retraining strategies. The removal of body weight would be most influential at heel-strike and toe-off. The unloading of the limbs occurring at toe-off could facilitate the activation of flexor muscles[29] in patients unable to control the unloading of their limbs. Training with BWS could facilitate not only flexion but also the gradual strengthening of extensor muscles during stance as BWS is decreased. Furthermore, an emphasis could be placed on the strengthening of muscle activity, at the appropriate joint angular position during gait, as suggested by Olney et al.[25]

An interactive BWS gait retraining strategy would allow patients lacking muscle control to develop the strength and coordination required to walk. Any BWS level below 70% can be beneficial for training, if the treadmill speed is adjusted to the patient's walking abilities and BWS condition. The amount of BWS can then be decreased and walking speed increased as the patient improves.

Summary and Conclusions

Normal gait under the influence of various BWS levels was compared with FWB gait to determine whether a BWS and treadmill stimulation strategy could be used in gait retraining. Adaptations to BWS were few and related to two factors.

First, modifications attributed directly to the removal of body weight were found. There was a reduced mean burst amplitude in muscles required for weight acceptance (ie, erector spinae and gluteus medius muscles) and push-off (ie, medial gastrocnemius muscle) and an increase in mean burst amplitude in the investigated muscle that is active during swing (ie, tibialis anterior muscle).

Second, alterations related to mechanical constraints of the BWS system and the dictated walking speed were revealed. These alterations were a raised center of gravity, leading to limited downward excursion, decreased percentage of stance, decreased total double-limb support time, decreased hip and knee angular displacement, and increased single-limb support time. These results led to the development of a BWS treadmill training scheme that may offer many clinical applications.

The training method used in this study did not produce abnormal gait. Furthermore, this interactive training protocol should provide an easier progression from stance to the swing phase of gait. Body-weight-supported gait appears to present temporal, kinematic, and EMG patterns that could be advantageous in gait retraining.

Acknowledgments

We recognize the contributions of Brenda Brouwer, Michael Wainberg, James Finch (experimental session), and Sylvain Bergeron (technical assistant).

Appendix. List of Equipment

Equipment	Company	Description
Treadmill	Collins (address not available)	1.15 m long × 0.37 m wide; speed range of 0–1.48 m·s^{-1}
Body-weight-support system	Industries Auteca Ltd, 3125 Bernard Pilon, St Mathieu de Balocil, Quebec, Canada J3G 4F5	See Barbeau et al[14]
Harness	In house	Modified Tyrolean climbing harness (see Barbeau et al[14])
Force transducer	Intertechnology Inc, 3675 Boul de Sources, DDO, Quebec, Canada H9B 2T6	Load cell 226.8 kg (500 lb) 363-D3-500-20P3 ti
Force readout	Intertechnology Inc	Indicator DRS 353000 KXLBO
FM tape recorder	Honeywell, Test Instruments Div, 740 Ellesmere Rd, Scarborough, Ontario, Canada N1H 2W4	Model 101; 14 channels; frequency response of 2,500 Hz at 9.5 cm/s (3.75 in/s)
Video recorder	Panasonic, 8270 Mayrand St, Montreal, Quebec, Canada H4P 2C5	Compatible with 1.9-cm (0.75-in) video tapes
Video camera	Panasonic	Rotary shutter video camera (exposure time of 2 ms)
Remote search controller	Panasonic	For field-to-field viewing; temporal resolutions of 16.7 ms, accuracy of ±2 fields
Video monitor	Panasonic	35.6-cm (14-in), high-resolution, black and white monitor
Footswitches	Nortel Manufacturing Ltd, 2000 Ellesmere Rd, Scarborough, Ontario, Canada N1H 2W4	Can be adapted with differential voltages to produce compound signal from collective footswitches
Electromyographic electrodes	Graphic Controls Canada Ltd, 215 Herbert St, PO Box 5500, Gananoque, Ontario, Canada K7G 2X1	Medi-Trace pellet electrodes; silver-silver chloride disposable electrodes
Amplifiers and preamplifiers	Centre de Recherche en Sciences Neurologiques de l'Université de Montréal, Montreal, Quebec, Canada H3C 3J7	Band-pass filtered 10–1,000 Hz (with notch filter at 60 Hz) and common mode rejection rate of 80 dB
Computer	Digital Equipment Co, 100 Herzberg Rd, PO Box 13000, Kanata, Ontario, Canada K2K 2A6	PDP 11/34A

References

1 Knutsson E, Richards C. Different types of disturbed motor control in gait of hemiparetic patients. *Brain*. 1979;102:405–430.

2 Eke-Okoro ST, Larsson L. A comparison of the gaits of paretic patients with the gait of control subjects carrying a load. *Scand J Rehabil Med*. 1984;16:151–158.

3 Carlsoo S, Dahllof AG, Holm J. Kinematic analysis of the gait in patients with hemiparesis and in patients with intermittent claudication. *Scand J Rehabil Med*. 1974;6:166–179.

4 Dickstein R, Nissan M, Pillar T, Scheer D. Foot-ground pressure pattern of standing hemiplegic subjects: major characteristics and patterns of improvement. *Phys Ther*. 1984;64:19–23.

5 Winstein CJ, Gardner ER, McNeal DR. Standing balance training: effect on balance and locomotion in hemiparetic adults. *Arch Phys Med Rehabil*. 1989;70:755–762.

6 Dietz V, Quintern S, Berger W. Electrophysiological studies of gait in spastic and rigidity. *Brain*. 1981;104:431–449.

7 Rossignol S, Barbeau H, Julien C. Locomotion of the adult chronic spinal cat and its modifications by monoaminergic agonists and antagonists. In: Golberger M, Gorio M, Murray M, eds. *Development and Plasticity of the Mammalian Spinal Cord*. Padua, Italy: Liviana Editrice SpAt; 1986:323–346.

8 Barbeau H, Rossignol S. Recovery of locomotion after chronic spinalization in the adult cat. *Brain Res*. 1967;412:84–95.

9 Finch L, Barbeau H. Influences of partial weight bearing on normal human gait: the development of a gait retraining strategy. *Can J Neurol Sci*. 1985;12:183. Abstract.

10 Ghori GM, Luckwill RG. Responses of the lower limb to load carrying in walking man. *Eur J Appl Physiol*. 1985;54:145–150.

11 Medeiros JM. *Investigation of Neuronal Mechanisms Underlying Human Locomotion: an EMG Analysis*. Iowa City, Iowa: The University of Iowa; 1978. Doctoral dissertation.

12 Neumann DA, Cook TM. Effect of load and carrying position on the electromyographic activity of the gluteus medius muscle during walking. *Phys Ther*. 1985;65:305–311.

13 Hewes DE, Spady AA, Harris RL. *Comparative Measurement of Man's Walking and Running Gaits in Earth and Simulated Lunar Gravity*. NASA report TDN-3363. 1967:1–33.

14 Barbeau H, Wainberg M, Finch L. Description and application of a system for locomotion rehabilitation. *Med Biol Eng Comput*. 1987;25:341–344.

15 Visintin M, Barbeau H. The effects of body weight support on the locomotor pattern of spastic paretic patients. *Can J Neurol Sci.* 1989;16:315–325.

16 Visintin M, Finch L, Barbeau H. Progressive weight bearing and treadmill stimulation during gait retraining of hemiplegics: a case study. *Phys Ther.* 1988;68:807. Abstract.

17 Zomlefer MR, Provencher J, Blanchette G, Rossignol S. Electromyographic study of lumbar muscles in acute high decerebrate and in low spinal cats. *Brain Res.* 1984;290:249–260.

18 Arsenault AB, Winter DA, Marteniuk RG. Characteristics of muscular function and adaptation in gait: a literature review. *Physiotherapy Canada.* 1987;39:5–12.

19 Snedecor GW, Cochran WG. *Statistical Method.* 7th ed. Ames, Iowa: Iowa State University Press; 1982.

20 Murray MP, Spurr GB, Gardner GM, Mollinger LA. Treadmill vs floor walking: kinematic electromyogram and heart rate. *J Appl Physiol.* 1985;59:87–91.

21 Margaria R, Cavagna GA. Human locomotion in subgravity. *Aerospace Medicine.* 1966;35:1140–1146.

22 Smidt GL. Hip motion and related factors in walking. *Phys Ther.* 1971;51:9–21.

23 Murray MP, Kory RC, Clarkson BH, Sepic SB. A comparison of free and fast speed walking pattern in normal men. *Am J Phys Med.* 1966;45:8–24.

24 Winter DA. Biomedical motor patterns in normal walking. *Journal of Motor Behavior.* 1983;15:302–330.

25 Olney SJ, Jackson VG, George SR. Gait reeducation guidelines for strike hemiplegic using mechanical energy and power analysis. *Physiotherapy Canada.* 1987;40:242–248.

26 Battye CK, Joseph J. An investigation by telemetry of the activity of some muscles in walking. *Med Biol Eng.* 1966;4:125–135.

27 Thorstensson A, Carlson H, Zomlefer MR, Nilsson J. Lumbar back muscle activity in relation to trunk movement during locomotion in man. *Acta Physiol Scand.* 1982;116:13–20.

28 Yang JF, Winter DA. Surface EMG profiles during different walking cadences in humans. *Electroencephalogr Clin Neurophysiol.* 1985;60:485–491.

29 Herman R, Weiter Z, Brampton S, Finley FR. Human solution for locomotion, I: single limb analysis. In: Herman R, Grillner S, Stein P, Stuart DG, eds. *Neural Control of Locomotion.* New York, NY: Plenum Publishing Corp; 1976:413–450.

30 Grillner S. Control of locomotion in bipeds, tetrapods and fish. In: *Handbook of Physiology, II: The Nervous System.* New York, NY: American Physiological Society; 1981:1179–1234.

31 Pearson K. The control of walking. *Sci Am.* 1976;235:72–86.

32 Thorstensson A, Nilsson J, Carlson H, Zomlefer MR. Trunk movements in human locomotion. *Acta Physiol Scand.* 1984;121:9–22.

33 Monster A. Loading reflexes during two types of voluntary muscle contraction. In: Herman R, Grillner S, Stein R, Stuart DG, eds. *Neural Control of Locomotion.* New York, NY: Plenum Publishing Corp; 1976:347–361.

34 Duysens J, Pearson KG. Inhibition of flexor burst generation by loading ankle extensor muscles in walking cats. *Brain Res.* 1980; 187:321–332.

Commentary

The concept of practicing skills under unusual loaded conditions is not unprecedented. Situations in which increased loads are used include the batter warming up with a weighted bat, the runner wearing a weighted vest or holding hand weights, and the sprinter charging up steps. Decreased loads are usually relegated to individuals with injuries; for example, the power lifter working out with only the bar, the runner avoiding hills, and the physical therapy patient exercising in water. Gait, in particular, was practiced by many of us in an unloaded state through time spent in a baby walker.

One of the concerns of performing motor skills under nonstandard loads is that inappropriate motor patterns may develop.[1] Some dedicated sprinters refuse to slowly run long distances because they are convinced that this practice will make them slower runners. If a karate student practices a block thousands of times, but in slow motion, will that technique be successful in deflecting a full-speed assault? Proponents of isokinetic dynamometers insist that proper speed is vital during rehabilitation.

Gait is a particularly complex motor pattern dependent on neurological control,[2] which may be disturbed by traumatic and pathological events. Physical therapists are often faced with the task of helping patients redevelop gait function while avoiding inappropriate gait patterns. Finch and colleagues have proposed that the more normal a gait pattern utilized during training, the more normal a gait pattern will result after rehabilitation is over. Of particular concern is that patients with neurological deficits walking with inadvertently abnormal gait patterns may take longer to recover gait function and may retain inappropriate gait patterns over a long period of time.

"Patients" with artificially induced neurological deficits in the form of spinal cats have been shown to regain near-normal gait patterns when walking with unloaded limbs.[3] For this reason, the authors believe that human patients with neurological deficits may benefit from practicing gait under unloaded conditions. Previous work by these authors and colleagues have examined gait patterns in a limited number of patients with neurological deficits.[4] These patients demonstrated improved gait patterns when walking under partially unloaded conditions.

Part of the difficulty with interpreting the results of gait studies (and evaluating gait, in general) is defining what is meant by "normal" gait. Many therapists would avoid defining normalcy in scientific terms, but would claim they "know it when they see it." This claim has not been supported by scientific studies.[5] The challenge faced by the scientist is to describe a number of quantifiable variables that can be used to characterize this motor skill. Gait consists of cyclical movement of a large number of joints, and this joint movement requires synchronization of action from many muscles. Both force output and timing are important. At the same time, infinite variations of these muscle actions may be possible to still achieve successful ambulation with the human structure. Clearly, a range of values must be used to define a normal gait pattern.[6-8]

Finch and colleagues have begun the task of establishing some values to represent partially unloaded gait in subjects without neurological disorders. These values could then provide a framework for comparison when the gait patterns of patients are studied. Although I am confident that these researchers will continue in their efforts to quantify both normal and pathological gait patterns, I would like to encourage others to join in these efforts. There are enough variations in gait patterns to keep many biomechanics laboratories busy for many years.

Weight unloading has added another source of variation in gait studies. Consider the differences between unloading by removing weight and unloading by decreasing the effect of gravity. During unloading by removal of weight, as demonstrated in this study, ground contact forces and associated reflexes are decreased, but airborne accelerations and decelerations are much less affected. These authors used a fairly rigid support system.[9] Possible variations that could be studied include freely translating counterweight support and incorporation of elastic elements in the support frame. Unloading by decreasing gravitational effects would include underwater walking. Of course, viscous resistance would then become an important factor. More fancifully, gait studies might be relegated to the space-shuttle bay or a future space station for an unqualified decrease in gravity.

The authors are to be congratulated for initiating investigations into an interesting, and hopefully fruitful, avenue for functional rehabilitation of patients with neurological deficits. Their rationale makes sense, and the early results are encouraging. Perhaps others who are involved in gait and neurological research will join in these efforts.

Jerome V Danoff, PhD, PT
Associate Professor
Department of Physical Therapy
College of Allied Health Sciences
Howard University
Washington, DC 20059

References

[1] Barnett ML, Ross D, Schmidt RA, Todd B. Motor skills learning and the specificity of training principle. *Research Quarterly*. 1973;44:440–447.

[2] Craik RL. Biomechanics: a neural control perspective. *Phys Ther*. 1984;64:1810–1811.

[3] Barbeau H, Rossignol S. Recovery of locomotion after chronic spinalization in the adult cat. *Brain Res*. 1967;412:84–95.

[4] Visintin M, Barbeau H. The effects of body weight support on the locomotor pattern of spastic paretic patients. *Can J Neurol Sci*. 1989;16:315–325.

[5] Eastlack ME, Arvidson J, Snyder-Mackler L, et al. Interrater reliability of videotaped observational gait-analysis assessments. *Phys Ther*. 1991;71:465–472.

[6] Winter DA. Kinematic and kinetic patterns in human gait: variability and compensating effects. *Human Movement Science*. 1984;3:51–76.

[7] Trueblood PR, Walker JM, Perry J, Gronley JK. Pelvic exercise and gait in hemiplegia. *Phys Ther*. 1989;69:18–26.

[8] Winter DA, Patla AE, Frack JS, Walt SE. Biomechanical walking pattern changes in the fit and healthy elderly. *Phys Ther*. 1990;70:340–347.

[9] Barbeau H, Wainberg M, Finch L. Description and application of a system for locomotor rehabilitation. *Med Biol Engr Comput*. 1987;25:341–344.